STEELS:
Heat Treatment and Processing Principles

STEELS:
Heat Treatment and Processing Principles

GEORGE KRAUSS
AMAX Foundation Professor
Department of Metallurgical Engineering
Colorado School of Mines

MATERIALS PARK, OHIO 44073

Copyright © 1990
ASM INTERNATIONAL®
All rights reserved

First printing, December 1989
Second printing, May 1993
Third printing, May 1994
Fourth printing, February 1995
Fifth printing, December 1997
Sixth printing, November 2000

No part of this book may be reproduced, stored in a retrieval system, or transmitted, in any form or by any means, electronic, mechanical, photocopying, recording, or otherwise, without the prior written permission of the publisher.

Nothing contained in this book is to be construed as a grant of any right of manufacture, sale, or use in connection with any method, process, apparatus, product or composition, whether or not covered by letters patent or registered trademark, nor as a defense against liability for the infringement of letters patent or registered trademark.

Library of Congress Cataloging-in-Publication Data

Krauss, George, 1933—
   Steels: heat treatment and processing principles / George Krauss.
     p.   cm.
   Includes bibliographical references.
   ISBN 0-87170-370-X
   1. Steel—Heat treatment. 2. Steel—Metallurgy. I. Title.
TN751.K73   1989   89-18136
672.3′6—dc20   CIP

PRINTED IN THE UNITED STATES OF AMERICA

To the next generation

Matt, Jon, Ben, and Tom

Preface to the 1980 Edition

This book is a completely rewritten version of *Principles of Heat Treatment* by M.A. Grossmann and E.C. Bain. It is a pleasure to acknowledge the contributions of these authors. Much of their work, especially that concerning the development of hardenability concepts, is incorporated here. *Principles of Heat Treatment* covered developments between 1935 and 1964, and Grossmann, Bain, and their contemporaries did their work so well that the heat treatment and metallurgy of carbon steels was almost taken for granted. Steels, however, are wonderfully complex, and continued effort in the last twenty years has brought deeper understanding of their response to thermal and mechanical treatments. New theoretical approaches to diffusion-controlled and martensitic transformations, the characterization of fine structure by transmission electron microscopy, fractography with the scanning electron microscope, new electron beam microanalysis techniques, fracture toughness testing, continued examination of hardenability, and the relationship of microstructure and fine structure to strength, toughness, and ductility are all areas, highly developed only in the last twenty years, that I have attempted to build onto the solid foundations of steel heat treatment developed by earlier workers. My approach has been to develop the structure-property-processing relationships that underlie the many heat treatments applied to steels. The origin and characterization of microstructures are emphasized because they are so often forgotten as the source of the handbook graphs and tables of processing parameters and properties.

The give-and-take of many conferences and the contributions of many investigators to the literature have been the basis for our growing understanding of steels and their behavior. I have drawn widely from published sources and through the cited references hope to recognize at least some university and industrial scientists and their contributions. The reference lists are by no means complete, but every paper opens up an area by listing tens or even hundreds of additional references.

Especially rewarding has been my association with other investigators. Professor Morris Cohen initiated my interest in steel, and the enthusiasm of my colleagues and students has sustained that interest. I have learned from every thesis investigation, and examples of the work of my students are shown throughout this book.

In recent years I have benefited much from association with the heat treatment activities of the American Society for Metals, first through membership on the Heat Treatment Technical Division Council, and more recently as editor of the *Journal of Heat Treating*. These associations have made me aware of the scope and sophistication of new approaches to the heat treatment of steels, and I am especially grateful to Norman Kates, Dale Breen, Jon Dossett, and Joe Riopelle for their introduction to the demanding, practical world of heat treatment.

A university research effort is very much dependent on outside support. I gratefully acknowledge the Army Research Office, the National Science Foundation, the Bethlehem Steel Corporation, the AMAX Foundation, and the American Iron and Steel Institute for the support that has enabled me and my students to remain actively involved in research on the behavior of steels. I acknowledge also fruitful discussions with Professors Glenn Edwards, Tom Bell, and Norman Breyer concerning parts of the text and am grateful for micrographs supplied by Professors R.W.K. Honeycombe, Robert Hehemann, and Marvin Wayman. Mark Geib deserves special mention for his help with some of the figures. Finally, I am deeply grateful to my wife, Ruth, for her competent assistance, her unwavering support, and the typing of the manuscript—all given between a busy schedule of rehearsals and performances of the Central City Opera and the Evergreen Chorale.

George Krauss

Evergreen, Colorado
September 25, 1979

Preface to the 1990 Edition

The 1980s have been a dynamic period for manufacturing, and it is appropriate that *Principles* be expanded to describe a broader selection of ferrous alloys used in manufacturing. Not only is deeper understanding of the performance of conventionally treated steels now available, but new alloys and new processes also have been developed. For example, new alloys under active development or brought to market in the '80s include duplex stainless steels, microalloyed bar and forging steels, ultrahigh-nitrogen stainless steels, low-cobalt maraging steels, steels with low manganese and silicon that are resistant to temper embrittlement, and austempered ductile cast irons. The success of these new alloys, as well as that of improved conventional steels, is often directly coupled to advances in melting, and the 1980s have seen the widespread adoption of ladle metallurgy and other special steelmaking techniques.

The most dramatic changes in processing have come in the area of surface modification, ranging from improvements in induction heating and gas carburizing to the development of plasma, physical vapor deposition, and laser heating processes. Thermochemical modifications, coatings, solid-state transformation hardening, and rapidly solidified, thin-surface layers are all possible with the new techniques. Thus exciting possibilities exist for manufacturing surfaces with special properties and engineered materials systems incorporating ferrous alloys.

For the revised edition of *Principles of Heat Treatment of Steel* I have added chapters on new surface modification techniques, stainless steels, tool steels, and cast irons, and have expanded four of the original chapters. Thus the revised text covers many aspects of alloying, processing, and microstructure evolution beyond those involved in conventional heat treatment of carbon steels. Also, the new surface modification techniques are often directed to producing engineered composite systems quite different from traditionally processed steels. In order to reflect the broader scope of the present edition, the title *Steels: Heat Treatment and Processing Principles* was selected. This new title moves steels to a prominent position, and recognizes the importance of processing other than heat treatment.

The principles of microstructure development, and the effects of microstructure on properties and performance, within the context of alloying, phase equilibria, and processing, remain the dominant theme of this book. About 110 new figures have been added, many of them selected to illustrate characteristic and special microstructural features of ferrous alloys. While heat treatment and thermal processing are still of prime importance, solidification, thermomechanical, mechanical, and surface deposition processing are also recognized as major factors which establish structure-property relationships in a broad spectrum of ferrous alloys.

Selected literature is cited throughout the text in order to lead readers to in-depth sources of information regarding topics of special interest. Unfortunately, the references cannot recognize all who have contributed to the vast field of processing, heat treatment, and performance of steels. Handbook and manufacturing literature must be referred to for processing details and property tabulations which cannot be included here.

The Army Research Office and National Science Foundation have continued to support steel research at the Colorado School of Mines into the 1980s, and I am grateful for continuing support of the AMAX Foundation for my professorship. In 1984, the Advanced Steel Processing and Products Research Center (ASPPRC), a cooperative industry-university research center, was established at the Colorado School of Mines with a seed grant from the National Science Foundation. This Center has made possible a renewed effort to deepen understanding of steel as a vital manufacturing material. I acknowledge with gratitude the support and interest of the following organizations who were sponsors of ASPPRC at the time of the writing of this second edition: Army Materials Technology Laboratory, Bethlehem Steel Corporation, Carpenter Technology Corporation, Caterpillar Incorporated, Chaparral Steel Company, Chrysler Corporation, Dofasco, Eaton Corporation, Ford Motor Company, Inland Steel Company, Lake Ontario Steel Corporation, National Institute for Standards and Technology, LTV Steel Company, Lukens Steel Company, North Star Steel Company, Rouge Steel Company, Stelco Incorporated, The Timken Company, and United States Steel Division, USX.

With pleasure I acknowledge helpful discussions and contributions of micrographs from a broadened list of colleagues: Tohru Arai (Toyota Research Laboratories), M. Grace Burke (Westinghouse Electric Company), J.R.T. Branco (CSM), R.H. Barkalow and R.W. Kraft (Lehigh University), Tom Bell (University of Birmingham), Scott Diets (CSM), David Hoffmann (Ford Motor Company), A.S. Korhonen (Helsinki University of Technology), Tom Majewski (Caterpillar Incorporated), Jeff McClain (CSM), Bob McGrew (CSM), Tadashi Maki and Imao Tamura (Kyoto University), Eric Mittemeijer (Delft University of Technology), Hisaki Okamoto (Tottori University), Mike Rigsbee (University of Illinois), Mike Shea (General Motors Corporation), Pan Jei (CSM), Steve Thompson (CSM), George Vander Voort (Carpenter Technology Corporation), Abdul Wahid (CSM), and Shen Yun (CSM).

I thank Scott Diets for his help with the figures, and I am especially grateful for the collaboration and support of my colleague, David K. Matlock. We have shared and accomplished much together in the 1980s. My wife Ruth supported this effort in many ways, including the word processing of many revisions of the final manuscript, and I thank her deeply for her help.

George Krauss

Evergreen, Colorado
July 30, 1989

Contents

1 Phases and Structures 1
The Iron-Carbon Equilibrium Diagram 1
Crystal Structures of Iron 4
Effects of Carbon 6
Crystal Structures in Fe-C Alloys 7
Effects of Alloying Elements 10
Critical Temperatures 11
References 15

2 Pearlite, Ferrite, and Cementite 17
Eutectoid Transformation 17
Structure of Pearlite 19
Pearlite Transformation Kinetics 21
Interphase Precipitation 31
Proeutectoid Phases 34
Proeutectoid Phase Formation 36
References 41

3 Martensite and Bainite 43
General Considerations 43
Martensitic Transformation Kinetics 47
Crystallography of Martensitic Transformation 57
Morphology of Ferrous Martensites 61
Plate Martensite 63
Lath Martensite 67
Bainite 78
References 84

4 Isothermal and Continuous Cooling Transformation Diagrams 89
Isothermal Transformation Diagrams 89
Continuous Cooling Transformation Diagrams 92
Continuous Cooling Transformation and Bar Diameter 104
Summary 105
References 105

5 Heat Treatments To Produce Ferrite and Pearlite 107
Full Annealing and Homogenizing 107
Normalizing 112
Spheroidizing 114
Process and Recrystallization Annealing 118
Cold Rolled and Annealed Sheet Steels 121
Strain and Quench Aging 125
Stress Relieving 130
Mechanical Properties of Ferrite-Pearlite Microstructures 133
Microalloyed Bar and Forging Steels 139
References 142

6 Hardness and Hardenability 145
Hardness and Carbon Content 145
Martensite Strength 149
Definitions of Hardenability 152
Hardness Distribution 152
Factors Affecting Cooling Rates 156
Severity of Quench 159
Quantitative Hardenability 163
Determination of Ideal Size 167
Jominy Test for Hardenability 169
Recent Developments 173
References 177

7 Austenite in Steels 179
Austenite and Properties 179
Austenite Formation 182
Austenite Grain Size 188
Austenite Grain Size Control 193
References 201

8 Tempering of Steel 205
Mechanical Property Changes 205
Alloying Elements and Tempering 212
Structural Changes on Tempering 218
Embrittlement Phenomena and Tempering 229
Tempered Martensite Embrittlement 231
Temper Embrittlement 236
Aluminum Nitride Embrittlement 239
Liquid Metal Embrittlement 241
Hydrogen Embrittlement 241

Overheating of Forgings   244
Flow and Fracture of Tempered Steels   245
References   255

9 Special Heat Treatments   263
Residual Stress, Distortion, and Quench Cracking   263
Martempering   266
Austempering   267
Thermomechanical Treatments   269
Intercritical Heat Treatment   274
References   279

10 Surface Hardening   281
Flame Hardening   281
Induction Heating   282
Carburizing: Processing Principles   285
Carburizing: Properties and Structure   291
Carburizing: Fatigue and Fracture   302
Nitriding   305
Carbonitriding   310
Ferritic Nitrocarburizing   312
References   315

11 Surface Modification   319
Introduction   319
Plasma Nitriding   320
Plasma Carburizing   322
Ion Implantation and Ion Mixing   325
Physical Vapor Deposition: Processing   328
Physical Vapor Deposition: Microstructures   331
Chemical Vapor Deposition   337
Salt Bath Coating Process   338
Laser and Electron Beam Surface Modification   339
Summary   345
References   346

12 Stainless Steels   351
Alloy Design and Phase Equilibria   351
Austenitic Stainless Steels   359
Intergranular Carbides in Austenitic Stainless Steels   361
Martensite Formation in Austenitic Stainless Steels   366
Other Phases in Austenitic Stainless Steels   370
Other Austenitic Stainless Steels   371
Heat Treatment of Austenitic Stainless Steels   373
Ferritic Stainless Steels   375

Intermetallic Phases in Ferritic Stainless Steels   377
475 °C (885 °F) Embrittlement in Ferritic Stainless
   Steels   379
Martensitic Stainless Steels   382
Precipitation-Hardening Stainless Steels   386
Duplex Stainless Steels   390
Summary   395
References   396

13 Tool Steels   401
Introduction   401
Classification of Tool Steels   401
Tool Steel Alloy Design   405
Primary Processing of Tool Steels   410
Annealing of Tool Steels   413
Stress Relief of Tool Steels   416
Hardening of Tool Steels   416
Preheating and Austenitizing   416
Hardenability and Martensite Formation   419
Grain Boundary Carbide Formation   422
Tempering of Tool Steels   424
Retained Austenite Transformation and Double Tempering
   in Tool Steels   427
Summary   428
References   428

14 Cast Irons   431
Phase Relationships   431
Gray Cast Irons   435
Ductile Cast Iron   438
Malleable Cast Iron   441
White Cast Irons   444
Austempered Ductile Irons   446
Cast Iron Surface Modification   450
Summary   451
References   452

Appendix 1: Glossary of Selected Terms   453

Appendix 2: Temperature Conversions   469

Index   473

CHAPTER 1

Phases and Structures

Steel can be heat treated to produce a great variety of microstructures and properties. The desired results are accomplished by heating *in* temperature ranges where a phase or combination of phases is stable (thus producing changes in the microstructure or distribution of phases), and/or heating or cooling *between* temperature ranges in which different phases are stable (thus producing beneficial phase transformations). The iron-carbon equilibrium phase diagram is the foundation on which all heat treatment of steel is based. This diagram defines the temperature-composition regions where the various phases in steel are stable, as well as the equilibrium boundaries between phase fields. This chapter describes the iron-carbon diagram and the phases found in steels and iron-carbon alloys.

The Iron-Carbon Equilibrium Diagram

The iron-carbon (Fe-C) diagram is a map that can be used to chart the proper sequence of operations for a given heat treatment. The iron-carbon diagram should be considered only a guide, however, because most steels contain other elements that modify the positions of phase boundaries. The effects of alloying elements on the phase relations shown in the iron-carbon diagram are described later in this chapter. Use of the iron-carbon diagram is further limited because some heat treatments are specifically intended to produce nonequilibrium structures whereas others barely approach equilibrium. Nevertheless, knowledge of the changes that take place in a steel as equilibrium is approached in a given phase field, or of those that result from phase transformations, provides the scientific basis for the heat treatment of steels.

2 / STEELS: HEAT TREATMENT AND PROCESSING PRINCIPLES

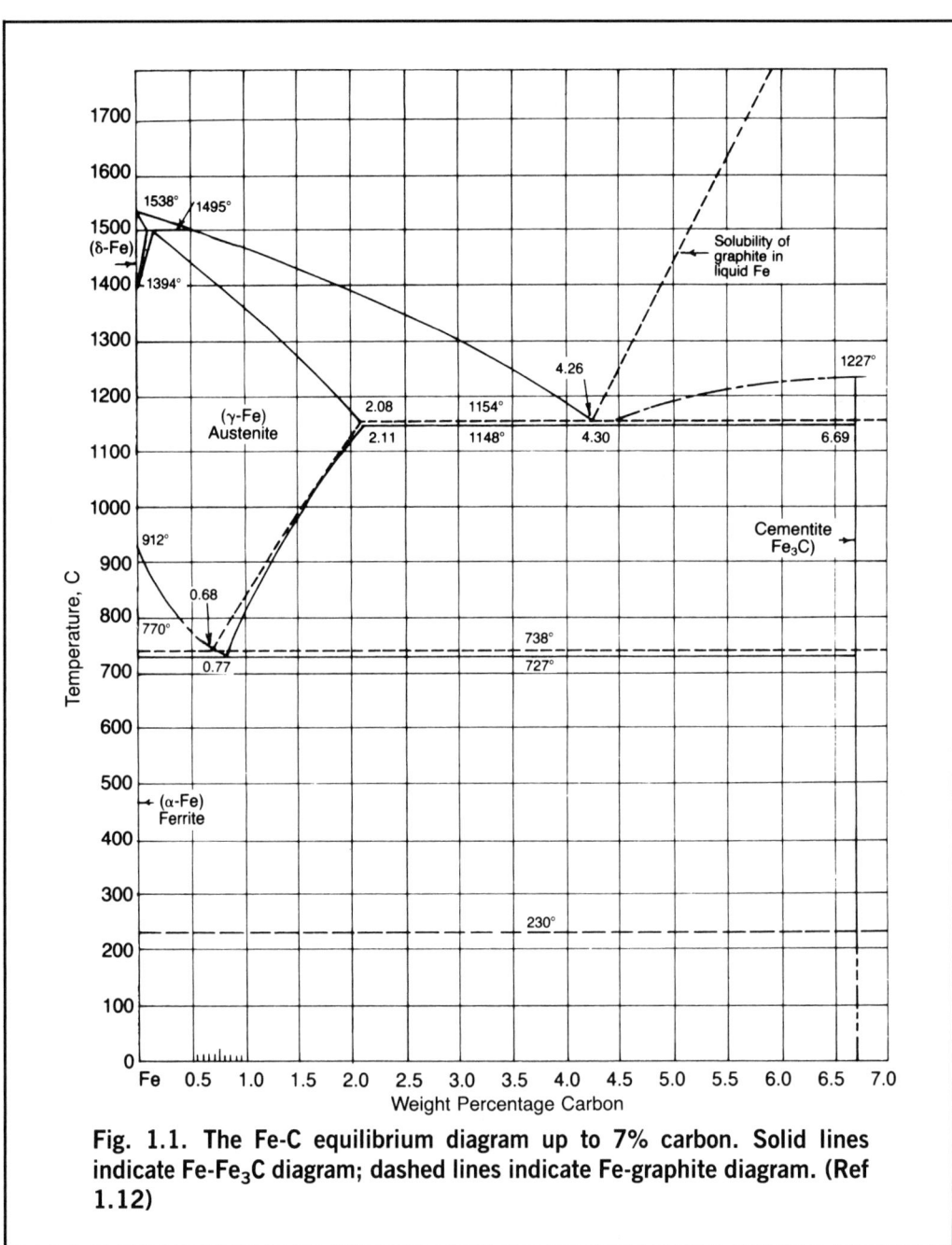

Fig. 1.1. The Fe-C equilibrium diagram up to 7% carbon. Solid lines indicate Fe-Fe$_3$C diagram; dashed lines indicate Fe-graphite diagram. (Ref 1.12)

Figure 1.1 shows the Fe-C equilibrium diagram for carbon contents up to 7%. Steels are alloys of iron, carbon, and other elements that contain less than 2% carbon—most frequently 1% or less. Therefore, the portion of the diagram below 2% carbon is of primary interest for steel heat treatment. Alloys containing more than 2% carbon are classified as cast irons. Actually, two diagrams are shown in Fig. 1.1: the solid lines show the equilibrium between Fe_3C and the several phases of iron, whereas the dashed lines show the equilibrium between graphite and the other phases. Graphite is a more stable form of carbon than Fe_3C and, given very long periods of time, Fe_3C will decompose to graphite. Graphitization, however, rarely occurs in steels, and thus the Fe-Fe_3C diagram is the more pertinent for understanding the heat treatment of steel. In cast irons, high carbon content and the usual high silicon additions promote graphite formation, and accordingly cast iron technology is based much more on the Fe-graphite diagram.

The diagram in Fig. 1.1 is strictly valid only at a pressure of one atmosphere. At very high pressures, the boundaries shift and new phases appear. For example, in pure iron a close-packed hexagonal crystal form of iron, epsilon iron, has been produced at high pressures (Ref 1.1). The triple point in pure iron between alpha iron, gamma iron, and epsilon iron occurs at 770 K and 110 kbars (11 GPa).

Compositions of the Fe-C alloys and phases represented by the Fe-C diagram are conventionally given in weight percent. The percent symbol (%), unless otherwise identified, is understood to represent weight percent, a convention that is followed in this text. Sometimes it is useful to determine compositions in atomic percent. Conversion from weight percent to atomic percent carbon in an Fe-C alloy is accomplished by the following equations:

$$\text{at.\% C} = \frac{\text{C atoms}}{\text{C atoms + Fe atoms}} \times 100 \qquad \text{(Eq 1.1)}$$

or

$$\text{at.\% C} = \frac{\dfrac{\text{wt.\% C}}{\text{at. wt. C}}}{\dfrac{\text{wt.\% C}}{\text{at. wt. C}} + \dfrac{\text{wt.\% Fe}}{\text{at. wt. Fe}}} \times 100 \qquad \text{(Eq 1.2)}$$

Application of this calculation to an Fe-0.4C alloy shows that 0.4% carbon is equivalent to 1.8 atomic percent carbon, a reflection of the much lighter atomic weight of carbon (12) compared with that of iron (56). Conversion to atomic percent for steels containing elements other than Fe and C requires an additional term in the denominator of Eq 1.1 or Eq 1.2 for each of the other elements present.

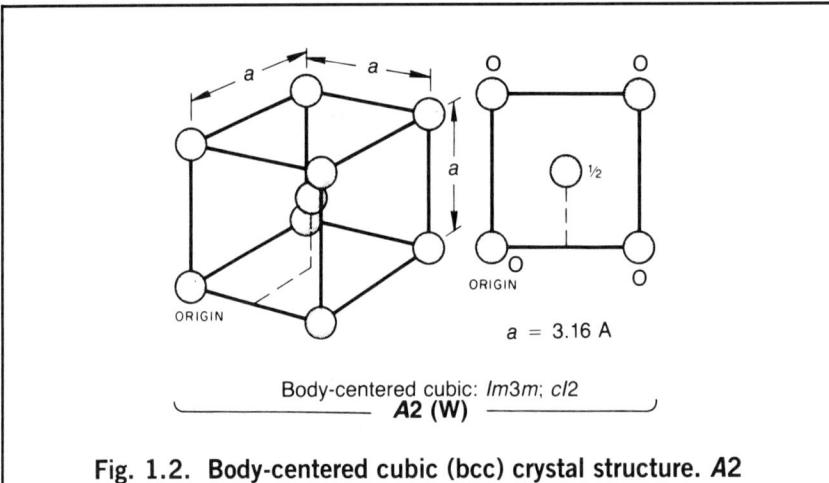

Fig. 1.2. Body-centered cubic (bcc) crystal structure. A2 is structure (Strukturbericht) symbol, and W is prototype metal with bcc structure. Ferrite in steel is bcc. (Ref 1.12)

The art and science of steel heat treatment is based on the existence of the austenite phase field in the Fe-C system. Controlled transformation of austenite to other phases is responsible for the great variety of microstructures and properties attainable in steels. Hot working of heavy sections into useful shapes and sizes by rolling or forging is also accomplished at temperatures where austenite is the stable phase.

Iron is an allotropic element: at atmospheric pressure, it may exist in more than one crystal form depending on the temperature. Alpha iron (ferrite) exists up to 912 °C (1674 °F); gamma iron (austenite) exists between 912 and 1394 °C (1674 and 2541 °F); and delta iron (delta ferrite) exists from 1394 °C (2541 °F) to the melting point of pure iron, 1538 °C (2800 °F). The temperature ranges in which the various crystal forms of iron are stable make up the left vertical boundary (the pure iron end) of the Fe-C phase diagram shown in Fig. 1.1.

Crystal Structures of Iron

The crystal structure of ferrite is characterized by the unit cell shown in Fig. 1.2. Ferrite belongs to the cubic crystal system—all three axes of the unit cell are of the same length a and are mutually perpendicular. The space lattice of ferrite is body-centered cubic (bcc). There are a total of two atoms per unit cell—the body-centered atom with coordinates $a/2$, $a/2$, $a/2$, and the atom at the origin of the unit cell with coordinates 0, 0,

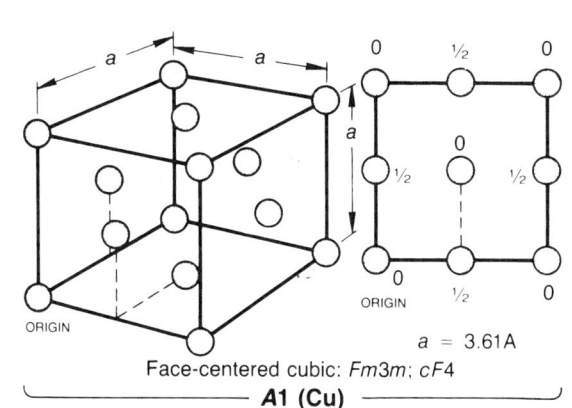

Fig. 1.3. Face-centered cubic (fcc) crystal structure. A1 is structure (Strukturbericht) symbol, and Cu is prototype metal with fcc structure. Austenite in steel is fcc. (Ref 1.12)

0. The latter atom represents all of the equivalent corner atoms of the unit cell, each of which is shared by eight unit cells that come together at a corner. The one-eighth atom per corner times the eight corners of the unit cell therefore accounts for one of the two atoms in a bcc unit cell.

The lattice parameter of alpha iron at room temperature is 2.86 Å, or 0.286 nm (Ref 1.2). The body diagonals of the unit cell, corresponding to $\langle 111 \rangle$ directions, are the directions in which the iron atoms are in contact in the bcc structure. Figure 1.2 shows that the body-centered atom has eight nearest neighbor atoms at a center-to-center distance of one-half a body diagonal, or $a\sqrt{3}/2$. Crystal structures in which the atoms are packed as closely together as possible have twelve nearest neighbor atoms, and therefore the bcc form of iron is a more open or less dense structure than the gamma iron structure described below. The difference in atomic packing between alpha and gamma is responsible for the volume expansion that occurs when the higher density gamma iron transforms to alpha iron on cooling.

The unit cell of gamma iron or austenite is shown in Fig. 1.3. Austenite also belongs to the cubic crystal system, but has a face-centered cubic lattice. There are a total of four atoms per unit cell with coordinates 0, 0, 0; 0, $a/2$, $a/2$; $a/2$, $a/2$, 0; $a/2$, 0, $a/2$, corresponding to a corner atom and an atom in the center of each face of the unit cell. Each face atom is shared by two adjacent unit cells; the six faces of the cubic cell thus contribute three atoms. As described above for the bcc cell, the eight corners together contribute only one atom.

The lattice parameter of austenite, about 3.56 Å (0.356 nm), is larger than that of ferrite. However, the close-packed structure and the 4 atoms per unit cell make the density of austenite greater than that of ferrite. The face diagonals, corresponding to ⟨110⟩ directions, are the close-packed directions in the fcc structure, and establish the center-to-center atom spacing of the 12 nearest neighbor atoms as $a\sqrt{2}/2$.

Austenite also may be characterized as a structure made up of planes of closest atomic packing stacked in a sequence that repeats every three layers. The orientation of the close-packed {111} planes relative to the unit cell may be readily identified because each {111} plane is defined by three face diagonals of the unit cell. The close-packed planes in austenite are extremely important: the dislocation motion that makes mechanical deformation of austenite possible occurs on {111} planes, and microstructural features within grains known as twins have {111} planes as boundaries. Twins are characterized by mirror symmetry of atoms across the planes separating the twins and the adjacent matrix (Ref 1.3). In austenite, twins frequently form as a result of growth accidents in the stacking of {111} planes—accidents caused by recrystallization and grain growth during heating or annealing in the temperature range where austenite is stable.

Finally, the third phase that may form in pure iron is delta ferrite, a body-centered cubic structure that is crystallographically identical to that of alpha iron. Delta ferrite forms only at temperatures close to the melting point of iron. It is generally only of academic interest in the heat treatment of carbon steels because it is replaced at lower temperatures by austenite, the usual starting structure for commercial heat treatment. However, because delta ferrite is the first phase to form during solidification of iron and steel ingots and welds, it may be associated with interdendritic segregation patterns or concentration gradients of alloying and/or impurity elements (Ref 1.4). Hot working and homogenizing steels in the austenite range generally significantly reduce the segregation produced during solidification, and some degree of segregation may be tolerated in many applications. Occasionally, however, a problem that arises during heat treatment may be traced back to the segregation produced as delta ferrite first formed from the molten steel.

Effects of Carbon

The addition of carbon to iron produces several important changes in the phases and phase equilibria just described. Differences in the ability of ferrite and austenite to accommodate carbon result not only in important characteristics of the Fe-C diagram but also in the formation of Fe₃C. The crystal structures of the bcc ferrite and fcc austenite are modified by introducing carbon atoms into the interstices or interstitial sites between iron atoms. Austenite and ferrite in Fe-C alloys and steels are, therefore, interstitial solid solutions.

Carbon is an element that stabilizes austenite and thereby increases the range of austenite formation in steels. Figure 1.1 shows that, with the addition of carbon, the austenite field greatly expands from 912 to 1394 °C (1675 to 2540 °F)—the range in pure iron—to a wide range of temperatures and compositions. The maximum solubility of carbon in austenite reaches 2.11% at 1148 °C (2018 °F). Ferrite has a much lower ability to dissolve carbon than does austenite: the solubility decreases continuously from a maximum of only 0.02% at 727 °C (1340 °F). The limited solubility of carbon in ferrite is emphasized by the very small ferrite field (see Fig. 1.1). An expanded portion of the low-carbon end of the Fe-C diagram showing the temperature-composition range of ferrite and the decreasing solubility of carbon in ferrite with decreasing temperature is shown in Fig. 1.4. The room temperature solubility of carbon in ferrite is almost negligible.

When the solubility limit for carbon in austenite is exceeded, a new phase—iron carbide or cementite—forms in iron-carbon alloys and steels. Cementite assumes many shapes, arrangements, and sizes that together with ferrite contribute to the great variety of microstructures found in steels. The various forms of cementite depend directly on thermal history or heat treatment. The crystal structures of cementite and of ferrite and austenite solid solutions will be discussed below, and the association and formation of those phases to produce characteristic microstructures will be discussed in later chapters.

Crystal Structures in Fe-C Alloys

The major difference between the structures of ferrite and austenite in steel and the corresponding phases in pure iron is the introduction of carbon atoms. There are two types of interstitial voids that may become sites for carbon atoms in bcc and fcc structures. Figures 1.5 and 1.6 show the octahedral and tetrahedral voids in the fcc and bcc structures, respectively. The two types of voids derive their names from the number of sides of the polyhedron formed by the iron atoms that surround a given site. A carbon atom has six nearest neighbor iron atoms if in an octahedral site and four if in a tetrahedral site.

The sizes of the different voids vary considerably. In austenite, assuming spherical iron atoms in contact, an octahedral site could accommodate an atom 0.052 nm in radius, but a tetrahedral site could

8 / STEELS: HEAT TREATMENT AND PROCESSING PRINCIPLES

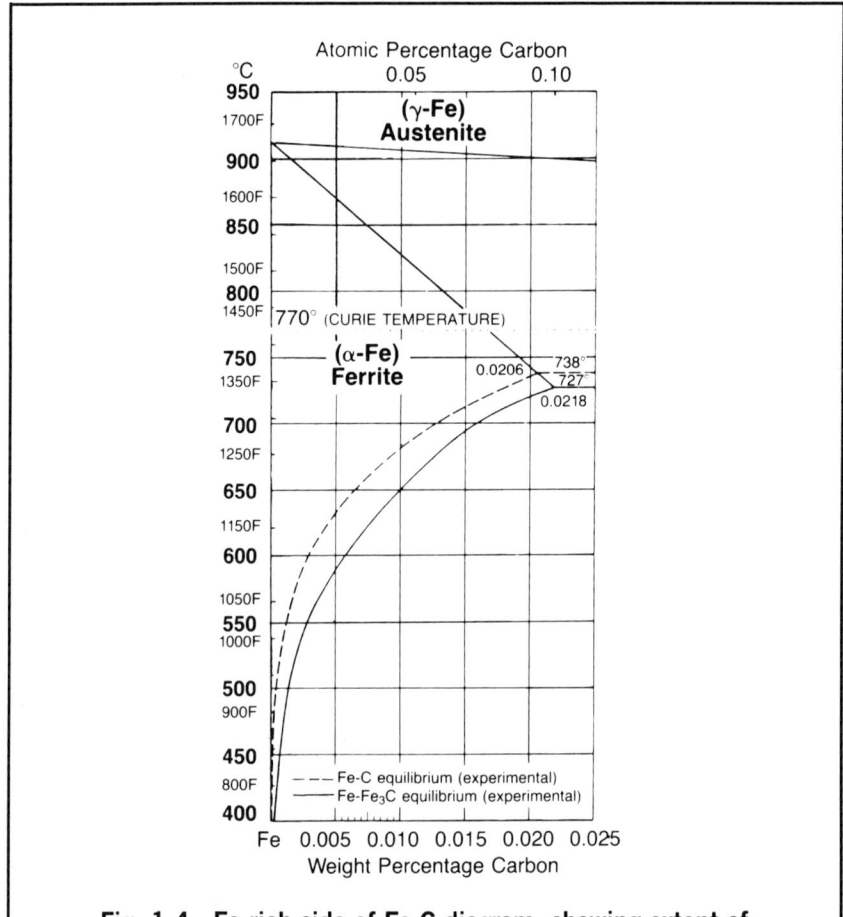

Fig. 1.4. Fe-rich side of Fe-C diagram, showing extent of ferrite phase field and decrease of carbon solubility with decreasing temperature. (Ref 1.12)

only accommodate an atom 0.028 nm in radius (Ref 1.5). Carbon atoms have radii of 0.07 nm, and are therefore more readily accommodated in the octahedral voids even though some lattice expansion is required.

In ferrite the interstitial sites are much smaller, thus explaining the very limited solubility of carbon. A tetrahedral site in ferrite could accommodate an interstitial atom 0.035 nm in radius and an octahedral site, an atom only 0.019 nm in radius. The octahedral sites in ferrite, however, are not symmetrical (see Fig. 1.6) and a carbon atom would severely displace only the two atoms at a distance of $a/2$, not those at a distance $a/\sqrt{2}$. Carbon atoms appear to prefer the octahedral sites in ferrite (Ref 1.5) and do produce a severe distortion of the lattice

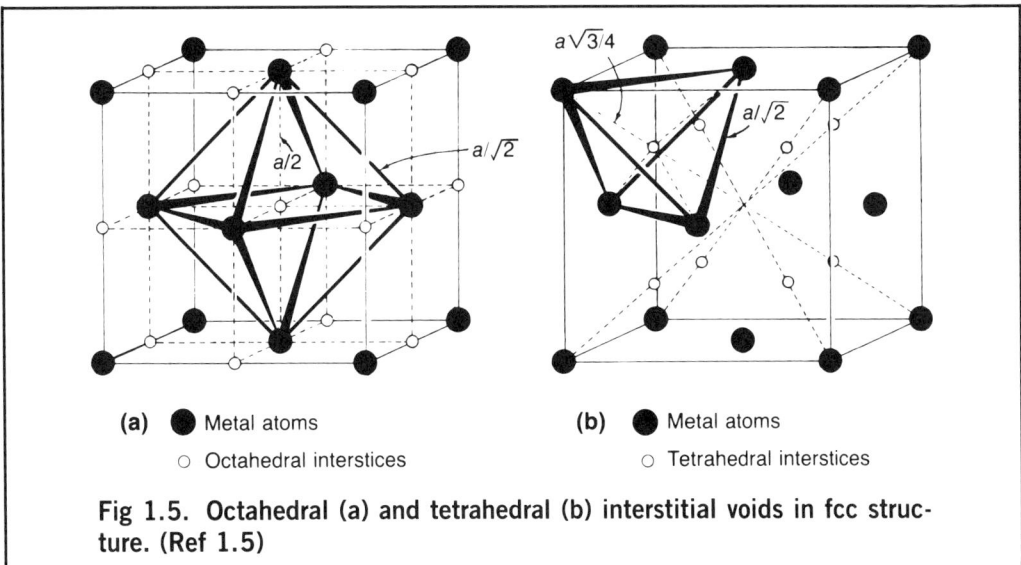

Fig 1.5. Octahedral (a) and tetrahedral (b) interstitial voids in fcc structure. (Ref 1.5)

in ⟨100⟩ directions. In ferrite, because of the limited number of carbon atoms that can be accommodated, the lattice remains essentially cubic. If large numbers of carbon atoms present in austenite are trapped in

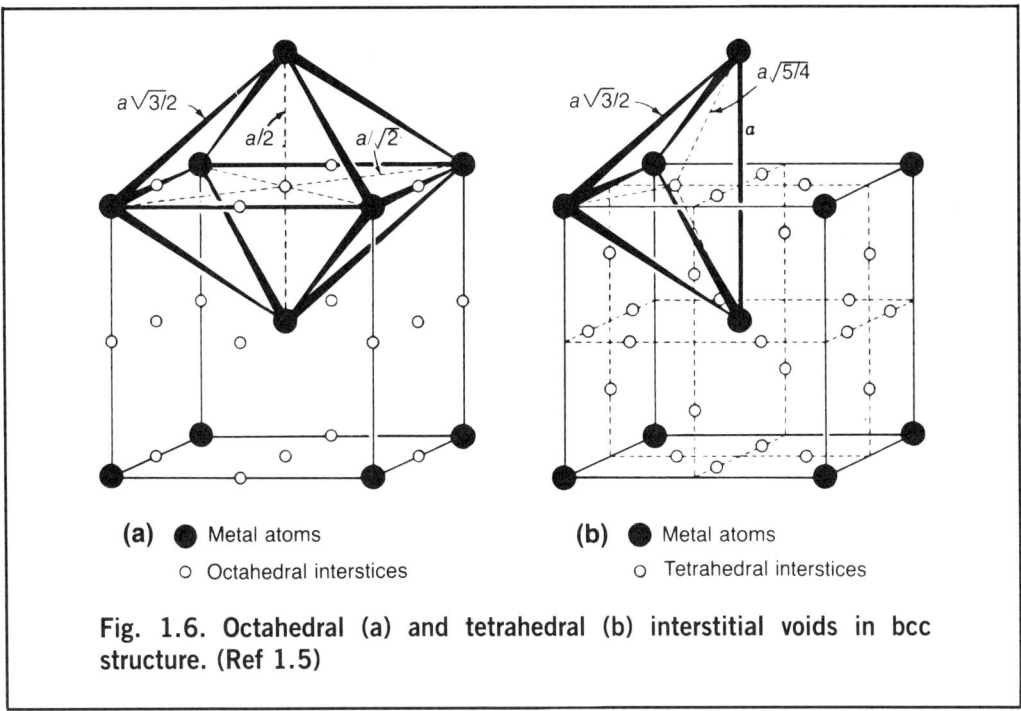

Fig. 1.6. Octahedral (a) and tetrahedral (b) interstitial voids in bcc structure. (Ref 1.5)

bcc octahedral sites by rapid cooling, the cubic structure may actually become tetragonal. The latter structure typifies the phase "martensite", and its formation is the object of the very important hardening heat treatments to be described in later chapters.

Cementite, the phase that forms when the solubility of carbon in ferrite and austenite is exceeded, is a significantly different phase from the interstitial solid solutions described above. Cementite is a compound with a specific ratio of one carbon atom to three iron atoms, and is frequently referred to as Fe_3C. Cementite contains 6.67% carbon and could exist alone only in an alloy at that composition, in contrast to ferrite or austenite, which may exist as single phases over a range of alloy carbon content.

Cementite is orthorhombic, with lattice parameters a = 0.452 nm, b = 0.509 nm, and c = 0.674 nm. Its unit cell contains 12 iron atoms and 4 carbon atoms. The positions of iron and carbon atoms relative to the unit cell axes of cementite (Ref 1.5) are shown in Fig. 1.7.

Effects of Alloying Elements

Up to this point, only the binary Fe-C diagram and the crystal structure of the phases that form in Fe-C alloys have been described. Steels, however, contain alloying elements and impurities that must be incorporated into austenite, ferrite, and cementite. Incorporation is usually by replacement of iron atoms if the alloy or impurity atoms are roughly the same size as iron atoms, but sometimes the atoms go into interstitial sites if they are significantly smaller than iron, as is nitrogen. In some cases, if sufficient quantities of alloying elements are present, solubility limits are exceeded and phases other than those already discussed may form. For example, small additions of chromium to Fe-C alloys at 890 °C (1634 °F) maintain the cementite structure, M_3C (M standing for a combination of chromium and iron atoms); larger additions cause the carbide M_7C_3 to form; and still larger additions produce the carbide $M_{23}C_6$ (Ref 1.6).

Some of the elements present in steels are austenite stabilizers (manganese and nickel, for instance), some are ferrite stabilizers (silicon, chromium, and niobium), and some are strong carbide formers (titanium, niobium, molybdenum, and chromium, if present in sufficient quantity). Ferrite and austenite stabilizers expand the respective phase fields. One measure of the effect of an alloying element on the Fe-C phase diagram is whether or not the eutectoid temperature (indicated by the horizontal line at 727 °C [1340 °F] in Fig. 1.1) is raised or lowered by an alloying addition. Austenite stabilizers lower

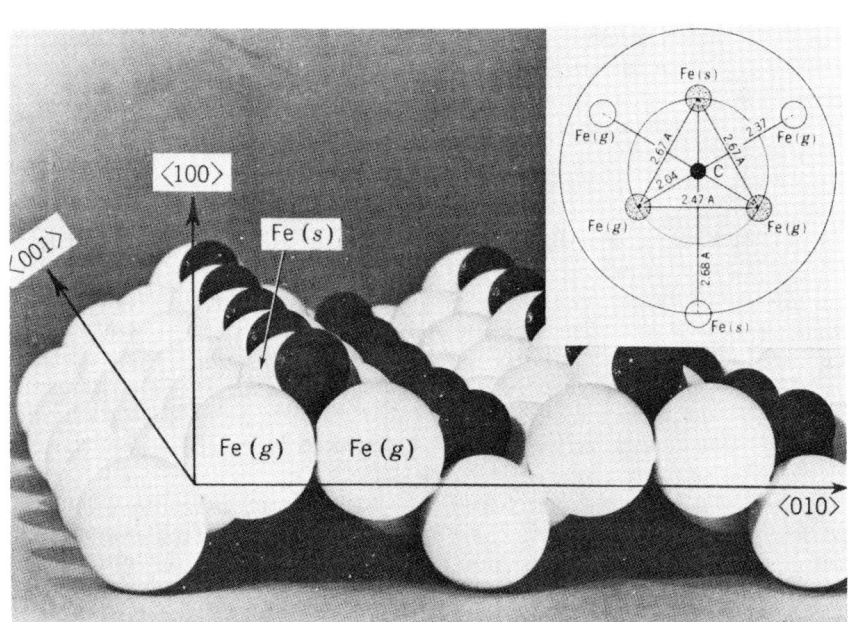

Fig. 1.7. Model of cementite structure that forms in steel. Insert is stereogram of iron nearest and next-nearest neighbor atoms around a carbon atom. (Ref 1.5)

the eutectoid temperature and thereby expand the temperature range over which austenite is stable. Figure 1.8 shows the change in eutectoid temperature with increasing amounts of several common alloying elements (Ref 1.7). Figure 1.9 shows a related effect of alloying elements on the Fe-C phase diagram: the decrease in carbon content of austenite of eutectoid composition. The type of evidence on which Fig. 1.8 and 1.9 were based is shown in Fig. 1.10 for the Fe-Cr-C system. Here, the austenite phase field shrinks with increasing chromium content. Associated changes in eutectoid composition and temperature are also graphically represented. The effect of alloying elements on the austenite phase field is usually determined by experimental techniques, but thermodynamic data have been used to calculate ranges of austenite stability in ternary alloys (Ref 1.8).

Critical Temperatures

The boundaries between phase fields of the Fe-C diagram shown in Fig. 1.1 identify temperatures for the various phase transforma-

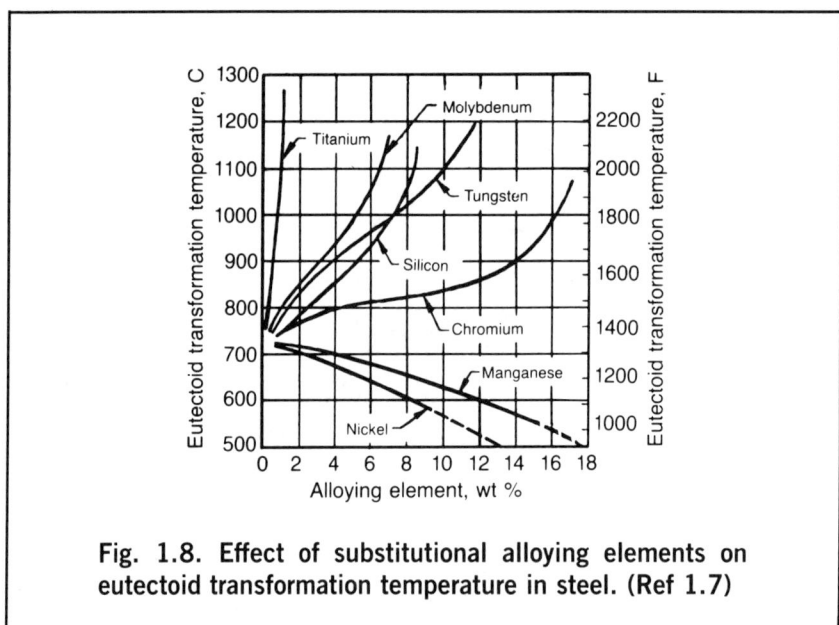

Fig. 1.8. Effect of substitutional alloying elements on eutectoid transformation temperature in steel. (Ref 1.7)

tions that may occur in Fe-C alloys. For example, if an Fe-0.5C alloy were heated from room temperature at an extremely low rate, some of the ferrite and all of the cementite would transform to austenite at

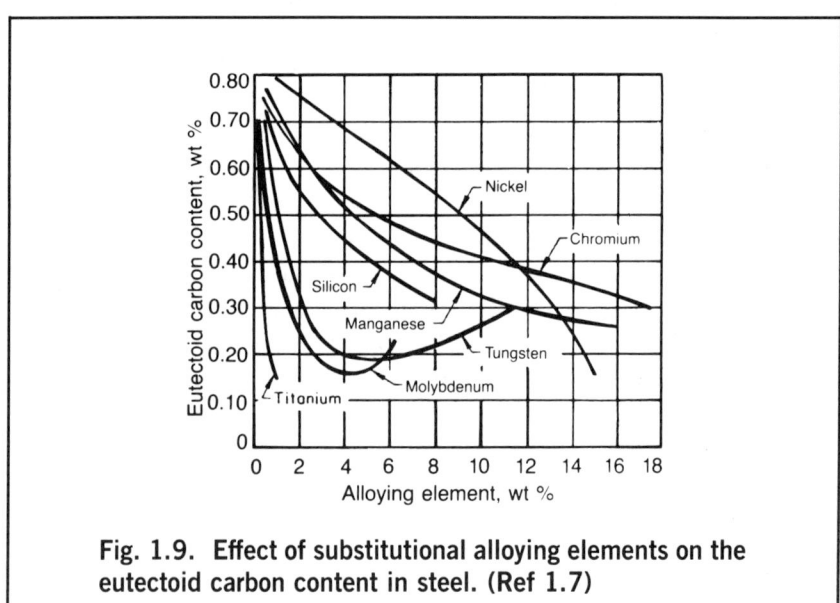

Fig. 1.9. Effect of substitutional alloying elements on the eutectoid carbon content in steel. (Ref 1.7)

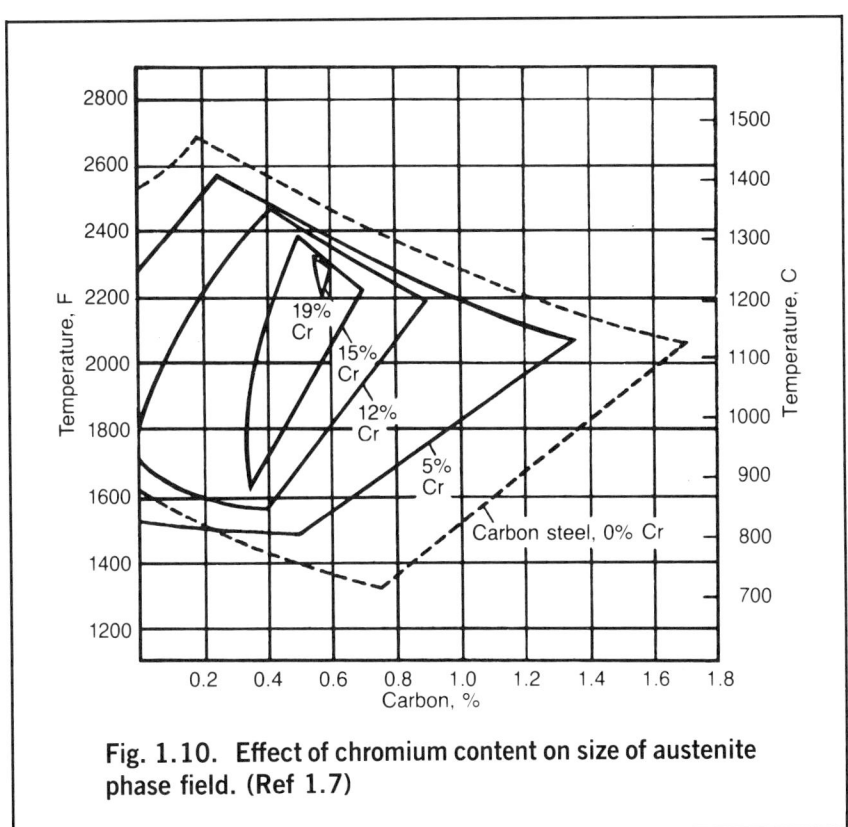

Fig. 1.10. Effect of chromium content on size of austenite phase field. (Ref 1.7)

727 °C (1340 °F), and at about 860 °C (1580 °F) the last bit of ferrite would be completely transformed to austenite.

The transformation temperatures are often referred to as critical temperatures and are observed by measuring changes in heat transfer or volume as specimens are heated or cooled. On heating, heat is absorbed and specimen contraction occurs as ferrite and cementite are replaced by the close-packed structure, austenite. On cooling, heat is evolved and specimen expansion occurs as austenite transforms to ferrite and cementite. The absorption or release of heat during phase transformation produces a change in slope, or "arrest", on a continuous plot of specimen temperature versus time. The letter "A" is the symbol for the thermal arrests that identify critical temperatures.

There are three critical temperatures of interest in the heat treatment of steel: the A_1, which corresponds to the boundary between the ferrite-cementite field and the fields containing austenite and ferrite or austenite and cementite; the A_3, which corresponds to the boundary between the ferrite-austenite and austenite fields; and the A_{cm}, which corresponds to the boundary between the cementite-austenite and the austenite fields. These temperatures assume equilibrium conditions—that is, extended periods of time at temperature or extremely slow rates of heating or cooling. Sometimes A_1, A_3, and A_{cm} are designated as Ae_1, Ae_3, and Ae_{cm}, respectively, the letter "e" indicating assumed equilibrium conditions.

The transformations that occur at A_1, A_3, and A_{cm} are diffusion controlled. Therefore, the critical temperatures are sensitive to composition and to heating and cooling rates. Rapid heating allows less time for diffusion and tends to increase the critical temperatures above those associated with equilibrium. Likewise, rapid cooling tends to lower the critical temperatures. The effect of heating or cooling rate is defined practically by a new set of critical temperatures designated "Aĉ" or "Ar̂" (for the arrests on heating or cooling, respectively). The terminology was developed by the French metallurgist, Osmond (Ref 1.9)—Ac stands for *arrêt chauffant* and Ar for *arrêt refroidissant*. As a result of heating and cooling effects, therefore, there are two other sets of critical temperatures: Ac_1, Ac_3, and Ac_{cm}, and Ar_1, Ar_3, and Ar_{cm}. These sets of critical temperatures are shown schematically in Fig. 1.11.

Generally, the critical temperatures for a given steel are determined experimentally. However, some empirical formulas that show the effects of alloying elements on the critical temperatures have been developed by regression analysis of large amounts of experimental data. For example, Andrews (Ref 1.10) has developed the following formulas for Ac_3 and Ac_1 in degrees Celsius:

$$Ac_3 = 910 - 203\sqrt{C} - 15.2Ni + 44.7Si + 104V$$
$$+ 31.5\ Mo + 13.1W \quad \text{(Eq 1.3)}$$
$$Ac_1 = 723 - 10.7Mn - 16.9Ni + 29.1Si + 16.9Cr$$
$$+ 290As + 6.38\ W \quad \text{(Eq 1.4)}$$

These formulas present another way of describing the effect of alloying elements on both the Fe-C diagram and the transformation behavior of steels. Elements that stabilize austenite lower the Ac_3 and Ac_1 as evidenced by their negative contributions to the corresponding equation, whereas elements that stabilize ferrite or carbide raise the Ac_3 and Ac_1 and make a positive contribution. The effect of alloying elements on the Ac_3 has also been determined by thermodynamic calculations (Ref 1.11).

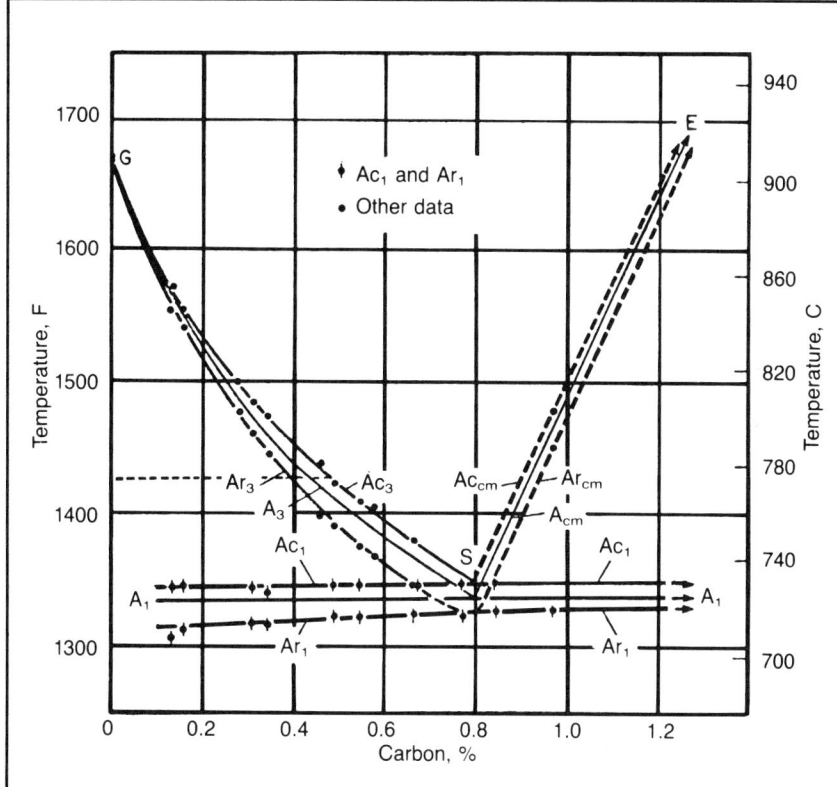

Fig. 1.11. Cooling (Ar), heating (Ac), and equilibrium (A) temperatures in Fe-C alloys. Heating and cooling at 0.125 °C/min. (Ref 1.7)

References

1.1 L. Kaufman and H. Bernstein, *Computer Calculation of Phase Diagrams*, Academic Press, New York, 1970
1.2 C.S. Roberts, Effect of Carbon on the Volume Fractions and Lattice Parameters of Retained Austenite and Martensite, *Trans TMS-AIME*, Vol 197, 1953, p 203
1.3 B.D. Cullity, *Elements of X-ray Diffraction*, Addison-Wesley, Reading, MA, 1956, p 55-60
1.4 M.C. Flemings, *Solidification Processing*, McGraw-Hill, New York, 1974
1.5 C.S. Barrett and T.B. Massalski, *Structure of Metals*, 3rd ed., McGraw-Hill, New York, 1966
1.6 L.R. Woodyatt and G. Krauss, Iron-Chromium-Carbon System at 870 °C, *Met Trans A*, Vol 7A, 1976, p 983-989
1.7 E.C. Bain and H.W. Paxton, *Alloying Elements in Steel*, 2nd ed., American Society for Metals, Metals Park, OH, 1961
1.8 J.S. Kirkaldy, B.A. Thomson, and E.A. Baganis, Prediction of Multicom-

ponent Equilibrium and Transformation Diagrams for Low Alloy Steels, in *Hardenability Concepts and Applications to Steel*, D.V. Doane and J.S. Kirkaldy (Eds.), TMS-AIME, Warrendale, PA, 1978

1.9 F. Osmond, *Transformation du Fer*, Baudoin and Co., Paris, 1888

1.10 K.W. Andrews, Empirical Formulae for the Calculation of Some Transformation Temperatures, *JISI*, Vol 203, 1965, p 721-727

1.11 J.S. Kirkaldy and E.A. Baganis, Thermodynamic Prediction of the Ae_3 Temperature of Steels with Additions of Mn, Si, Ni, Cr, Mo and Cu, *Met Trans A*, Vol 9A, 1978, p 495-501

1.12 *Metals Handbook*, 8th ed., Vol 8, American Society for Metals, Metals Park, OH, 1973, p 236, 275, 276

CHAPTER 2

Pearlite, Ferrite, and Cementite

Chapter 1 described the crystal structures of the phases that form in steels and the Fe-C phase diagram which defines the temperature-composition ranges over which these phases may exist. This chapter shows how various arrangements of phases or microstructures are produced by austenite transformation to ferrite and cementite. Alloy composition and the rate at which austenite is cooled profoundly affect which microstructure forms. The emphasis in this chapter is on the microstructures produced by the diffusion-controlled transformations that occur in carbon steels during relatively slow cooling from the austenite phase field.

Eutectoid Transformation

The Fe-C diagram introduced in Chapter 1 provides the basic framework for understanding the phase transformations and microstructures of concern in this chapter. Figure 2.1 is an enlarged section of the Fe-C diagram that includes the areas most pertinent to the transformation of austenite in slowly cooled steels. Consider first the Fe-0.77C alloy, which would be completely austenitic at all temperatures down to the A_1 temperature (727 °C or 1340 °F). If held for a very long period of time at this temperature, or cooled very slowly through A_1 (that is, under conditions approaching equilibrium), the phase diagram shows that the austenite must be replaced by a mixture of ferrite and cementite. A phase transformation in which one solid phase is replaced by two different solid phases is classified as a eutectoid transformation, and in the Fe-C system may be written in the form of the following reaction:

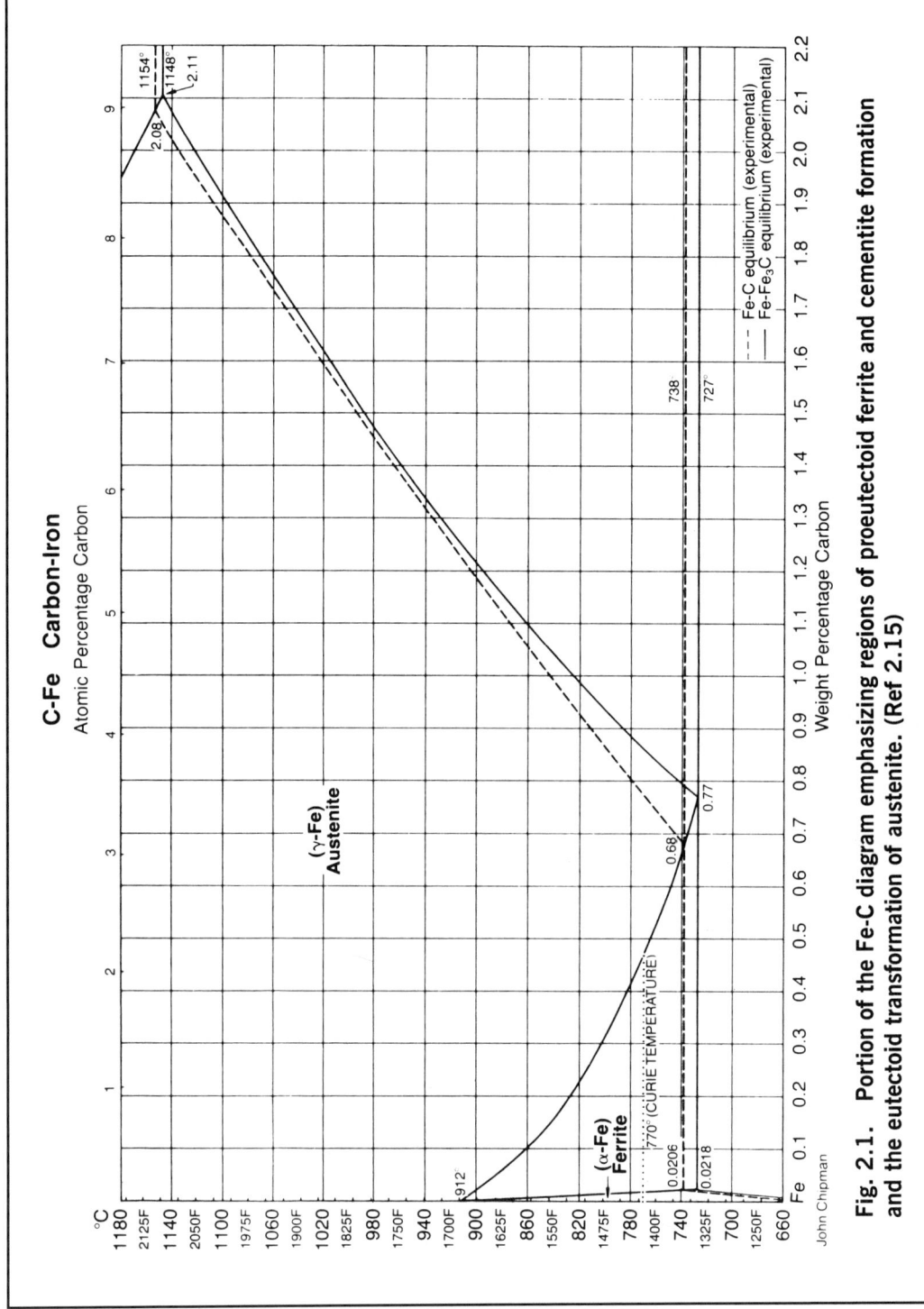

Fig. 2.1. Portion of the Fe-C diagram emphasizing regions of proeutectoid ferrite and cementite formation and the eutectoid transformation of austenite. (Ref 2.15)

$$\gamma\ (0.77\%\ C) \underset{\text{heating}}{\overset{\text{cooling}}{\rightleftharpoons}} \alpha\ (0.02\%\ C) + Fe_3C(6.67\%\ C) \quad \text{(Eq 2.1)}$$

This equation shows that the phases involved in the eutectoid reaction have fixed compositions and that the reaction is reversible depending upon whether heat is removed or added. Ideally the eutectoid reaction occurs isothermally at 727 °C (1340 °F). Equilibrium conditions, however, are rarely obtained in actual practice and the eutectoid reaction may in fact occur over a wide range of temperatures below A_1.

Structure of Pearlite

The eutectoid transformation in steels produces a unique microstructure termed "pearlite". Pearlite is made up of alternate lamellae of ferrite and cementite as shown in Fig. 2.2, a light micrograph of a furnace-cooled specimen of an Fe-0.75C alloy. Colonies of lamellae of various orientations and spacings characterize the microstructure. The spacing variations of the cementite lamellae in different areas may be partly due to differences in the angles that the lamellae make with the plane of polish, and partly due to the fact that the pearlite may have formed over a range of temperatures. Assuming all the pearlite was formed at about the same temperature, and that therefore all the lamellae had almost identical spacing, those colonies with lamellae perpendicular to the plane of polish would show the true spacing or closest spacing of the ferrite and cementite lamellae. Those lamellae at angles less than 90° would show a wider spacing. Determination of the true pearlite spacing from metallographically prepared specimens where the lamellae form a range of angles with the specimen surface requires special quantitative metallographic analyses (Ref 2.1, 2.2).

The origin of the term "pearlite" is related to the regular array of the lamellae in the colonies and the fact that etching attacks the ferrite phase more severely than the cementite. The raised and regularly spaced cementite lamellae of the colonies then act as diffraction gratings, and a pearl-like lustre is produced by diffraction of light of various wave lengths from the different colonies.

The amounts of cementite and ferrite in pearlite formed at 727 °C (1340 °F) can be determined by a calculation based on the lever rule. The lever rule can be applied to any two-phase field of a binary phase diagram to determine the amounts of the different phases present at a given temperature in a given alloy. A horizontal line, referred to as a tie line, represents the lever, and the alloy composition its fulcrum.

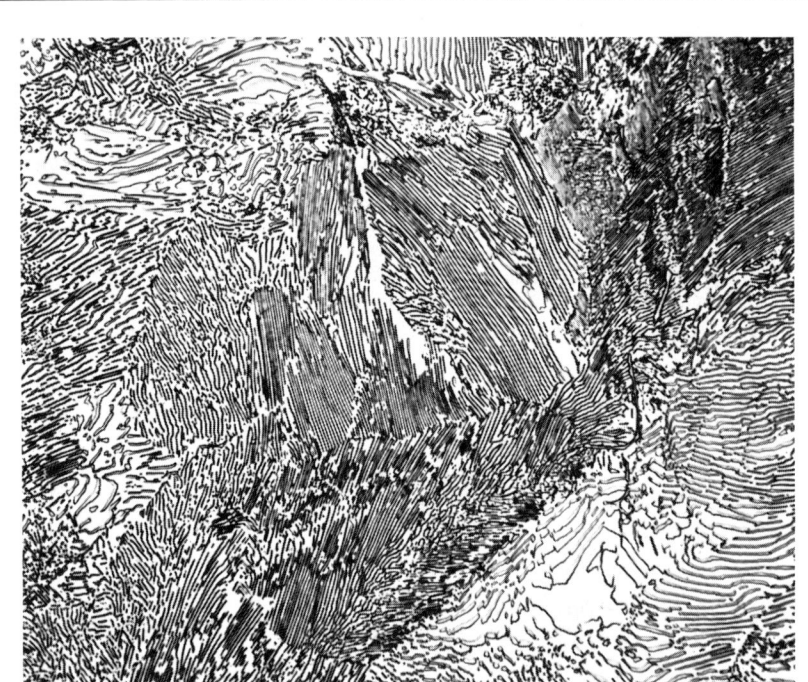

Fig. 2.2. Pearlite in a furnace-cooled Fe-0.75C alloy. Picral etch. Magnification, 500×. Courtesy of A.R. Marder and A. Benscoter, Bethlehem Steel Corp., Bethlehem, PA

The intersection of the tie line with the boundaries of the two-phase field fixes the compositions of the coexisting phases, and the amounts of the phases are proportional to the segments of the tie line between the alloy and the phase compositions. For pearlite, assume a tie line immediately below 727 °C (1340 °F) that spans the ferrite-cementite phase field (see Fig. 1.1). Application of the lever rule calculation for the Fe-0.77C alloy, the alloy that transforms entirely to pearlite, shows that:

$$\text{wt.\% Fe}_3\text{C in pearlite} = \frac{0.77 - 0.02}{6.67 - 0.02} \times 100 = 11\% \quad (\text{Eq 2.2})$$

By difference, the weight percent ferrite in pearlite is 89%. Therefore, whenever austenite containing 0.77% carbon transforms to pearlite at or close to 727 °C (1340 °F), ferrite and cementite form in the fixed weight percentages as shown above. The densities of ferrite and cementite, 7.87 and 7.70 g/cm³ respectively, are so close that the volume percentages of ferrite and cementite in pearlite are essentially the same as the weight percentages. Therefore, in Fe-C alloys, the amounts of phases calculated by the lever rule with compositions by weight should correlate well with the amounts of phases revealed in light micrographs. The amounts of phases visible in micrographs are related to area percentages, which in turn are directly related to their volume percentages if the phases are uniformly distributed.

The development of a pearlite colony has been shown to initiate from either ferrite or cementite crystals (Ref 2.3, 2.4). Originally, the lamellar structure was thought to develop only by sidewise nucleation of separate lamellae; however, branching of a single cementite crystal into parallel lamellae with spacing characteristic of a given transformation temperature has also been shown to produce the lamellar structure. According to the latter mechanism, all of the cementite in a given colony is interconnected, and a colony of pearlite may be regarded as two single crystals of ferrite and cementite. The latter structure was strikingly revealed in a serial sectioning experiment in which a pearlite colony was repeatedly photographed as successive layers were removed by polishing in 1-μm steps (Ref 2.3). All of the apparently separate cementite lamellae were shown to have a common origin. Once a pearlite colony is established by sidewise nucleation and/or branching of the ferrite and cementite, the lamellae are considered to grow by extension of their edges into the austenite, a process frequently referred to as edgewise growth (Ref 2.5).

Pearlite Transformation Kinetics

The preceding section described the lamellar structure of pearlite and its formation by a eutectoid reaction at or close to 727 °C (1340 °F). In actual practice, however, the formation of pearlite rarely occurs close to the A_1. Figure 2.3 shows an isothermal transformation diagram for eutectoid 1080 steel. Curves for the beginning and end of pearlite formation, obtained by cooling from the austenite phase field and holding at various temperatures below A_1, are shown. The beginning of transformation curve is asymptotic to the A_1, thus indicating that pearlite would not form at temperatures close to A_1 unless the steel were held at temperature for extended periods of time. In commercial heat treating practice, the slow rates of cooling that would permit pearlite formation close to the A_1 are approached only in very heavy sections or by furnace cooling. With increased undercooling below A_1, however, the time periods for the beginning and end of

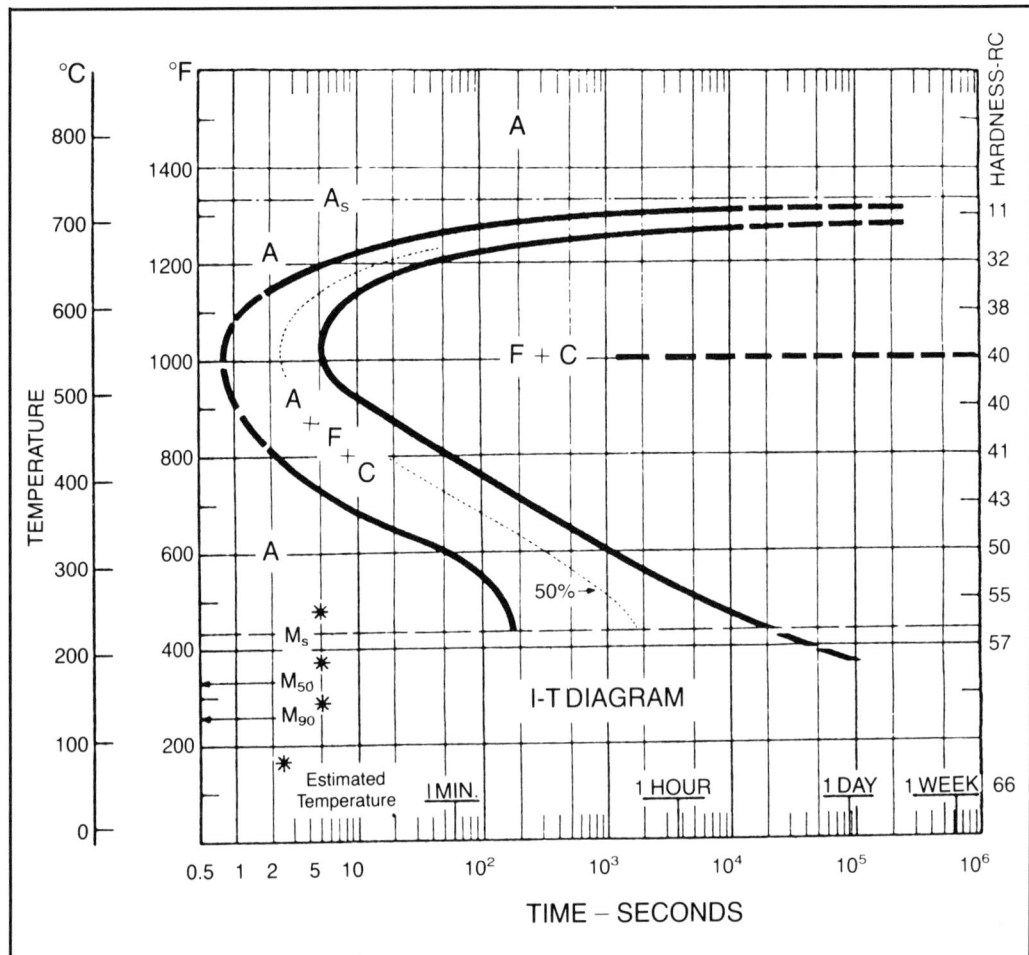

Fig. 2.3. Isothermal transformation diagram for 1080 steel containing 0.79% carbon and 0.76% manganese. Specimens were austenitized at 900 °C (1650 °F) and had an austenite grain size of ASTM No. 6. (Ref 2.16)

pearlite transformation are reduced substantially. At the nose of the transformation curve, 540 °C (1004 °F), the lowest temperature at which pearlite forms in this steel, only seconds are required for complete transformation. Below 540 °C (1004 °F), bainite, a nonlamellar microstructure of ferrite and cementite, is formed.

A number of factors influence the rate of pearlite formation. Perhaps most important is the fact that substantial carbon atom rearrangement must take place to accomplish the transformation of austenite (containing nominally 0.77% carbon) to low-carbon ferrite and high-carbon cementite according to Eq 2.1. The diffusion of carbon, as characterized by its diffusion coefficient, is temperature dependent. One equation that has been developed (Ref 2.6) to show the temperature dependence of carbon diffusion in austenite is

$$D_C^\gamma = 0.12 e^{-32,000/RT} \qquad \text{(Eq 2.3)}$$

where D_C^γ is the average diffusion coefficient (cm^2/s) of carbon in austenite, R is the gas constant (1.98 cal/g-mol/K), and T is the absolute temperature (°C + 273). Equation 2.3 shows that the diffusion coefficient decreases exponentially with decreasing temperature, a powerful effect that significantly lowers the diffusion coefficient for small decreases in temperature. At first glance, the temperature dependence of diffusion appears to contradict the experimentally established fact (see Fig. 2.3) that pearlite formation is faster at lower temperatures than at higher temperatures. This apparent anomaly is explained by the reduction of interlamellar spacing as the temperature of pearlite transformation decreases. Thus, the distance that carbon has to diffuse to distribute itself between the ferrite and cementite decreases, and despite the fact that diffusion becomes more sluggish at lower temperatures, the growth of pearlite colonies accelerates.

The interrelationships between diffusion and the lamellar structure of pearlite help to explain how the eutectoid transformation proceeds, but not why the transformation occurs. The stability of all phases and microstructures in metals and alloys is based on the principle of minimum free energy. If the free energy of a given microstructure or system is not a minimum, then either a phase transformation (for example, the austenite to pearlite transformation under consideration here) or microstructural rearrangement without a phase change (for example, grain growth or particle coarsening) would occur in order to lower free energy to the minimum possible value.

The free energy (G) per unit volume of a phase or combination of phases is defined in terms of other thermodynamic parameters, enthalpy or heat content (H), the absolute temperature (T), and entropy (S) as follows:

$$G = H - TS \qquad \text{(Eq 2.4)}$$

Enthalpy is the total energy of a phase (or microstructure composed of several phases) per unit volume of that structure. Entropy is a measure of the degree of order associated with a given structure at a given temperature. It may be influenced by the amplitude of atom vibration, the mixing of several component types of atoms and/or vacant lattice sites in a given phase, or the degree of order associated with a given solid or liquid structure. The TS term, therefore, is a measure of the energy associated with the order of a unit volume of a given structure at a given temperature, and is especially important in establishing phase stability at high temperatures. Equation 2.4 shows that the difference between the enthalpy and entropy terms defines free energy. A rigorous approach to the development of atomistic and classical thermodynamics is presented in the text by Swalin (Ref 2.7).

A helpful example of the application of the principle of minimum free energy in establishing phase stability is the melting of a solid crystal structure. With increasing temperature, H increases, but TS increases much more if the liquid, with its high degree of atomic disorder, replaces the ordered crystal structure. Therefore, above the melting point, because of its higher entropy, the liquid has the lower free energy and is the stable phase relative to the solid. Similar considerations apply to transformations between solid phases such as the transformation of austenite to ferrite and cementite.

At the Ae_1 temperature in the Fe-C system, the free energy of austenite is exactly equal to the free energy of ferrite and cementite and there is no incentive for transformation to occur, especially if interfaces or boundaries between the austenite and pearlite must be created. Interfaces accommodate structural and chemical discontinuities between phases, and therefore make positive contributions to or raise the energy of a system. However, with decreasing temperature below Ae_1, the free energy of a unit volume of a mixture of ferrite and cementite becomes much less than that of austenite. This free energy difference is frequently referred to as the driving force for transformation and increases with decreasing temperature or undercooling below the Ae_1. A larger driving force makes possible not only the development of more colonies of pearlite but also a finer lamellar spacing within a pearlite colony, structural changes that result in increased interfacial area of two types. A higher density of pearlite colonies results in increased austenite/pearlite interfacial area, and a reduced interlamellar spacing results in increased ferrite-cementite interfacial energy within the colonies. The high driving force at low temperatures offsets the positive energy contributions due to the various interfaces produced during the austenite transformation to pearlite.

Many relationships for the change in interlamellar spacing with undercooling have been proposed, but the one most closely related to the above considerations was developed by Zener and Hillert (Ref 2.8) as presented in the following equation:

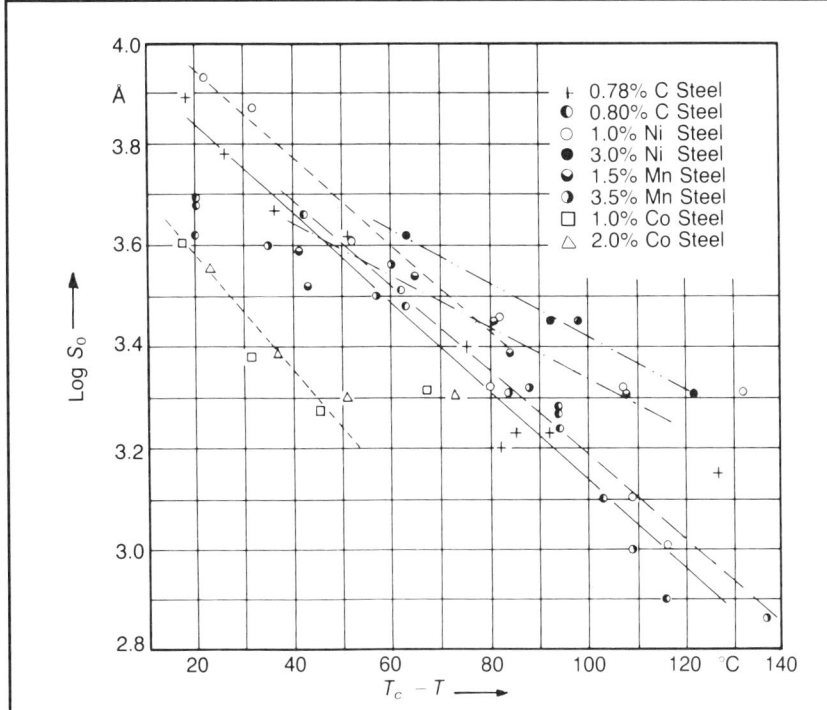

Fig. 2.4. Average true interlamellar spacings of pearlite, S_o, as a function of undercooling below Ae_1 for various steels as indicated. (Ref 2.1)

$$S = \frac{4\sigma_{\alpha/Fe_3C}T_E}{\Delta H_V \Delta T} \qquad \text{(Eq 2.5)}$$

where S is the interlamellar spacing defined by the combined width of the α and Fe_3C lamellae; σ_{α/Fe_3C} is the interfacial energy per unit area of α/Fe_3C boundary; T_E is the equilibrium temperature in degrees Kelvin (Ae_1 in the case of Fe-C alloys and steels); ΔH_V is the change in enthalpy per unit volume between austenite and the mixture of ferrite and cementite; and ΔT is the undercooling below Ae_1. Figure 2.4 shows a set of measurements illustrating the decrease in pearlite spacing with increasing undercooling for a variety of steels.

The isothermal transformation kinetics of the eutectoid transformation, i.e., the progress of pearlite formation as a function of time at a constant temperature, are based on the nucleation and growth rates of pearlite colonies. Figure 2.5 shows circular cross sections of pearlite colonies in an Fe-C alloy of eutectoid composition that has been

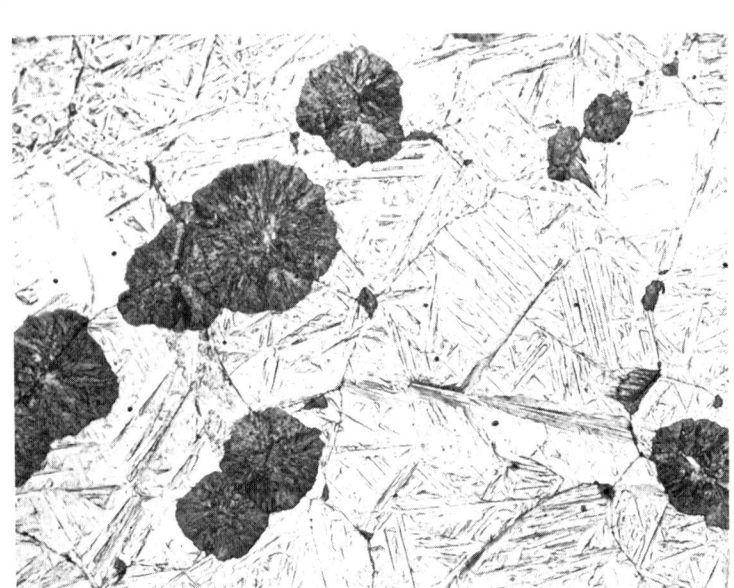

Fig. 2.5. Cross sections of spherical colonies of pearlite (dark) in eutectoid steel. Remainder of microstructure is martensite formed in austenite not transformed to pearlite at the reaction temperature. Magnification, 250×. Courtesy of A.R. Marder and B. Bramfitt, Bethlehem Steel Corp., Bethlehem, PA

partially transformed to pearlite. A number of colonies of pearlite have been nucleated and are in the process of growing into the austenite at the reaction temperature. In contrast to the pearlite shown in Fig. 2.2, the individual lamellae are too closely spaced to be resolved at the magnification of the micrograph, and the pearlite colonies have a dark appearance. The balance of the microstructure is white-etching martensite, formed in any untransformed austenite when it was quenched from the reaction temperature. Martensite and its formation are described in Chapter 3.

Johnson and Mehl (Ref 2.9), assuming that the pearlite colonies are spherical and randomly nucleated as a function of time, developed the following equation for isothermal pearlite formation:

$$f(t) = 1 - \exp[-\pi N G^3 t^4 / 3] \qquad \text{(Eq 2.6)}$$

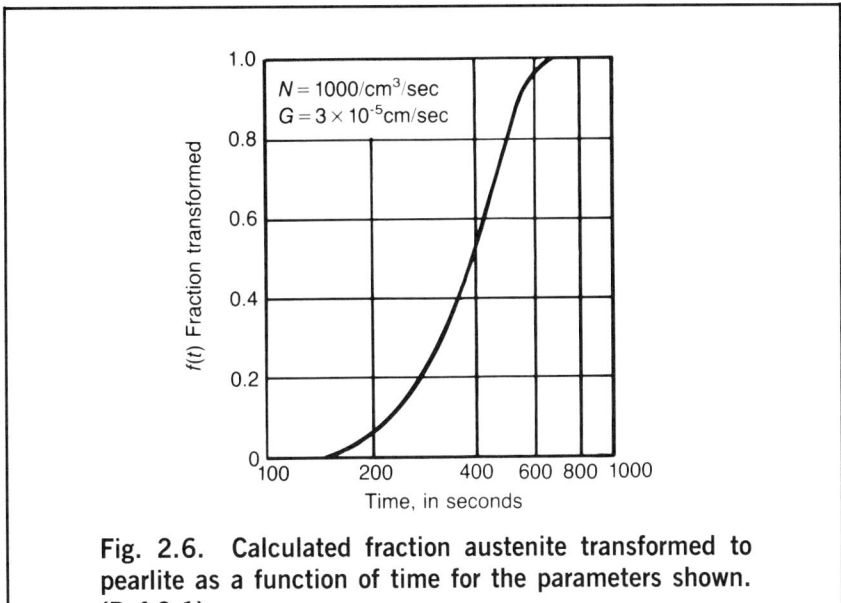

Fig. 2.6. Calculated fraction austenite transformed to pearlite as a function of time for the parameters shown. (Ref 2.1)

where f(t) is the volume fraction pearlite formed at any time t at a given temperature, N is the nucleation rate of the colonies, and G is the rate at which the colonies grow into the austenite. The Johnson-Mehl equation describes mathematically the rate at which austenite is converted to a pearlitic microstructure by the nucleation and growth of pearlite colonies. At any given temperature, f(t) versus time fits an "S-shaped" or sigmoidal curve as shown in Fig. 2.6. The initial transformation rate is quite low and is associated with what is referred to as an incubation period, the time when the first stable nuclei develop. As more and more nuclei develop and are in various stages of growth, the rate of transformation increases. Finally the colonies impinge and the rate of transformation again slows as the microstructure gradually approaches complete transformation. The elapsed time periods needed to initiate and complete the pearlite transformation are directly related to the beginning and end of transformation curves in isothermal transformation diagrams shown schematically in Fig. 2.7. The exact beginning and end of pearlite formation at any given temperature is of course dependent on the sensitivity of the experimental techniques used to follow the transformation, but generally the accuracy is on the order of 1%. Therefore, the beginning and end of transformation curves in Fig. 2.7 correspond to 1% and 99% transformation, respectively.

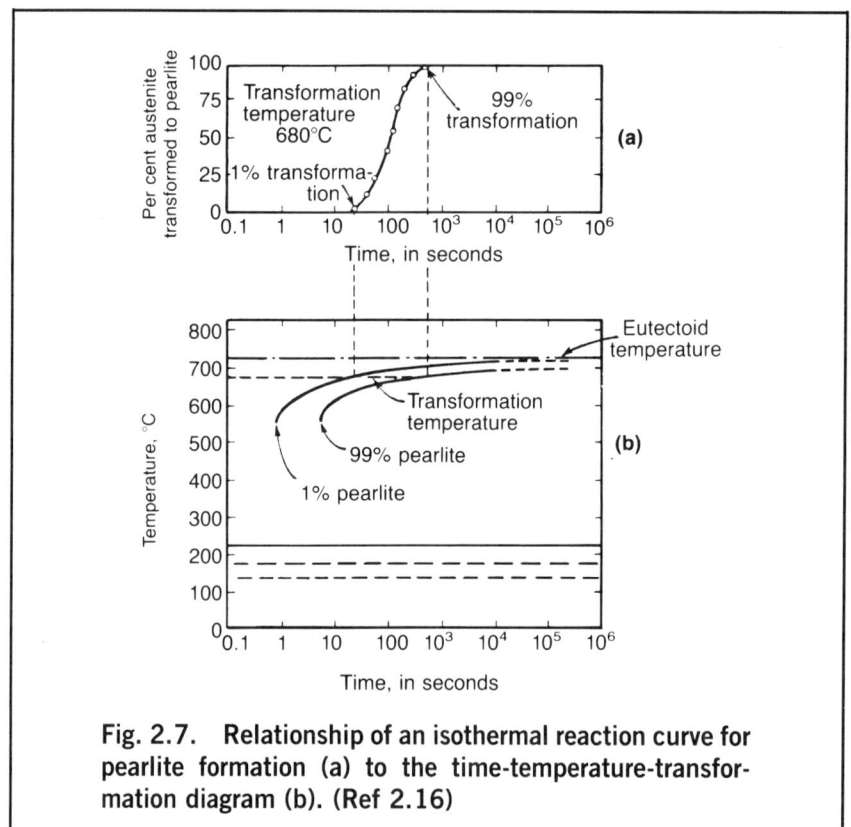

Fig. 2.7. Relationship of an isothermal reaction curve for pearlite formation (a) to the time-temperature-transformation diagram (b). (Ref 2.16)

The shape of the isothermal transformation curve for eutectoid steel (see Fig. 2.3) is explained by the temperature dependence of the nucleation and growth rates of the pearlite colonies. Figure 2.8 shows that both N and G in a 0.78% carbon steel increase with decreasing transformation temperature, thus, according to the Johnson-Mehl equation, accelerating the eutectoid transformation at lower temperatures. As discussed earlier, the greater driving force associated with increased undercooling produces more nuclei and smaller interlamellar spacing. The latter in turn increases the growth rate of the pearlite colonies by effectively reducing the distance over which carbon must diffuse at the austenite/pearlite interface.

The above discussion shows that the Johnson-Mehl equation offers a highly effective approach to characterizing the kinetics of pearlite transformation. The assumption that the pearlite colonies nucleate randomly in the austenite throughout the course of the transformation is not always valid, however. As shown in Fig. 2.5, the pearlite colonies invariably nucleate at austenite grain boundaries. Eventually, the

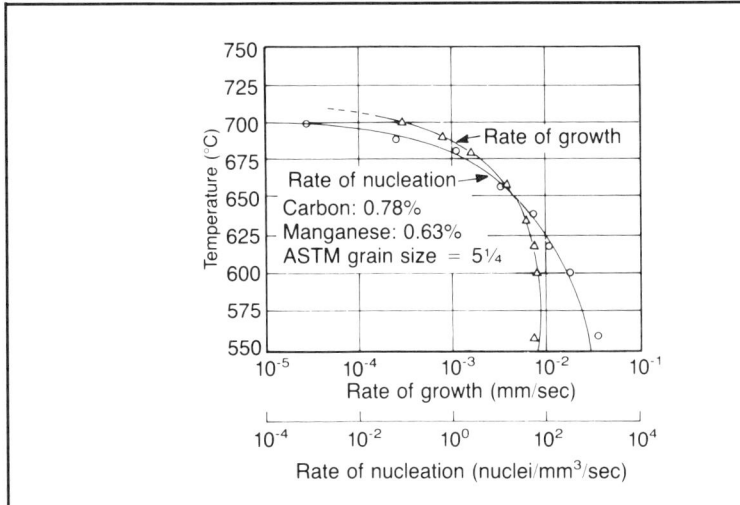

Fig. 2.8. Variation of nucleation and growth rates for pearlite formation as a function of temperature in a eutectoid steel. (Ref 2.17)

grain boundaries become saturated with nuclei, nucleation terminates, and the balance of the transformation is accomplished solely by growth of the grain boundary nucleated colonies into the austenite (Ref 2.10).

The mechanism of pearlite formation continues to receive theoretical and experimental attention. Perhaps the most active considerations involve the way in which carbon and other alloying elements distribute themselves between the ferrite and cementite lamellae. Earlier in this chapter it was tacitly assumed that the growth of a pearlite colony is dependent on the diffusion of carbon atoms through the austenite ahead of the pearlite interface. Such diffusion through a crystal phase is referred to as bulk or volume diffusion. Another possibility, however, is that the carbon diffuses along the advancing interface between the pearlite and the austenite (Ref. 2.5). Such interface or grain boundary diffusion occurs more rapidly than volume diffusion because of the more irregular or open packing of atoms at grain boundaries in comparison to the regular, close atom packing within a grain.

In ternary systems and steels the effects of alloying elements must also be considered. Puls and Kirkaldy (Ref 2.8) suggest that manganese and nickel do not partition themselves between the ferrite and cementite and that, therefore, pearlite formation in

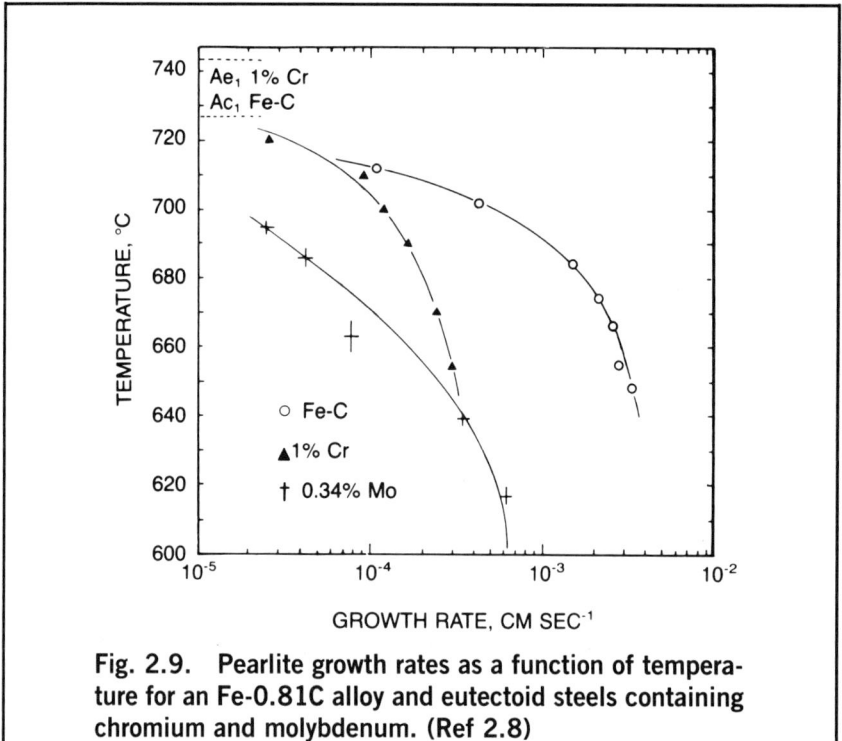

Fig. 2.9. Pearlite growth rates as a function of temperature for an Fe-0.81C alloy and eutectoid steels containing chromium and molybdenum. (Ref 2.8)

Fe-C-Mn and Fe-C-Ni alloys is dependent primarily on volume diffusion of carbon in austenite. Any reduction in the rate of pearlite growth in these systems is due to the effect of manganese and nickel on the diffusion of carbon in austenite. However, chromium and molybdenum, which are strong carbide-forming elements, are considered to partition to the carbide lamellae by interface diffusion. In the Fe-C-Cr and Fe-C-Mo systems, then, pearlite growth is retarded because chromium and molybdenum atoms must diffuse, a process that is much more sluggish than the diffusion of carbon because of the much larger size of the alloying element atoms compared to carbon atoms. Figures 2.9 and 2.10 show the effects of the various alloying elements on the growth rate of pearlite as a function of temperature. The alloying elements all slow the growth of pearlite, an effect that is extremely valuable when nonpearlitic microstructures are the desired objectives of heat treatment. The practical effects of alloying elements in retarding pearlite formation in steels are the basis of the topic of hardenability discussed in Chapter 6.

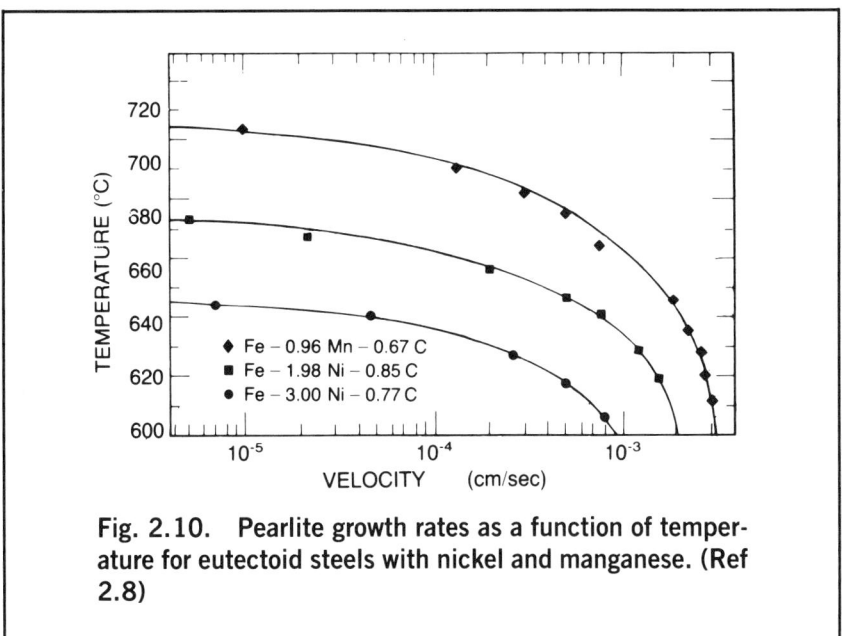

Fig. 2.10. Pearlite growth rates as a function of temperature for eutectoid steels with nickel and manganese. (Ref 2.8)

Interphase Precipitation

In alloy steels containing strong carbide-forming elements such as vanadium, niobium, titanium, molybdenum, and tungsten, a special type of austenite decomposition occurs. Honeycombe and his colleagues (Ref 2.4) have shown that rows of very fine alloy carbide particles form along the interface between the decomposing austenite and newly formed ferrite. In view of the fact that the carbides nucleate and grow at the interface between austenite and ferrite, the reaction is termed interphase precipitation. However, one solid phase transforms into two other solid phases, and the reaction may also be classified as a eutectoid transformation that produces a special microstructure much different from lamellar pearlite.

The alloy carbides that make up the rows left behind by the moving austenite-ferrite interface are frequently on the order of 100 Å (10 nm) or less and are too fine to be resolved in the light microscope. Figure 2.11(a) is a light micrograph that shows the initiation of interphase precipitation at austenite grain boundaries in an Fe-0.75V-0.15C alloy held for 10 s at 680 °C (1256 °F). The colonies of the interphase precipitate have curved interfaces and outline the austenite grain boundaries, but no structure is visible within the colonies. The balance of the microstructure is a low-carbon martensite formed on rapid cooling after the 10-s hold at 680 °C (1256 °F). The rows of fine

32 / STEELS: HEAT TREATMENT AND PROCESSING PRINCIPLES

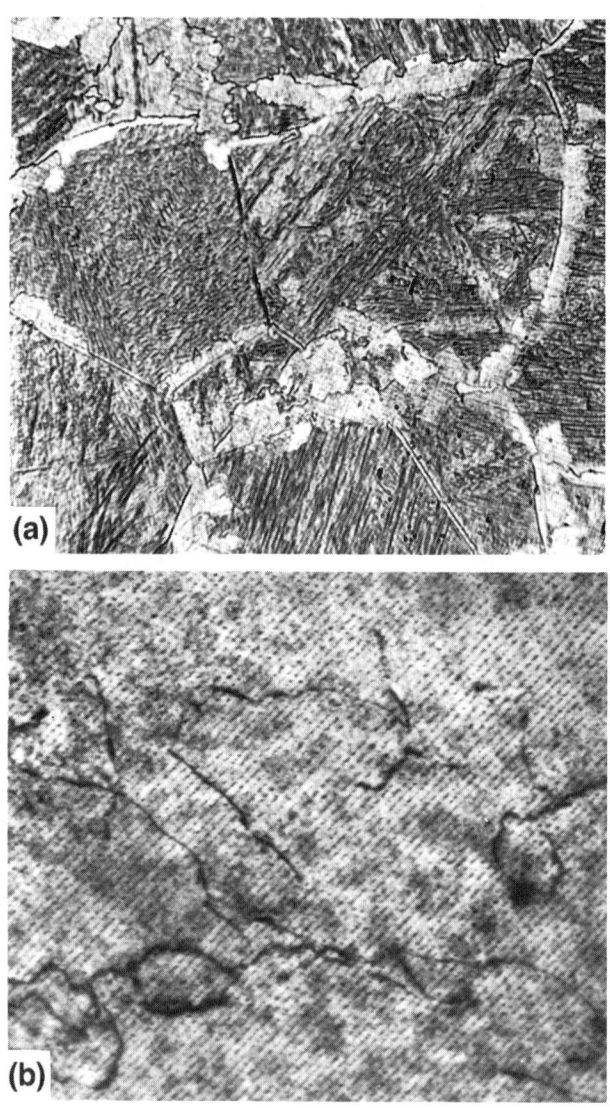

Fig. 2.11. (a) Colonies of interphase precipitation (light areas) nucleated at austenite grain boundaries of an Fe-0.75V-0.15C alloy held 10 s at 680 °C (1256 °F). Magnification, 125×. (b) Rows of fine alloy carbides within a colony of the same steel held 5 min at 725 °C (1340 °F). Magnification, 100 000×. Courtesy of R.W.K. Honeycombe, University of Cambridge, U.K.

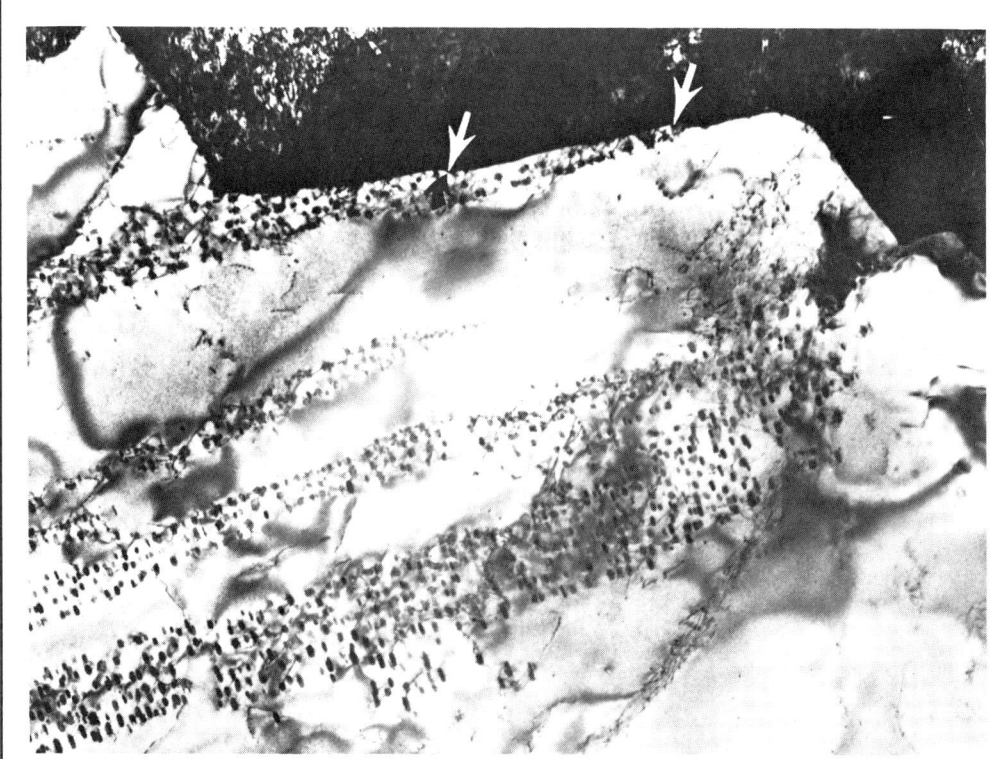

Fig. 2.12. Interphase precipitation and ledges in Fe-12Cr-0.2C steel isothermally transformed at 650 °C (1202 °F) for 36 min. Transmission electron micrograph. Magnification, about 70 000×. Courtesy of K. Campbell and R.W.K. Honeycombe, University of Cambridge, U.K.

precipitates present in the colonies of interphase precipitation are shown in Fig. 2.11(b), a transmission electron micrograph taken from the Fe-0.75V-0.15C steel held for 5 min at 725 °C (1335 °F).

An interesting aspect of the mechanism of interphase precipitation is the growth of the colonies by the extension of ledges in a direction parallel to the austenite-ferrite interface. As successively nucleated ledges complete their growth, a net extension of the colonies normal to the colony interface develops. Figure 2.12 shows examples of growth ledges in an Fe-12Cr-0.2C steel isothermally transformed for 36 min at 650 °C (1202 °F). Arrows point to the ledges. The planar interfaces left behind by the movement of the ledges are the sites for the carbide formation, while the ledges themselves are free of particles.

At this time, steels which transform solely by interphase precipitation are not used commercially, although alloying elements such as vanadium and niobium that form interphase carbides are being used increasingly in a new high-strength low-alloy (HSLA) or micro-alloyed steels. The very fine interphase alloy carbides and the high strength of microstructures with these particles are, however, quite attractive, and special applications may be developed in the future (Ref 2.4).

Proeutectoid Phases

In most steels—i.e., those not of eutectoid composition—austenite begins to transform well above the A_1 temperature. Figure 2.1 shows that ferrite forms below the A_3 temperature in steels that contain less than the eutectoid carbon content (hypoeutectoid steels), and cementite forms below the A_{cm} in steels containing more than the eutectoid carbon content (hypereutectoid steels). The ferrite and cementite that form prior to the eutectoid transformation are referred to as proeutectoid ferrite and cementite in order to indicate that they have formed by a mechanism other than the eutectoid transformation.

Proeutectoid ferrite and cementite are identical in crystal structure and composition to the ferrite and cementite of pearlite but are distributed quite differently in the microstructure than their lamellar arrangement in pearlite. Figure 2.13 shows a microstructure of proeutectoid ferrite and pearlite that formed in an Fe-0.4C alloy during slow cooling from the austenite phase field. The coarse network of white-etching proeutectoid ferrite is in marked contrast to the lamellar pearlite. Figure 2.1 shows that proeutectoid ferrite in a slowly cooled Fe-0.4C alloy begins to form just above 780 °C (1436 °F) and continues to grow until the A_1 temperature is reached. Tie lines through the ferrite-austenite field at successively lower temperatures and the applications of the lever rule to the Fe-0.4C alloy show that the amount of proeutectoid ferrite and the carbon content of the austenite increase continuously with decreasing temperature. The low solubility of carbon in ferrite requires that the carbon content build up in the austenite. At the A_1 temperature the carbon content of the austenite coexisting with the ferrite in the Fe-0.4C alloy, or any other hypoeutectoid steel for that matter, is 0.77%, which is just the right composition required for the eutectoid reaction, as written in Eq 2.1. Consequently any austenite coexisting with ferrite at the A_1 temperature transforms to pearlite on cooling, producing microstructures such as those shown in Fig. 2.13. The lever rule applied to the Fe-0.4C

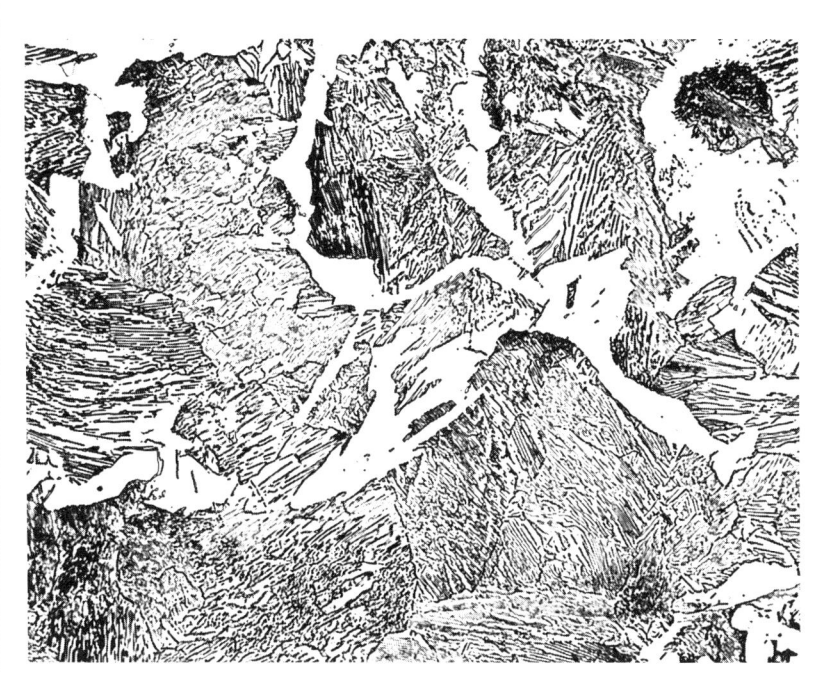

Fig. 2.13. Proeutectoid ferrite (white network) and pearlite in an Fe-0.4C alloy air cooled from the austenite field. Nital etch. Magnification, 500×. Courtesy of A.R. Marder and A. Benscoter, Bethlehem Steel Corp., Bethlehem, PA

alloy in the ferrite-austenite phase field at 727 °C (1340 °F) shows that there should be about 50% by weight proeutectoid ferrite in the microstructure according to the following calculation:

$$\text{wt.\% proeutectoid ferrite} = \frac{0.77 - 0.4}{0.77 - 0.02} \times 100 \qquad (\text{Eq 2.7})$$

Alloys or steels with less carbon than 0.4% would contain more proeutectoid ferrite; those with more carbon would contain more pearlite. Depending on the carbon content of the steel, then, it is possible to have microstructures consisting of 100% ferrite (if the carbon content is less than or equal to 0.02%) or 100% pearlite (if the carbon content is equal to 0.77%) or any combination of proeutectoid ferrite and pearlite between these extremes. Steels designated for applications that require good formability—for example, automotive panel parts—have microstructures that are predominantly ferrite, while steels selected for applications where hardness and wear resistance are most important—for example, railroad rails—have microstructures that are completely pearlitic. The properties of steels heat treated to have microstructures of ferrite and pearlite are described in Chapter 5.

Up to this point only the formation of proeutectoid ferrite has been considered. Steels with carbon content greater than the eutectoid compositions form proeutectoid cementite if slowly cooled through or held in the cementite-austenite phase field (see Fig. 2.1). As the cementite (containing 6.67% carbon) forms, the carbon content of the austenite must decrease. With decreasing temperature, the austenite composition follows the A_{cm} until at the eutectoid temperature the austenite contains 0.77% carbon, again just the right composition for the eutectoid reaction. The balance of the austenite then transforms to pearlite.

Figure 2.14 shows a network of proeutectoid cementite that has formed by holding an Fe-1.22C alloy at 780 °C (1436 °F) for 30 min. Some colonies of pearlite are also present, the dark circular patches, and the balance of the microstructure is martensite formed during quenching from 780 °C (1436 °F). The carbon content of steels rarely exceeds 1.2%; therefore, little proeutectoid cementite ever forms. Application of the lever rule in the austenite-cementite field to the 1.2% carbon alloy at 727 °C (1340 °F) shows that only about 7% proeutectoid cementite could form. However, even though there can never be a large amount of proeutectoid cementite, the presence of a proeutectoid cementite network is considered to be very detrimental to the workability and toughness of high-carbon steels. Normalizing and spheroidizing heat treatments designed to modify or eliminate the cementite networks are discussed in Chapter 5.

Proeutectoid Phase Formation

Two major types of proeutectoid phase morphology develop in steels. One, shown in Fig. 2.13 and 2.14, consists of ferrite or cementite grains that nucleate and grow along the austenite grain boundaries. In a morphological classification system developed by Dubé and extended by Aaronson (Ref 2.11), these crystals are referred to as grain boundary allotriomorphs. The second type of morphology consists of needle-like or plate-shaped grains that extend from the austenite grain boundaries. Examples of such grains of proeutectoid ferrite are shown in Fig. 2.15. In the Dubé classification system, the plate-shaped

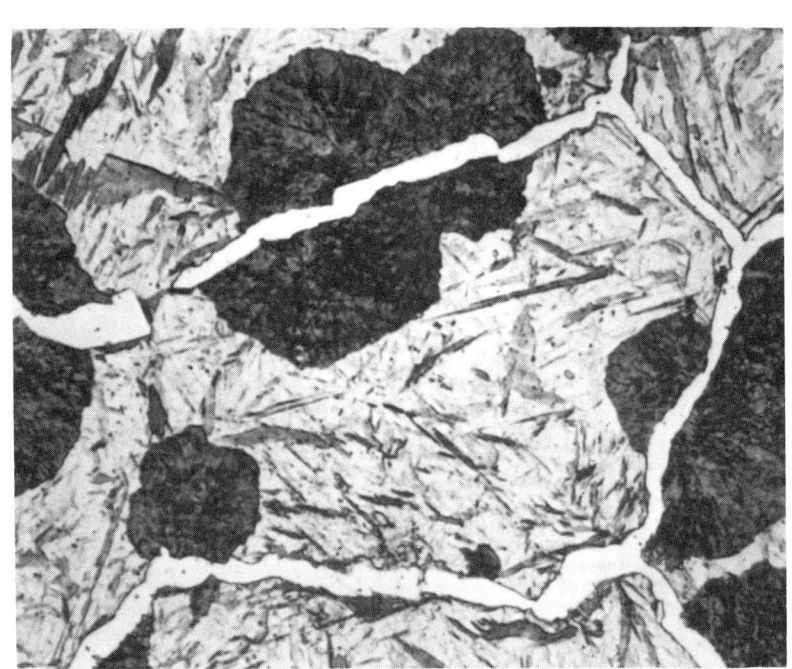

Fig. 2.14. Proeutectoid cementite (white network) formed at austenite grain boundaries in an Fe-1.22C alloy held at 780 °C (1436 °F) for 30 min. Dark patches are pearlite colonies and remainder of microstructure is martensite and retained austenite. Nital etch. Magnification, 600×. Courtesy of T. Ando, Colorado School of Mines, Golden

crystals are referred to as Widmanstätten side plates, the adjective "Widmanstätten" taken from the name of the French scientist, Alois de Widmanstätten. Widmanstätten is now widely used to describe any elongated or plate-shaped crystals that appear to form along specific crystallographic directions in a parent crystal.

Grain boundary allotriomorphs form under conditions approaching equilibrium, i.e., either by holding for extending periods of time in or by slow cooling through a two-phase field. Under these conditions, there is ample time for diffusion but the thermodynamic driving force is relatively low. Relatively large nuclei, therefore, form heterogeneously on the austenite grain boundaries, a process incorporating some of the existing austenite grain boundary and minimizing the interfacial energy increase. Nucleation of a proeutectoid phase within

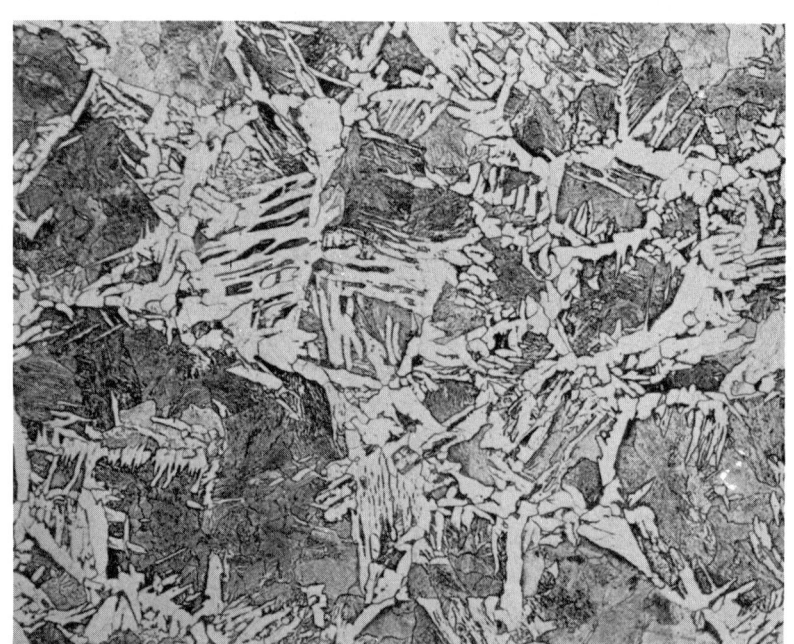

Fig. 2.15. Examples of proeutectoid ferrite with a Widmanstätten side plate morphology in an Fe-0.4C alloy. Nital etch. Magnification, 200×. Courtesy of M.D. Geib, Colorado School of Mines, Golden

an austenite grain is prohibitive because all the new parent-product interface would have to be created. Widmanstätten side plates, on the other hand, are formed when the steel is cooled more rapidly and/or substantially undercooled below the A_3 or A_{cm}. Greater driving force is therefore available for the transformation of austenite to a proeutectoid phase, but diffusion is more limited. The latter disadvantage is partially offset by the fact that carbon atoms can diffuse in all directions around the tip of a growing side plate and that the diffusion distances in the austenite at the tips of plates are relatively short (Ref 2.12). The increased driving force apparently causes a shift to a mechanism that favors a closer crystallographic coupling between the austenite and the proeutectoid phase and some shear or cooperative movement of the iron atoms into the product crystal arrangement.

C.S. Smith (Ref 2.13) discussed the role of the crystallographic relationship between a proeutectoid crystal nucleated on an austenite grain boundary and the two adjacent austenite grains. He proposed that the crystal structure of one austenite grain might closely match the atom arrangement in the ferrite grain, i.e., a definite crystallographic orientation relationship might exist between the two crystals, and the resulting interface would have a high degree of coherency. The relatively good packing of atoms at the interface, however, would make transfer of the atoms across the interface difficult and result in a boundary with low diffusional mobility. The atom arrangement between the ferrite and the other austenite grain might not match nearly as well; thus, an incoherent interface with a large degree of misfit would separate the two crystals. Atoms in such an interface would easily move from fcc packing to the bcc structure, thus producing a boundary with a high degree of diffusional mobility. At high transformation temperatures and low undercooling, the incoherent boundary could migrate and produce the typical grain boundary allotriomorphic morphology. With a high degree of undercooling, however, the migration of the incoherent boundary by diffusion would be restricted, and the high driving force would cause the ferrite with the coherent boundary in the other grain to propagate, resulting in a Widmanstätten side plate. The frequently observed growth of Widmanstätten side plates into only one grain is explained by the Smith hypothesis.

Figure 2.16 illustrates a microstructure that appears to have formed as a result of Smith's considerations. The microstructure was produced by water quenching an Fe-0.2C alloy with the intention of forming a completely martensitic structure. A number of proeutectoid ferrite crystals, however, nucleated at an austenite grain boundary prior to the martensite that makes up the bulk of the microstructure. The grains marked A and B have different orientations, as indicated by the grain boundary between them. The rapid cooling has prevented the motion of one interface of each grain, presumably because the interfaces were incoherent and the temperature was lowered too quickly for any significant amount of diffusional growth. Each ferrite grain, however, apparently had a coherent interface that could propagate into one of the austenitizing grains with sufficient undercooling. The fact that the Widmanstätten side plates are parallel to the fine martensite crystals and merge into the martensitic matrix supports the statement that some shear must be associated with Widmanstätten ferrite formation. Ferrite formation requires the rejection of carbon, but the shear formation of martensite at temperatures below those at which ferrite forms traps the carbon in the martensitic structure.

Two other mechanisms of ferrite formation may operate under certain conditions or in certain alloy compositions. One involves the

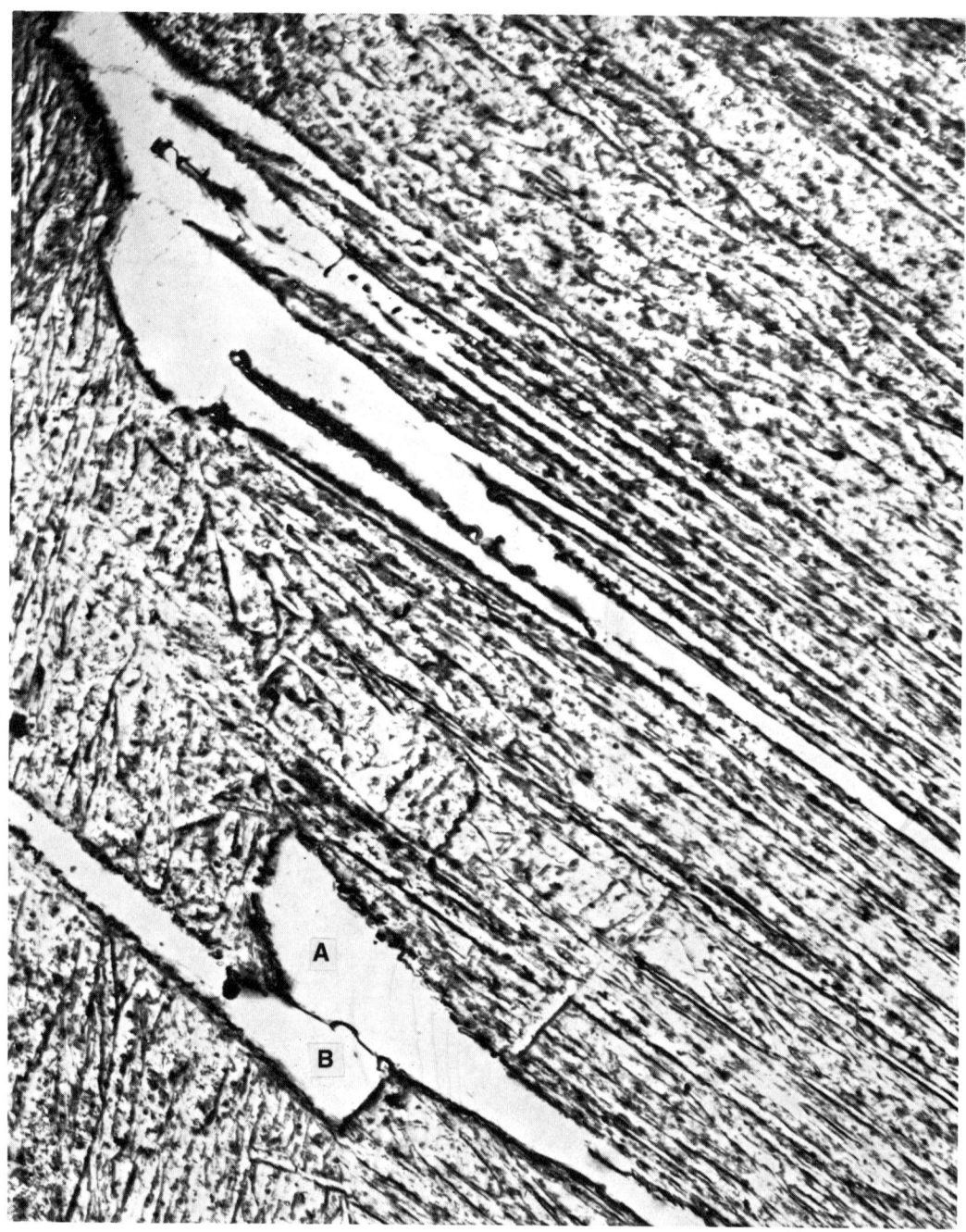

Fig. 2.16. Widmanstätten side plate formation in a quenched Fe-0.2C alloy. Grains A and B, separated by a grain boundary, have orientations that favor growth into different austenite grains. Electron micrograph from an extraction replica. Magnification, 7500×. Courtesy of R.N. Caron, Olin Corp., New Haven, CT

growth or thickening of both grain boundary allotriomorphs and Widmanstätten side plates by motion of ledges parallel to coherent interfaces (Ref 2.14). The ledges are considered to have incoherent interfaces with the austenite and therefore a high degree of mobility, as discussed relative to the Smith hypothesis. The other mechanism operates only in low-carbon steels containing less than 0.02% carbon. When these steels are cooled rapidly from the single-phase austenite field to the single-phase ferrite field, no change in composition is required. Carbon is, therefore, not rejected from the ferrite as is the case for higher carbon steels discussed earlier. The transformation is accomplished only by short range diffusion, on the order of one atomic distance, across the interface. This type of transformation has been classified as a massive or short range diffusion (SRD) transformation (Ref 2.15), the former term selected because massive, equiaxed grains are produced by this transformation mechanism.

References

2.1 R.F. Mehl and W.C. Hagel, The Austenite:Pearlite Reaction, in *Progress in Metal Physics,* B. Chalmers and R. King (Eds.), Vol 6, 1956, Pergamon Press, New York, p 74–134

2.2 D. Brown and N. Ridley, Rates of Nucleation and Growth and Interlamellar Spacing of Pearlite in a Low-Alloy Eutectoid Steel, *JISI,* Vol 204, 1966, p 811

2.3 M. Hillert, The Formation of Pearlite, in *Decomposition of Austenite by Diffusional Processes,* V.F. Zackay and H.I. Aaronson (Eds.), Interscience, New York, 1962, p 197–247

2.4 R.W.K. Honeycombe, Transformation from Austenite in Alloy Steels, *Met Trans A,* Vol 7A, 1976, p 915–936

2.5 B.E. Sundquist, The Edgewise Growth of Pearlite, *Acta Met,* Vol 16, 1968, p 1413–1427

2.6 C. Wells and R.F. Mehl, Rate of Diffusion of Carbon in Austenite in Plain Carbon, in Nickel and in Manganese Steels, *Trans AIME,* Vol 140, 1940, p 279

2.7 R.A. Swalin, *Thermodynamics of Solids,* John Wiley & Sons, New York, 1962

2.8 M.P. Puls and J.S. Kirkaldy, The Pearlite Reaction, *Met Trans,* Vol 3, 1972, p 2777–2796

2.9 W.A. Johnson and R.F. Mehl, Reaction Kinetics in Processes of Nucleation and Growth, *Trans AIME,* Vol 135, 1939, p 416–458

2.10 J.W. Cahn and W.C. Hagel, Theory of the Pearlite Reaction, in *Decomposition of Austenite by Diffusional Processes,* V.F. Zackay and H.I. Aaronson (Eds.), Interscience, New York, 1962, p 131–196

2.11 H.I. Aaronson, The Proeutectoid Ferrite and the Proeutectoid Cementite Reactions, in *Decomposition of Austenite by Diffusional Processes,* V.F.

Zackay and H.I. Aaronson (Eds.), Interscience, New York, 1962, p 387–548
2.12 P.G. Shewmon, *Transformations in Metals,* McGraw-Hill, New York, 1969, p 220
2.13 C.S. Smith, *Trans ASM,* 1953, Vol 45, p 533
2.14 H.I. Aaronson, C. Laird, and K.R. Kinsman, Mechanisms of Diffusional Growth of Precipitate Crystals, in *Phase Transformations,* American Society for Metals, Metals Park, OH, 1970, p 313
2.15 T.B. Massalski, Massive Transformation Structures, in *Metals Handbook,* Vol 8, 8th ed., American Society for Metals, Metals Park, OH, 1973, p 186–187
2.16 *Atlas of Isothermal Transformation and Cooling Transformation Diagrams,* American Society for Metals, Metals Park, OH, 1977, p 28
2.17 R.F. Mehl and A. Dubé, *Phase Transformations in Solids,* John Wiley & Sons, New York, 1951, p 545

CHAPTER 3

Martensite and Bainite

This chapter describes the diffusionless, shear-type transformation of austenite to martensite. Athermal transformation kinetics, crystallographic features, and the development of fine structure are all special characteristics of the martensitic transformation. These features are described and related to the major morphologies and microstructural arrangements of martensite, lath and plate, that form in steel. Rapid cooling or quenching is required to form martensite, primarily to avoid the diffusion-dependent transformations described in Chapter 2, but the exact cooling conditions that will result in martensite in a given steel are strongly dependent on carbon content, alloying, and austenitic grain size, factors which determine hardenability, a subject discussed in Chapter 6. At the end of this chapter, bainite, a product of transformation that involves both shear and diffusion, is described.

General Considerations

Martensite, named after the pioneering German metallurgist, Adolf Martens, has long been used to designate the hard microstructure found in quenched carbon steel (Ref 3.1). More recently, however, emphasis has been placed on the nature of the transformation itself rather than the product. Martensitic transformation also occurs in many nonferrous systems (Ref 3.2), the Cu-Al and Au-Cd systems to name just two metal systems, and in oxides such as SiO_2 (Ref 3.3) and ZrO_2 (Ref 3.4). In fact, ceramists and geologists have independently identified the characteristics of martensitic transformation in nonmetal systems and used the term "displacive" to describe transformations which would be called martensitic by metallurgists (Ref 3.3). Martensite then becomes any phase produced by a martensitic or

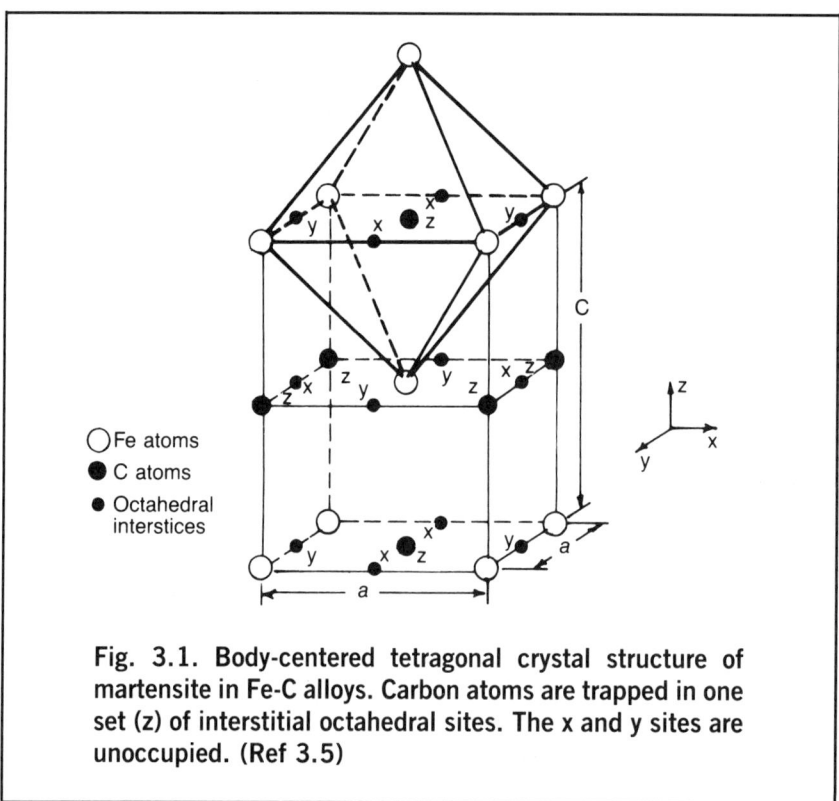

Fig. 3.1. Body-centered tetragonal crystal structure of martensite in Fe-C alloys. Carbon atoms are trapped in one set (z) of interstitial octahedral sites. The x and y sites are unoccupied. (Ref 3.5)

displacive transformation, even though that phase may have significantly different composition, crystal structure, and properties than does martensite in steels.

In Fe-C alloys and steels, austenite is the parent phase that transforms to martensite on cooling. The martensitic transformation is diffusionless, and, therefore, the martensite has exactly the same composition as does its parent austenite, up to 2% carbon (see Fig. 1.1 and 2.1), depending on the alloy composition. Since diffusion is suppressed, usually by rapid cooling, the carbon atoms do not partition themselves between cementite and ferrite (see Chapter 2) but instead are trapped in the octahedral sites of a body-centered cubic structure, thus producing a new phase, martensite. The solubility of carbon in a bcc structure is greatly exceeded when martensite forms; hence, martensite assumes a body-centered tetragonal (bct) unit cell (see Fig. 3.1) in which the c parameter of the unit cell is greater than the other two a parameters. With higher carbon concentration of the martensite, more interstitial sites are filled, and the tetragonality increases, as shown in Fig. 3.2.

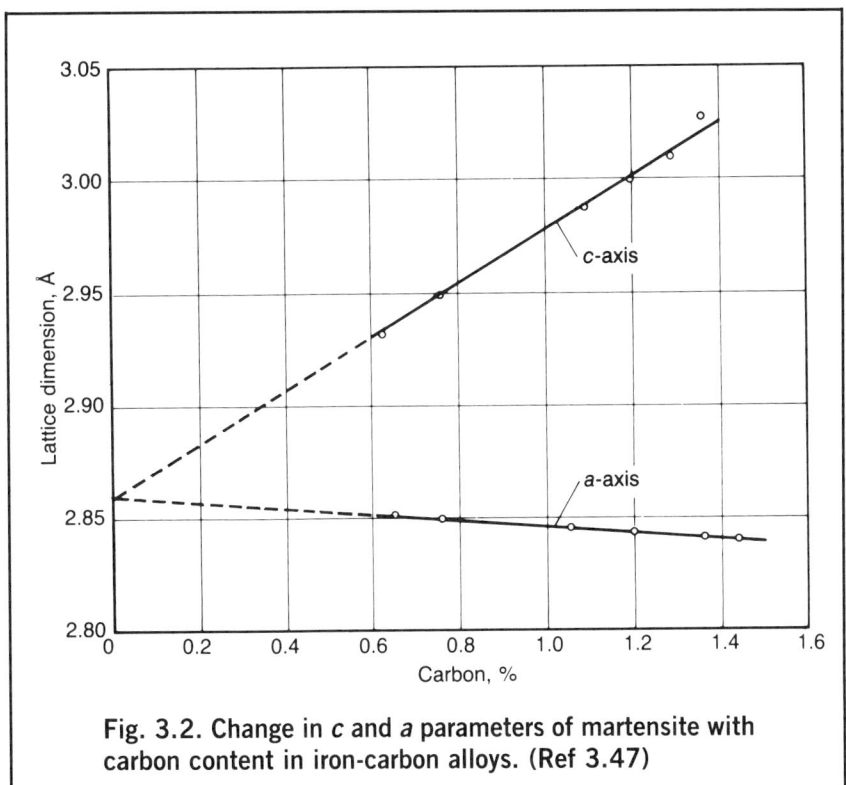

Fig. 3.2. Change in *c* and *a* parameters of martensite with carbon content in iron-carbon alloys. (Ref 3.47)

Martensite is a unique phase that forms in steels. It has its own crystal structure and composition and is separated by well-defined interfaces from other phases, but it is a metastable phase present only because diffusion has been suppressed. If the martensite is heated to a temperature where the carbon atoms have mobility, the carbon atoms diffuse from the octahedral sites to form carbides. As a result, the tetragonality is relieved, and martensite is replaced by a mixture of ferrite and cementite as required by the Fe-C phase diagram. The decomposition of martensite to other structures on heating is referred to as tempering and is the subject of Chapter 8.

Martensite forms by a shear mechanism. Many atoms move cooperatively and almost simultaneously to effect the transformation, a mechanism very much in contrast to atom-by-atom movement across interfaces during diffusion-dependent transformations. Figure 3.3 shows schematically a number of features of the shear or displacive transformation of austenite to martensite. The arrows point in the directions of shear on opposite sides of the plane on which the transformation was initiated. The martensite crystal formed is dis-

46 / STEELS: HEAT TREATMENT AND PROCESSING PRINCIPLES

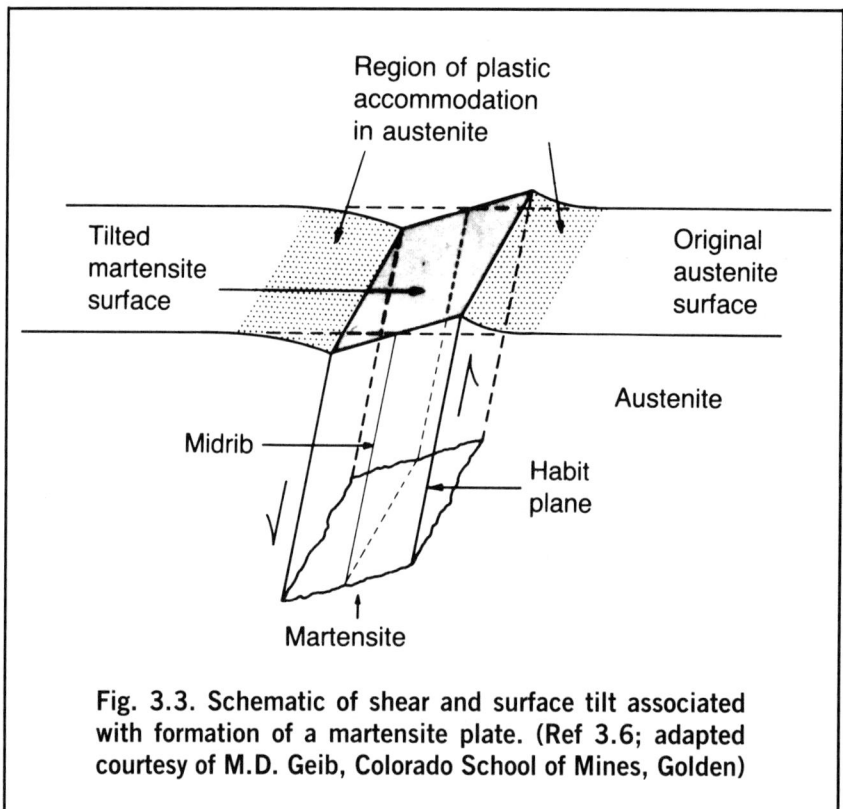

Fig. 3.3. Schematic of shear and surface tilt associated with formation of a martensite plate. (Ref 3.6; adapted courtesy of M.D. Geib, Colorado School of Mines, Golden)

placed partly above and partly below the surface of the austenite by the shear. Thus (as shown in Fig. 3.3) the originally horizontal surface of the parent phase is rotated or tilted into a new orientation by the shear transformation. Surface tilting is an important characteristic of shear-type or martensitic transformation. The atom-by-atom transfer across interfaces by which diffusion-controlled transformations proceed does not produce tilting but tends to produce surfaces of the product phase parallel to the surface of the parent phase.

Figure 3.3 also shows that considerable flow or plastic deformation of the parent austenite must accompany the formation of a martensite crystal. Eventually the constraints of the parent phase limit the width of a martensite lath, and further transformation can proceed only by the nucleation of new plates. If the parent austenite could not accommodate the shape change produced by the martensitic shears, separation or cracking at the martensite/parent phase interface would occur. Fortunately austenite in steels has sufficient ductility to accommodate martensite formation. However, in many ceramic systems, the parent phase cannot accommodate the shape change, and displacive transformations must be avoided.

Martensite crystals ideally have planar interfaces with the parent austenite (see Fig. 3.3). The preferred crystal planes of the austenite on which the martensite crystals form are designated as habit planes. The habit planes vary according to alloy composition, and some examples are presented in the section on morphology in this chapter. The midrib shown in Fig. 3.3 is generally considered to be the starting plane for the formation of a plate of martensite and may in fact have a different fine structure than other parts of the plate.

An example of surface relief and its relationship to martensitic microstructure is shown in Fig. 3.4. This series of light micrographs was obtained after a prepolished Fe-0.2C alloy specimen was austenitized and quenched in a hot stage microscope with an argon gas atmosphere. Figure 3.4(a) shows the surface relief associated with the formation of hundreds of martensite crystals. The surface tilting is emphasized in some areas by the dark shadows present on surfaces tilted away from the light source. In Fig. 3.4(b) the surface relief has been almost polished away, and in Fig. 3.4(c) the surface shown in (b) has been etched. Finally, Fig. 3.4(d) shows the microstructure after the surface has been polished to remove all relief and etched once again. Comparison of Fig. 3.4(c) and (d) with Fig. 3.4(a) shows the direct correspondence of the surface relief with the martensitic units in the polished and etched microstructure. In polished and etched sections, the individual crystals of martensite appear to be long and thin and are very often characterized as acicular or needlelike. In three dimensions, however, the crystals have a lath or plate shape with flat interfaces, as shown schematically in Fig. 3.3. The needlelike shapes visible on polished and etched surfaces, therefore, are cross sections through laths or plates.

Martensitic Transformation Kinetics

The conversion of an austenitic microstructure to a martensitic microstructure in many commercial steels takes place continuously with decreasing temperature during uninterrupted cooling. This mode of transformation kinetics is referred to as athermal (without thermal activation) in order to differentiate it from the isothermal kinetics that characterize thermally activated diffusion-controlled transformations. Pearlite formation, for example, occurs continuously as a function of time if austenite is held at a constant temperature below A_1. Martens-

48 / STEELS: HEAT TREATMENT AND PROCESSING PRINCIPLES

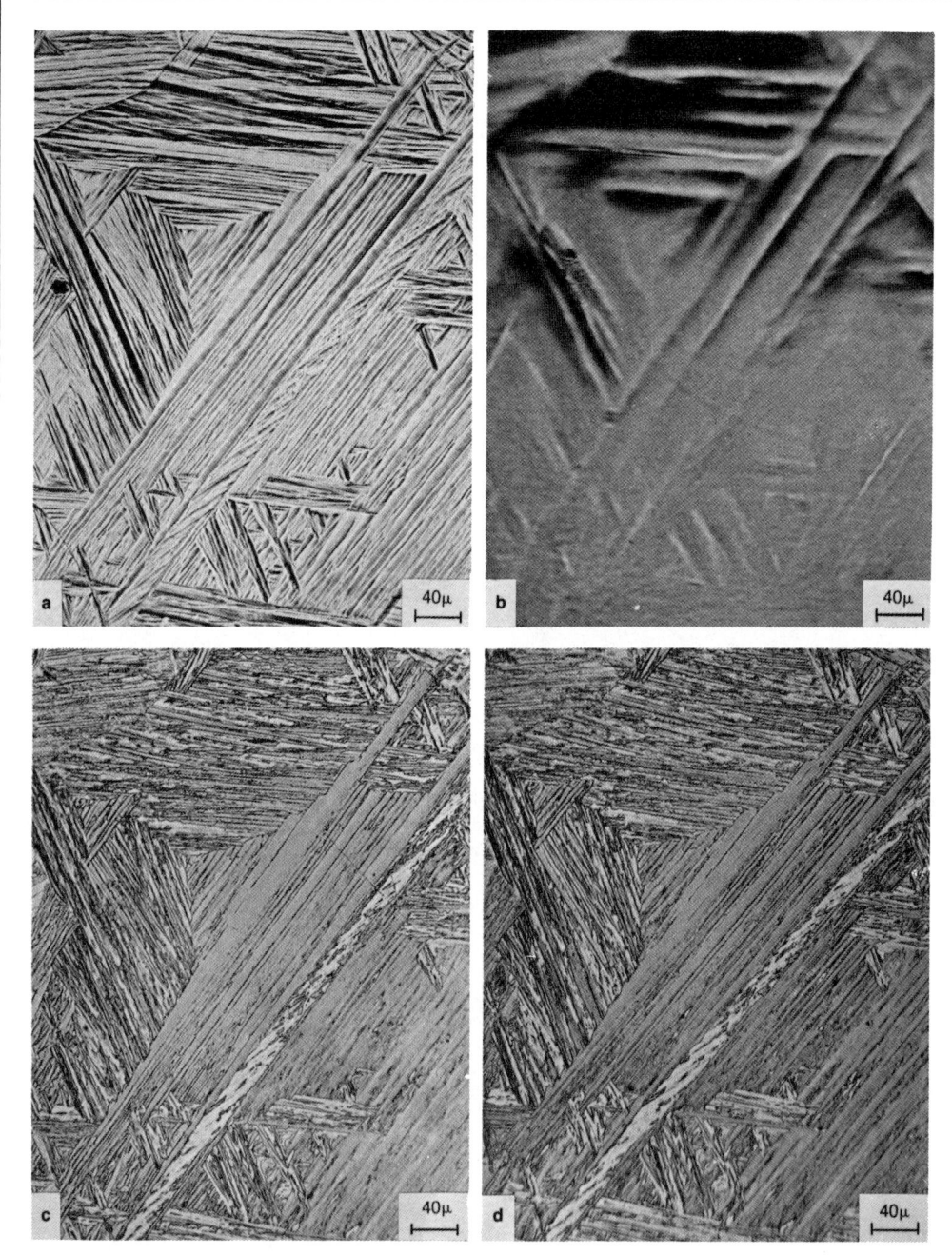

Fig. 3.4. Surface tilting and its relationship to martensitic structure in an Fe-0.2C alloy. (a) Surface tilting after quenching; (b) partially polished surface; (c) area in (b) after etching; (d) same area after polishing to remove all relief and re-etching. Nital etch. (Ref 3.7)

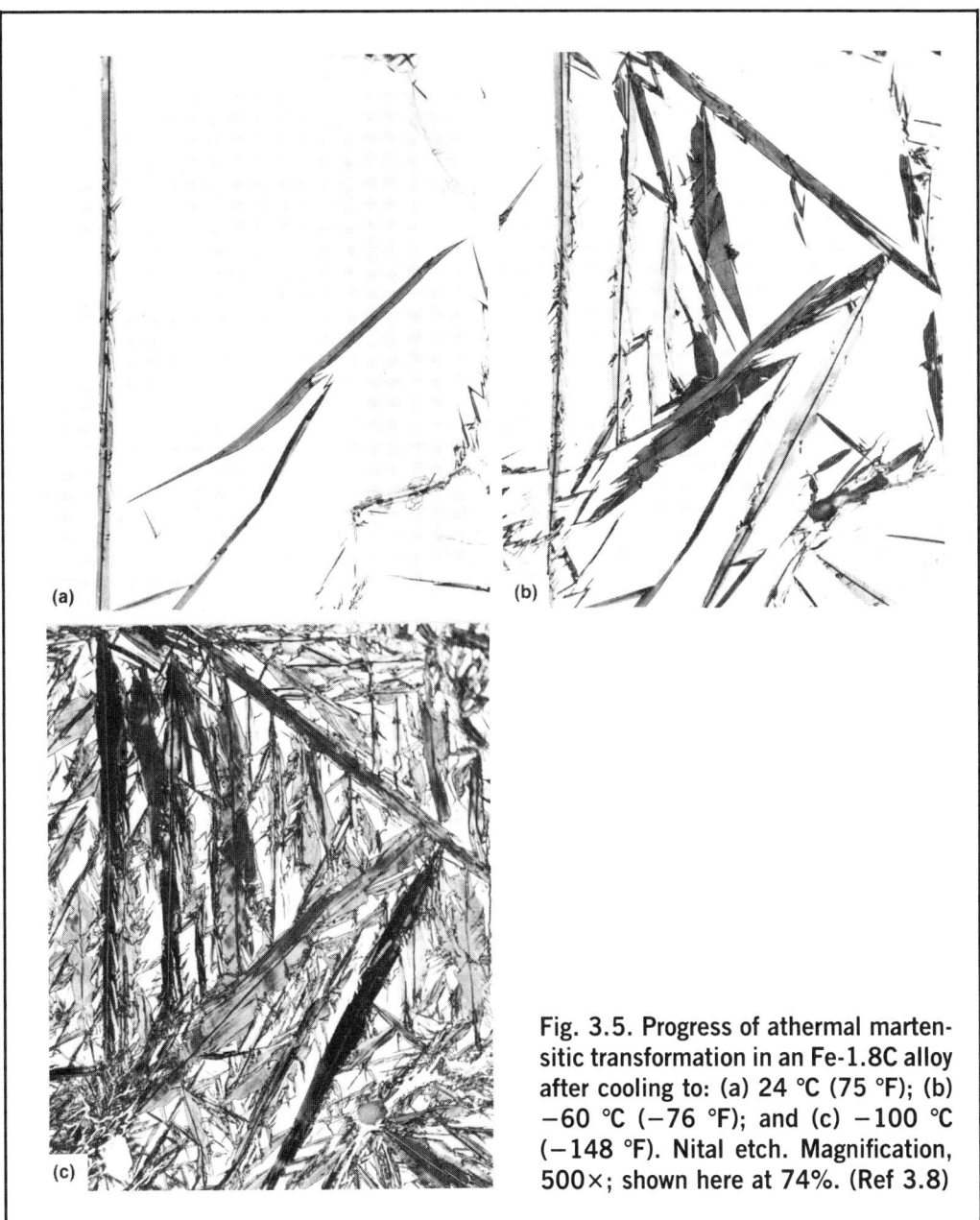

Fig. 3.5. Progress of athermal martensitic transformation in an Fe-1.8C alloy after cooling to: (a) 24 °C (75 °F); (b) −60 °C (−76 °F); and (c) −100 °C (−148 °F). Nital etch. Magnification, 500×; shown here at 74%. (Ref 3.8)

ite formation, however, is accomplished virtually as soon as a given temperature is reached; should cooling be stopped at that temperature, no further transformation to martensite will occur. Additional transformation, usually by means of the nucleation and rapid growth of new plates of martensite, is accomplished only by cooling to lower temperatures.

Figure 3.5 shows the progress of athermal transformation in an Fe-1.86C alloy. The austenite in this high-carbon alloy is quite stable, and martensitic transformation was initiated just above room temperature. Figure 3.5(a) shows a few very large plates of martensite that formed on cooling to room temperature. The balance of the microstructure is austenite. Figures 3.5(b) and (c) show how some of the austenite retained at room temperature is transformed to new plates on successive subzero cooling to −60 °C (−76 °F) and −100 °C (−148 °F), respectively. The new plates have nucleated within the framework of the initially formed plates (see Fig. 3.5a), and the parent austenite has been subdivided into smaller and smaller units with increasing amounts of martensitic formation. Clearly, the martensitic transformation effectively ceases on reaching a given temperature, and only additional undercooling drives the transformation further.

In contrast to the Fe-1.86C alloy, most hardenable steels transform to martensite at temperatures well above room temperature. Figure 3.6 shows the transformation of austenite to martensite in an Fe-1.94Mo alloy, an alloy in which austenite transforms to martensite in the same manner as in low- and medium-carbon steels. Hot stage cinephotomicrography was required to follow the high-temperature formation of the martensite in the continuously cooled Fe-1.94Mo alloy (Ref 3.8). Figure 3.6 shows a sequence of frames taken from a film of the transformation sequence. Frame 1 shows several austenite grains that are largely untransformed and the succeeding frames show the step-by-step formation of the martensitic microstructure. The martensite plates in Fig. 3.6 are visible only because of the surface tilting associated with transformation; it was obviously impossible to polish and etch (as in the case of the Fe-1.86C alloy) between frames during the cooling of the Fe-1.94Mo alloy. Figure 3.6 also shows that an important characteristic of the athermal transformation of the Fe-Mo alloy, and low- and medium-carbon steels that behave similarly, is the development of parallel groups of plates or laths by nucleation and growth of new plates parallel and adjacent to existing plates.

The temperature at which martensite starts to form in a given alloy is designated as the martensite start temperature (M_s). The M_s reflects the amount of thermodynamic driving force required to initiate the shear transformation of austenite to martensite. Figure 3.7 shows that the M_s decreases significantly with increasing carbon content in Fe-C alloys and carbon steels. Carbon in solid solution increases the strength or shear resistance of the austenite and, therefore, greater

MARTENSITE AND BAINITE / 51

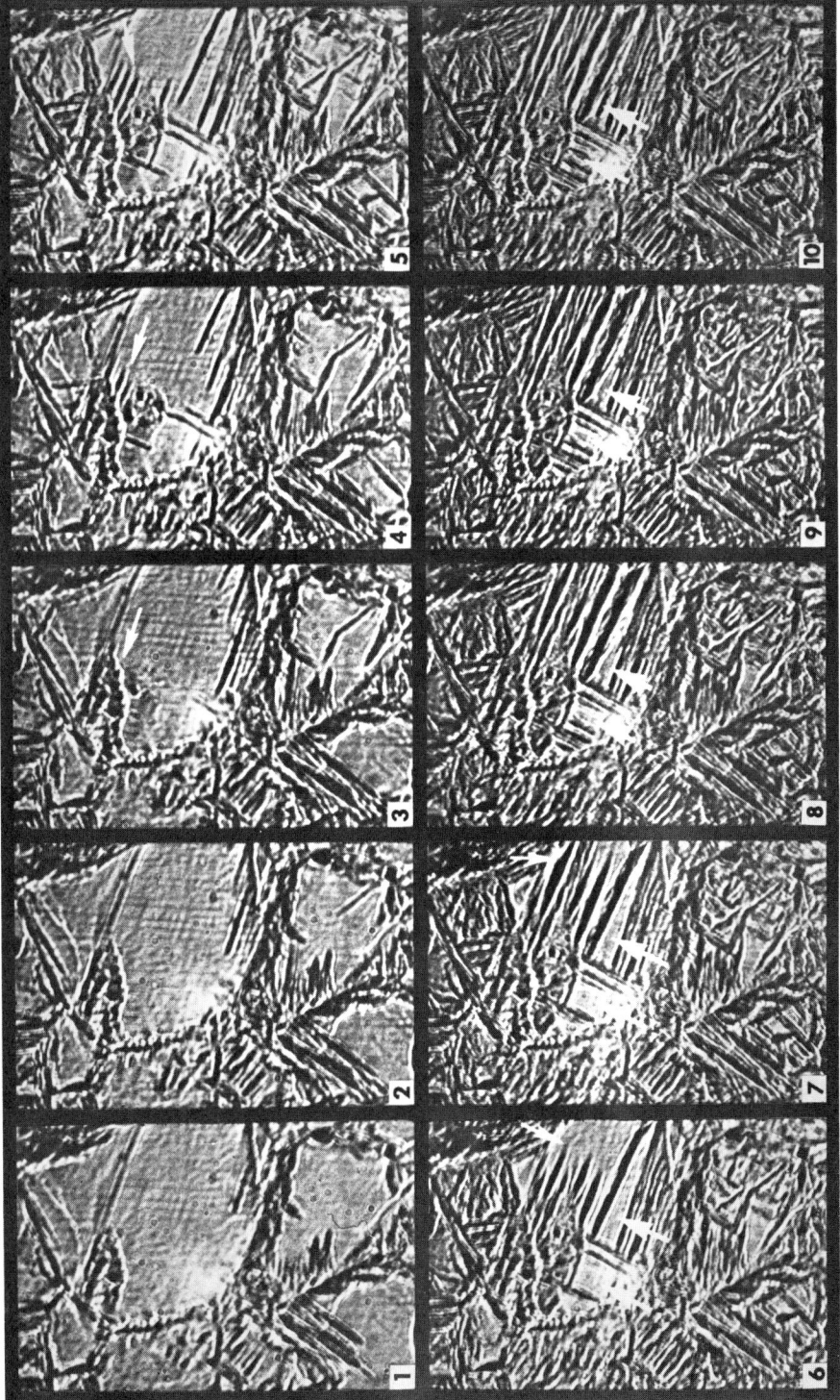

Fig. 3.6. Progress of athermal martensitic transformation in an Fe-1.94Mo alloy. Successive exposures taken of surface relief on a hot stage microscope. Magnification, 105×; shown here at 85%. (Ref 3.8)

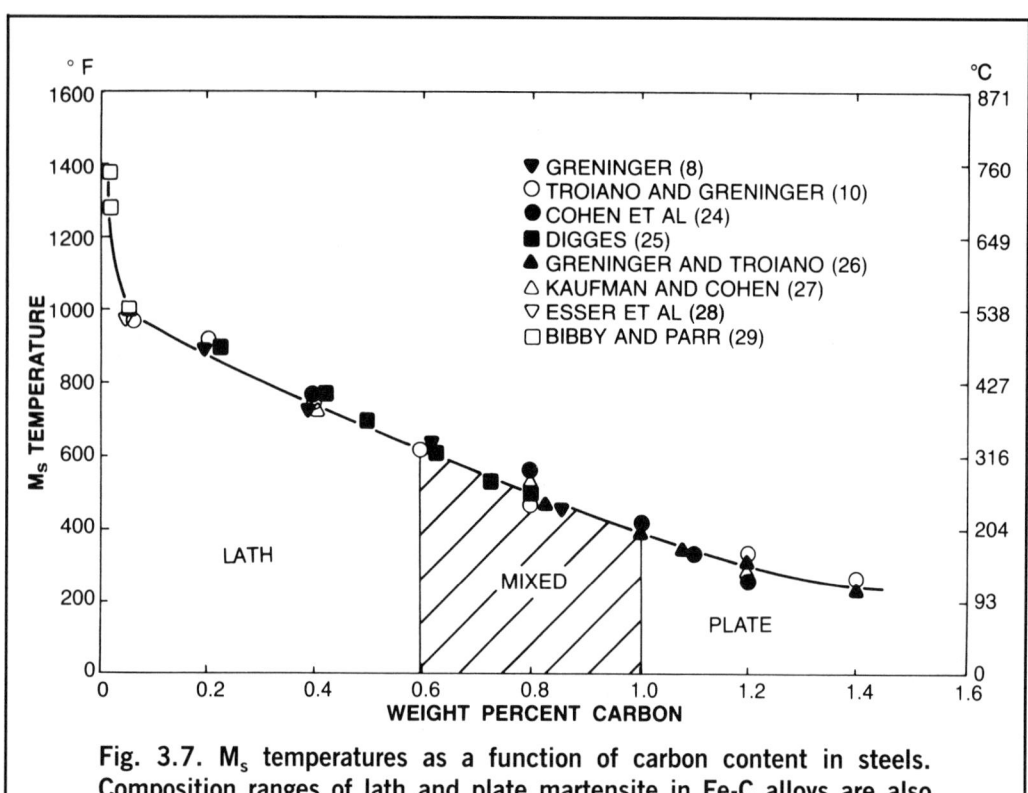

Fig. 3.7. M_s temperatures as a function of carbon content in steels. Composition ranges of lath and plate martensite in Fe-C alloys are also shown. (Ref 3.11; investigations indicated are identified by their numbers in this reference)

undercooling or driving force is required to initiate the shear for martensite formation in higher carbon alloys. The martensite finish temperature (M_f), or the temperature at which the martensite transformation is complete in a given alloy, is also a function of carbon content. The detection of the last small amounts of untransformed austenite is experimentally difficult (Ref 3.9); therefore, the M_f curves based on results of early investigations are only approximate. The M_f drops below room temperature in alloys containing more than about 0.3% carbon. Therefore, significant amounts of untransformed austenite, especially in high-carbon steels, may be present with martensite at room temperature. Figure 3.8 shows that this is actually the case. Retained austenite content, measured by x-ray diffraction techniques (Ref 3.10, 3.11) at room temperature, is as high as 30 to 40% in Fe-C alloys containing 1.2 to 1.4% carbon. Even in alloys containing only 0.3 to 0.4% carbon, some small

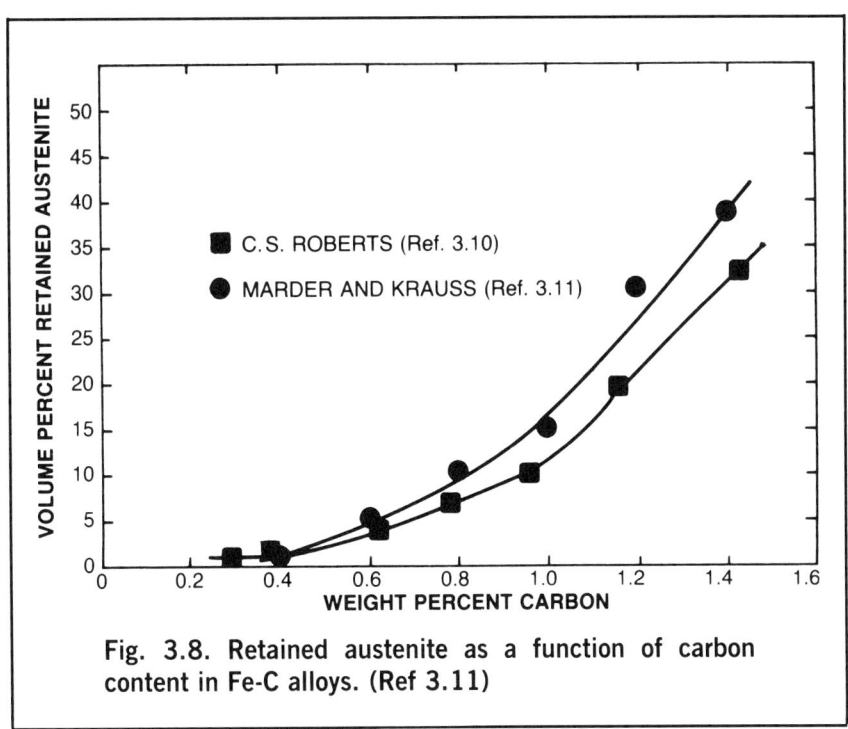

Fig. 3.8. Retained austenite as a function of carbon content in Fe-C alloys. (Ref 3.11)

amount of austenite is retained. Alloying elements that stabilize austenite increase the amount of retained austenite at any given carbon level and temperature.

Table 3.1. List of Formulas for M_s Calculation From Alloy Composition

Investigators	Date	(Ref)	Equation
Payson and Savage	1944	(3.12)	M_s (°F) = 930 − 570C − 60Mn − 50Cr − 30Ni − 20Si − 20Mo − 20W
Carapella	1944	(3.13)	M_s (°F) = 925 × (1−0.620C)(1−0.092Mn)(1−0.033Si) (1−0.045Ni)(1−0.070Cr)(1−0.029Mo) (1−0.018W)(1+0.120Co)
Rowland and Lyle	1946	(3.14)	M_s (°F) = 930 − 600C − 60Mn − 50Cr − 30Ni − 20Si − 20Mo − 20W
Grange and Stewart	1946	(3.15)	M_s (°F) = 1000 − 650C − 70Mn − 70Cr − 35Ni − 50Mo
Nehrenberg	1946	(3.16)	M_s (°F) = 930 − 540C − 60Mn − 40Cr − 30Ni − 20Si − 20Mo
Steven and Haynes	1956	(3.17)	M_s (°C) = 561 − 474C − 33Mn − 17Cr − 17Ni − 21Mo
Andrews (linear)	1965	(3.18)	M_s (°C) = 539 − 423C − 30.4Mn − 12.1Cr − 17.7Ni − 7.5Mo
Andrews (product)	1965	(3.18)	M_s (°C) = 512 − 453C − 16.9Ni + 15Cr − 9.5Mo + 217(C)^2 − 71.5(C)(Mn) − 67.6(C)(Cr)

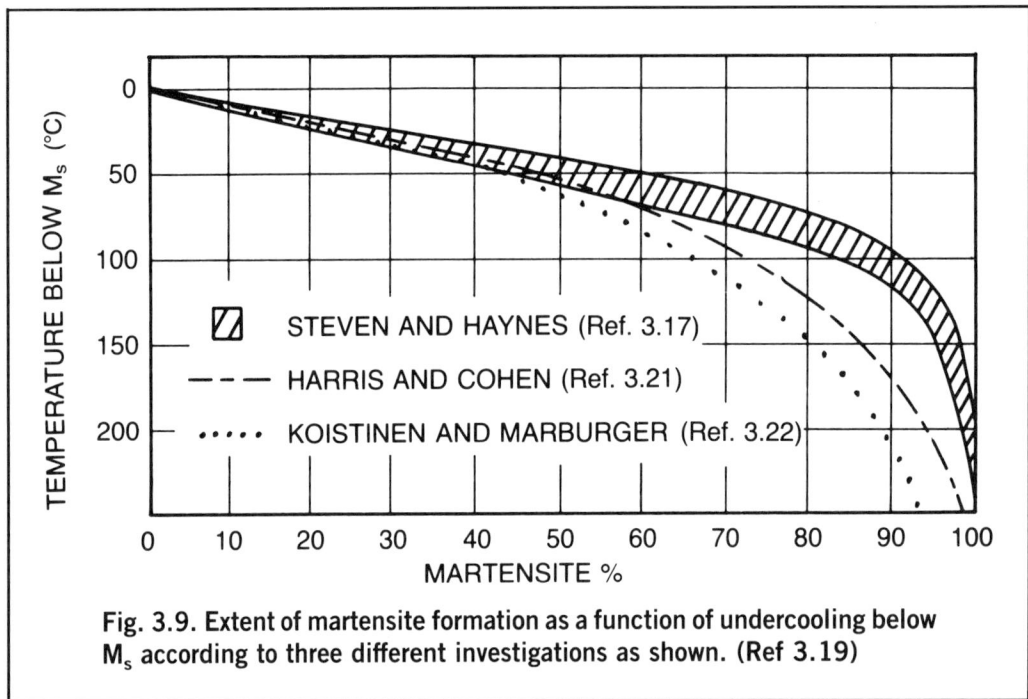

Fig. 3.9. Extent of martensite formation as a function of undercooling below M_s according to three different investigations as shown. (Ref 3.19)

Alloying elements also influence the M_s temperatures of steels, and a number of equations have been developed to relate M_s to steel composition. Table 3.1 lists various equations that have been developed over the years. All alloying elements, except cobalt, lower M_s temperatures. The equations developed by Andrews (Ref 3.18) are based on measurements of M_s temperatures and compositions of a large number of steels of British, German, French, and American manufacture with maximum carbon content of 0.6%, manganese up to 4.9%, chromium up to 5%, nickel up to 5%, and molybdenum up to 5.4%. Andrews showed that 92 and 95% of measured M_s temperatures for the steels were within ±25 °C of the M_s temperatures calculated from their compositions according to the linear and product equations, respectively. A test of Andrews' equations with M_s measurements and steel compositions published in the 1970s shows that Andrews' equations continue to give good agreement between measured and calculated M_s values with the ±25 °C limits (Ref 3.19). A more recent evaluation of the M_s temperature equations recommends only slight changes in the Stevens and Haynes and Andrews linear equations, and incorporates the effects of cobalt and silicon (Ref 3.20).

Once the M_s of a steel is known, the extent of the athermal transformation of austenite to martensite is dependent only on the amount of undercooling below the M_s temperature. Two equations have been developed to describe the athermal transformation kinetics of martensite formation:

$$f = 1 - 6.96 \times 10^{-15} (455 - \Delta T)^{5.32} \qquad \text{(Eq 3.1)}$$
$$f = 1 - \exp - (1.10 \times 10^{-2} \Delta T) \qquad \text{(Eq 3.2)}$$

where f is the volume fraction of martensite and ΔT is the undercooling below M_s in degrees Centigrade. Equation 3.1 was developed by Harris and Cohen (Ref 3.21) for steels containing 1.1% carbon, and Eq 3.2 was developed by Koistinen and Marburger (Ref 3.22) from Fe-C alloys containing between 0.37 to 1.1% carbon. The data used to develop Eq 3.1 and Eq 3.2 together with measurements of Steven and Haynes (Ref 3.17) from hardenable steels containing 0.32 to 0.44% carbon are shown in Fig. 3.9. All sets of data agree closely for small amounts of undercooling, but they diverge significantly as undercooling increases. For example, at 100 °C (212 °F) below M_s, the Koistinen and Marburger data show only about 65% transformation while the Steven and Haynes data show 90% transformation. The discrepancy is probably due to experimental difficulties in determining the amount of austenite remaining at any given temperature. Koistinen and Marburger used x-ray analysis while the other investigators used light microscopy to determine the amount of retained austenite. The detection of small amounts of retained austenite by the latter technique is difficult in high-carbon steels and virtually impossible in medium-carbon steels. Therefore, the Koistinen and Marburger equation, based on the most accurate technique for determining small amounts of retained austenite, is considered to give the best representation of martensite transformation over the entire range of undercooling.

During the course of the athermal martensite formation discussed up to this point, two types of anomalies may develop: bursting and stabilization. The burst phenomenon occurs in Fe-Ni and Fe-Ni-C alloys with subzero M_s temperatures. Large numbers of martensite plates, sometimes enough to transform 70% of the austenite, form in a "burst" at a temperature designated the M_B (Ref 3.23). This transformation behavior is related to the ability of plates of martensite to nucleate other plates of martensite, a process called autocatalysis. The stimulus to nucleation is the stress, generated at the tips of plates, that helps to initiate the shear transformation process on other favorably oriented variants of the habit plane (Ref 3.24). The habit plane variants that are activated are generally not parallel to that of the

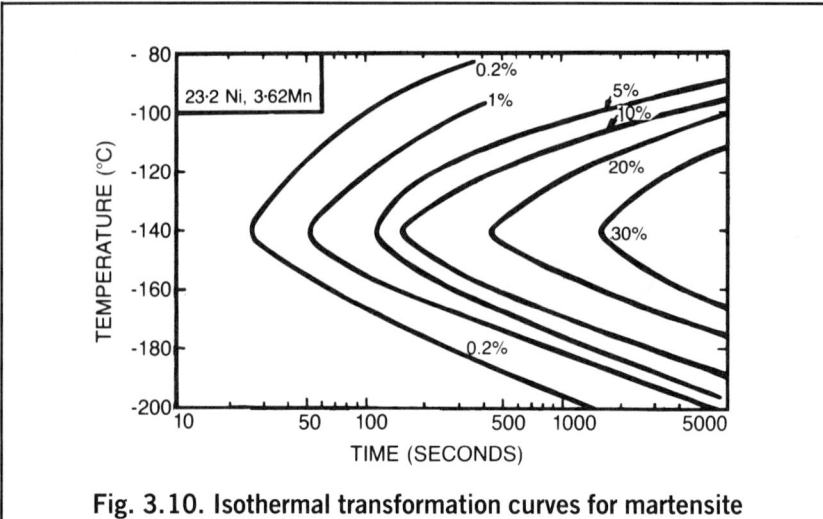

Fig. 3.10. Isothermal transformation curves for martensite formation in an Fe-23Ni-3.6Mn alloy. Curves are identified by the percentage of martensite formed. (Ref 3.26)

initiating plate, and frequently zig-zag arrays of martensite plates are observed in alloys susceptible to autocatalytic nucleation or bursting.

Stabilization, a phenomenon that reduces the ability of austenite to transform into martensite, occurs during slow cooling or interruption of cooling before complete transformation. For example an oil-quenched steel may contain more retained austenite than the same steel water quenched, and if transformation of a steel is interrupted by holding at some temperature between M_s and M_f, no martensite transformation may occur when cooling is resumed until substantial undercooling below the hold temperature is accomplished (Ref 3.25). One explanation of stabilization assumes that carbon segregates to potential embryos, or sites of martensitic nucleation, during slow cooling or on holding of a partially transformed specimen at a constant temperature. Once segregated, the carbon atoms increase the shear resistance of the austenite, thereby effectively stabilizing the austenite.

Although athermal martensite transformation kinetics are the dominant mode of transformation in heat treatable carbon steels, isothermal transformation has been observed in Fe-Ni-Mn and Fe-Ni-Cr alloys. The isothermal transformation is time-dependent and occurs at subzero temperatures; plotting this transformation frequently forms C-curves. Figure 3.10 shows a time-temperature-transformation diagram developed for isothermal martensite formation in an Fe-23Ni-3.7Mn alloy (Ref 3.26). Mathematical modeling of isothermal transformation kinetics has made possible a separation of the effects of preexisting nucleation sites or embryos and those produced by autocatalysis (Ref 3.23, 3.27). Also, the studies of isothermal transformation have shown a relationship between embryo size and kinetic mode of transformation. Alloy systems with large embryos or lattice sites predisposed to transformation require little or no thermal activation and therefore transform to martensite athermally. Systems with smaller embryos require thermal activation to produce martensite nuclei of size sufficient to initiate transformation, a process which leads to the time-dependent isothermal kinetics. Also, it has been shown that the activation energy for isothermal martensitic nucleation in Fe-Ni-Mn alloys is inversely proportional to the chemical driving force for the transformation, i.e., the greater the driving force the lower the activation energy (Ref 3.27, 3.28).

Crystallography of Martensitic Transformation

The diffusionless, shear mechanism of martensitic transformation requires good crystallographic coupling between the parent and product phases. Two important crystallographic parameters or characteristics emphasize this interrelationship between austenite and martensite in ferrous alloys. One is the orientation relationship between the crystal structure of the parent and the product martensite. The orientation relationship specifies planes and directions of the parent phase and the planes and directions in the product martensite to which they are parallel. Two well-known orientation relationships have been determined in ferrous alloy systems by means of x-ray diffraction techniques (Ref 3.29). The Kurdjumov-Sachs orientation relationship:

$$\{111\}_A \parallel \{101\}_M$$

$$\langle 110 \rangle_A \parallel \langle 111 \rangle_M$$

is valid for high-carbon steels with $\{225\}_A$ habit planes. The other orientation relationship, which was determined by Greninger and Troiano and is also attributed to Nishiyama, is:

$$\{111\}_A - \{011\}_M$$

$$\langle 112 \rangle_A \parallel \langle 011 \rangle_M$$

This relationship is observed in alloys where the martensite plates have $\{259\}_A$ habit planes.

The other crystallographic parameter that emphasizes the interrelationship of the parent and product phases is the habit plane, already mentioned in the discussion of Fig. 3.3. In steels, the habit plane is the plane in the parent austenite on which the martensite forms and grows. When the martensitic transformation is complete, ideally the habit plane is the planar interface between any retained austenite present and the martensite crystals. In actual fact, however, the interfaces between martensite and austenite in steels might be quite irregular, and the habit plane may be truly planar only at the midrib or point of origin of a martensite crystal. The habit plane is important not only because of its association with the initiation and progress of the transformation, but also because it affects the microstructural arrangements in the parent austenite grains of the many martensite plates that make up a hardened microstructure. The habit plane is a function of alloy composition, especially carbon content, and the various habit planes that characterize martensite in steels are presented in the section on morphology in this chapter.

The orientation relationship and habit plane in a given steel are parameters that relate the crystallography of austenite to martensite after transformation. Crystallography is also important in describing the martensitic transformation itself. A crystallographic theory of martensitic transformation was developed in the 1950s by Wechsler, Lieberman, and Read (Ref 3.30) and by Bowles and MacKenzie (Ref 3.31). The significance of the crystallographic theory is the understanding it provides for the origin of the internal fine structure found within any martensite crystal. The fine structure may consist of dislocations, twins, or a mixture of the two, depending on alloy composition. The presence of fine structure is a unique result of martensitic transformation, but occurs to some extent in any transformation where shear and diffusion are required to form the new phase. Widmanstätten ferrite and bainite formation are examples of the latter type of transformation.

The crystallographic theory of martensite formation is based on two important microstructural (macroscopic relative to atomic dimensions) observations: that the habit plane is unrotated and undistorted, and that the shape change that produces the surface tilting in Fig. 3.3 is homogeneous and a result of plane strain. The plane strain may be visualized as shear or displacement on planes parallel to the habit plane. Greninger and Troiano (Ref 3.32) first noted that the shape change could not be produced merely by the lattice deformation (i.e., in steels, the change in lattice from fcc austenite to bct martensite).

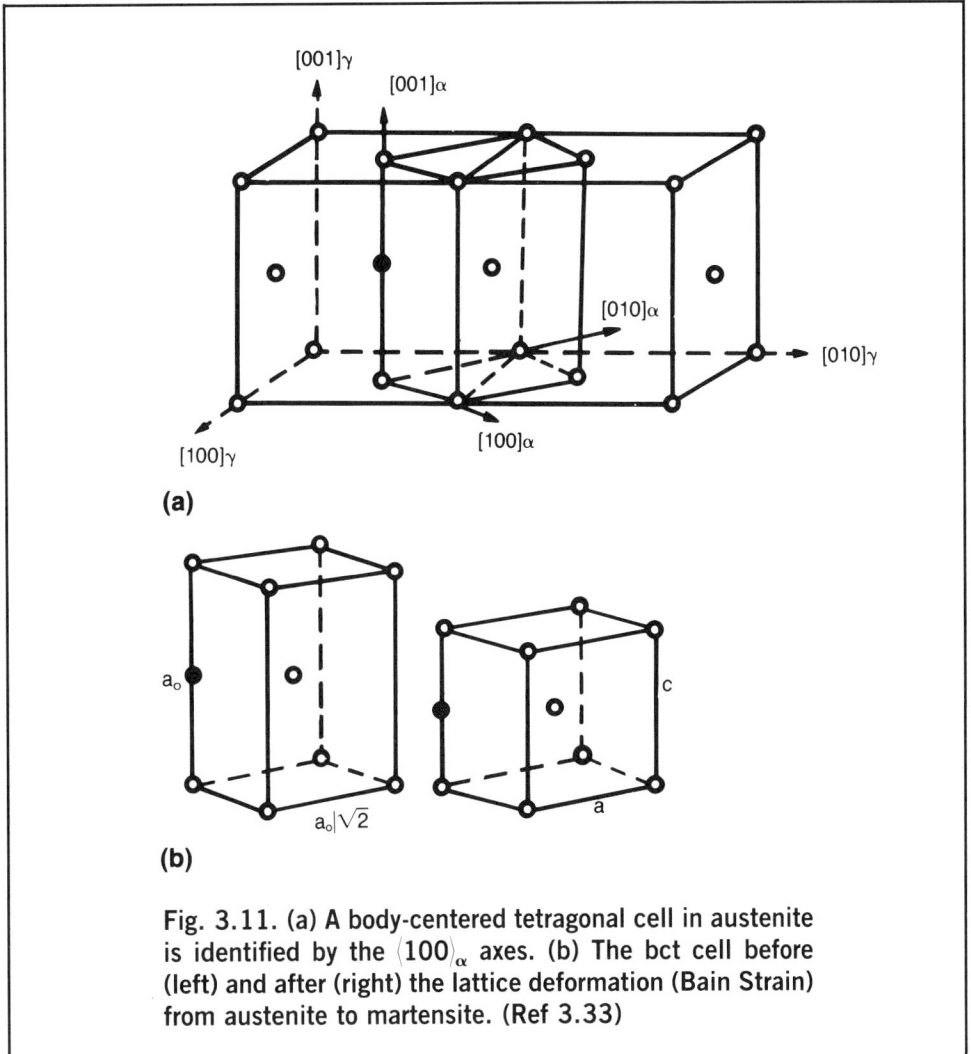

Fig. 3.11. (a) A body-centered tetragonal cell in austenite is identified by the $\langle 100 \rangle_\alpha$ axes. (b) The bct cell before (left) and after (right) the lattice deformation (Bain Strain) from austenite to martensite. (Ref 3.33)

Another deformation was necessary to satisfy the requirement of plane strain and the undistorted habit plane. This additional deformation, lattice invariant deformation, involves deformation of the bct martensite crystal by twinning or slip but not a change of the lattice or crystal structure itself.

The major elements of the crystallographic theory for martensite formation in steels are shown schematically in Fig. 3.11 and 3.12. Figure 3.11(a) shows two adjacent fcc unit cells of austenite in which a bct unit cell has been identified. This identification of a set of atoms in the parent phase that will transform to a set of atoms in the product

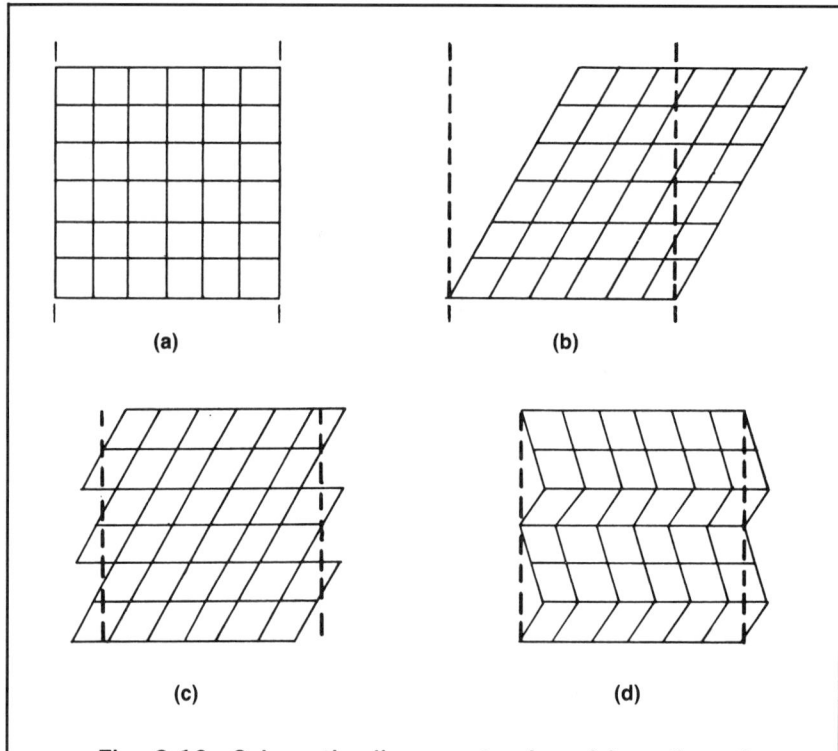

Fig. 3.12. Schematic diagrams to show (a) portion of parent crystal; (b) new lattice (martensite) produced by lattice deformation; and lattice invariant deformation by (c) slip and (d) twinning to make martensite conform to original position of parent crystal (a). (Ref 3.6)

phase is referred to as the lattice correspondence. The atoms identified in Fig. 3.11(a) have been isolated in the unit cell schematic on the left of Fig. 3.11(b). At this stage, the dimensions of the bct cell are still those derived from the austenitic lattice parameter. The unit cell on the right of Fig. 3.11(b) is that of martensite with lattice parameters a and c, corresponding to given carbon content (see Fig. 3.2). A lattice deformation was required to produce the martensite from austenite. In steels, the lattice correspondence shown in Fig. 3.11(a) was first identified by Bain and the lattice deformation from fcc to bct is referred to as the "Bain strain." Figure 3.11 shows that the Bain strain produces a contraction along the c axis and an expansion along the a axes.

In general, the lattice deformation will cause rotation away from the habit plane, as shown in Fig. 3.12(a) and (b), where it is assumed a number of cells of a parent crystal (a) are transformed to a new lattice

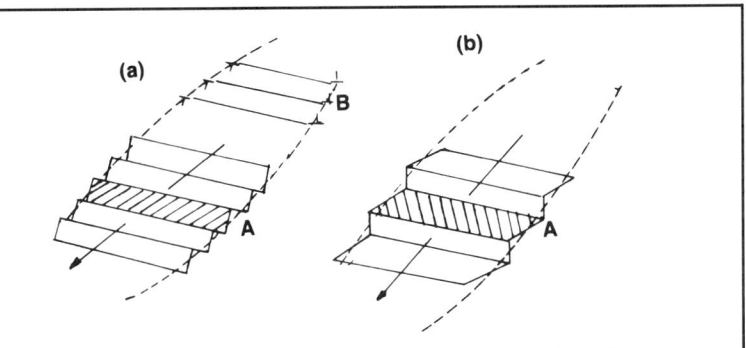

Fig. 3.13. Schematic representations of lattice invariant deformation by (a) slip and (b) twinning in martensite plates. (Ref 3.6)

(b). The vertical dashed lines represent the unrotated, undistorted habit plane. The constraints of the surrounding parent phase, however, cause the martensite unit to accommodate or deform by a lattice invariant deformation to the original boundaries (a) as required by the crystallographic theory. Figures 3.12(c) and (d) show, respectively, the martensite deformed by slip (dislocation movement) or twinning to satisfy on a macroscopic scale the requirement of an unrotated, undistorted habit plane. Figure 3.13 is another schematic representation of the slip (a) and twinning (b) modes of lattice invariant deformation with martensite plates. These sketches, of course, are idealized to demonstrate the concept of the lattice invariant shear. In actual crystals, when slip is the mechanism of accommodation, not only are dislocations introduced at the austenite-martensite interface, but also a high dislocation density remains in the fine structure within the plates. Examples of the latter are presented in the next section of this chapter.

The crystallographic theory of the martensitic transformation is well developed mathematically and has successfully predicted crystallographic parameters in a number of alloys. For example, if the lattice and lattice invariant deformations are specified, the habit plane may be predicted. For development of the theory and its application see Bilby and Christian (Ref 3.6) and Wayman (Ref 3.2). Successes and limitations of the theory are reviewed by Dunne and Wayman (Ref 3.34).

Morphology of Ferrous Martensites

Two major morphologies of martensite, lath and plate, develop in heat treatable carbon steels. Figure 3.14 shows the carbon ranges of

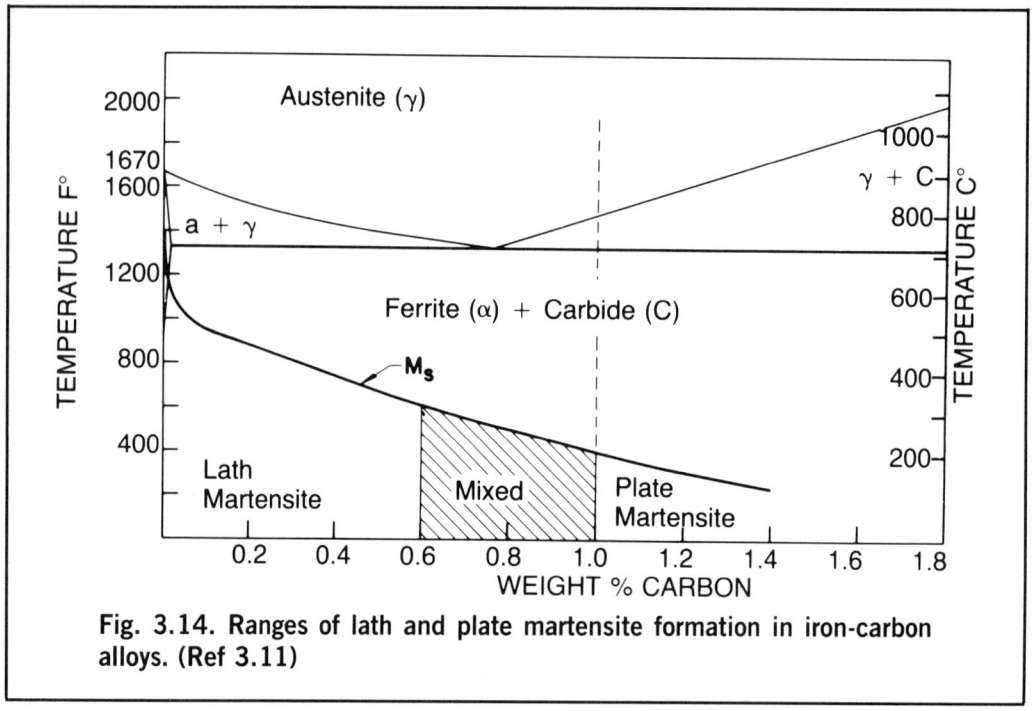

Fig. 3.14. Ranges of lath and plate martensite formation in iron-carbon alloys. (Ref 3.11)

formation and M_s temperature of the two morphologies. The boundaries of the various regions are based on characterization of high-purity Fe-C alloys and may shift in alloy steels. The designations of the two morphologies originate from the shape of the individual units of martensite. The lath designation is used to describe the board-shaped units of martensite that form in low- and medium-carbon steels, while the plate designation accurately describes the shape of the martensite units that form in high-carbon steels. The terms lath and plate, therefore, refer to the three-dimensional shapes of individual martensite crystals. In metallographic specimens, sections through the martensite laths or plates are revealed by polishing and etching. Generally, these cross sections will appear to be needlelike or acicular, and the latter adjectives are often used to describe martensite microstructures. Other terms based on one or another feature of the different forms of martensite have been used to describe the two morphologies of martensite, but the terms lath and plate are preferred (Ref 3.8).

Until the advent of the electron microscope, the plate martensites, which could be readily resolved by light microscopy, received the most emphasis in the literature. The units of plate martensite are well within the size range resolvable in the light microscope, and frequently the retained austenite that coexists with the martensite in high-carbon alloys helps to sharply define the plates in the light microscope. On the other hand, as will be demonstrated, many of the individual units of lath martensite are below the resolution of the light microscope, and any retained austenite present is also too fine to be resolved. Although plate martensites are important in some heat treated applications (such as the case microstructure of carburized steels), most hardenable steels have low- or medium-carbon content and, therefore, microstructures composed of lath martensite. As a result, lath martensites have overwhelming industrial significance. Microstructures with plate martensite for engineering applications are found in tool steels and the high-carbon case structures of carburized steels. In order to follow the historical development of the understanding of martensitic microstructures, plate martensite will be described first in the following sections. The characteristics of individual units and the arrangement of the units to produce the microstructures that are put into service as a result of good heat treatment practice will be described for both lath and plate martensite.

Plate Martensite

Many other ferrous systems show the same transition from lath to plate martensite (see Fig. 3.14) with increasing alloying as does the Fe-C system (Ref 3.8). Figure 3.15 shows plate martensite that was produced by cooling a single crystal of Fe-33.5Ni austenite in liquid nitrogen (−196 °C or −321 °F). Subzero cooling was required because the high nickel content had lowered the M_s to −30 °C (−22 °F). The specimen was not polished or etched after the liquid nitrogen treatment, and therefore all features shown in Fig. 3.15 are due to the surface relief generated by the martensitic transformation. On the scale shown, the surface tilting is indeed quite homogeneous except for small dark bands visible in some of the martensite plates. These bands are deformation twins formed in response to the constraints of the austenite matrix. The deformation twins, however, are micron sized and irregularly distributed in contrast to the much finer and more regularly distributed fine structure that results from the lattice invariant deformation.

Figure 3.16 is a transmission electron micrograph of the fine structure that formed in a single plate of martensite in the Fe-33.5Ni alloy. Fine transformation twins (small dark bands), dislocations (the fine linear features), and a large deformation twin band are present. Figure 3.17 shows the dislocation fine structure at a higher magnifi-

Fig. 3.15. Plate martensite formed in an austenitic single crystal of an Fe-33.5Ni alloy by cooling to −196 °C (−321 °F). Plates are visible only because of surface relief generated by martensitic transformation. Magnification, 200×. (Ref 3.35)

cation. Relatively straight dislocation lines in two directions are visible. By selected area diffraction techniques (Ref 3.35), the preferred directions were shown to correspond to $\langle 111 \rangle$ directions and, therefore, the dislocations are mostly screw dislocations. This type of dislocation array is characteristic of those formed by deformation of bcc iron at low temperatures and/or high strain rates, and illustrates one type of fine structure formed by the lattice invariant deformation of bcc martensite as a result of the austenitic constraints. Figure 3.18 shows another type of fine structure, very fine transformation twins, about 100 Å in thickness, in Fe-33.5Ni plate martensite. The twins lie on $\{112\}_m$ planes and represent another plastic deformation mode that forms in bcc crystals deformed at low temperatures and high strain rates. Also shown in Fig. 3.18 is a larger deformation twin across which the fine transformation twins have changed their orientation to a $\{112\}_m$ orientation in the twin (Ref 3.36).

Examples of plate martensite in Fe-C alloys are shown in Fig. 3.5 and 3.19. Many different orientations of the martensite plates are apparent in the microstructures shown. This characteristic appearance of a plate martensitic microstructure is directly related to the habit planes of the plate martensite and the tendency of adjacent plates to assume different variants of the habit plane. Plate martensites have

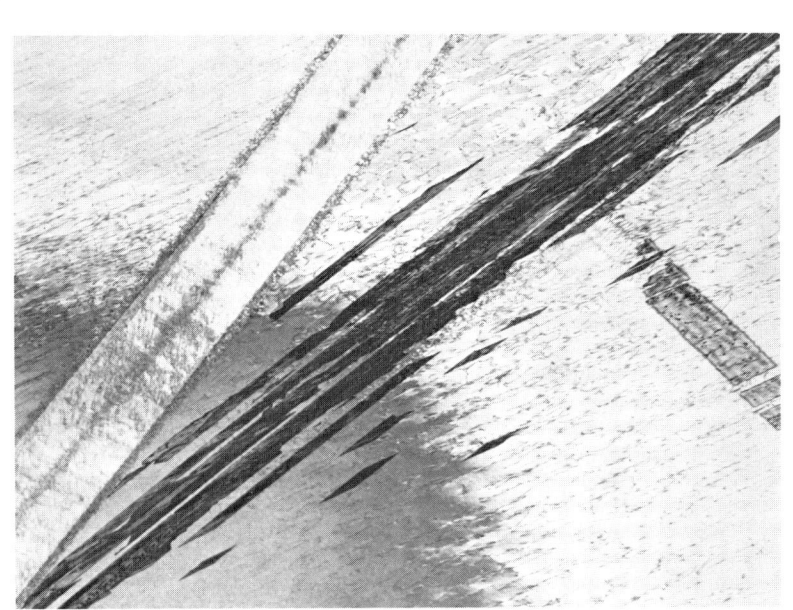

Fig. 3.16. Fine structure within martensite plates shown in Fig. 3.15. A deformation twin, fine transformation twins, and dislocations are shown. Transmission electron micrograph; magnification, 20 000×, shown here at 75%. (Ref 3.35)

irrational habit planes, i.e., the planes are not defined by low number indices such as (100) or (111). Early work by Greninger and Troiano (Ref 3.37) (see Fig. 3.20) showed that Fe-1.78C alloys had habit planes best characterized as $\{259\}_A$, and lower carbon alloys, containing 0.92% and 1.4% carbon, had $\{225\}_A$ habit planes. The Fe-Ni plate martensites have been the subject of extensive crystallographic studies. Figure 3.21 shows that there is considerable scatter of the habit plane in Fe-Ni alloys containing 29.0 to 35.0% nickel. The scatter may be due to compositional variations, the habit plane shifting toward $\{225\}_A$ with decreasing nickel content, or to mixtures of lattice invariant deformations such as combinations of twinning and dislocations in a given alloy (Ref 3.35).

The many orientations of martensite plates in the microstructures shown in Fig. 3.5 and 3.19 are due to the many variants of the irrational habit planes. A variant is merely a different orientation of a given {hkl} plane as defined by a different arrangement of the same hkl indices. For example, $(925)_A$, $(592)_A$, and $(952)_A$ are all variants of the

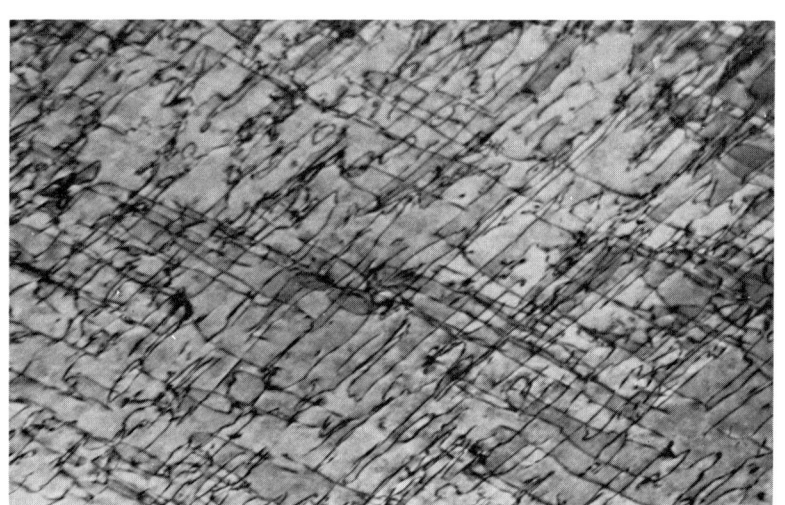

Fig. 3.17. Dislocation fine structure in martensite plates shown in Fig. 3.15. Transmission electron micrograph magnification, 20 000×, shown here at 75%. (Ref 3.35)

$\{259\}_A$ plane. Any plane where h, k, l are all different, as is the case for $\{259\}_A$, has 24 different variants, and a plane with two indices equal, such as the $\{225\}_A$ plane, has 12 variants. Thus, the plate martensite microstructures, because of the large number of variants possible, and the fact that adjacent plates assume different variants, appear quite haphazardly arranged, despite the fact that there is only a single habit plane for all the plates in a given alloy.

An important consequence of the nonparallel plate formation in Fe-C alloys is the development of microcracks in the martensite plates as a result of the impingement of plates of different habit plane variants (Ref 3.39). Figure 3.22 shows an example of the microcracks in the plate martensite of an Fe-C alloy. The microcracks tend to form in the largest martensite plates and therefore are not present to any great extent in steels where austenite grain size, and accordingly martensite plate size, are fine (Ref 3.41). Also in lower carbon steels, the morphology shift to lath martensite eliminates the impingements and the development of microcracks (Ref 3.42). The high carbon plate martensites are quite brittle and sensitive to microcracking. However, in Fe-Ni alloys where the martensite is much more ductile, the impingement of martensite plates is accommodated by deformation twinning rather than cracking.

Fig. 3.18. Fine transformation twins in plate martensite of an Fe-33.5Ni alloy. Note change in orientation of fine twins in large deformation twin. Transmission electron micrograph; magnification, 15 000×, shown here at 75%. (Ref 3.36)

Lath Martensite

Light micrographs of lath martensite in Fe-C alloys are shown in Fig. 3.23. The lath martensite units tend to be quite fine, but the characteristic acicularity of a martensitic microstructure is apparent. An important microstructural characteristic of lath martensites is the tendency of many laths to align themselves parallel to one another in large areas of the parent austenite grain. These regions of parallel lath alignment are referred to as packets and tend to develop most prominently in lower carbon alloys, as shown in Fig. 3.23(a). The packets are delineated because of the different etching characteristics of the different variants or orientations of the laths in the various packets.

Figure 3.24 shows the transition in martensite morphology that develops as the carbon content of the Fe-C alloys increases from 0.67 to 1.00% carbon. With increasing carbon content, more plates of martensite, differentiated from lath martensite by their larger size and their tendency to microcrack, are discernible in the microstructure.

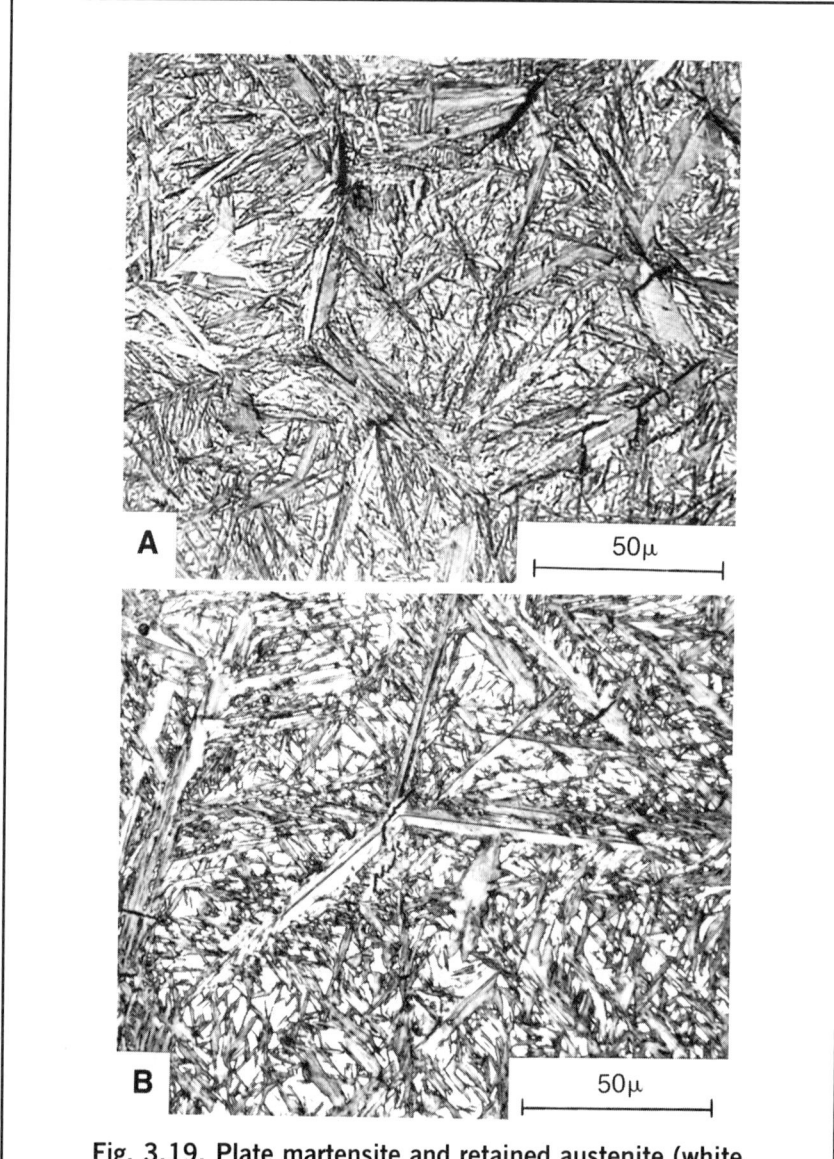

Fig. 3.19. Plate martensite and retained austenite (white patches) in (A) Fe-1.22C and (B) Fe-1.4C alloys. Light micrographs. (Ref 3.11)

A transmission electron micrograph of the lath martensite in an Fe-0.2C alloy is shown in Fig. 3.25. All of the laths, even the very thin ones, are resolved, in contrast to the light micrograph of the same structure in Fig. 3.23(a) where many units are not clearly defined.

MARTENSITE AND BAINITE / 69

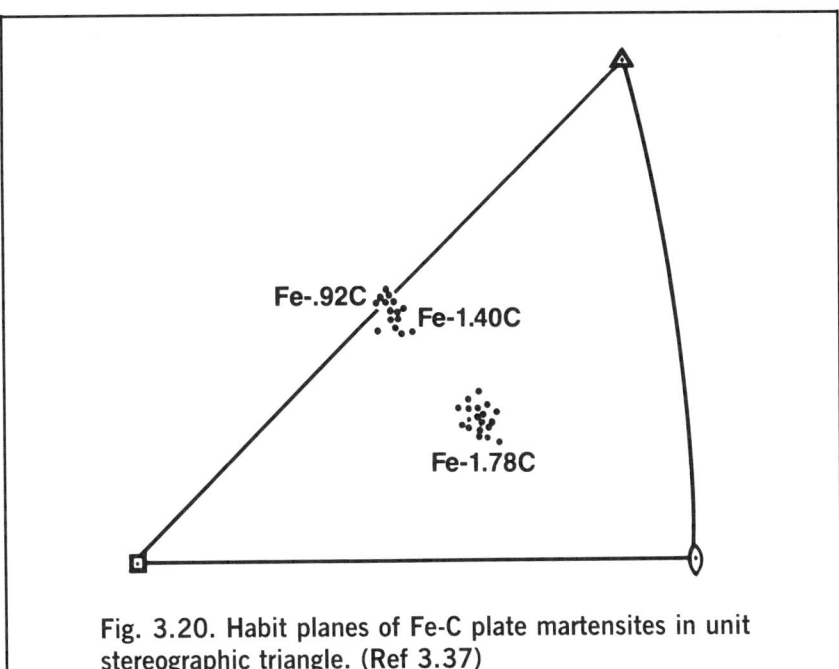

Fig. 3.20. Habit planes of Fe-C plate martensites in unit stereographic triangle. (Ref 3.37)

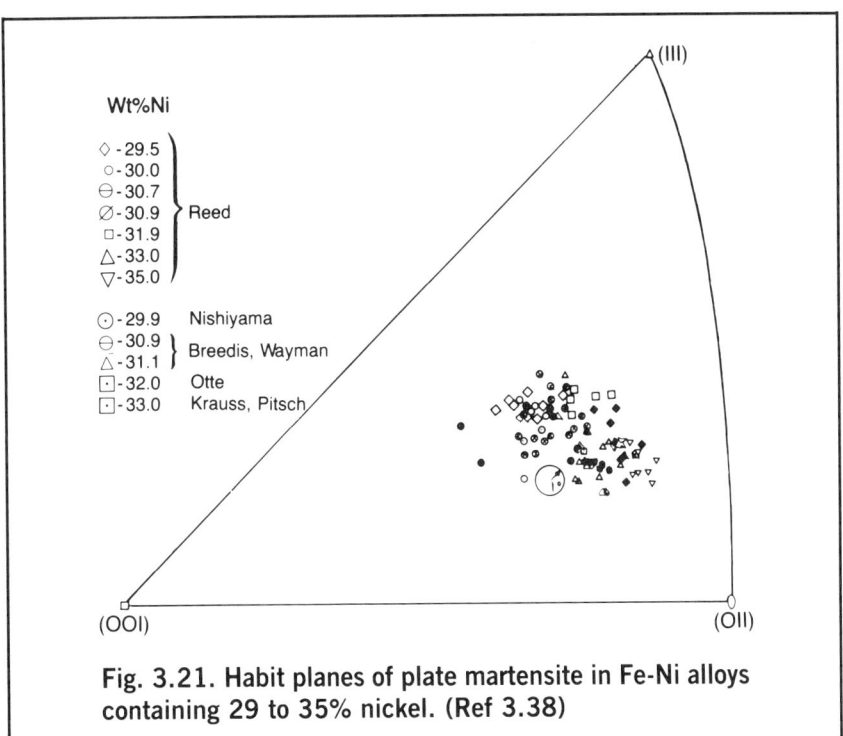

Fig. 3.21. Habit planes of plate martensite in Fe-Ni alloys containing 29 to 35% nickel. (Ref 3.38)

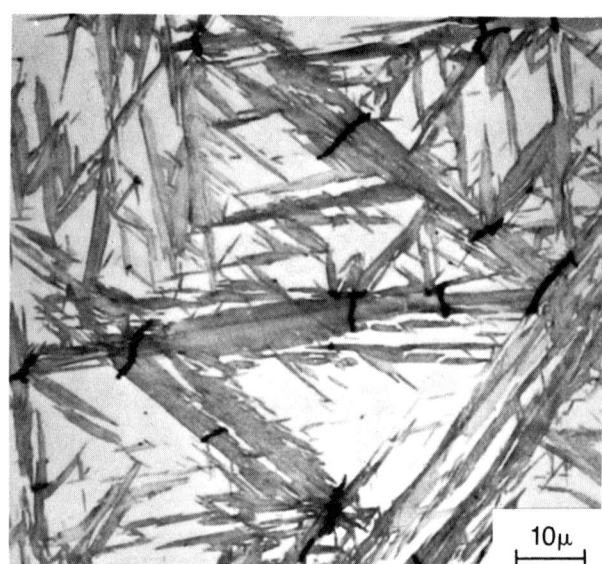

Fig. 3.22. Microcracks in plate martensite of an Fe-1.4C alloy. (Ref 3.40)

Parts of two packets are shown. In each packet there appear to be two major orientations or variants of the martensite laths, and there are many very fine laths. Figure 3.26 shows the distribution of lath widths obtained by measurements from electron micrographs obtained from thin foils and replicas of polished and etched metallographic specimens (Ref 3.43). The important result shown in Fig. 3.26 is that most of the laths have widths smaller than 0.5 μm, the resolution limit of the light microscope, and therefore cannot possibly be revealed by light metallography. There are some laths with widths up to almost 2 μm, and these larger laths would, of course, be visible in the light microscope as some are in Fig. 3.23. It is the very fine size of most of the laths in a packet of low- or medium-carbon martensite that have over the years made the light metallographic characterization of lath martensite difficult.

The habit plane of lath martensite, $\{557\}_A$, is irrational, a plane close to $\{111\}_A$, as shown in Fig. 3.27. There are three $\{557\}_A$ variants clustered about each of the four $\{111\}_A$ planes, and the angle between these variants is only 16°. Laths of different orientation within packets (see Fig. 3.25) frequently are observed to make angles of about 16° with each other, leading to the conclusion that the variants

MARTENSITE AND BAINITE / 71

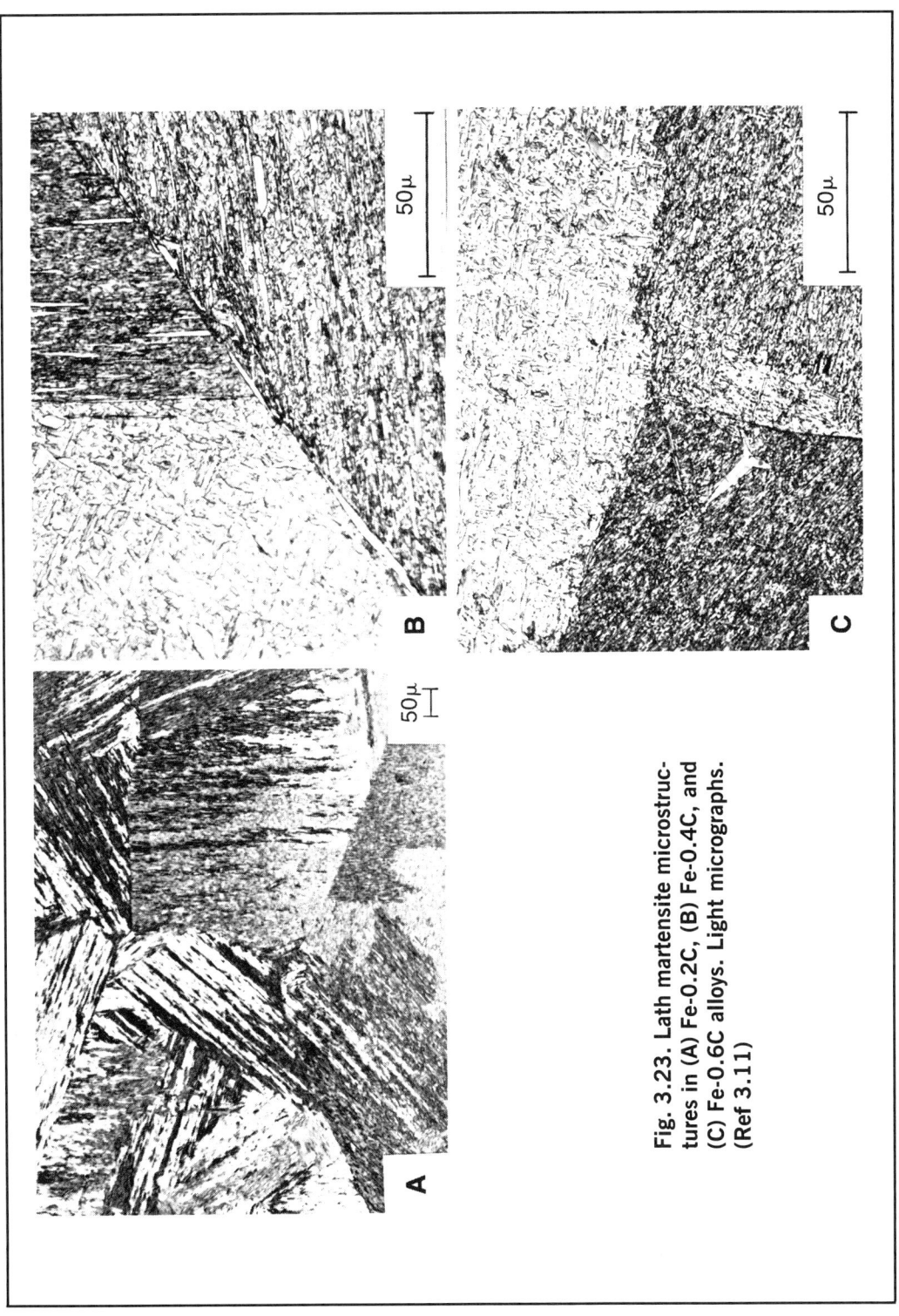

Fig. 3.23. Lath martensite microstructures in (A) Fe-0.2C, (B) Fe-0.4C, and (C) Fe-0.6C alloys. Light micrographs. (Ref 3.11)

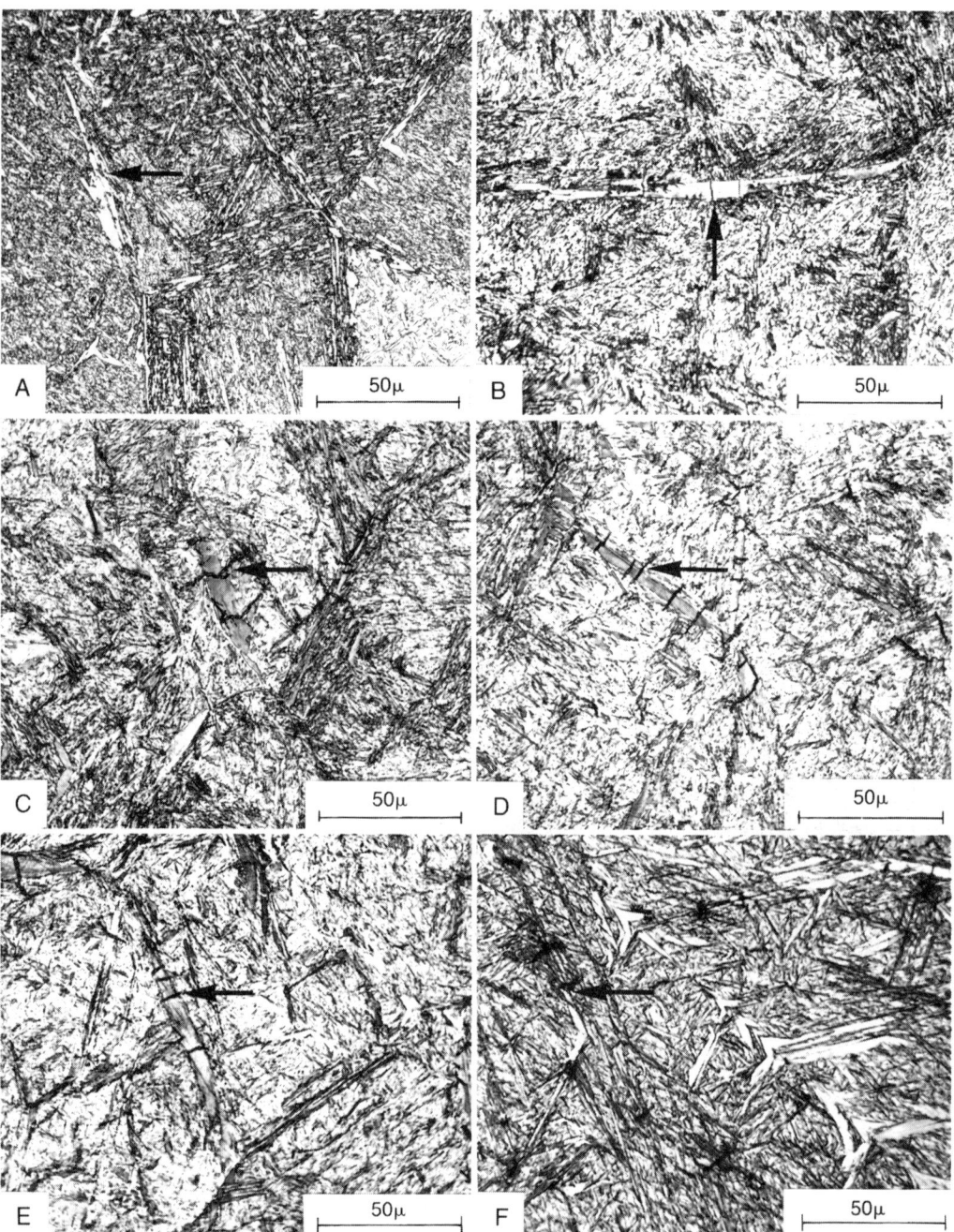

Fig. 3.24. Transition from lath to plate martensite microstructures in Fe-C alloys between 0.67 and 1.00% carbon. A: 0.67% carbon; B: 0.75%; C: 0.82%; D: 0.85%; E: 0.93%; F: 1.00%. Light micrographs. Sodium bisulfite etch. (Ref 3.11)

Fig. 3.25. Lath martensite in an Fe-0.2C alloy. Two packets, each with two variants of laths, are shown. Transmission electron micrograph. (Ref 3.43)

in a given packet all have variants close to the same $(111)_A$ plane. This coupling of variants, the small angles between variants, and the fine size of the laths give the microstructural impression that lath martensite has a $\{111\}_A$ habit with only four variants. Lath marten-

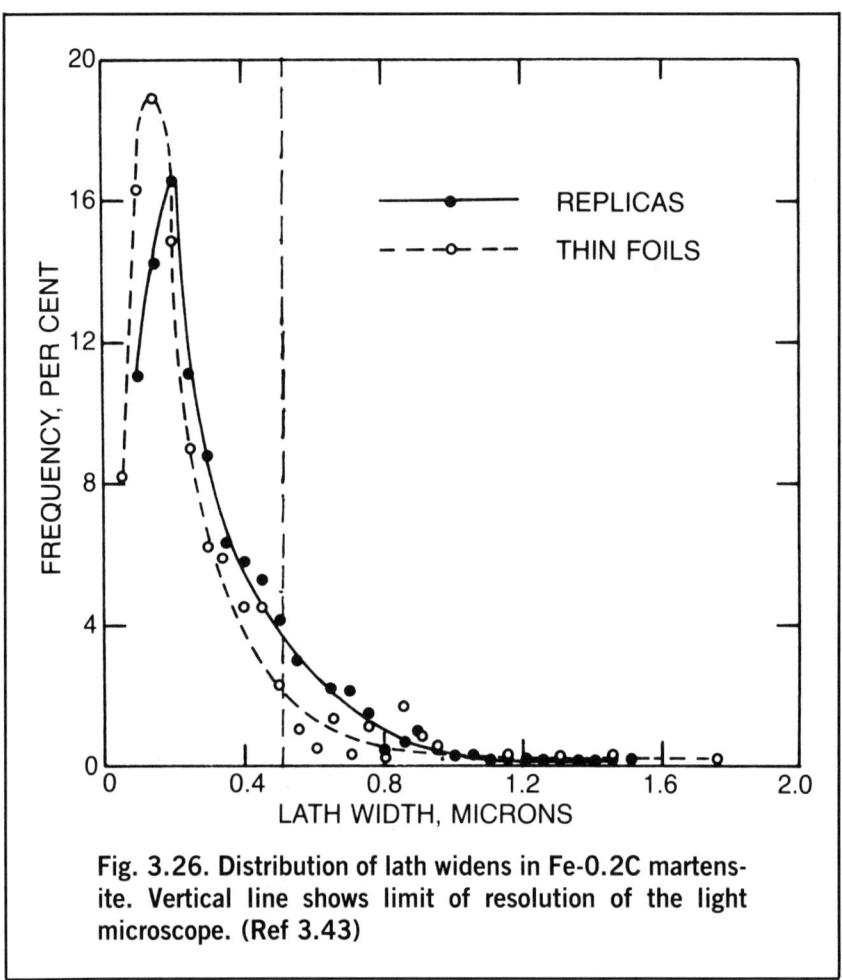

Fig. 3.26. Distribution of lath widens in Fe-0.2C martensite. Vertical line shows limit of resolution of the light microscope. (Ref 3.43)

sitic microstructures, therefore, appear much more orderly (see Fig. 3.23) than do the plate martensitic microstructures (see Fig. 3.19) with as many as 24 variants. The $[557]_A$ habit plane has also been measured in an extensive study of lath martensite in an Fe-20Ni-5Mn alloy (Ref 3.44).

Although there may be several variants of laths in a packet of lath martensite, one variant tends to be dominant (Ref 3.43). This characteristic of a packet means that most of the laths, separated by low-angle boundaries or perhaps retained austenite, have the same crystal orientation and that a packet may be considered as a single grain or crystal, albeit a grain divided by many low-angle boundaries and containing a fine structure of many dislocations. The packet

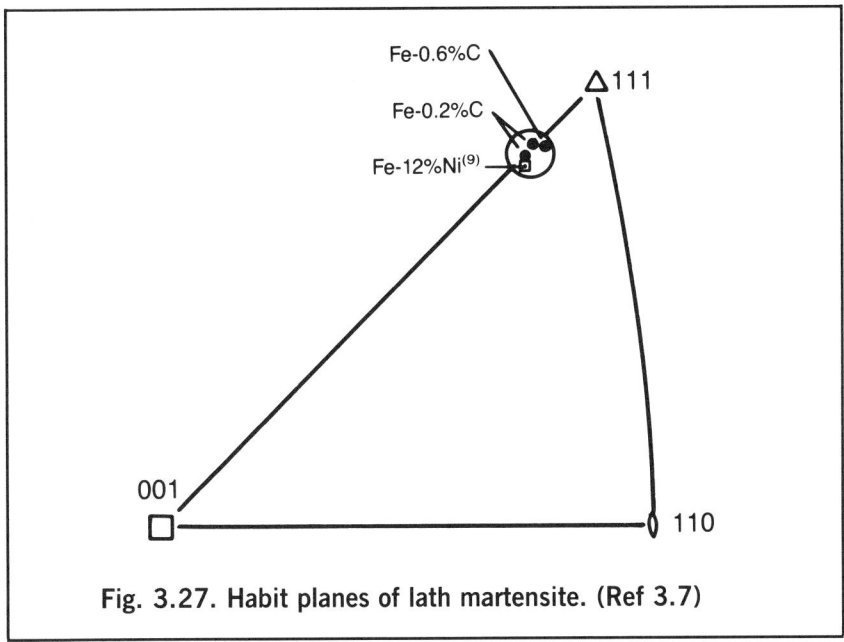

Fig. 3.27. Habit planes of lath martensite. (Ref 3.7)

structure of lath martensite has no counterpart in plate martensite and is important in determining mechanical properties and fracture behavior of the martensite that forms in low- and medium-carbon steels.

The fine structure of lath martensite consists predominantly of a very high density of dislocations, too high to be resolved even by electron microscopy of thin foils. However, Speich (Ref 3.45) was able to determine, indirectly by electrical resistivity measurements, a dislocation density of almost 10^{12} dislocations per square centimeter in low-carbon lath martensite.

An example of the fine structure of lath martensite in an Fe-0.2C alloy is shown in Fig. 3.28. The dislocations are tangled and arranged in incipient dislocation cells, a structure much different from the essentially straight, uniformly distributed dislocations of the Fe-Ni plate martensite shown in Fig. 3.17. The dislocation tangles are a result of a plastic deformation mode consistent with the high M_s and high-temperature range of formation of the low-carbon lath martensite (see Fig. 3.14) whereas, as already noted, the straight dislocations of the Fe-Ni martensite are consistent with low-temperature deformation of bcc iron alloys. Another consequence of the high M_s temperatures of lath martensite formed in low-carbon steels is autotempering or quench tempering, the precip-

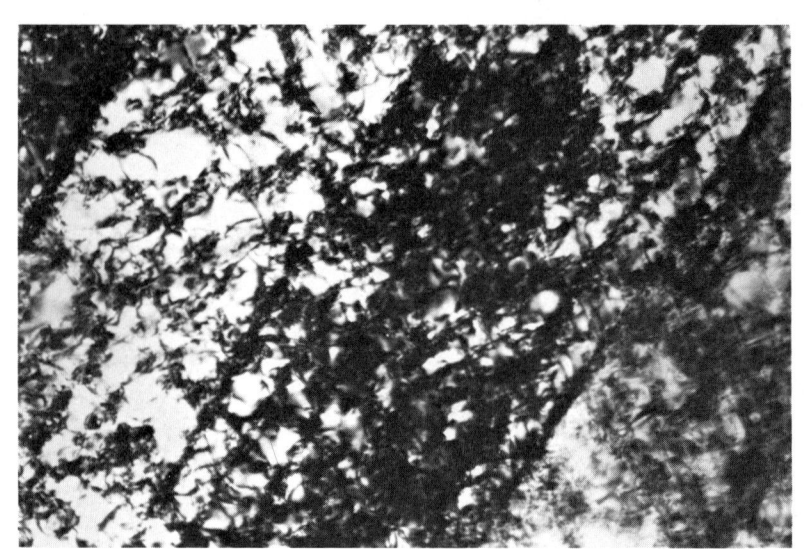

Fig. 3.28. Dislocation fine structure in lath martensite of an Fe-0.2C alloy. Transmission electron micrograph; magnification, 82 500×. (Ref 3.46)

itation of cementite in martensite during quenching. Aborn (Ref 3.47) presents evidence of autotempering in an early study of structure and properties of low-carbon martensites.

Although dislocations are the major fine structural component in lath martensite, fine transformation twins, a low temperature mode of plastic accommodation, are also found to some extent in Fe-C lath martensite. The amount of fine twinning increases in accord with the decreasing M_s temperatures and lower athermal transformation ranges of lath martensite formation as carbon content increases.

The major change in morphology of martensite in Fe-C alloys and steels is the change from lath to plate morphologies which begins in alloys containing about 0.6% carbon. However, there is a gradual change in morphology within the lath range, as indicated in Fig. 3.23 and more clearly shown in Fig. 3.29 (Ref 3.48). In the alloys with 0.43% and 0.55% carbon, although the martensite units still appear to be largely parallel and quite fine, the packet structure is more difficult to define. Also, on the scale resolvable with the electron microscope, more adjoining laths assume nonparallel variants (Ref 3.48).

MARTENSITE AND BAINITE / 77

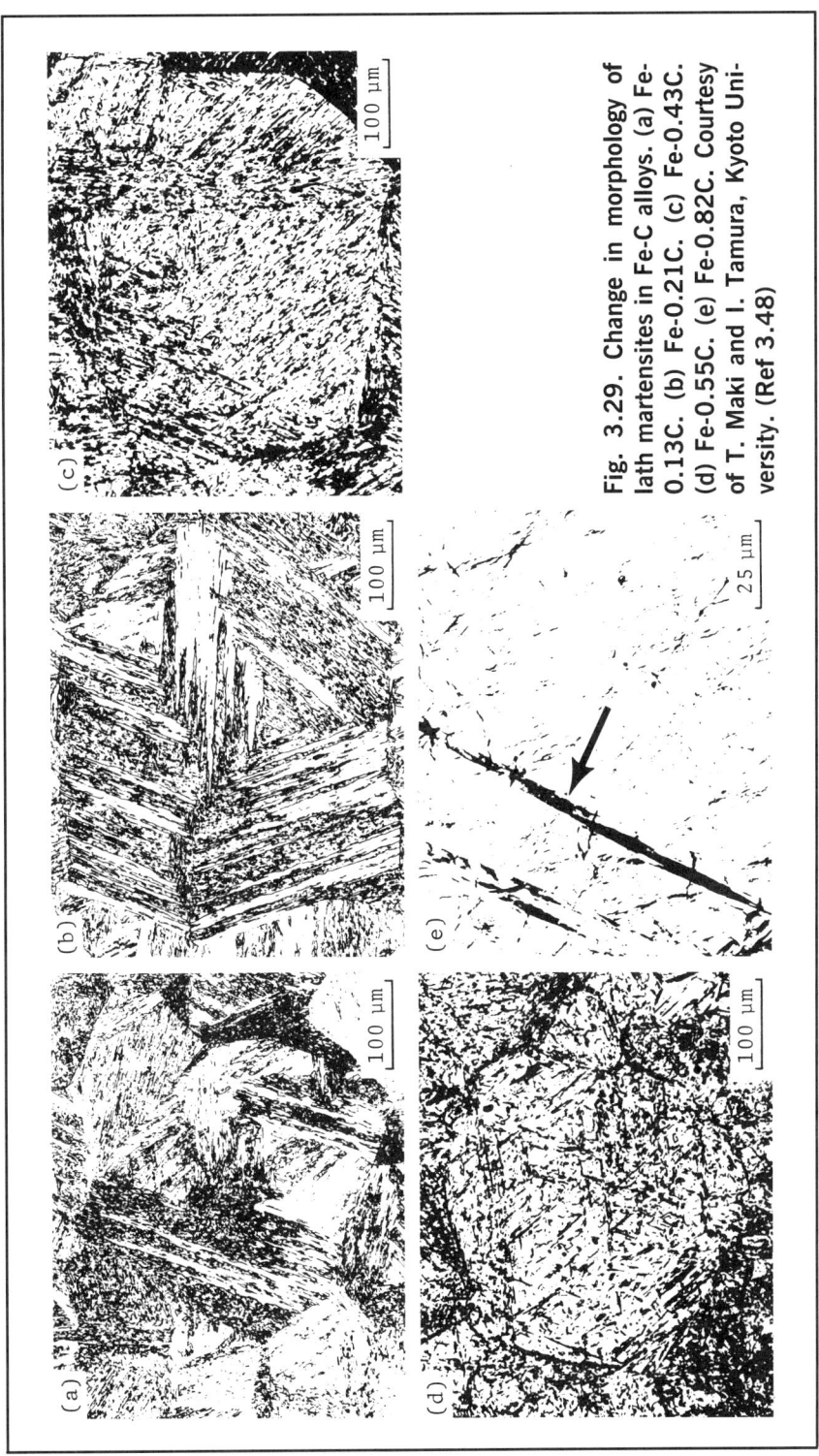

Fig. 3.29. Change in morphology of lath martensites in Fe-C alloys. (a) Fe-0.13C. (b) Fe-0.21C. (c) Fe-0.43C. (d) Fe-0.55C. (e) Fe-0.82C. Courtesy of T. Maki and I. Tamura, Kyoto University. (Ref 3.48)

Bainite

The formation and structure of the various microstructural constituents that form during the heat treatment of steels (in particular, proeutectoid ferrite and cementite, pearlite, and martensite) have already been discussed. The description of bainite, the last microstructural constituent to be discussed, has been deferred to this point because it forms under continuous cooling or isothermal transformation conditions intermediate to those of pearlite and martensite formation and because there are transformation and structural similarities to both pearlite and martensite.

Similar to pearlite, bainite is a mixture of the phases ferrite and cementite, and is therefore dependent on the diffusion-controlled partitioning of carbon between ferrite and cementite. However, unlike pearlite, the ferrite and cementite are present in nonlamellar arrays whose characteristics are dependent on alloy composition and transformation temperature. Similar to martensite, the ferrite of bainite may be in the form of laths or plates containing a dislocation structure, and to some extent, therefore, the mechanism of bainite formation involves shear as well as diffusion.

There are two major morphologies or forms of bainite: upper bainite that forms in the temperature range just below that of pearlite formation, and lower bainite which forms at temperatures closer to M_s. Figure 3.30 shows upper bainite in a 4360 steel partially transformed at 410 °C (770 °F) and at 495 °C (923 °F). In Fig. 3.30 the bainite is dark and the white-etching matrix is martensite formed from untransformed austenite on quenching. The bainite appears dark (i.e., has low reflectivity) because of roughness produced by etching around the carbide particles, but the carbides themselves in this example are too fine to be resolved in the light microscope. The feathery appearance of the clusters of ferrite laths is clearly shown in the light micrographs and is an important identifying feature of upper bainite. The carbide particles of upper bainite tend to be elongated and form between ferrite laths, areas to which carbon has been rejected during the growth of the ferrite. Figure 3.31 is a thin foil electron micrograph which shows the interlath cementite in a 4360 steel transformed to bainite at 495 °C (923 °F). In certain steels, in particular those with high silicon content, the carbon-enriched austenite between the bainite laths is quite stable and is retained between the ferrite laths at room temperature. Figure 3.32 shows such retained austenite, marked A, in bainite formed at 400 °C (752 °F) in a steel containing 0.6% carbon and 2.0% silicon.

MARTENSITE AND BAINITE / 79

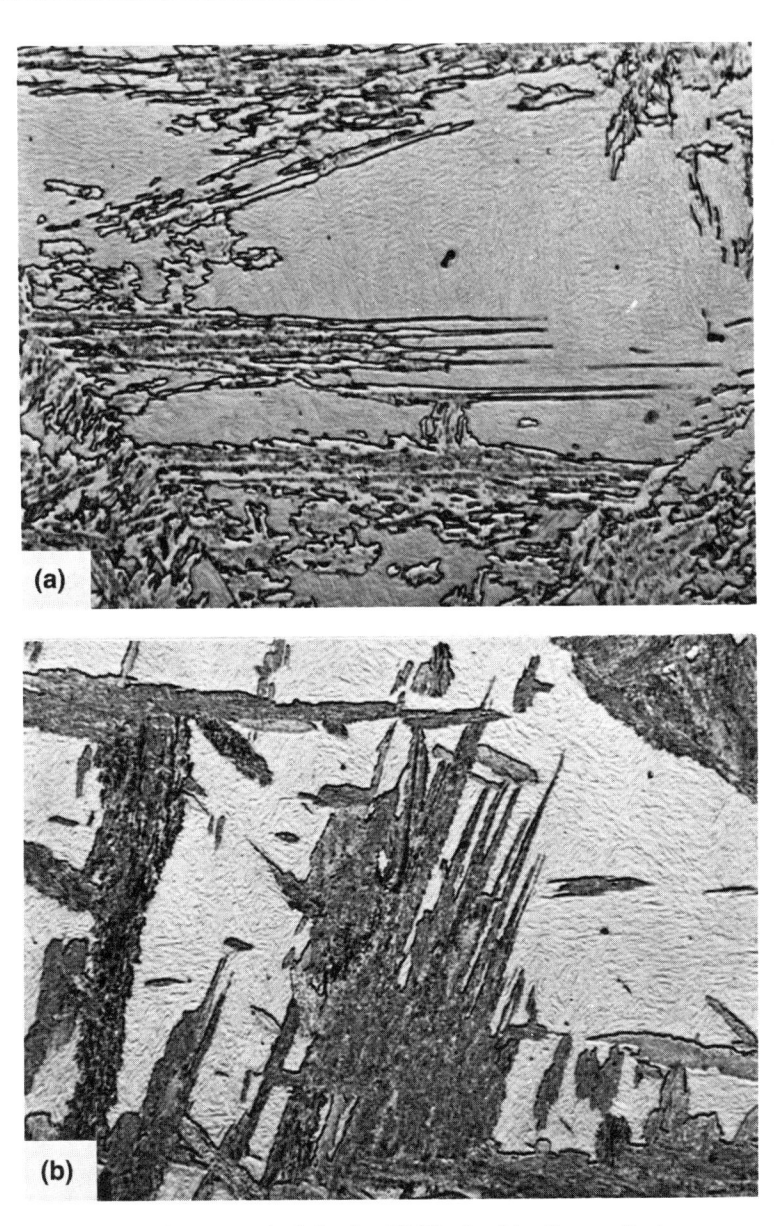

Fig. 3.30. Upper bainite in 4360 steel isothermally transformed at (a) 495 °C (923 °F) and (b) 410 °C (770 °F). Light micrographs. Picral etch. Magnification, 750×. (Ref 3.49)

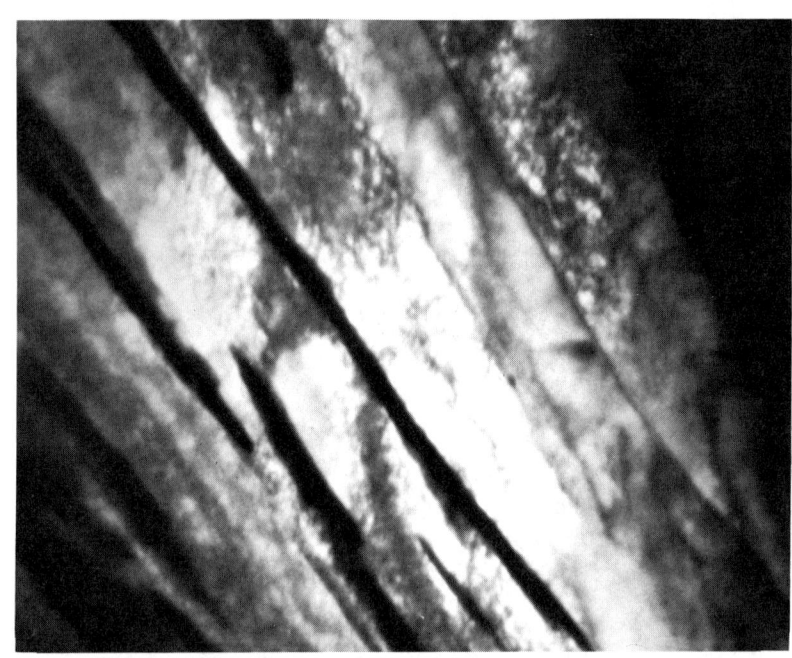

Fig. 3.31. Carbides (dark) formed between laths of upper bainite in 4360 steel transformed at 495 °C (923 °F). Transmission electron micrograph; magnification, 25 000×. (Ref 3.49)

Lower bainite in 4360 steel partially transformed at 300 °C (572 °F) and quenched is shown in Fig. 3.33. Again the bainite etches dark and the white-etching matrix is martensite formed from austenite untransformed at 300 °C (572 °F). The lower bainite is composed of large plates and, analogously to plate martensitic microstructures, is often characterized as acicular. The carbides in lower bainite are responsible for its dark etching appearance but are much too fine to be resolved in the light microscope. Figure 3.34 shows the carbides in lower bainite formed in the 4360 steel transformed at 300 °C (572 °F). In contrast to upper bainite, fine carbides have formed within the plates, rather than between plates, and are significantly finer than the interlath carbides of upper bainite. In some lower bainites the fine carbides have been identified as a transition carbide, epsilon carbide, that forms prior to cementite at lower temperatures. The transition carbide is especially stable in steels containing aluminum and silicon. Okamoto and Oka (Ref 3.50) have reported a new type of

Fig. 3.32. Retained austenite between laths of upper bainite in 0.6% carbon steel containing 2.0% silicon transformed at 400 °C (752 °F). Retained austenite is identified by letter A. Transmission electron micrograph; magnification, 40 000×. (Ref 3.49)

lower bainite which forms in hypereutectoid steels and is termed "lower bainite with midrib" (Fig. 3.35). This bainite forms isothermally at lower temperatures (150 to 200 °C or 300 to 390 °F) than the temperatures at which conventional lower bainite forms (200 to 350 °C or 390 to 660 °F). The midrib is an isothermally formed thin plate of martensite which provides the interface at which subsequently the two phase carbide-ferrite lower bainitic structure forms (Fig. 3.35b).

Several studies have characterized crystallographic parameters such as habit planes and orientation relationships in the bainite of Fe-Cr-C alloys (Ref 3.51), and the mechanism and characterization of bainite has been the subject of continuing debate (Ref 3.52-3.54). Although the carbon atoms have high mobility in the bainite transformation range, the iron atoms do not, and therefore a shear mechanism is required to effect the formation of bainite, at least at the start of the transformation. Carbon diffusion is then necessary to form the

Fig. 3.33. Lower bainite in 4360 steel transformed at 300 °C (572 °F). Light micrograph. Picral etch. Magnification, 750×. (Ref 3.49)

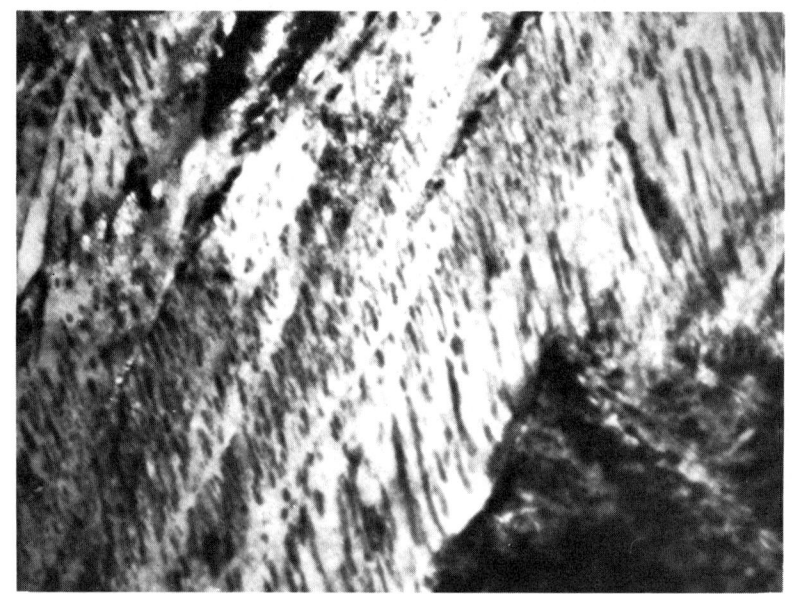

Fig. 3.34. Lower bainite with fine carbides within plates in 4360 steel transformed at 300 °C (572 °F). Transmission electron micrograph; magnification, 24 000×. (Ref 3.49)

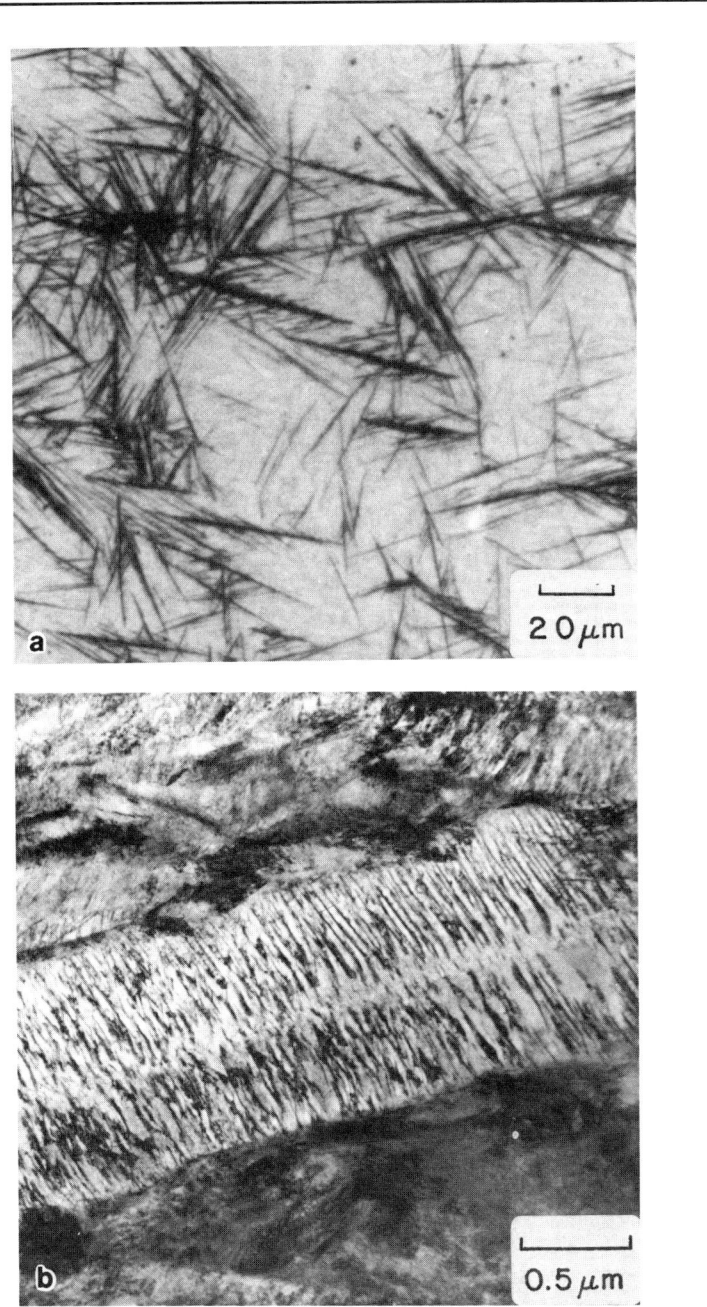

Fig. 3.35. Lower bainite with midribs in a 1.10% carbon steel transformed at 190 °C (374 °F) for 5 h. (a) Light micrograph. (b) TEM micrograph. Courtesy of H. Okamoto, Tottori University. (Ref 3.50)

carbides within two-phase bainitic structure. Aaronson and his coworkers, as reviewed in Ref 3.53 and 3.54, propose a ledge mechanism for the growth of bainite.

The trend to producing steels with low carbon content but high strength and toughness has resulted in the development of steels with bainite-like microstructures quite different from the classical two-phase, nonlamellar ferrite-carbide bainites formed in medium-carbon steels. These structures are formed on continuous cooling in low-carbon alloys, often containing manganese and molybdenum, and are mixtures of ferrite and austenite, with some of the austenite sometimes transformed to martensite (Ref 3.55). There are many morphologies of these unique structures, and the related transformation mechanisms, classification, and terminology are currently under active discussion (Ref 3.56, 3.57).

References

3.1 F. Osmond, Méthode générale pour l'analyse micrographique des aciers au carbone, *Bulletin de la societe d'Encouragement pour l'Industrie National*, Vol 10, 1895, p 480

3.2 C.M. Wayman, *Introduction to the Crystallography of Martensite Transformations*, MacMillan, New York, 1964

3.3 W.D. Kingery, *Introduction to Ceramics*, John Wiley & Sons, New York, 1960

3.4 G.K. Bansal and A.H. Heuer, On a Martensitic Phase Transformation in Zirconia (ZrO_2)—I. Metallographic Evidence, *Acta Met*, Vol 20, 1972, p 1281-1289

3.5 M. Cohen, The Strengthening of Steel, *Trans TSM-AIME*, Vol 224, 1962, p 638-656

3.6 B.A. Bilby and J.W. Christian, The Crystallography of Martensite Transformations, *JISI*, Vol 197, 1961, p 122-131

3.7 A.R. Marder and G. Krauss, The Formation of Low-Carbon Martensite in Fe-C Alloys, *Trans ASM*, Vol 62, 1969, p 957-964

3.8 G. Krauss and A.R. Marder, The Morphology of Martensite in Iron Alloys, *Met Trans*, Vol 2, 1971, p 2343-2357

3.9 G. Thomas, Retained Austenite and Tempered Martensite Embrittlement, *Met Trans A*, Vol 9A, 1978, p 439-450

3.10 C.S. Roberts, Effect of Carbon on the Volume Fractions and Lattice Parameters of Retained Austenite and Martensite, *Trans AIME*, Vol 197, 1953, p 203-204

3.11 A.R. Marder and G. Krauss, The Morphology of Martensite in Iron-Carbon Alloys, *Trans ASM*, Vol 60, 1967, p 651-660

3.12 P. Payson and C.H. Savage, Martensite Reactions in Alloy Steels, *Trans. ASM*, Vol 33, 1944, p 261-275

3.13 L.A. Carapella, Computing A^{11} or M_s (Transformation Temperature on Quenching) from Analysis, *Metal Progress*, 1944, Vol 46, p 108

3.14 E.S. Rowland and S.R. Lyle, The Application of M_s Points to Case Depth Measurement, *Trans ASM*, 1946, Vol 37, p 27-47
3.15 R.A. Grange and H.M. Stewart, The Temperature Range of Martensite Formation, *Trans AIME*, 1946, Vol 167, p 467-490
3.16 A.E. Nehrenberg, *Trans AIME*, 1946, Vol 167, p 494-498
3.17 W. Steven and A.G. Haynes, The Temperature of Formation of Martensite and Bainite in Low-alloy Steel, *JISI*, Vol 183, 1956, p 349-359
3.18 K.W. Andrews, Empirical Formulae for the Calculation of Some Transformation Temperatures, *JISI*, Vol 203, 1965, p 721-727
3.19 G. Krauss, Martensitic Transformation, Structure and Properties in Hardenable Steels, in *Hardenability Concepts with Applications to Steel*, AIME, Warrendale, PA, 1978, p 229-248
3.20 C.Y. Kung and J.J. Rayment, An Examination of the Validity of Existing Empirical Formulae for the Calculation of M_s Temperature, *Met Trans A*, Vol 13A, 1982, p 328-331
3.21 W.H. Harris and M. Cohen, Stabilization of the Austenite-Martensite Transformation, *Trans AIME*, Vol 180, 1949, p 447-470
3.22 D.P. Koistinen and R.E. Marburger, A General Equation Prescribing the Extent of the Austenite-Martensite Transformation in Pure Iron-Carbon Alloys and Plain Carbon Steels, *Acta Met*, Vol 7, 1959, p 59-60
3.23 A.R. Entwisle, The Kinetics of Martensite Formation in Steel, *Met Trans*, Vol 2, 1971, p 2395-2407
3.24 J.C. Bokros and E.R. Porter, The Mechanism of the Martensite Burst Transformation in Fe-Ni Single Crystals, *Acta Met*, Vol 11, 1963, p 1291-1301
3.25 K.R. Kinsman and J.S. Shyne, Thermal Stabilization of Austenite in Iron-Nickel-Carbon Alloys, *Acta Met*, Vol 15, 1967, p 1527-1543
3.26 C.H. Shih, B.L. Averbach, and M. Cohen, Some Characteristics of the Isothermal Martensitic Transformation, *Trans AIME*, Vol 203, 1955, p 183-187
3.27 V. Raghavan and M. Cohen, Measurement and Interpretation of Isothermal Martensitic Kinetics, *Met Trans*, Vol 2, 1971, p 2409-2418
3.28 S.R. Pati and M. Cohen, Nucleation of the Isothermal Martensitic Transformation, *Acta Met*, Vol 17, 1969, p 189-199
3.29 E.R. Petty, in *Martensite, Fundamentals and Technology*, Longman, London, 1979, p 6
3.30 M.S. Wechsler, D.S. Lieberman, and T.A. Read, On the Theory of the Formation of Martensite, *Trans AIME*, Vol 197, 1953, p 1503-1515
3.31 J.S. Bowles and J.K. MacKenzie, The Crystallography of Martensite Transformations, *Acta Met*, Vol 2, 1954, p 129-137, 138-147, 224-234
3.32 A.B. Greninger and A.R. Troiano, The Mechanism of Martensite Formation, *Trans AIME*, Vol 185, 1949, p 590-598
3.33 J.W. Christian, in *Martensite Fundamentals and Technology*, E.R. Petty (Ed.), Longman, London, 1970, p 13
3.34 D.P. Dunne and C.M. Wayman, The Crystallography of Ferrous Martensites, *Met Trans*, Vol 2, 1971, p 2327-2341
3.35 G. Krauss and W. Pitsch, The Fine Structure and Habit Planes of Martensite in an Fe-33 wt pct Ni Single Crystal, *Trans TMS-AIME*, Vol 233, 1965, p 919-926

3.36 G. Krauss and W. Pitsch, Deformation Twins in Martensite, *Acta Met*, Vol 12, 1964, p 278-279

3.37 A.B. Greninger and A.R. Troiano, Crystallography of Austenite Decomposition, *Trans AIME*, Vol 140, 1940, p 307-336

3.38 H.M. Ledbetter and R.P. Reed, On the Martensite Crystallography of Fe-Ni Alloys, *Mater Sci Eng*, Vol 5, 1969-70, p 341-349

3.39 A.R. Marder and A.O. Benscoter, Microcracking in Fe-C Acicular Martensite, *Trans ASM*, Vol 61, 1968, p 293-299

3.40 A.R. Marder, A.O. Benscoter, and G. Krauss, Microcracking Sensitivity in Fe-C Plate Martensite, *Met Trans*, Vol 1, 1970, p 1545-1549

3.41 R.P. Brobst and G. Krauss, The Effect of Austenite Grain Size on Microcracking in Martensite of an Fe-1.22 C Alloy, *Met Trans*, Vol 5, 1974, p 457-462

3.42 M.G. Mendiratta, J. Sasser, and G. Krauss, Effect of Dissolved Carbon on Microcracking in Martensite of an Fe-1.39 C Alloy, *Met Trans*, Vol 3, 1972, p 351-353

3.43 C.A. Apple, R.N. Caron, and G. Krauss, Packet Microstructure in an Fe-0.2% C Martensite, *Met Trans*, Vol 5, 1974, p 593-599

3.44 B.P.J. Sandvik and C.M. Wayman, Characteristics of Lath Martensite, *Met Trans A*, Vol 14A, 1983, Part I, p 809-822; Part II, p 823-834; Part III, p 835-843

3.45 G.R. Speich, Tempering of Low-Carbon Martensite, *Trans TMS-AIME*, Vol 245, 1969, p 2552-2564

3.46 T. Swarr and G. Krauss, The Effect of Structure on the Deformation of As-quenched and Tempered Martensite in an Fe-0.2% C Alloy, *Met Trans*, Vol 7A, 1976, p 41-48

3.47 R.H. Aborn, Low Carbon Martensites, *Trans ASM*, Vol 48, 1956, p 51-85

3.48 T. Maki, K. Tsuzaki, and I. Tamura, The Morphology of Microstructure Composed of Lath Martensites in Steels, *Trans Iron Steel Inst Japan*, Vol 20, 1986, p 207-214

3.49 R.F. Hehemann, Ferrous and Nonferrous Bainitic Structures, in *Metals Handbook*, 8th ed., Vol 8, American Society for Metals, Metals Park, OH, 1973, p 194-196

3.50 H. Okamoto and M. Oka, Lower Bainite with Midrib in Hypereutectoid Steels, *Met Trans A*, Vol 17A, 1986, p 1113-1120

3.51 G.R. Srinivasan and C.M. Wayman, The Crystallography of the Bainite Transformation, *Acta Met*, Vol 16, 1969, p 621-636

3.52 R.F. Hehemann, K.R. Kinsman, and H.I. Aaronson, A Debate on the Bainite Reaction, *Met Trans*, Vol 3, 1972, p 1077-1094

3.53 J.W. Christian and D.V. Edmonds, The Bainite Transformation, in *Phase Transformations in Ferrous Alloys*, A.R. Marder and J.I. Goldstein (Eds.), TMS-AIME, Warrendale, PA, 1984, p 293-325

3.54 S.K. Liu, W.T. Reynolds, Jr., H. Hu, G.J. Shiflet, and H.I. Aaronson, Discussion of the Bainite Transformation in a Silicon Steel, and reply by H.K.D.H. Bhadeshia and D.V. Edmonds, *Met Trans A*, Vol 16A, 1985, p 457-468

3.55 S.W. Thompson, D.J. Colvin, and G. Krauss, On the Bainitic Structure Formed in a Modified A710 Steel, *Scripta Metall*, Vol 22, 1988, p 1069-1074

3.56 T. Araki, K. Shibata, and M. Enomoto, Reviewed Concepts on the Microstructural Identification and Terminology of Low Carbon Ferrous Bainites, *ICOMAT '89*, to be published

3.57 B.L. Bramfitt and J.G. Speer, A Perspective on the Morphology of Bainite, *Met Trans A*, to be published

CHAPTER 4

Isothermal and Continuous Cooling Transformation Diagrams

This chapter describes the transformation diagrams that have been developed to define the progress of diffusion-controlled phase transformations of austenite to various mixtures of ferrite and cementite. Both isothermal and continuous cooling transformation diagrams are described, and references to atlases containing collections of these diagrams for a variety of steels are given. The availability of these diagrams makes possible the selection of steels and the design of heat treatments that will either produce desirable microstructures of ferrite and cementite or avoid diffusion-controlled transformations, and thereby produce martensitic microstructures of maximum hardness.

Isothermal Transformation Diagrams

Diagrams that define the transformation of austenite as a function of time at constant temperatures are referred to as isothermal transformation (IT) diagrams or time-temperature-transformation (TTT) diagrams. An IT diagram for 1080 steel has already been presented in Fig. 2.3 in connection with the description of the nucleation and growth kinetics of pearlite formation. The IT diagram for eutectoid steel with negligible alloy content is quite straightforward. Only pearlite forms above the nose of the IT diagram, and only bainite forms below the nose. The curves defining the beginning and end of pearlite or bainite formation are the major features of the diagram.

Steels with carbon content above or below the eutectoid composition and alloy steels have more complex transformation diagrams.

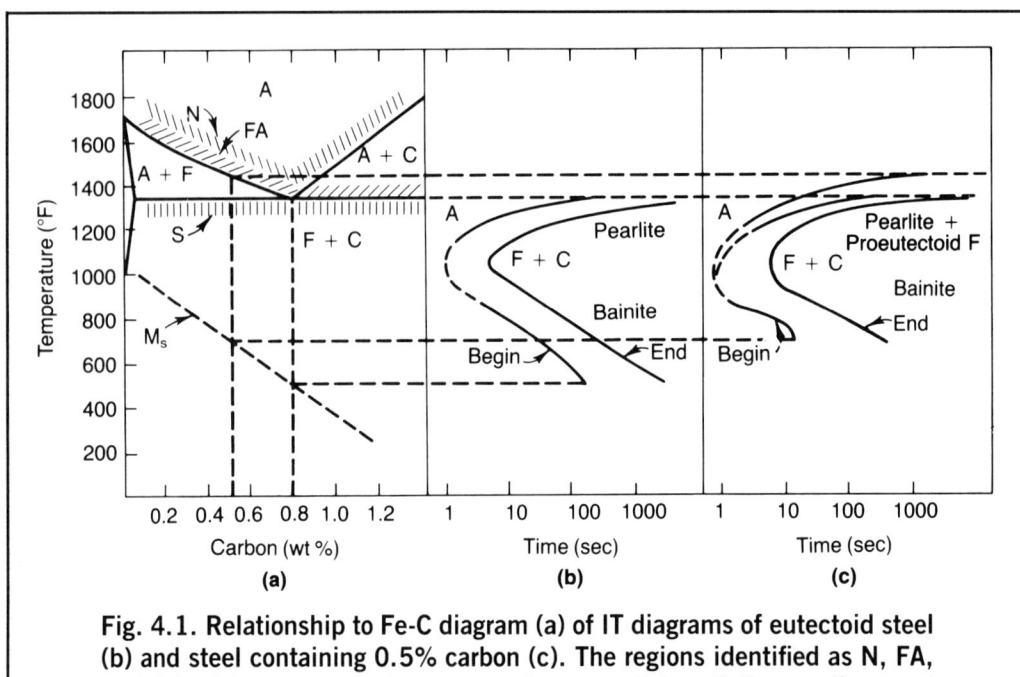

Fig. 4.1. Relationship to Fe-C diagram (a) of IT diagrams of eutectoid steel (b) and steel containing 0.5% carbon (c). The regions identified as N, FA, and S in (a) are temperature ranges for normalizing, full annealing, and spheroidizing heat treatments, respectively, as discussed in Chapter 5. (Ref 4.14)

Figure 4.1 shows schematic IT diagrams for eutectoid steel and a hypoeutectoid plain carbon steel containing nominally 0.5% carbon. Also shown is their relationship to the iron-carbon diagram. The beginning and ending curves for pearlite formation approach the Ae_1 temperature at very long transformation times and move to shorter times with decreasing transformation temperature for reasons discussed in Chapter 2. The IT diagram for the hypoeutectoid steel has an extra curve to mark the beginning of proeutectoid ferrite formation. As indicated in Fig. 4.1, the latter curve approaches the Ae_3 temperature for the 0.5% carbon steel with increasing transformation time. Hypoeutectoid steels with lower carbon contents would have higher Ac_3 temperatures and, therefore, expanded regions of proeutectoid ferrite coexistence with austenite. Similarly, hypereutectoid steels would have IT diagrams with curves for the beginning of proeutectoid cementite formation.

Figure 4.1 shows other differences between the IT diagrams for eutectoid and hypoeutectoid steels. One difference is in M_s temperatures: the lower the carbon content, the higher the M_s temperature.

Another difference is the acceleration of austenite transformation to proeutectoid ferrite with decreasing carbon content, as shown by the position of the nose of the hypoeutectoid steel at shorter times relative to that of the eutectoid steel. The dotted lines in Fig. 4.1(b) and (c) reflect experimental uncertainty in the exact positions of the beginning of transformation curves.

IT diagrams are usually produced by metallographic examination of series of specimens held for various times at various temperatures between Ae_3 or A_{cm} and M_s. More than a hundred specimens are often required to determine a complete IT diagram for a given steel (Ref 4.1). The procedure used is to heat the metallographic specimens in the single-phase austenite field for a sufficient time, usually 1 h, to produce a homogeneous austenite. The austenitizing treatment sets the austenite grain size and the extent of carbide solution. Both of the latter microstructural factors may influence the course of isothermal transformation of austenite, and therefore it is necessary to record the austenitizing temperature used to determine the IT. Once austenitizing is complete, a series of specimens is cooled rapidly, usually by immersion in a molten salt bath, to a given isothermal transformation temperature. The specimens are held for various times and then quenched to room temperature. The specimens held for the shortest times will transform completely to martensite on cooling because there is insufficient time at the hold temperature for any diffusion-controlled transformation. The austenite in specimens held for longer periods of time would transform to ferrite, cementite, pearlite, and/or bainite, depending on temperature and the composition of the steel. The detection of the first small amounts of these phases in specimens largely transformed to martensite establishes the time for the beginning of transformation at a given temperature. With longer holding times at the transformation temperature, more and more of the austenite transforms to ferrite, cementite, or mixtures of ferrite and cementite, and less of the specimen is martensitic after quenching to room temperature. Finally, after holding for a sufficiently long period of time at temperature, transformation of the austenite is complete prior to quenching, and the time for the end of transformation is established. When the process is repeated for a number of temperatures, the complete IT diagram is established.

Although metallographic examination of specimens isothermally held for various times is the most accurate method of determining IT diagrams (particularly with respect to differentiating regions of proeutectoid ferrite, cementite formation, and pearlite or bainite formation), other experimental techniques are also useful. Hardness measurements, for example, reflect the phases present in transformed specimens. A list of the phases in the order of increasing hardness would include ferrite, pearlite, bainite, and martensite. Hardness would, therefore, be a maximum for microstructures produced by quenching after short isothermal holding times to a minimum for specimens held long enough for complete isothermal austenite transformation. Beginning and end of transformation could therefore be established by following hardness changes as a function of isothermal holding time. Dilatometry, an experimental technique that measures changes in length of specimens, has also been used to determine IT diagrams. The application of this technique is possible because of the expansion that accompanies the transformation of austenite to ferrite or ferrite-carbide mixtures, as discussed in Chapter 1. Dilatometry has been used by German investigators (Ref 4.2) for IT diagram determination. By cross checking with metallographic examination, dilatometry has been found to indicate the beginning of transformation after about 3% of the austenite has been transformed, as compared to the ability of microstructural examination to reveal the first 1% of austenite transformation. Figure 4.2 compares IT diagrams determined by dilatometry and metallography and shows the greater sensitivity of the latter technique for IT diagram determination.

Continuous Cooling Transformation Diagrams

Many of the heat treatments performed on steel are carried out by continuous cooling rather than by isothermal holding, and as a result diagrams that represent the transformation of austenite on cooling at various rates have been developed. The latter type of diagram for a given steel is referred to as a continuous cooling (CC) diagram or cooling transformation (CT) diagram. Generally, continuous cooling shifts the beginning of austenite transformation to lower temperatures and longer times. Figure 4.3 shows a derived (i.e., not experimentally determined) CT diagram for eutectoid steel and its relationship to the IT diagram (Ref 4.1 and 4.4). Also shown in the top part of Fig. 4.3 is a Jominy specimen. The latter specimen is water quenched only at one end, and therefore the cooling rate is a maximum at that end and drops with increasing distance into the specimen. The cooling rates at various locations of a Jominy specimen have been measured by attachment of thermocouples, and four of these cooling rates have been superimposed on the lower part of Fig. 4.3. With decreasing cooling rate or increasing distance from the quenched end of the Jominy specimen, the austenite transforms to microstructures containing increasingly greater quantities of pearlite. The decreased hardness

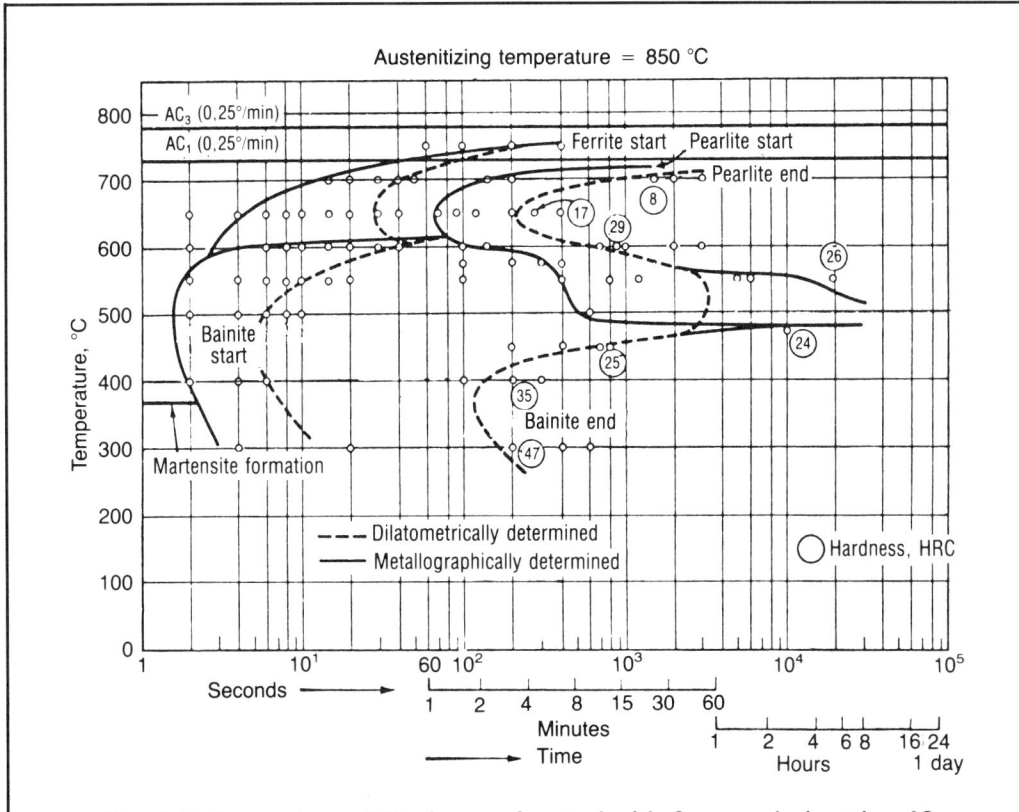

Fig. 4.2. Comparison of IT diagram for steel with German designation 42 CrMo 4 (0.38% C, 0.99% Cr, and 0.16% Mo) determined by dilatometry (dashed lines) and metallography (continuous lines). (Ref 4.2)

associated with the replacement of martensite by pearlite with decreasing cooling rate is also shown in the top part of Fig. 4.3.

In general, especially for hardenable alloy steels, attempts to derive CT diagrams from IT diagrams without experimental verification have proven unsatisfactory (Ref 4.3). For example, the bainite transformation range is dominated by pearlite formation in eutectoid steel and has not been included in the derived curve of Fig. 4.3. The following list of CT characteristics with no IT counterparts has been published (Ref 4.4):

(a) the depression of the M_s temperature at slow cooling rates
(b) the tempering of martensite that takes place on cooling from the M_s temperature to about 204 °C (400 °F)

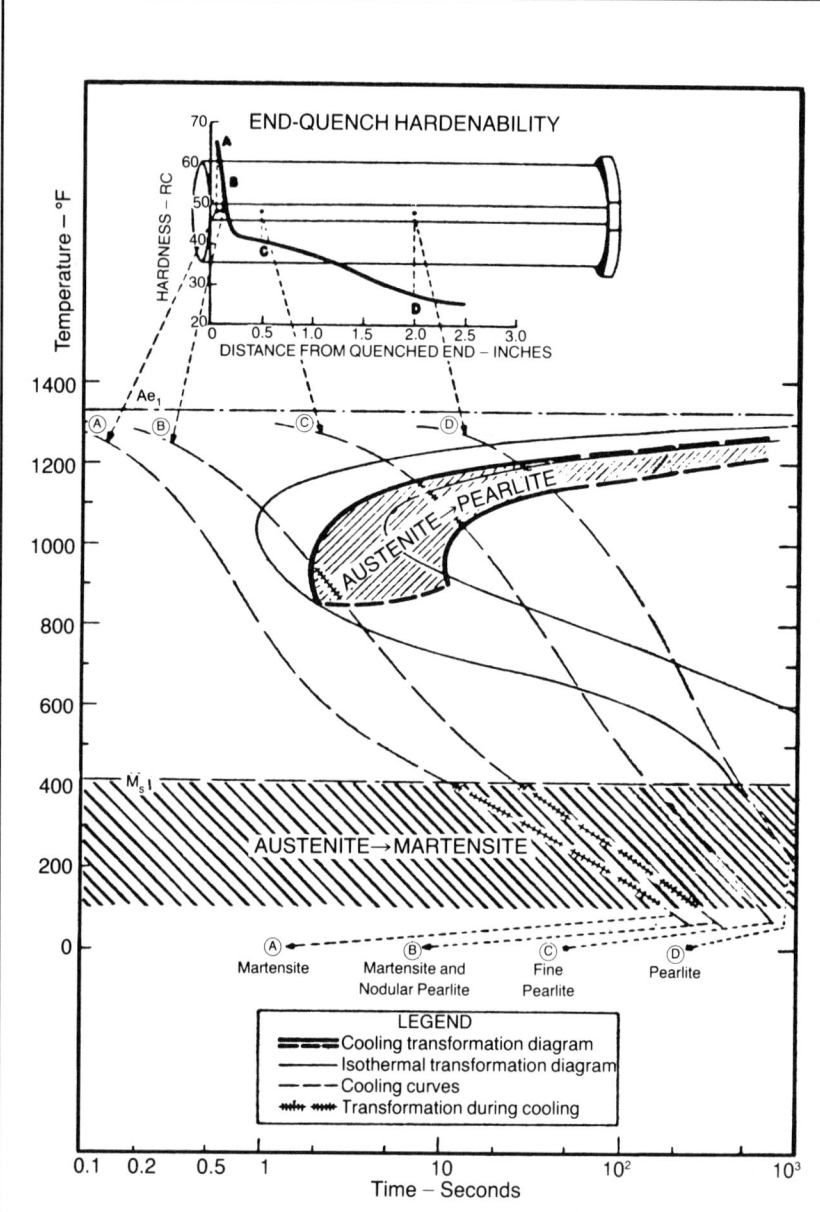

Fig. 4.3. Relationship of CT (heavy lines) and IT (light lines) diagrams of eutectoid steel. Also shown are Jominy end-quench specimen and four cooling rates from different positions on the specimen superimposed on the CT diagram. (Ref 4.4)

(c) the prevalence of bainite as a transformation product
(d) the extraordinary variety of microstructures encountered
(e) the unexpected occurrence of ferrite in a high-carbon steel such as AISI 52100

The following comments expand on the observations in the above list. The depression of the M_s temperature with decreasing cooling rate in a given steel is due to the rejection of carbon into austenite as ferritic or bainitic structures form on cooling. The untransformed austenite therefore has higher carbon concentration and a lower M_s temperature, as discussed in Chapter 3. The tempering of martensite on cooling is referred to as autotempering and is most common in low-carbon steels with high M_s temperatures. The latter situation results in the presence of martensite over a large temperature range on cooling. During this period of the quench, carbon has sufficient mobility to form the carbides characteristic of tempered martensite. Bainite formation, item (c), is promoted by certain alloying elements, in particular molybdenum, and by the more rapid cooling rates that favor shear transformation over diffusion-controlled transformation. The complexity of microstructures is due to the increasing fineness and intermixing of the austenite transformation products as transformation proceeds at successively lower temperatures on cooling. Finally, proeutectoid ferrite is sometimes observed in high-carbon steels where normally proeutectoid cementite would be expected because not all of the carbides may be dissolved during austenitizing. As a result, some of the carbon is tied up in carbide particles, and the austenite has a lower than expected carbon content approaching that of a hypoeutectoid steel.

In addition to the above differences between IT and CT diagrams, frequently there is a gap noted in CT diagrams. This gap represents a temperature range where apparently no transformation occurs on cooling and may be due to carbon enrichment of austenite on cooling as high-temperature ferrite forms and/or changes in incubation times for pearlite and bainite nucleation on cooling (Ref 4.3).

As a result of the differences between isothermal and continuous cooling transformation, CT diagrams are determined primarily by experiment, although there is still some interest in calculating CT diagrams from IT diagrams (Ref 4.3). The use of quenching dilatometers, in which changes in length and temperature with time of a standard specimen are simultaneously recorded, is now well established as the major approach to experimental determination of CT diagrams. The changes in specimen length due to the expansion associated with austenite transformation can therefore be related to

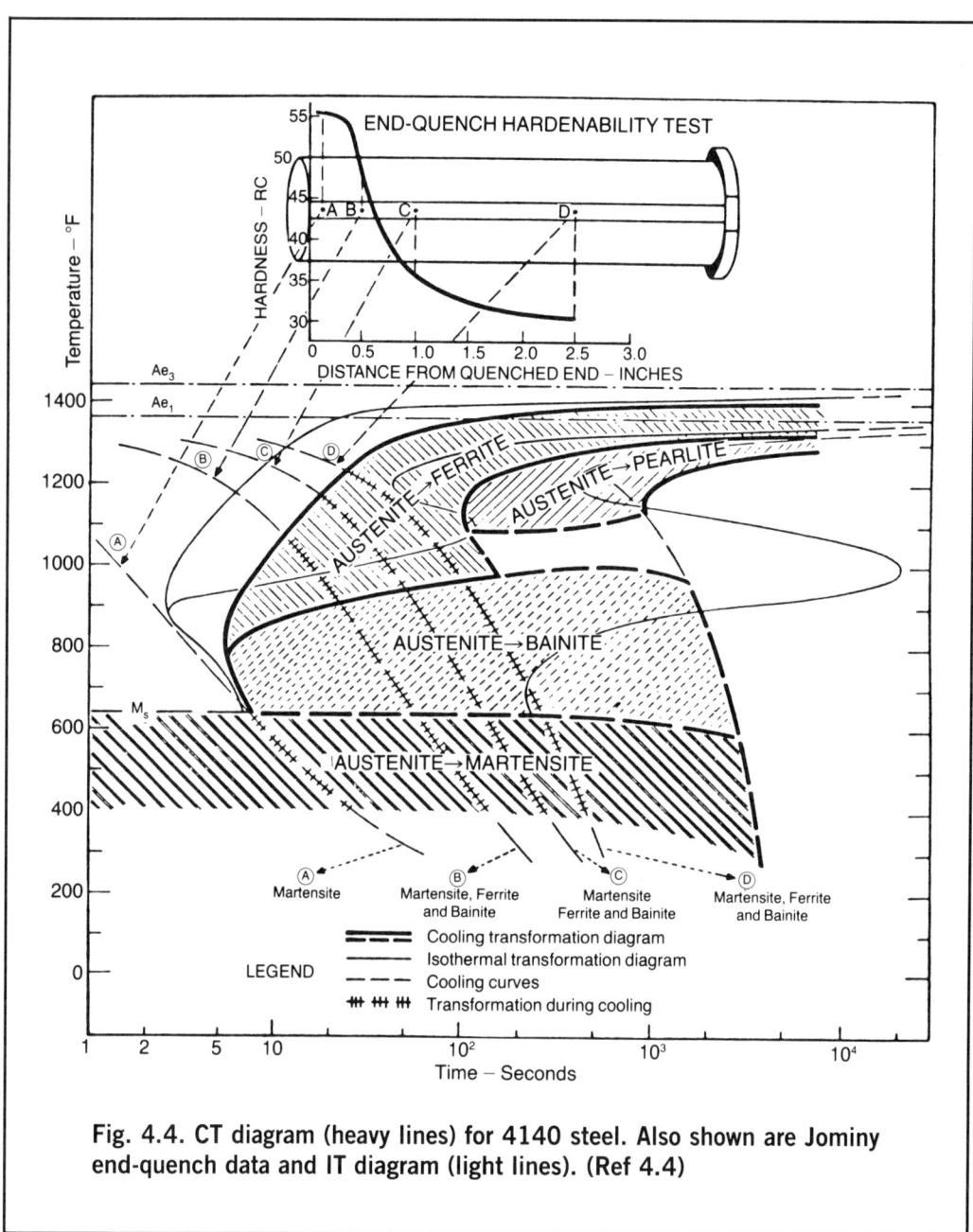

Fig. 4.4. CT diagram (heavy lines) for 4140 steel. Also shown are Jominy end-quench data and IT diagram (light lines). (Ref 4.4)

points on a series of cooling curves. Metallographic examination of the transformed specimens then establishes the microstructure produced by a given cooling sequence. Experimental and instrumental details of the dilatometric approach are given in published atlases of

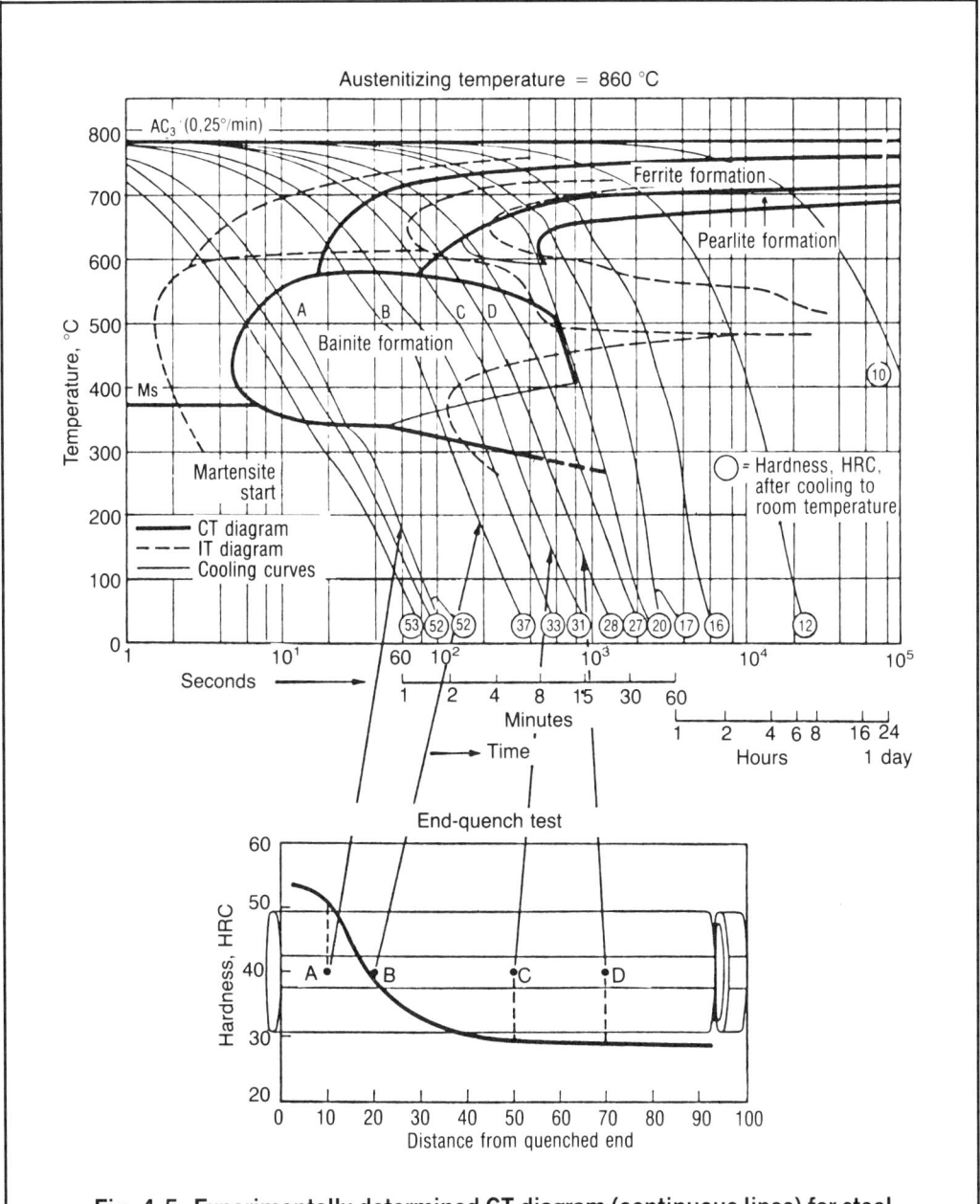

Fig. 4.5. Experimentally determined CT diagram (continuous lines) for steel with German designation 42 CrMo 4 (0.38% C, 0.99% Cr, and 0.16% Mo) for comparison with that derived for a similar steel, 4140, as shown in Fig. 4.4. IT diagram is also shown (dashed lines). (Ref 4.2)

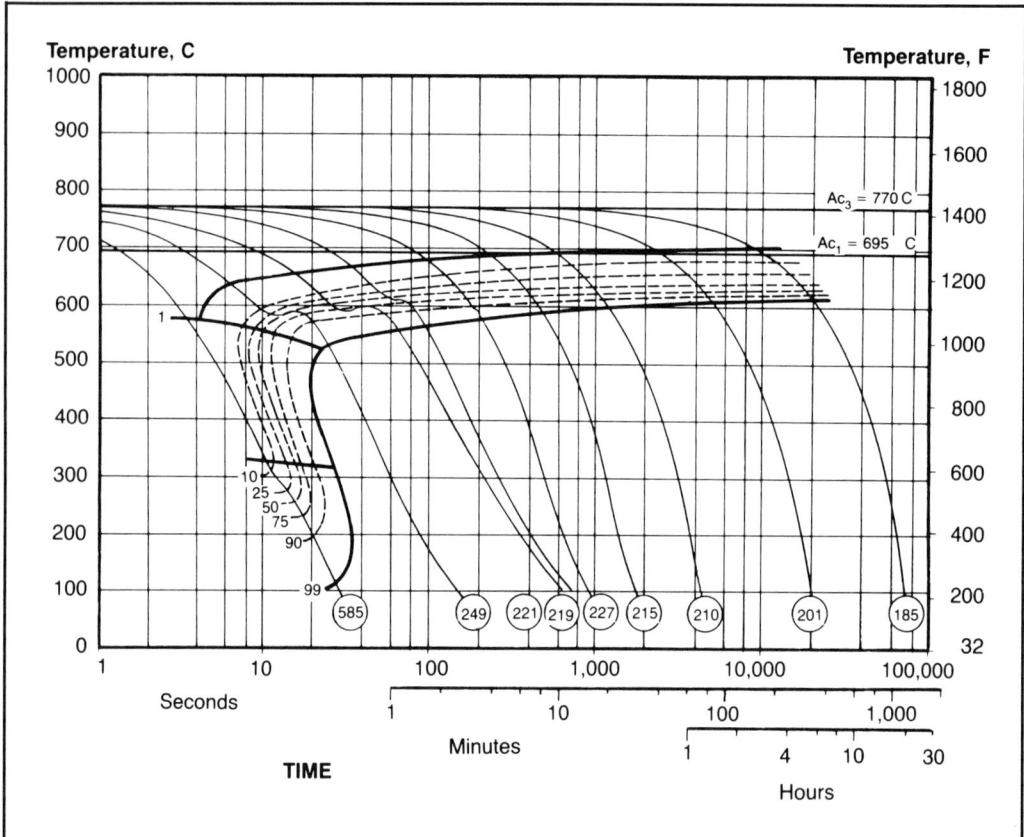

Fig. 4.6. CT diagram for a steel containing 0.37% C, 0.36% Si, 0.85% Mn, 1.44% Ni, and 0.02% Mo. The steel was austenitized at 800 °C (1470 °F) for 20 min. The circled numbers correspond to DPH hardness of microstructures produced by cooling at the rates shown. (Ref 4.5)

CT diagrams (Ref 4.2, 4.5), and Eldis (Ref 4.6) has critically reviewed the relationship of dilatometry to the construction of CT diagrams.

CT diagrams for alloy steels are more complicated than that shown in Fig. 4.3 for eutectoid steel. Figures 4.4 and 4.5 show IT and CT diagrams for SAE 4140 steel and 42 CrMo 4 steel determined by U.S. Steel and Max-Planck Institut für Eisenforschung investigators, respectively. The steels are quite comparable in composition and contain nominally 0.4% carbon, 1% chromium, and 0.2% molybdenum as the major alloy additions. The CT diagram in Fig. 4.4 was derived from the IT diagram and that in Fig. 4.5 was experimentally determined. In the case of the 4140-type steel there is relatively good

ISOTHERMAL AND CONTINUOUS COOLING TRANSFORMATION DIAGRAMS / 99

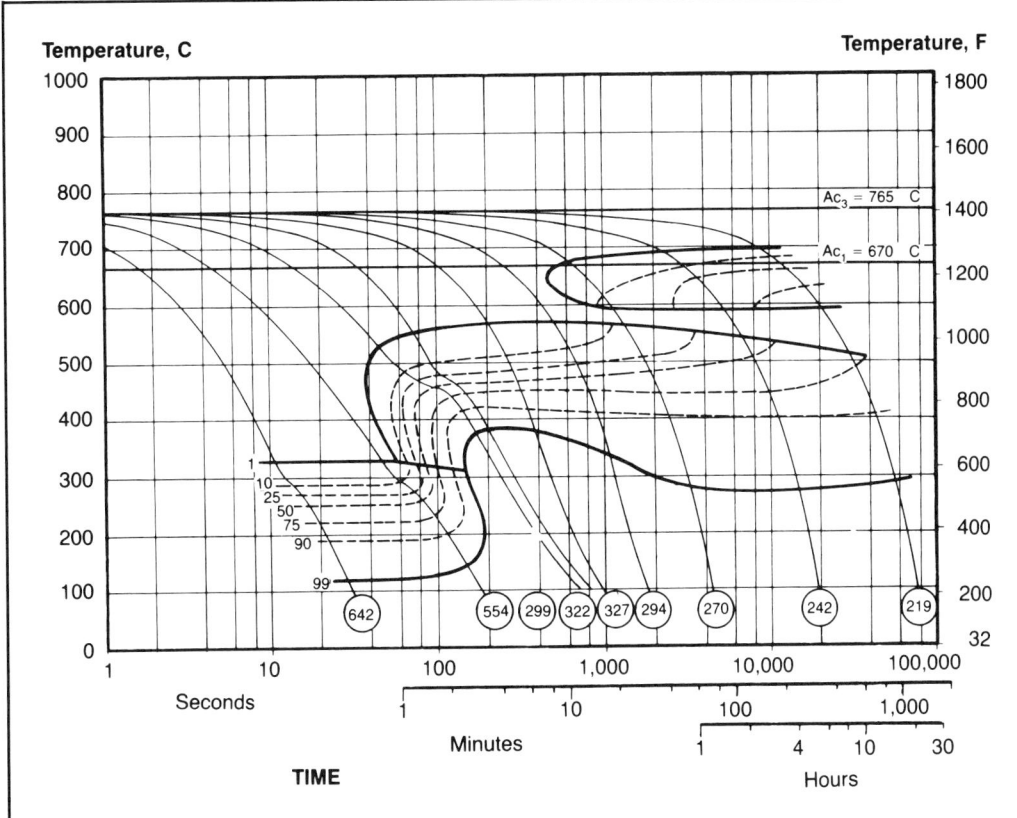

Fig. 4.7. CT diagram for a steel containing 0.37% C, 0.36% Si, 0.84% Mn, 1.40% Ni, and 0.47% Mo. The steel was austenitized at 795 °C (1465 °F) for 70 min. The circled numbers correspond to DPH hardness of microstructures produced by cooling at the rates shown. (Ref 4.5)

agreement between the two methods of CT diagram determination, and the diagrams show the dominance of ferrite and bainite formation at intermediate rates of cooling.

Figures 4.6 and 4.7 are CT diagrams that show a number of effects of alloying on cooling transformation. Two steels, containing 1.4% nickel, 0.36% silicon, and 0.85% manganese and differing only in molybdenum content, are compared. The diagrams have been selected from an atlas that systematically characterizes the effects of molybdenum, chromium, nickel, and silicon on CT diagrams of 0.4% carbon steels (Ref 4.5). The microstructures resulting from selected cooling curves from Fig. 4.6 and 4.7 are shown in Fig. 4.8 and 4.9, respectively.

100 / STEELS: HEAT TREATMENT AND PROCESSING PRINCIPLES

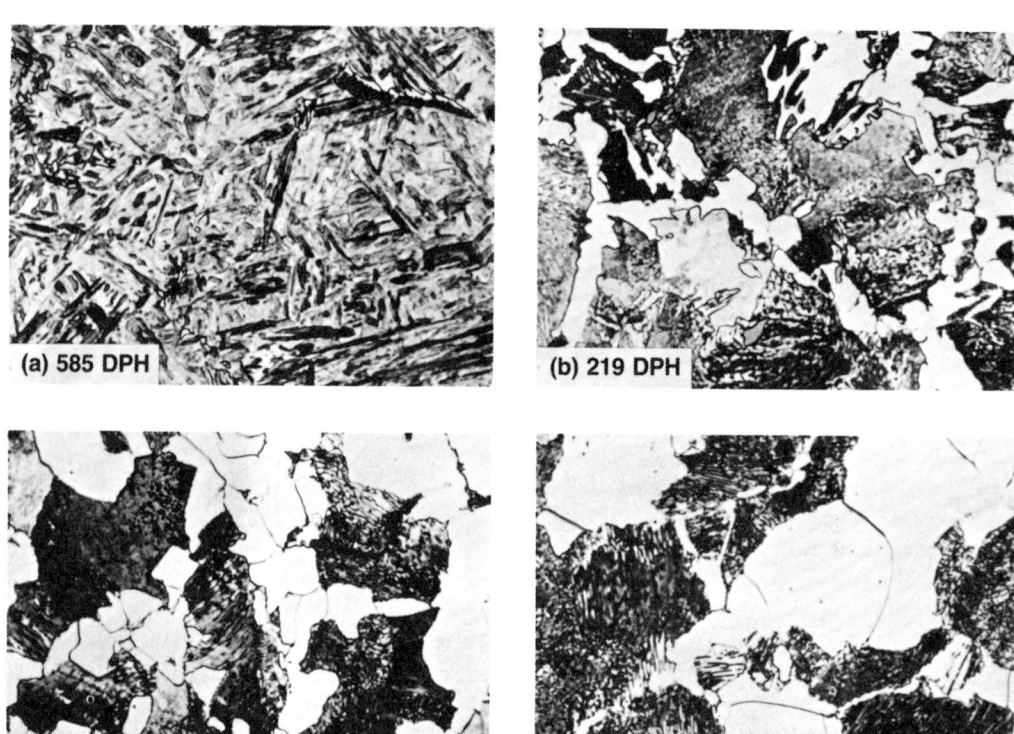

Fig. 4.8. Microstructures produced by cooling steel shown in Fig. 4.6 at four rates as identified by DPH hardness in Fig. 4.6. 2% nital etch. Magnification, 1000×; shown here at 75%. (Ref 4.5)

Each cooling curve and microstructure is identified by the DPH hardness of the microstructure produced by that cooling sequence.

Figures 4.6 and 4.8 show that nickel depresses the Ac_3 and Ac_1 temperatures in accord with its role as an austenite stabilizer in steels and increases hardenability (i.e., the ability to form martensite on cooling) primarily by shifting the proeutectoid and pearlite transformation to longer time periods. Although the austenite-ferrite and austenite-pearlite regions are not differentiated in Fig. 4.6, the microstructures in Fig. 4.8 show that equiaxed proeutectoid ferrite and pearlite are the transformation products for continuous cooling that produces DPH hardnesses of 219, 210, and 185 [see Fig. 4.8(b), (c), and (d), respectively]. Kirkaldy (Ref 4.7) has attributed the improvement in hardenability due to elements such as nickel, copper, and manganese to the lowering of the transformation temperatures and the attendant lower rates of diffusion.

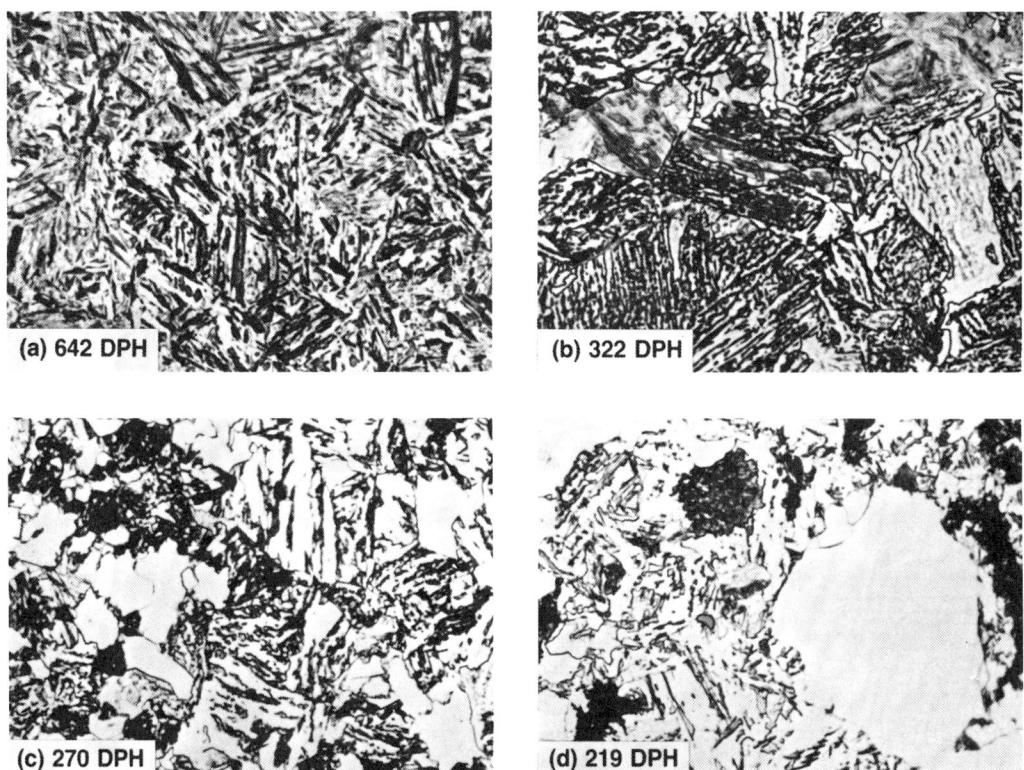

Fig. 4.9. Microstructures produced by cooling steel shown in Fig. 4.7 at four rates as identified by DPH hardness in Fig. 4.7. 2% nital etch. Magnification, 1000×; shown here at 75%. (Ref 4.5)

Figures 4.7 and 4.9 show that the addition of about 0.5% molybdenum to the 1.4% nickel steel produces significant changes in cooling transformation characteristics and microstructure. Hardenability is greatly improved, pearlite and equiaxed proeutectoid ferrite formation is severely retarded, and the bainite transformation becomes quite prominent. The gap that sometimes forms between two mechanisms of transformation is also apparent. The strong effect of molybdenum and similar ferrite stabilizers such as chromium and silicon has been attributed (Ref 4.7) to the fact that molybdenum must diffuse or partition during pearlite formation. Since molybdenum diffuses very sluggishly below Ae_1, the pearlite transformation is significantly retarded. Ferrite formation by a shear mechanism, on the other hand, requires no such partitioning of substitutional elements, and as a result the lower nose for Widmanstätten ferrite and bainite (which is nucleated by ferrite) are prominent features of the CT diagram for the

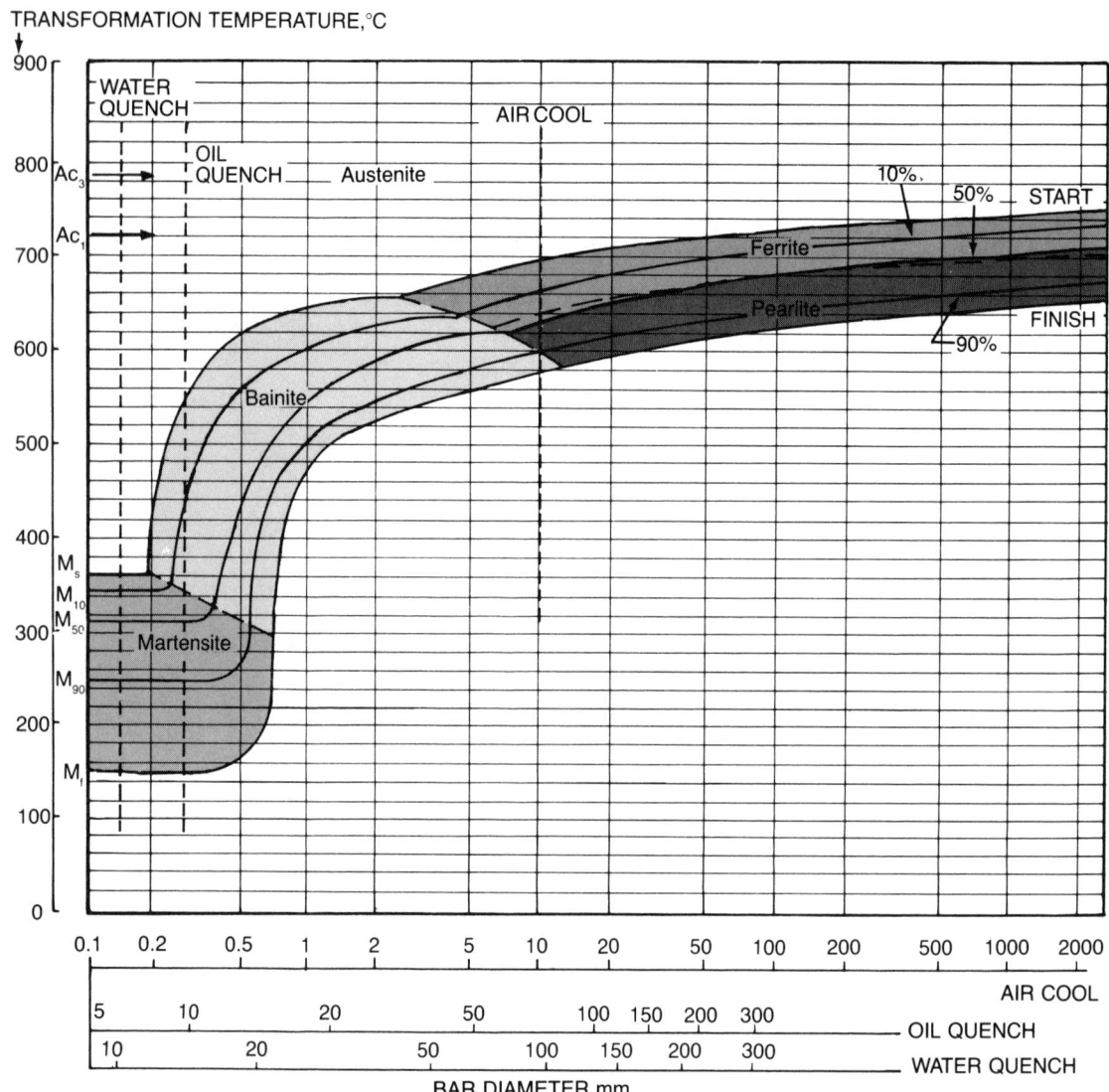

Fig. 4.10. CT diagram for plain carbon steel containing 0.38% C and 0.70% Mn. Transformation and microstructures are plotted as a function of bar diameter. (Ref 4.8)

steel containing molybdenum. Of course, the excellent hardenability shown in Fig. 4.7 is due to the combination of both the nickel and molybdenum alloying effects.

Figures 4.6 through 4.9 show other examples of the characteristics of continuous cooling transformation discussed earlier. The M_s tem-

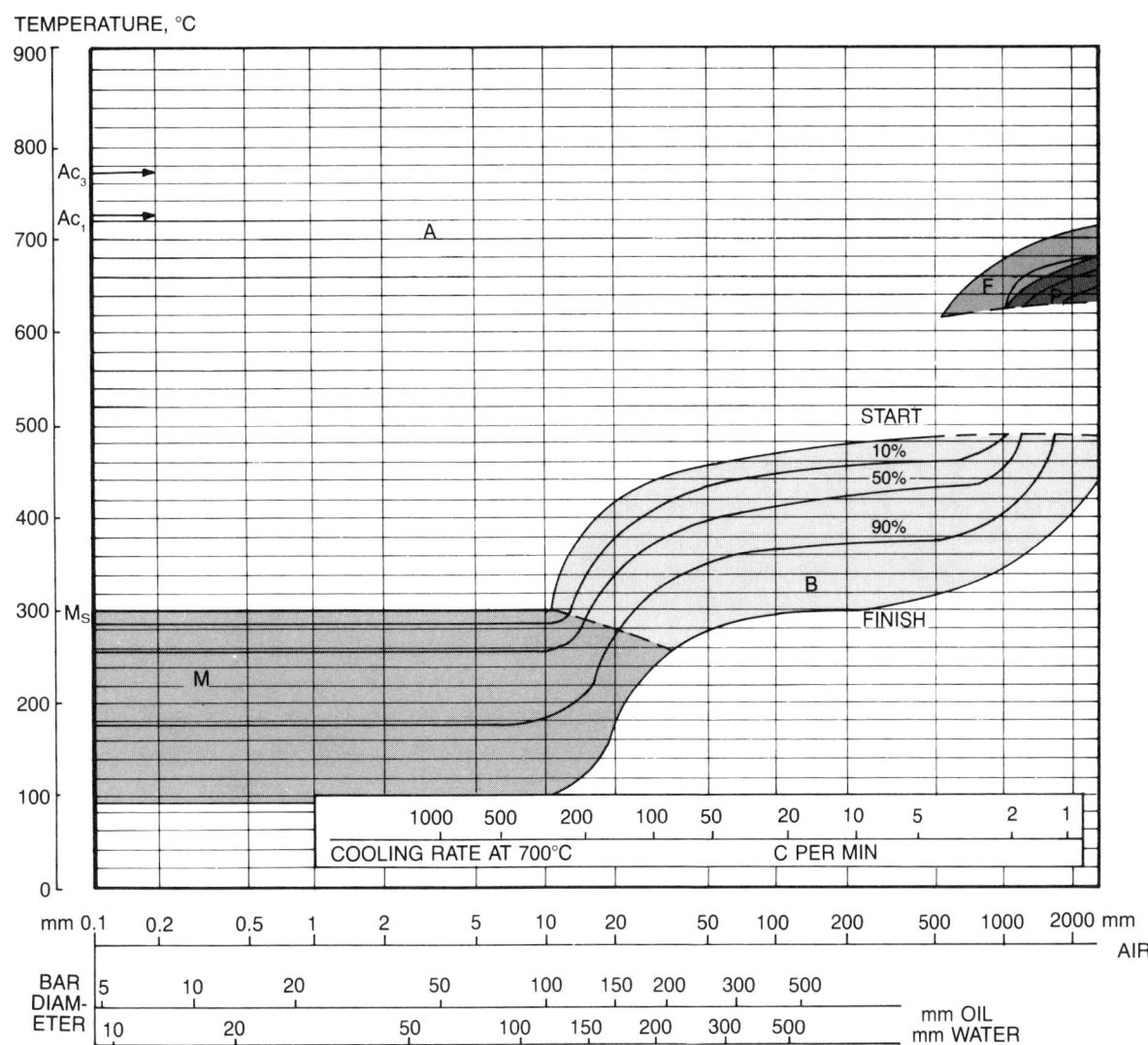

Fig. 4.11. CT diagram for an alloy steel with 0.40% C, 1.50% Ni, 1.20% Cr, and 0.30% Mo, plotted as a function of bar diameter. Steel was austenitized at 850 °C (1562 °F); previous treatment: rolling, then softening at 650 °C (1202 °F). (Ref 4.8)

peratures are slightly lowered once some bainite has formed and the microstructures, especially those produced in the Ni-Mo steel (see Fig. 4.9), are quite complex. Figures 4.6 and 4.7 also show the effects of the evolution of the heat of formation of the austenite transformation products on the cooling curves. This phenomenon is referred to as "recalescence" and causes changes in slopes of the cooling curves and sometimes even temperature increases on transformation. Figures 4.6 and 4.7 show that recalescence is most prominent at high and intermediate cooling rates; at slow cooling rates the specimen has sufficient time to dissipate to its surroundings the heat generated by transformation. Also, measurable recalescent effects lag the dilatometric detection of transformation and therefore are not suitable for accurate CT diagram determination.

Continuous Cooling Transformation and Bar Diameter

An atlas that presents continuous cooling transformation as a function of bar diameter rather than time has recently been prepared by the British Steel Corporation (Ref 4.8). This atlas is especially valuable to the heat treater because it permits an estimate of the microstructure that will form in the center of a bar of a given diameter for a large number of engineering steels in air-cooled, oil-quenched, and water-quenched conditions. Accompanying each of the CT diagrams is a plot of hardness as related to the bar diameters in the as-cooled condition and sometimes the hardness after tempering. The bar diameter characterization is essentially a representation of hardenability, a subject that is developed more fully in Chapter 6 but is included here because of its direct relationship to continuous cooling transformations.

Figure 4.10 shows the CT diagram for a plain carbon steel containing 0.38% carbon, 0.20% silicon, and 0.70% manganese. The abscissa is plotted as bar diameter associated with air cooling, oil quenching, and water quenching. Vertical lines associated with a given diameter show the microstructures to be expected in the center of a bar of that diameter. For example, the vertical dashed line identified as "Air Cool" shows that a microstructure of ferrite, pearlite, and a small amount of bainite is expected in a 10-mm diameter bar that has been air cooled. Likewise, the vertical dashed lines marked "Water Quench" and "Oil Quench" indicate that martensite and bainite plus martensite, respectively, would be expected for 10-mm diameter bars quenched in the two different media. Figure 4.11 shows the CT diagram for a more highly alloyed 0.40% carbon steel. The diagram shows that a 10-mm bar of this steel, even if air cooled, would be entirely martensitic and that oil-quenched bars up to 100 mm in diameter would be fully hardened.

The CT diagrams relating microstructure to bar diameter, therefore, permit a direct assessment not only of the possibility of producing maximum hardness in a bar of given diameter, but also of the ability to produce air-cooled or normalized structures with ferrite-pearlite microstructures of low hardness. It should be remembered that the microstructures are those that would be present at the center of the bars, and that microstructure and hardness gradients may exist between the center and surface of the bars because of cooling rate variations between those points. Some variation in microstructures from those represented in the CT diagrams must also be expected because of chemistry variations within specification limits of a given grade of steel, variations in austenitizing, and/or different degrees of agitation during water or oil quenching.

Summary

A wealth of information characterizing the isothermal (Ref 4.1, 4.4, 4.9, 4.10) and continuous cooling (Ref 4.2, 4.4, 4.5, 4.8, 4.10 to 4.13) transformation behavior of a large number of steels is now available in the form of atlases. There is a strong trend to the use of CT diagrams because they better represent the large number of heat treatments that are based directly on continuous cooling. The latter trend is in turn related to the ready availability of CT diagrams, a situation that has been significantly aided by the development of experimentally convenient and accurate dilatometric techniques for CT diagram determination.

References

4.1   *Atlas of Isothermal Transformation Diagrams*, 2nd ed., United States Steel Corp., Pittsburgh, 1951
4.2   *Atlas zur Wärmebehandlung der Stähle*, Vol 1-4, Max-Planck-Institut für Eisenforschung, in cooperation with the Vereins Deutscher Eisenhüttenleute, Verlag Stahleisen, M.B.H., Düsseldorf, 1954-1976
4.3   A.K. Cavanagh, *Met Trans A*, Vol 10A, 1979, p 129-132
4.4   *Atlas of Isothermal Transformation and Cooling Transformation Diagrams*, American Society for Metals, Metals Park, OH, 1977
4.5   W.W. Cias, *Phase Transformation Kinetics and Hardenability of Medium-Carbon Alloy Steels*, Climax Molybdenum Co., Greenwich, CT, 1972
4.6   G.T. Eldis, A Critical Review of Data Sources for Isothermal and Continuous Transformation Diagrams, in *Hardenability Concepts with Applications to Steel*, AIME, Warrendale, PA, 1978, p 126-153
4.7   J.S. Kirkaldy, Prediction of Alloy Hardenability from Thermodynamic and Kinetic Data, *Met Trans*, Vol 4, 1973, p 2327-2333
4.8   M. Atkins, *Atlas of Continuous Cooling Transformation Diagrams for*

Engineering Steels, British Steel Corp., Sheffield, 1977; revised U.S. edition published by American Society for Metals, Metals Park, OH, 1980

4.9 *Supplement to Atlas of Isothermal Transformation Diagrams*, United States Steel Corp., Pittsburgh, 1953

4.10 A. Schrader and A. Rose, Structure of Steels, Vol 2, *De Ferri Metallographia*, Verlag Stahleisen MBH, Düsseldorf, 1966

4.11 *Transformation Diagrams of Steels Made in France*, Vol 1-4, I.R.S.I.D., St. Germaine-en-Laye, 1953-1960

4.12 M. Economopoulos, L. Harbraleen, and N. Lambert, *Transformation Diagrams of Steels Made in Benelux Countries*, C.N.R.M., Brussels, 1967

4.13 A.A. Popov and L.E. Popova, *Isothermal and Thermokinetic Diagrams of the Breakdown of Supercooled Austenite*, Metalurgiya, Moscow, 1965

4.14 G. Krauss and J.F. Libsch, *Phase Diagrams in Ceramic, Glass and Metal Technology*, A.M. Alper (Ed.), Academic Press, New York, 1970

CHAPTER 5

Heat Treatments To Produce Ferrite and Pearlite

This chapter describes heat treatments such as full annealing, normalizing, and spheroidizing that have been developed to produce uniformity in microstructure, improve ductility, reduce residual stresses, and/or improve machinability of steels. The microstructures that are produced by these heat treatments consist of various distributions of ferrite and cementite, and therefore are produced by relatively long holding times at temperature and slow cooling rates. The latter conditions permit the diffusion-controlled formation of desirable ferrite and cementite microstructures. The heat treatments are described in terms of the Fe-C diagram and the various transformation diagrams discussed in earlier chapters. Relationships for the mechanical properties of mixtures of ferrite and pearlite as influenced by steel composition are presented in the last section of this chapter.

Full Annealing and Homogenizing

The term annealing has been used in its broadest sense to refer to any heat treatment that has as its objective the development of a nonmartensitic microstructure of low hardness and high ductility. This understanding of annealing is much too broad, however, and a number of more specific annealing heat treatments have been developed and defined. Full annealing is a heat treatment accomplished by heating steels into the single-phase austenite field and slowly cooling, usually in a furnace, through the critical transformation ranges. When

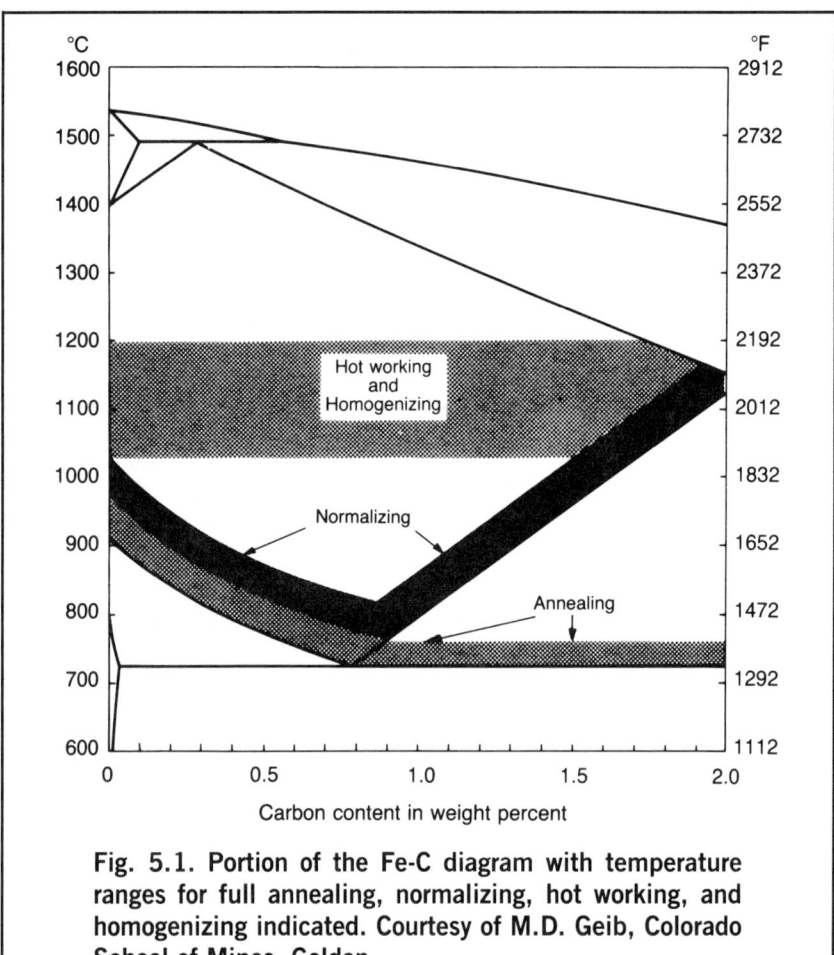

Fig. 5.1. Portion of the Fe-C diagram with temperature ranges for full annealing, normalizing, hot working, and homogenizing indicated. Courtesy of M.D. Geib, Colorado School of Mines, Golden

the term annealing is used without an adjective in reference to carbon steels, full annealing is the implied heat treatment practice (Ref 5.1).

Figure 5.1 shows the temperature ranges for several heat treatments involving austenitizing superimposed on the Fe-C diagram. As shown, the temperature for full annealing is a function of the carbon content of the steel, staying just above the A_3 temperature for hypoeutectoid steels and above the A_1 for hypereutectoid steels. The critical temperatures will vary somewhat with the alloy content of the steel, but the objective of heating into the single-phase austenite field for low- and medium-carbon steels and into the austenite-cementite field for high-carbon steels remains the same no matter what the steel composition.

The reason for heating the hypereutectoid steels in the two-phase field is to agglomerate or spheroidize the proeutectoid cementite. If such steels are heated above A_{cm}, proeutectoid cementite would form on slow cooling at the austenite grain boundaries, as discussed in Chapter 2. The resulting network of carbides on the austenitic grain boundaries provides an easy fracture path and renders the steel brittle to forming or service stresses. Figure 5.2(a) shows a carbide network developed in SAE 52100 steel, a high-carbon bearing steel containing nominally 1% carbon and 1.5% chromium. Figure 5.2(b) shows how fracture produced by impact loading has followed the carbide network along prior austenite grain boundaries in a microstructure similar to that shown in Fig. 5.2(a). In Fig. 5.2(b), the steel has been hardened by quenching from the austenite-cementite field and martensite coexists with the carbide network. A carbide network formed on slow cooling from above A_{cm} in a 52100 steel is shown in Fig. 5.3(a). Pearlite instead of martensite has formed within the austenite grains. The object of full annealing high-carbon steels in the austenite-carbide field, then, is to break up such continuous carbide networks by agglomeration into separated, spherical carbide particles. The driving force for this process is the reduction in austenite/cementite interface area and thus the reduction in interfacial energy that accompanies spheroidization. Figure 5.3(b) shows the partial spheroidization of a cementite network. Although the structure was formed during austenitizing for hardening, the austenitizing temperature ranges for hardening and full annealing are identical in high-carbon steels.

Not only is the temperature range of heating an important part of full annealing, but the slow cooling rate associated with full annealing is also a vital part of the process. Figure 5.4 compares schematic temperature-time schedules for full annealing and the normalizing heat treatments discussed in the next section of this chapter. The cooling rates are superimposed on a schematic CT diagram for a hypoeutectoid steel. The slow cooling rates characteristic of furnace cooling ensure that the austenite transforms first to proeutectoid ferrite and then to pearlite at temperatures approaching the equilibrium A_3 and A_1 temperatures. As a result the ferrite will be equiaxed and relatively coarse-grained, and the pearlite will have a coarse interlamellar spacing. The latter microstructural characteristics lower hardness and strength and increase ductility, the major objectives of the full annealing treatment. Once the austenite has fully transformed to ferrite and pearlite, the cooling rate could be increased to reduce the time of annealing and thereby improve productivity. A number of additional rules for developing optimum full annealing practices and properties are given in Ref 5.1.

110 / STEELS: HEAT TREATMENT AND PROCESSING PRINCIPLES

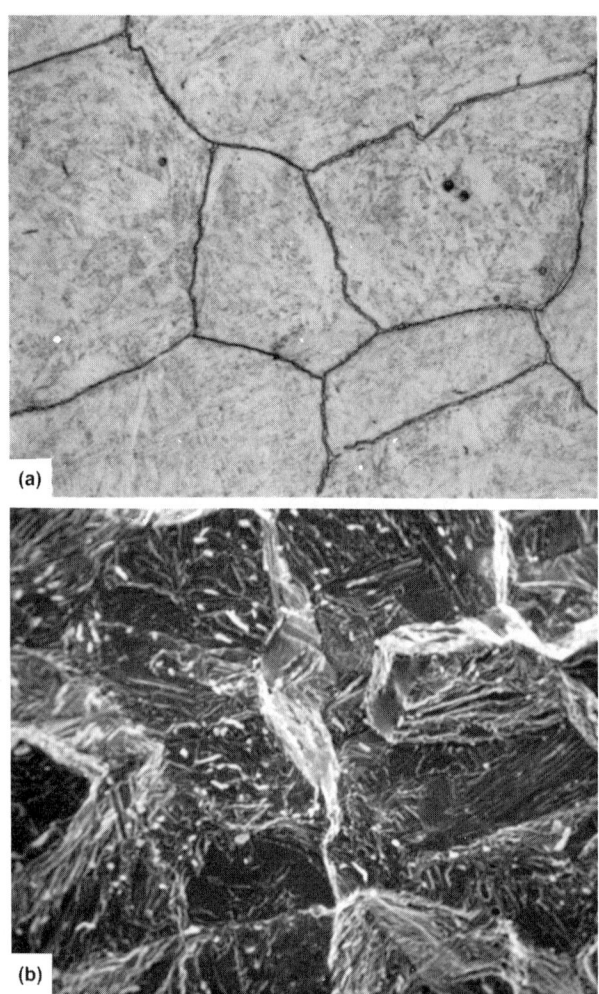

Fig. 5.2. (a) Carbide network at prior austenite grain boundaries in 52100 steel. Light micrograph. Nital etch. Magnification, 600×; shown here at 75%. (b) Fracture along grain boundary carbides in 52100 steel. Scanning electron micrograph. Magnification, 415×; shown here at 75%. Courtesy of T. Ando, Colorado School of Mines, Golden

Figure 5.1 also shows the temperature range for homogenizing, a type of annealing treatment usually performed in earlier stages of steel processing prior to hot rolling or forging, working operations that

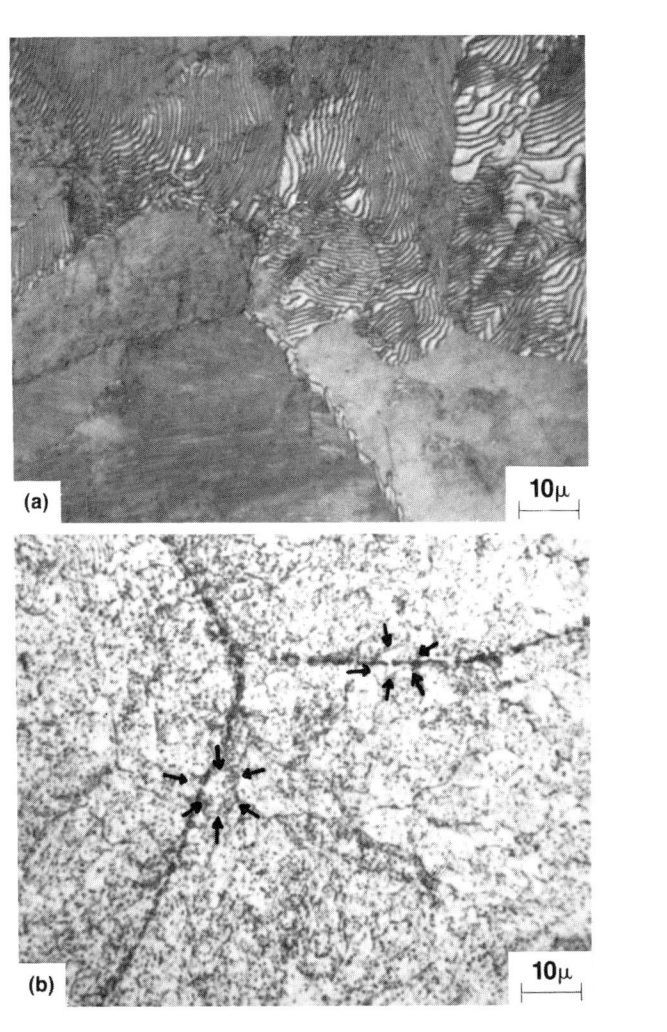

Fig. 5.3. (a) Proeutectoid cementite network in normalized 52100 steel. (b) Residual cementite network after austenitizing structure in (a) at 850 °C (1562 °F) for hardening. Very fine particles are from spheroidization of cementite in original pearlite matrix. Arrows point to fine austenite grains that have formed on austenitizing. (Ref 5.2)

are also performed in the same temperature range. Homogenizing is performed at high temperatures in the austenite phase field to speed the diffusion-controlled reduction of segregation or chemical concen-

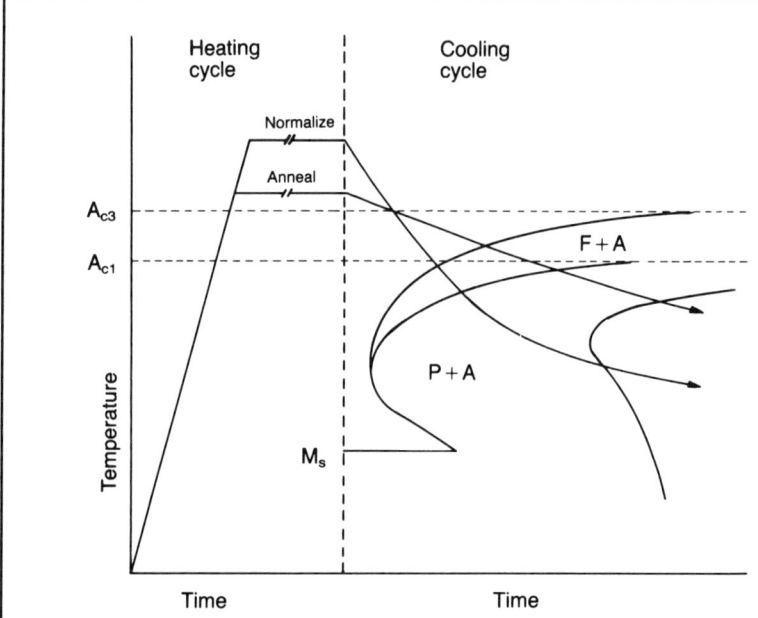

Fig. 5.4. Schematic time-temperature cycles for normalizing and full annealing. The slower cooling of annealing results in higher temperature transformation to ferrite and pearlite and coarser microstructures than does normalizing. Courtesy of M.D. Geib, Colorado School of Mines, Golden

tration gradients that are produced by ingot solidification. In addition, second phases such as carbides are dissolved as fully as possible. The resulting uniformity or homogeneity of the austenite not only improves hot workability but also contributes to uniformity in the response of a steel to subsequent annealing or hardening operations.

Normalizing

Normalizing is a heat treatment which, similar to full annealing, produces a uniform microstructure of ferrite and pearlite. There are, however, several important differences between normalizing and annealing. Normalizing in hypoeutectoid steels is performed at temperatures somewhat higher than those used for annealing, while in hypereutectoid steels the heating temperature range is above the A_{cm} (see Fig. 5.1). In normalizing, heating is followed by air cooling, in contrast to the slower furnace cooling of full annealing.

The somewhat higher austenitizing temperatures used for normalizing, as compared to those used for annealing hypoeutectoid steels, in effect produce greater uniformity in austenitic structure and composition similar to a homogenizing treatment, although at a much lower temperature and for shorter times than those used for homogenizing. Another of the major objectives of normalizing is to refine the grain size that frequently becomes very coarse during hot working at high temperatures or that is present in as-solidified steel castings. As such hot worked or cast products are heated through the Ac_1 and Ac_3 temperatures, new austenite grains are nucleated and, if the austenitizing temperature is limited to the range shown in Fig. 5.1, a uniform fine-grained austenitic structure is produced. Exceeding the indicated temperature range might result in excessive austenitic grain size, as discussed in Chapter 7. Normalizing, then, produces a uniform, fine-grained austenite grain structure that in hypoeutectoid steels transforms to ferrite-pearlite microstructures on air cooling. The resulting microstructure may have good uniformity and desirable mechanical properties for a given application or may be reaustenitized for final hardening by quenching to martensite.

In hypereutectoid steels, normalizing is performed above the A_{cm} not only to refine austenitic grain size but also to dissolve carbides and carbide networks that may have developed during prior processing. The normalized structures that result respond more readily to the spheroidizing treatments for good machinability as described below and/or provide better response to a subsequent and final hardening heat treatment. There is the possibility that continuous carbide networks may develop on cooling from a normalizing temperature above A_{cm}, and that as a result a somewhat brittle normalized microstructure might develop. On subsequent austenitizing for hardening, however, the network carbides agglomerate or spheroidize somewhat (see Fig. 5.3b), and fracture toughness is in fact improved relative to a microstructure without the partially spheroidized network (Ref 5.2).

The air-cooling step of a normalizing treatment produces subtle but significant differences in microstructures compared to those produced by full annealing. Figure 5.4 shows schematically that air cooling lowers the temperature range over which proeutectoid ferrite and pearlite form compared to the transformation range in full annealing. As a result, both the ferrite grain size and the pearlite interlamellar spacing are reduced compared to those in the same steel in the fully annealed condition. The finer microstructure of a normalized steel in turn has higher strength and hardness and slightly lower ductility than a fully annealed steel.

The actual mechanical properties of any normalized or annealed steel are determined by a number of factors, the most important being carbon content. The higher the carbon content, the more pearlite that forms and the higher the strength and hardness of the steel. Quantitative relationships for the contributions of carbon and other parameters to the mechanical properties of ferrite-pearlite steels are discussed in a later section of this chapter.

It is also important to realize that the air cooling associated with a normalizing heat treatment produces a range of cooling rates depending on the section size. Heavier sections air cool at much lower rates than do light sections because of the added time required for thermal conductivity to lower the temperature of central portions of the work piece. Two important consequences follow from the effect of section size on cooling rate. In very heavy sections, the surface may cool at significantly higher rates than the interior, thus producing residual stresses. In very light sections, especially in alloy hardenable steels, air cooling may actually be rapid enough to produce bainitic or martensitic microstructures instead of ferrite and pearlite. The British Steel Corporation atlas (Ref 5.3) which plots cooling transformation as a function of air-cooling section size (see Chapter 4) enables this effect to be evaluated. Other aspects of normalizing carbon steels are discussed in Ref 5.1.

Spheroidizing

The most ductile, softest condition of any steel is associated with a microstructure that consists of spherical carbide particles uniformly dispersed in a ferrite matrix. Figure 5.5 shows a spheroidized microstructure in a 0.66C-1Mn steel. The high ductility of such a microstructure is directly related to the continuous ductile ferrite matrix; pearlite (see Fig. 2.2) with its fine lamellar carbides separating the ferrite, more effectively hinders deformation and, therefore, increases hardness and lowers ductility compared to a spheroidized structure. The good ductility of spheroidized microstructures is extremely important for low- and medium-carbon steels that are cold formed, and the low hardness of spheroidized structures is important for high-carbon steels that undergo extensive machining prior to final hardening.

Spheroidized microstructures are the most stable microstructures found in steels and will form in any prior structure heated at temper-

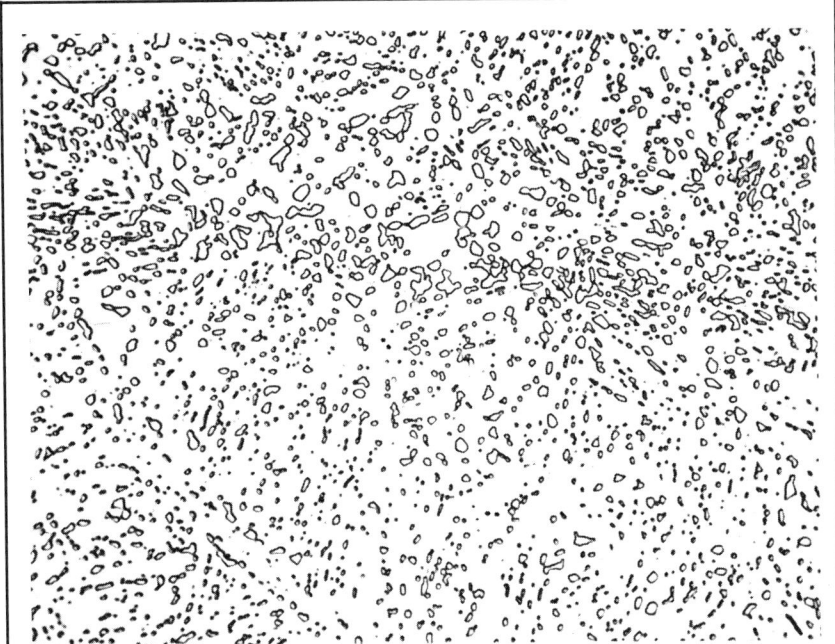

Fig. 5.5. Spheroidized microstructure in an Fe-0.66C-1Mn alloy formed by heating martensite at 704 °C (1300 °F) for 24 h. Picral etch. Magnification, 1000×. Courtesy of A.R. Marder and A. Benscoter, Bethlehem Steel Corp., Bethlehem, PA

atures high enough and times long enough to permit the diffusion-dependent development of the spherical carbide particles. As a result, there are many different heat treatment approaches for producing spheroidized microstructures. The slowest spheroidizing is associated with pearlitic microstructures, especially those with coarse interlamellar spacings. Figure 5.6 shows the percent of carbides that have spheroidized in fine to coarse pearlites produced by isothermally transforming a 0.74-0.71Si steel between 700 and 580 °C (1292 and 1076 °F), followed by annealing at 700 °C (1292 °F) (Ref 5.4). Many hundreds of hours are required to spheroidize the pearlitic microstructures. Spheroidizing is more rapid if the carbides are initially in the form of discrete particles, as in bainite, and even more rapid if the starting structure is martensite. Spheroidizing of martensitic microstructures is most frequently performed on highly alloyed tool steels that form martensite on air cooling as shown schematically in Fig. 5.7.

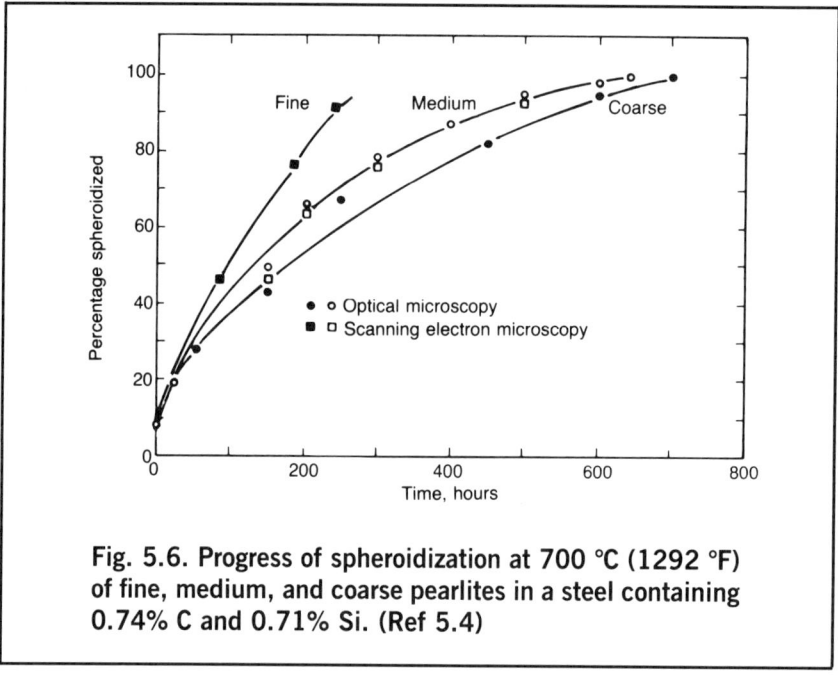

Fig. 5.6. Progress of spheroidization at 700 °C (1292 °F) of fine, medium, and coarse pearlites in a steel containing 0.74% C and 0.71% Si. (Ref 5.4)

Spheroidizing at rates much faster than those shown in Fig. 5.6 is accomplished by either complete or partial austenitizing, and then holding just below Ac_1, cooling very slowly through the Ac_1, or cycling above and below Ac_1 (Ref 5.1, 5.5). These temperature ranges for spheroidizing are shown in Fig. 5.8. It is important to limit the austenitizing temperature in order to retain a degree of heterogeneity in the austenite, especially since undissolved carbide particles appear to promote the transformation of the austenite to spheroidized microstructures. As noted earlier, homogenized austenite free of undissolved carbides as produced by normalizing or full annealing promotes the formation of pearlitic structures rather than spheroidized structures.

Spheroidized microstructures are stable because the ferrite is generally strain-free and because the spherical shape of the cementite particles is one of minimum interfacial area per unit volume of particle. Lamellar cementite particles, as present in pearlite, have a very large interfacial area per unit volume of particle and therefore high interfacial energy. In order to reduce the interfacial energy, cementite lamellae or plates break up into smaller particles that eventually assume spherical shapes. Figure 5.9 shows a representation of the breakup process of a single plate as determined by serial sectioning of a specimen annealed for 150 h at 700 °C (1292 °F) (Ref 5.4). Once the lamellae have broken up, the small spherical particles

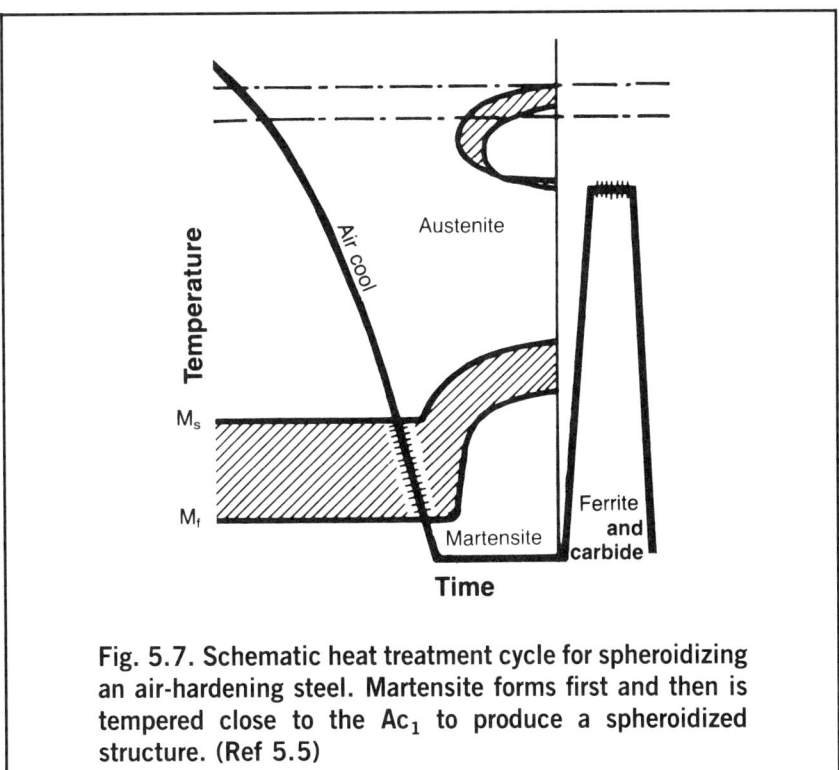

Fig. 5.7. Schematic heat treatment cycle for spheroidizing an air-hardening steel. Martensite forms first and then is tempered close to the Ac_1 to produce a spheroidized structure. (Ref 5.5)

dissolve, and the larger particles grow, again driven by the reduction in interfacial energy. The following equation (Ref 5.4, 5.7) describes the rate of coarsening of a spheroidized microstructure:

$$\frac{dr}{dt} = \frac{2\gamma \, V_{Fe_3C}^2 \, X_c \, D_c^{eff}}{V_{Fe} \, RT r_1} \left(\frac{1}{\bar{r}} - \frac{1}{r_1} \right) \quad \text{(Eq 5.1)}$$

where γ is the interfacial energy; V_{Fe_3C} and V_{Fe} are the molar volumes of cementite and ferrite; X_c is the mole fraction of carbon in equilibrium with cementite in ferrite; D_c^{eff} is the effective carbon diffusion coefficient; R is the gas constant; T is the absolute temperature; r_1 is the radius of newly created particles; and $\bar{r}$ is the mean size of the already spheroidized particles. Equation 5.1 shows that the rate of spheroidization is directly related to the diffusion of carbon in ferrite and decreases as the average size of particles in a spheroidized microstructure increases. Alloying elements slow the rate of carbon diffusion in ferrite, and therefore the spheroidization process. Also if

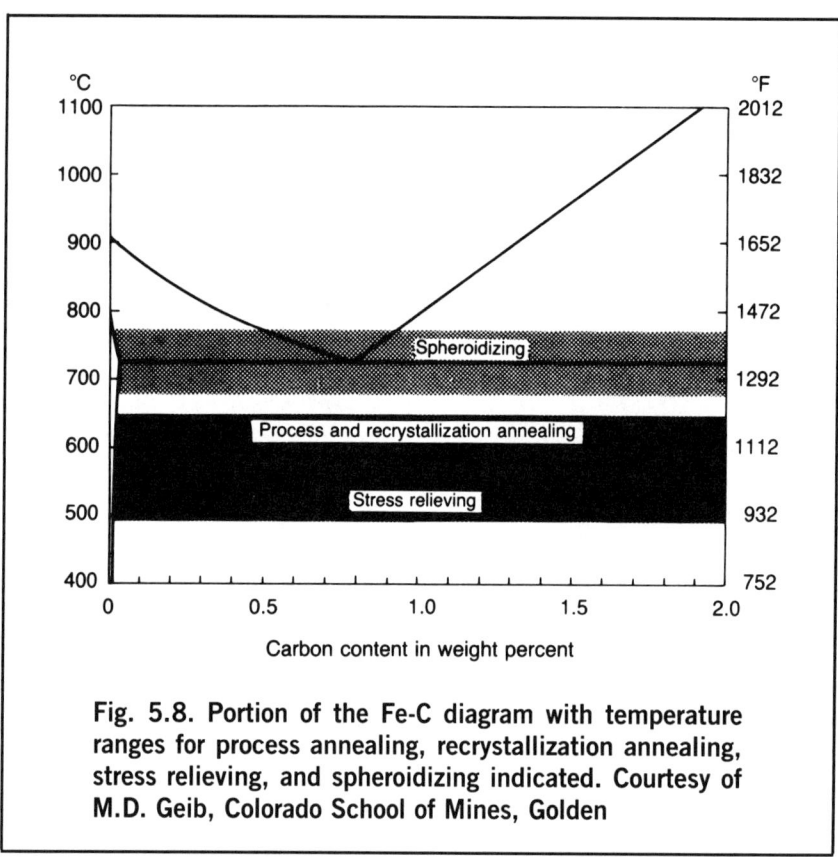

Fig. 5.8. Portion of the Fe-C diagram with temperature ranges for process annealing, recrystallization annealing, stress relieving, and spheroidizing indicated. Courtesy of M.D. Geib, Colorado School of Mines, Golden

present, strong carbide-forming elements would have to diffuse for alloy carbide coarsening, and thus would greatly reduce the rate of spheroidization.

Process and Recrystallization Annealing

Process and recrystallization annealing are similar subcritical annealing treatments usually applied to restore ductility to cold worked steel products of a variety of shapes. Since these heat treatments are performed in the ferrite and cementite two-phase field of the Fe-C diagram (see Fig. 5.8), no phase transformation accompanies the microstructural changes produced by the treatments. Generally, the microstructure of the low- and medium-carbon steels prior to cold work is spheroidized or largely ferritic with small amounts of pearlite, both highly ductile microstructures. The ferrite in these microstructures is

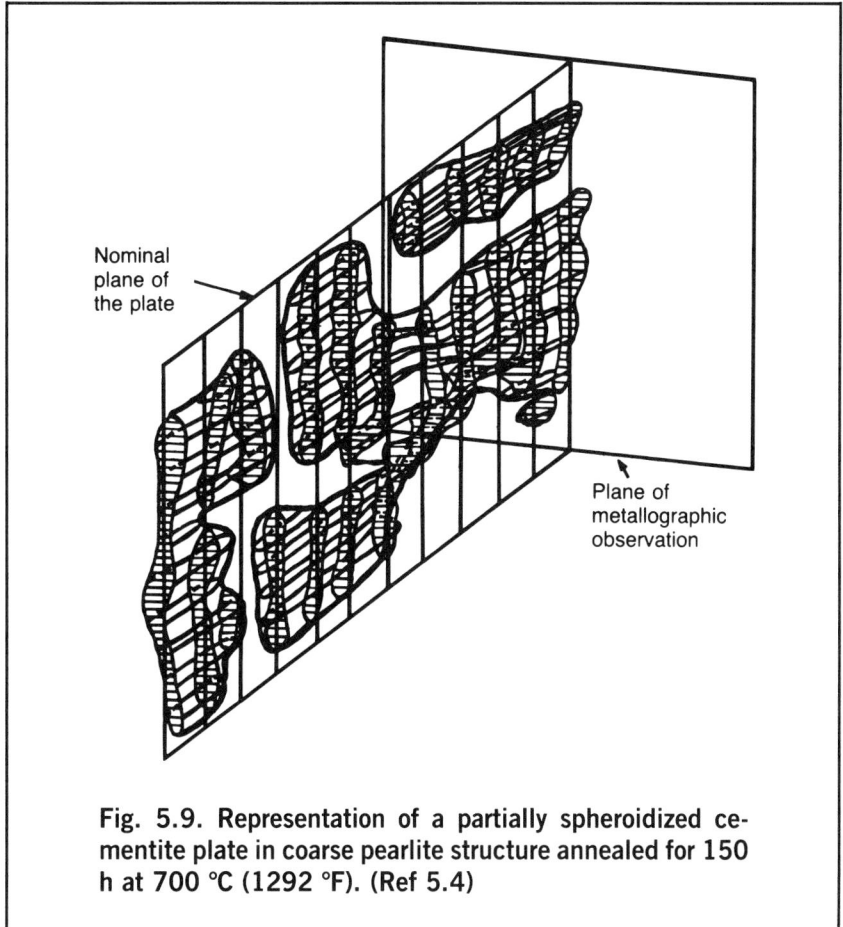

Fig. 5.9. Representation of a partially spheroidized cementite plate in coarse pearlite structure annealed for 150 h at 700 °C (1292 °F). (Ref 5.4)

equiaxed and strain-free. Cold working deforms or work hardens the ferrite, tending to elongate the ferrite grains in the direction of working and introducing a high density of crystal imperfections within the grains. On heating, high strain energy of the deformed ferrite at first drives recovery, a mechanism by which some of the crystal imperfections are eliminated or rearranged into new configurations (Ref 5.8), and eventually drives recrystallization, a process where new, strain-free equiaxed grains nucleate and grow in the deformed ferrite (Ref 5.8). The end result of the subcritical annealing process is a restoration of the ductile, spheroidized microstructure which is again capable of undergoing significant cold deformation.

Figure 5.10 shows the effects of cold work and recrystallization annealing on a very low-carbon (0.003%) sheet steel (Ref 5.9). Because of the low carbon content no carbides are visible and the only

120 / STEELS: HEAT TREATMENT AND PROCESSING PRINCIPLES

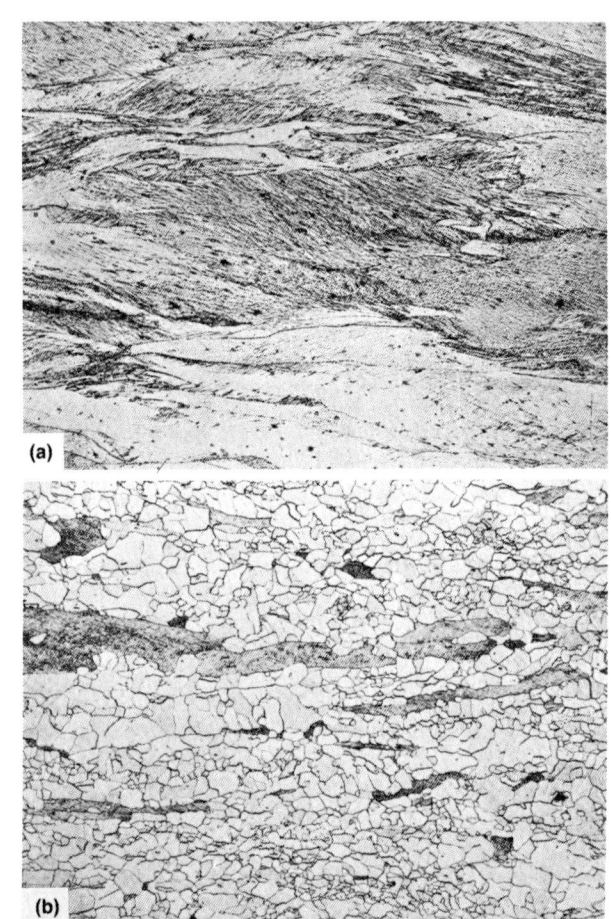

Fig. 5.10. (a) Microstructure of an Fe-0.003C alloy cold rolled 60%. (b) Microstructure of an Fe-0.003C alloy after annealing at 538 °C (1000 °F) for 2 h. About 80% of structure has recrystallized to fine equiaxed grains. Light micrographs. Nital etch. Magnification, 100×; shown here at 75%. Courtesy of D.A. Witmer, Bethlehem Steel Corp., Bethlehem, PA

second-phase particles present are oxide and sulfide inclusions. Figure 5.10(a) shows the deformed microstructure produced by a 60% reduction in sheet thickness by cold rolling. The elongated grain structure and etching effects associated with the deformation within the grains are clearly visible. Figure 5.10(b) shows the same steel after about 80%

recrystallization has been produced by annealing at 538 °C (1000 °F) for 2 h. The recrystallized ferrite grains are quite fine and equiaxed, and etching shows no evidence of deformation.

A third stage of the annealing process, grain growth or coarsening, has not yet developed in the example shown in Fig. 5.10(b). Prolonged heating beyond the point of recrystallization would cause such grain growth to occur, a process driven by the reduction in grain boundary or interfacial energy made possible by an increase in grain size. Reference 5.8 describes the mechanisms of recovery and recrystallization stages of annealing in detail.

Cold Rolled and Annealed Sheet Steels

The previous section has introduced the topic of annealing of cold worked steels. This section describes in more detail the processing of cold rolled and annealed sheet steels used in large tonnages for the automotive and appliance industries. These steels must have excellent formability, high-quality surfaces, and good spot weldability, requirements which are best satisfied by low-carbon steels with largely ferritic microstructures.

The steels for sheet products are cast as ingots and rolled to slabs or continuously cast as slabs. The slabs are then reduced in hot strip mills through sets of roughing and finishing rolls and coiled. Hot strip mill processing is important not only to produce the proper strip dimensions, but also to establish structures which will ultimately provide good properties in cold rolled and annealed sheet. For example, aluminum-killed steels which develop crystallographic textures good for deep drawing must be finish rolled at relatively high temperatures, about 890 °C (1635 °F), and rapidly cooled by water sprays to low coiling temperatures, on the order of 580 °C (1075 °F) (Ref 5.10). This practice keeps aluminum and nitrogen in solid solution throughout coiling and subsequent cold rolling. During annealing of the cold rolled sheet AlN particles precipitate and promote the formation of a (111)[110] cube-on-corner recrystallization texture. With most of the grains in this orientation, the sheet steel resists thinning and through-thickness fracture in subsequent deep-drawing forming operations.

Low-carbon ferritic steels are quite ductile at room temperature, and screw dislocations readily cross slip and intersect in response to cold rolling (Ref 5.10). As the sheet is cold rolled, a dislocation substructure characterized by dislocation walls and intervening dislocation-free cells develops. With increasing cold work, the cells decrease in size and the walls become better defined. Figure 5.11 shows the dislocation substructure formed in an aluminum-killed steel after 5 and 10% tensile strain, and Fig. 5.12 shows cell size as a function of tensile strain (Ref 5.11). The cells decrease in size at a lower rate with increasing strain and approach a constant size. The data of Langford and Cohen (Ref 5.12), obtained from highly strained iron wire, continues the trend established in the tensile tests. The approach to a minimum cell size is attributed to dynamic recovery and dislocation annihilation at high strains.

Figure 5.13 shows the microstructure of a 0.08C-1.45Mn-0.2Si steel after cold rolling 50% and after annealing at 700 °C (1292 °F) for 20 min (Ref 5.13). The ferrite-pearlite microstructure which had formed on cooling of the hot rolled strip is deformed and elongated by the cold rolling. The annealing has caused the deformed ferrite to recrystallize as strain-free, equiaxed grains, and the deformed pearlite colonies have spheroidized into arrays of fine, spherical particles.

Recrystallization kinetics for the 50% cold rolled steel annealed at several temperatures are shown in Fig. 5.14. The kinetics are characterized by an incubation period, during which strain-free grains nucleate, but in this case recrystallization proceeded so rapidly that only at the lowest annealing temperature, 650 °C (1202 °F), was a sigmoidal-shaped curve observed. The activation energy determined from the times required to achieve 50% recrystallization at the various temperatures was 226 kJ/mol (54 kcal/mol), close to the activation energy for the self-diffusion of iron in bcc ferrite (Ref 5.13). This observation is consistent with the mechanism of iron atom transfer from strained grains to strain-free grains across the interface between the two structures. An annealing temperature of 760 °C (1400 °F) is in the two-phase austenite-ferrite field. Thus austenite formation as well as recrystallization occurs. Figure 5.15 shows a partially recrystallized specimen of the 50% cold rolled C-Mn-Si steel which has been held at 760 °C (1400 °F) for 10 s. Austenite nucleates at the boundaries of the deformed ferrite grains and eventually at carbide particles in recrystallized ferrite grains (Ref 5.13).

The above annealing experiments were performed in a laboratory salt bath. In large-scale industrial practice two methods of annealing are used: box annealing and continuous annealing. These two approaches have been reviewed by Mould (Ref 5.14), and a diagram comparing the two annealing methods is shown in Fig. 5.16. In the box annealing process a furnace is placed around stacks of

HEAT TREATMENTS TO PRODUCE FERRITE AND PEARLITE / 123

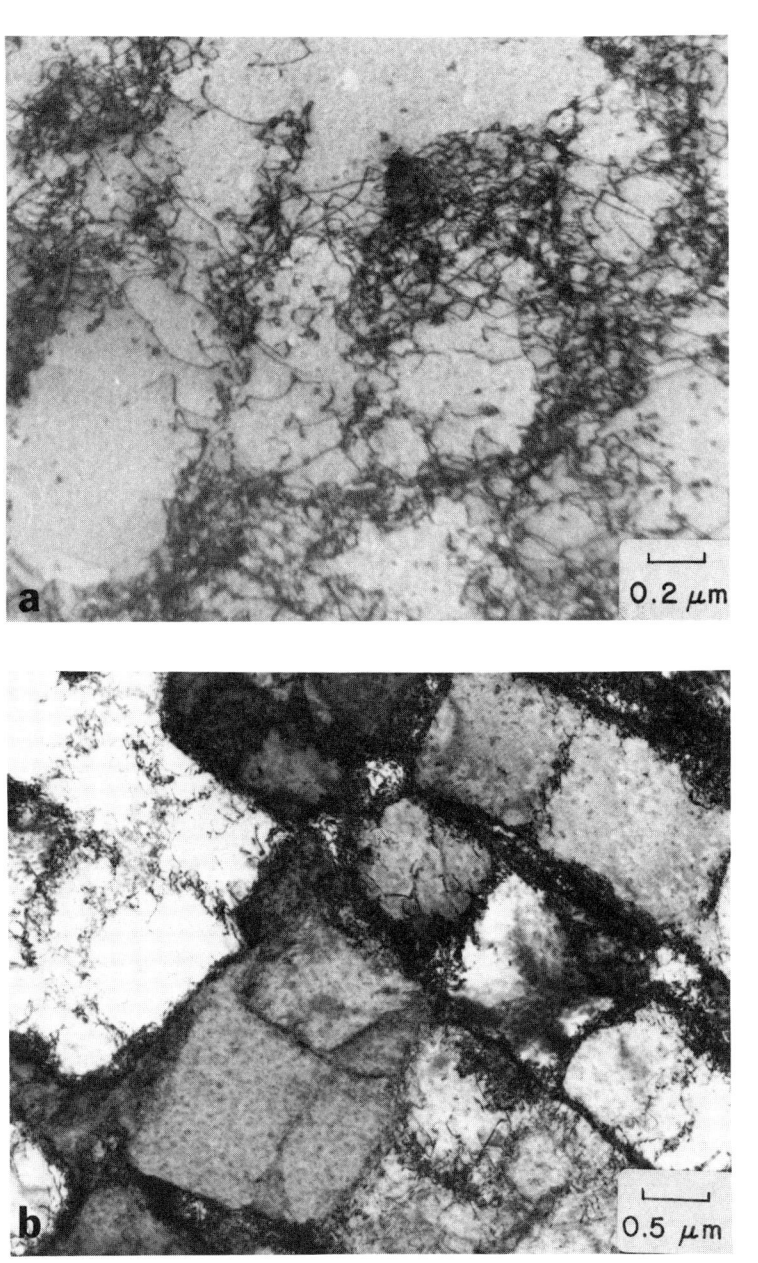

Fig. 5.11. Dislocation substructure in an aluminum-killed steel containing 0.05C-0.29Mn-0.03Al. (a) Strained 5% in tension. (b) Strained 10% in tension. Transmission electron micrographs. Courtesy of J. Pan, Colorado School of Mines, Golden. (Ref 5.11)

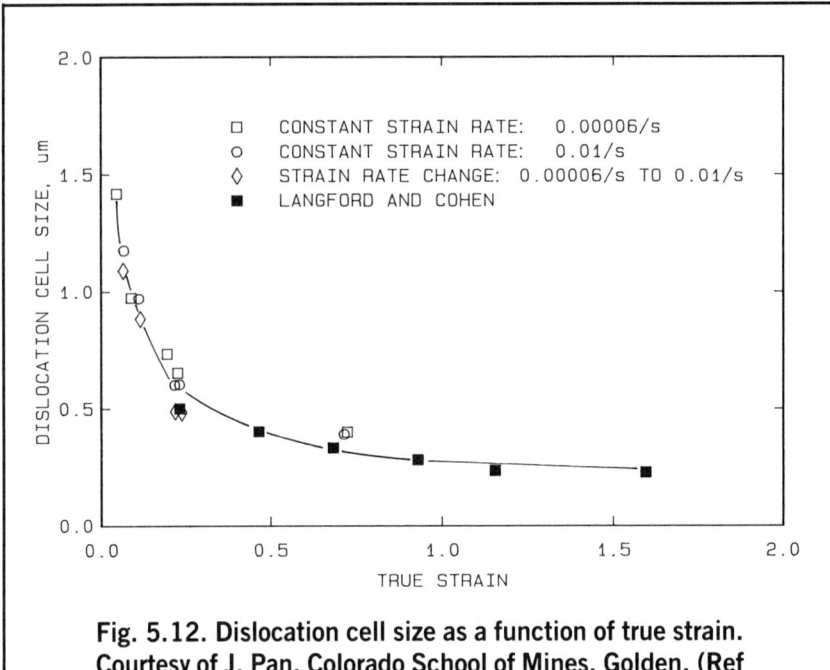

Fig. 5.12. Dislocation cell size as a function of true strain. Courtesy of J. Pan, Colorado School of Mines, Golden. (Ref 5.11)

coils. The very large mass of steel heats and cools very slowly, and the process requires several days for completion. Annealed grain sizes are coarse and the slow cooling rates ensure that all carbon dissolved during annealing precipitates on cooling. Thus excellent ductility results, although some nonuniformity develops because inside and outside parts of a coil experience different thermal histories.

During continuous annealing the coils of cold rolled steels are unwound and the steel is passed through a two-stage furnace for times on the order of a few minutes. The first stage heats the steel and accomplishes recrystallization, while the second stage heats at a lower temperature to effectively overage the steel and remove carbon from solution. Without this step, the thin sheet would cool too rapidly and retain carbon in solution. This carbon would eventually cause strain or quench aging (as described below) and reduce sheet formability. Several processing approaches to overaging are shown in Fig. 5.17. Continuous annealing lines can be used to heat either subcritically or intercritically, and thereby produce higher strength dual-phase steel as well as annealed steel.

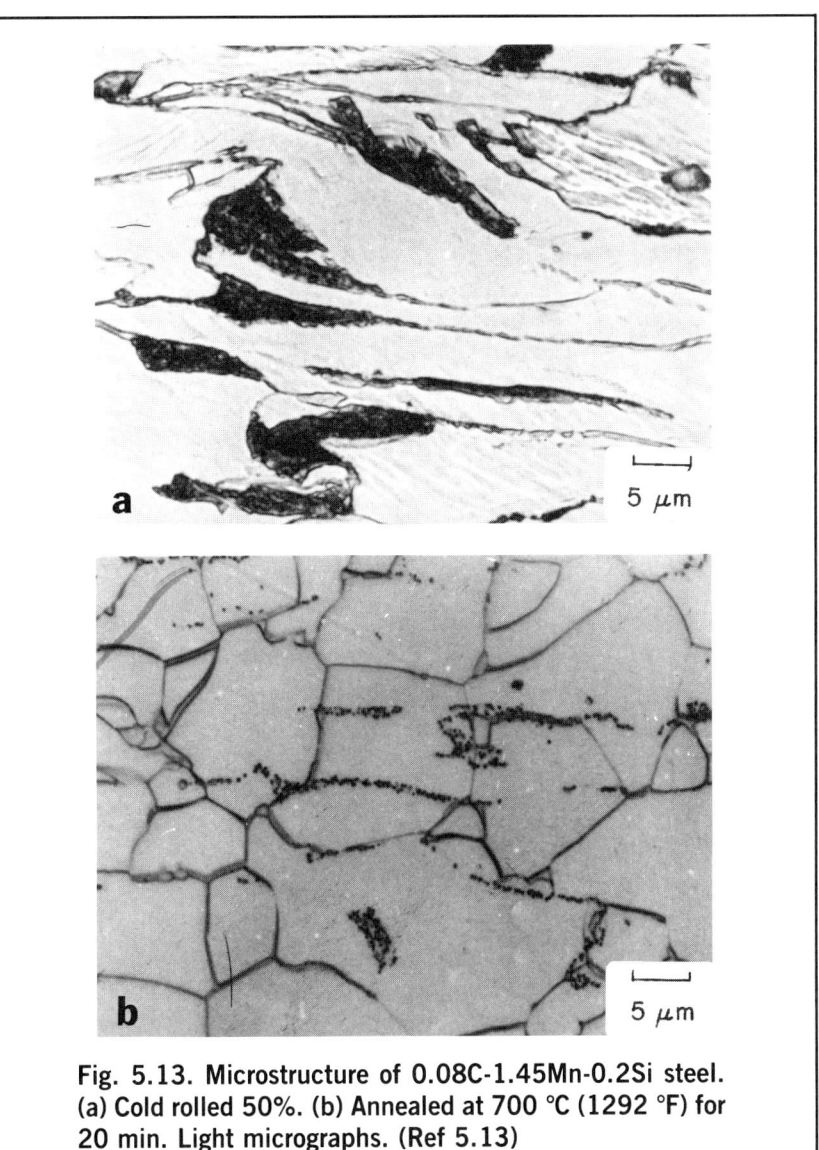

Fig. 5.13. Microstructure of 0.08C-1.45Mn-0.2Si steel. (a) Cold rolled 50%. (b) Annealed at 700 °C (1292 °F) for 20 min. Light micrographs. (Ref 5.13)

Strain and Quench Aging

The enlarged section of the iron-rich portion of the Fe-C diagram (Fig. 1.4) shows that the solubility of carbon in ferrite drops from a maximum of 0.0218 wt.% at 727 °C (1340 °F) to a negligible amount at room temperature. Similarly, a maximum of 0.093 wt.% nitrogen dissolves in ferrite at 585 °C (1085 °F) and the solubility of nitrogen in ferrite also decreases with decreasing temperature to a negligible

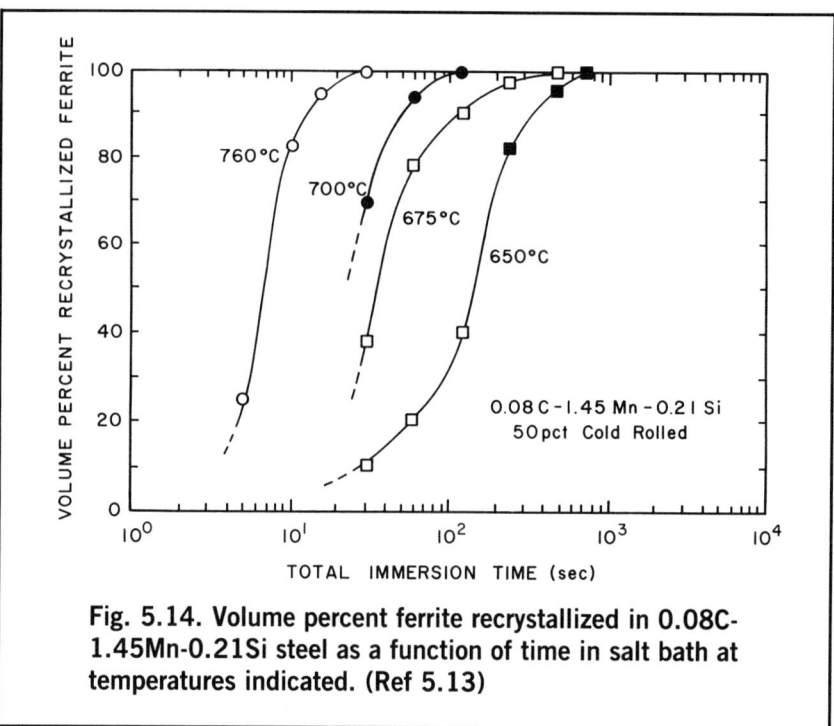

Fig. 5.14. Volume percent ferrite recrystallized in 0.08C-1.45Mn-0.21Si steel as a function of time in salt bath at temperatures indicated. (Ref 5.13)

amount at room temperature (Ref 5.10). Thus during heating, carbon and nitrogen dissolve in ferrite, and during cooling, depending on cooling rate, may be retained in solid solution, segregate to grain boundaries or dislocations, or precipitate as carbide and nitride particles. Strain aging is associated with the segregation of interstitial atoms to the strain fields of dislocations. The strong interaction of the atoms with the dislocations prevents their motion when stress is applied, and the dislocations are said to be pinned or locked (Ref 5.15). Quench aging is caused by the precipitation of carbides and nitrides from ferrite supersaturated with interstitial atoms quenched-in by rapid cooling from high temperatures. Both aging phenomena cause strengthening, decreased ductility, and discontinuous yielding.

Figure 5.18 shows schematically the effect of quench aging on the early stages of yielding in an intercritically annealed and quenched low-carbon steel. The intercritical annealing converts part of the microstructure to islands of austenite, which on quenching transform to martensite. The ferrite which is retained during the intercritical annealing dissolves carbon and on quenching becomes supersaturated with carbon atoms. The resulting microstructure then consists of martensite and ferrite. Steels in this condition are referred to as

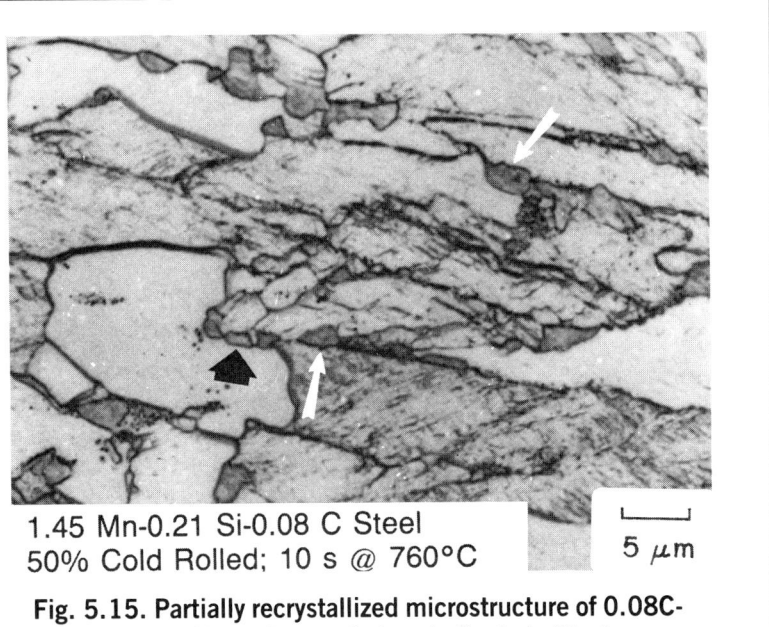

1.45 Mn-0.21 Si-0.08 C Steel
50% Cold Rolled; 10 s @ 760°C   5 μm

Fig. 5.15. Partially recrystallized microstructure of 0.08C-1.45Mn-0.21Si steel annealed as indicated. Black arrow points to austenite adjacent to recrystallized grain, and white arrows point to austenite formed on boundaries between deformed ferrite grains. Light micrograph. (Ref 5.13)

dual-phase steels (Ref 5.16, 5.17), and their microstructures and properties are discussed in more detail in Chapter 9. The ferrite immediately surrounding the martensite islands contains a high density of dislocations generated by the transformation of austenite to martensite. These dislocations are present in high density, are not pinned, and are free to glide at low stresses. The low yield strengths, high initial strain hardening rates, and continuous yielding typical of dual-phase steels are shown schematically for the quenched specimen in Fig. 5.18.

When the intercritically annealed and quenched steel is aged at temperatures close to room temperature, carbide precipitation pins the dislocations. As a result, higher stresses are required to generate new dislocations for plastic deformation. The sudden formation of a high density of new dislocations generates a sudden local thinning on a tensile specimen or steel sheet. The local or discontinuous thinning is referred to as a Lüders band and the associated strain as the Lüders strain. With continued stressing the Lüders band propagates over the

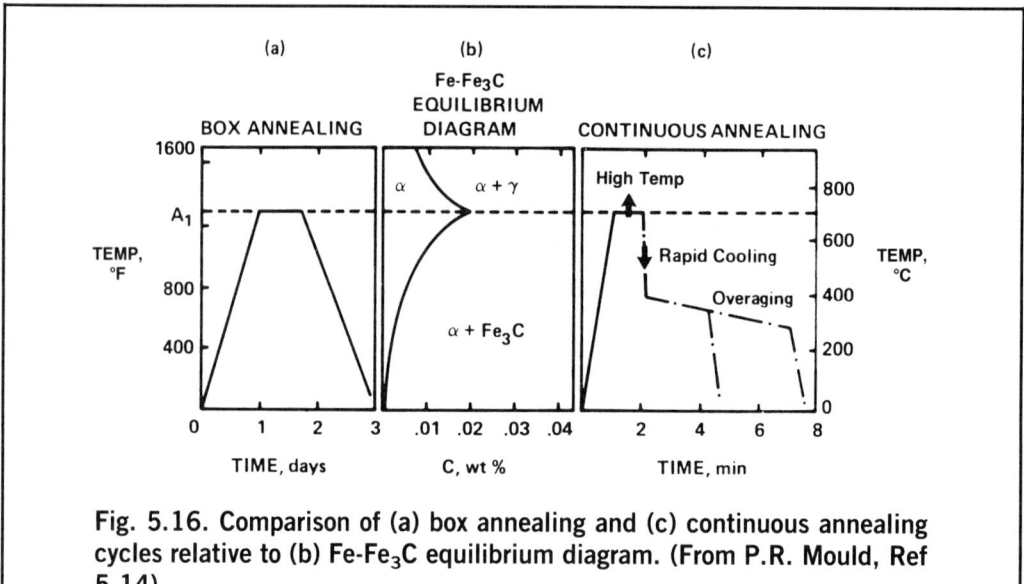

Fig. 5.16. Comparison of (a) box annealing and (c) continuous annealing cycles relative to (b) Fe-Fe$_3$C equilibrium diagram. (From P.R. Mould, Ref 5.14)

gage length of a specimen at a constant stress called the lower yield stress, and when the entire gage length is strained by an amount equal to the Lüders strain, strain hardening begins, as shown in Fig. 5.18, uniformly over the entire specimen gage length.

Figure 5.19 shows how discontinuous yielding develops with aging

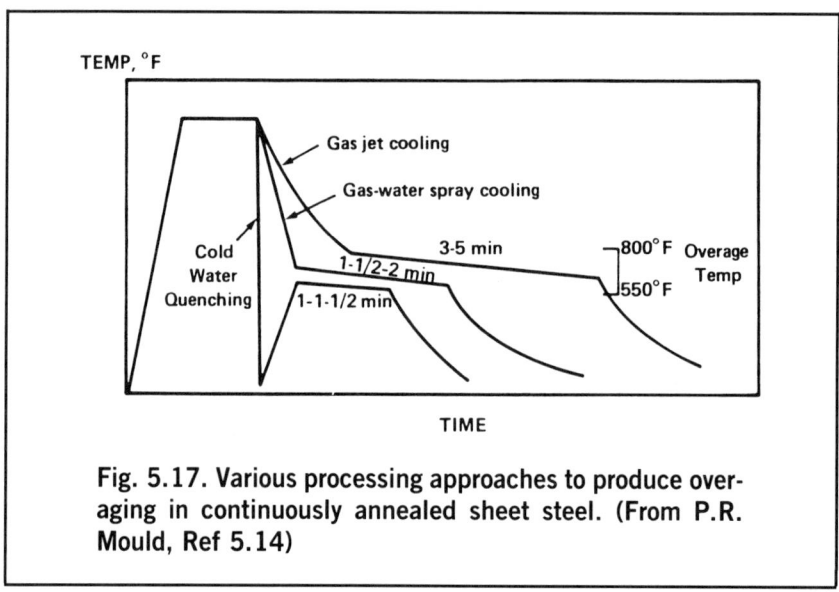

Fig. 5.17. Various processing approaches to produce overaging in continuously annealed sheet steel. (From P.R. Mould, Ref 5.14)

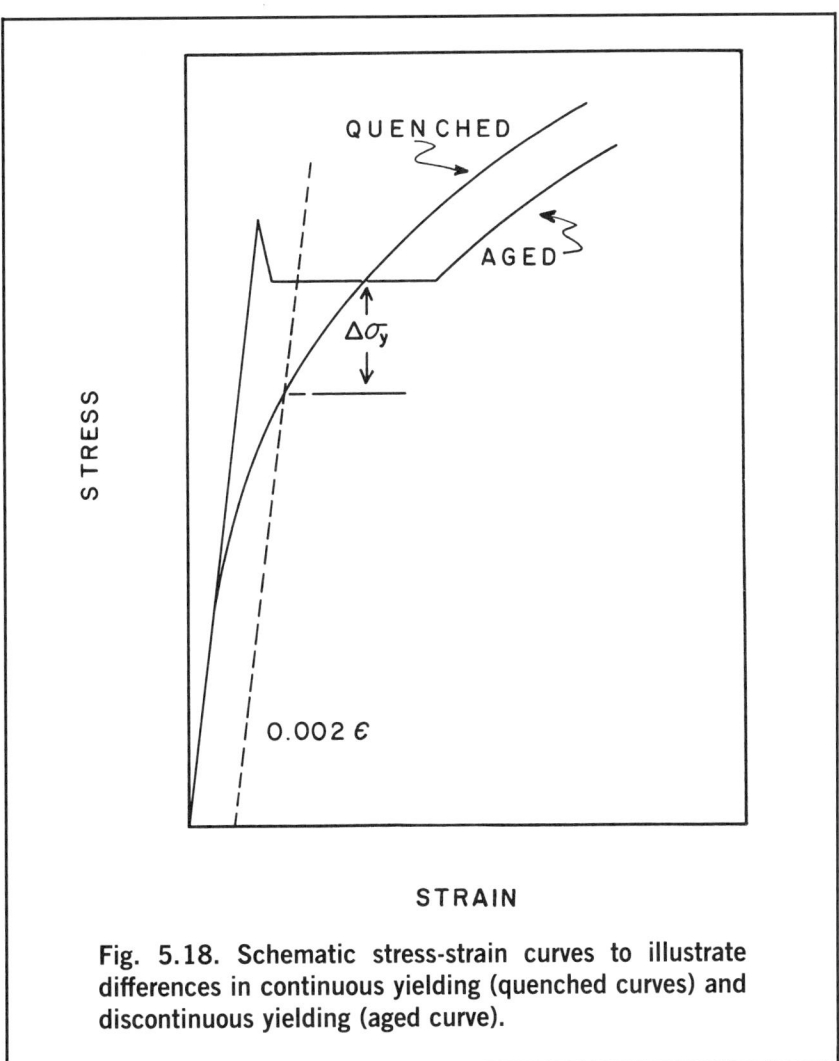

Fig. 5.18. Schematic stress-strain curves to illustrate differences in continuous yielding (quenched curves) and discontinuous yielding (aged curve).

time at 120 °C (248 °F) in a 0.08C-1.45Mn-0.2Si steel intercritically annealed at 760 °C (1400 °F) after cold rolling (Ref 5.18). Yield strengths and the extent of discontinuous yielding gradually increase with aging time. Figure 5.20 shows various precipitate dispersions which develop on quench aging. At low temperatures, fine precipitates form throughout the matrix and on dislocation lines (Ref 5.20, 5.21). The fine precipitates are believed to be a metastable carbide and to form on vacancy clusters. At higher temperatures, cementite forms in platelet and dendritic morphologies (Ref 5.20, 5.21).

Cold rolled and annealed sheet steels invariably yield discontinuously, whether by strain aging of the low density of dislocations

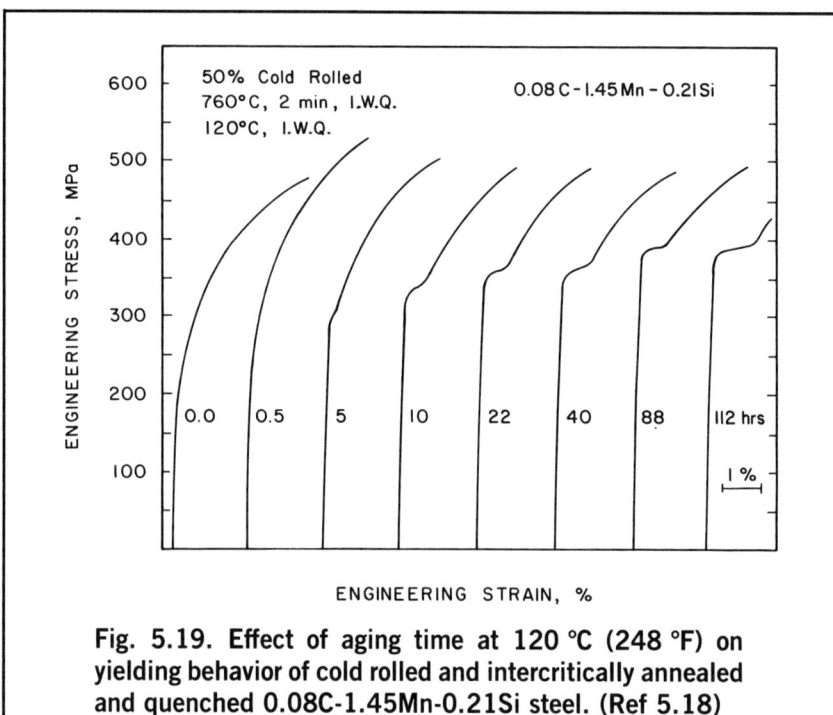

Fig. 5.19. Effect of aging time at 120 °C (248 °F) on yielding behavior of cold rolled and intercritically annealed and quenched 0.08C-1.45Mn-0.21Si steel. (Ref 5.18)

present after annealing or by quench aging. The Lüders bands, also called stretcher strains, are undesirable if formed during shaping of exposed automotive body panels. The susceptibility to Lüders band formation is eliminated by a light cold-rolling operation called temper rolling. Just enough deformation is applied to strain the sheet into the uniform deformation portion of the stress-strain curve. In time, however, carbon or nitrogen atoms may diffuse to the new dislocations generated during temper rolling and re-establish the conditions for discontinuous yielding. This renewed strain aging will add an increment of strength to the flow stress produced by straining and also reduce ductility (Ref 5.10).

Stress Relieving

A number of thermal and mechanical processes produce residual stresses that might be detrimental to the performance of fabricated steel parts or assemblies. The residual stresses may cause distortion, cracking during heat treatment or processing, or failure below design stresses in service. One source of residual stresses is the cooling of

HEAT TREATMENTS TO PRODUCE FERRITE AND PEARLITE / 131

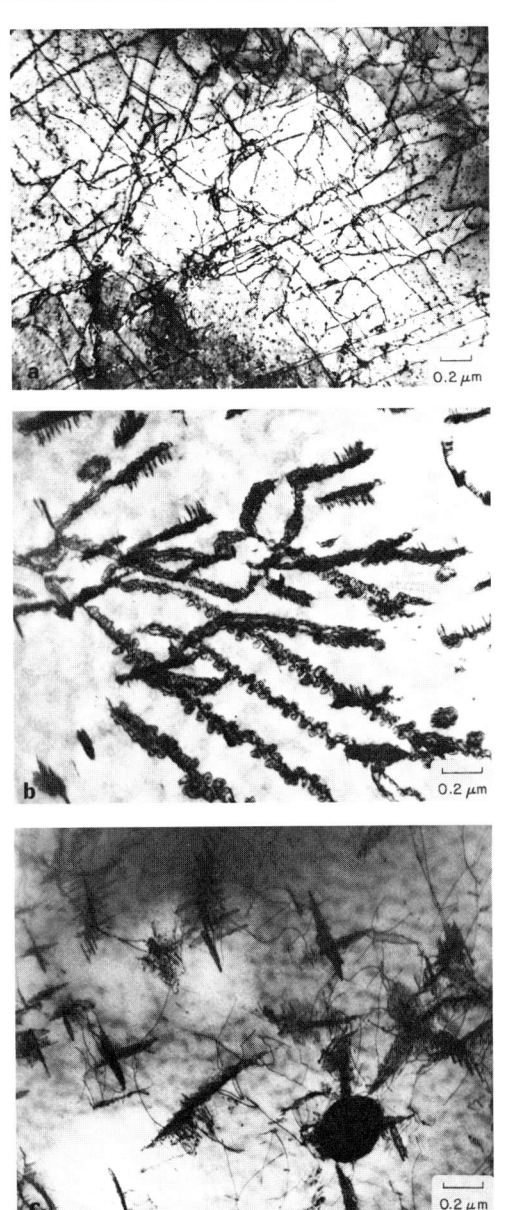

Fig. 5.20. Precipitate dispersions in quench aged steels. (a) 0.052C-0.93Mn-0.004N steel aged at 97 °C (207 °F) for 20 min. (b) 0.077C-0.63Mn-0.018N steel aged at 97 °C (207 °F) for 115 h. (c) 0.052C-0.93Mn-0.004N steel aged at 138 °C (280 °F) for 10 h. Courtesy of E. Indacochea, University of Illinois at Chicago. (Ref 5.19)

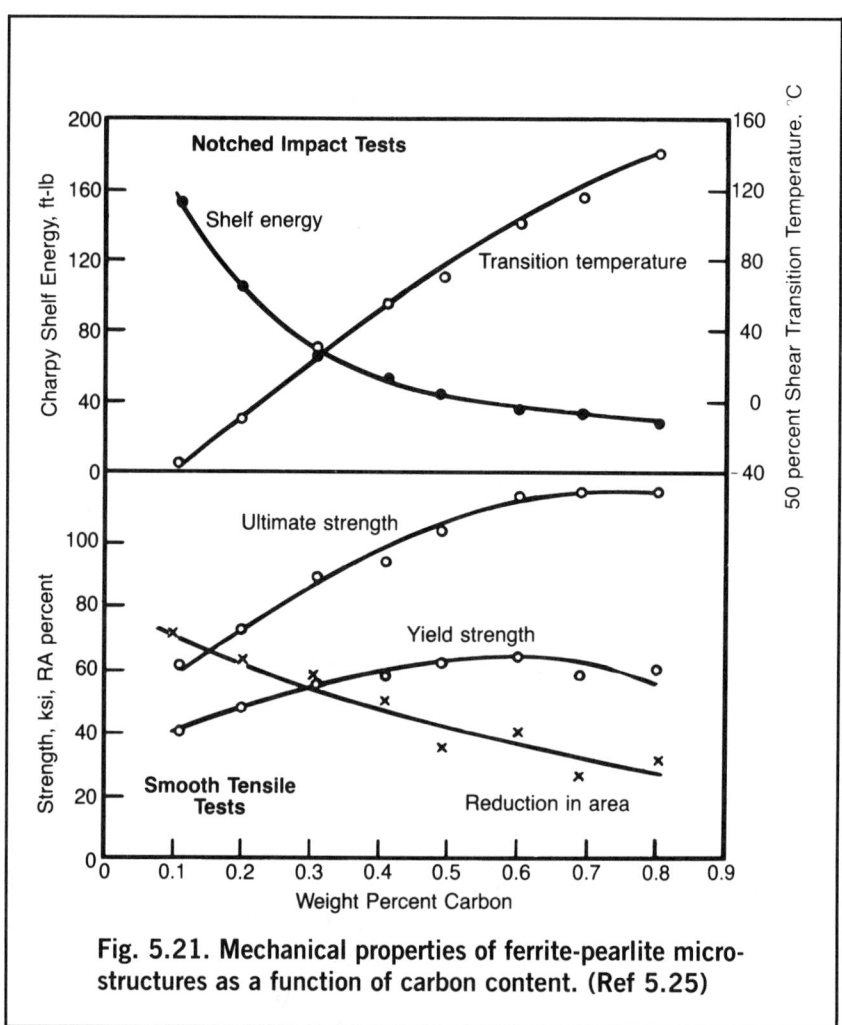

Fig. 5.21. Mechanical properties of ferrite-pearlite microstructures as a function of carbon content. (Ref 5.25)

heavy sections after austenitizing. Even during air cooling, the surface of a heavy section may transform to ferrite and cementite well before the center. When the center eventually transforms, the volume expansion associated with the ferrite formation is restrained by the cooler, already transformed surface. As a result the center is compressed and the surface put in tension. Quenching to form martensite produces a similar but even more severe residual stress problem, even in smaller sections, and is one reason why hardenable steels are alloyed to permit martensite formation at lower cooling rates. Martensitic steels, as discussed in Chapter 8, are invariably tempered, a process that reduces residual stresses and increases ductility and toughness. Machining and cold work also may introduce residual stresses in steel, due to differences in the amount of deformation between the surface and interior regions of a part. Welding is another process which produces residual tensile stresses. As the weld metal solidifies and contracts, it is restrained by the adjacent base metal. As a result, stress-relieving treatments are frequently specified for welded assemblies.

Residual stresses are reduced or eliminated by subcritical heat treatments performed at temperatures either somewhat below or overlapping those used for process or recrystallization annealing (see Fig. 5.8). Heating to and cooling from the stress-relief temperature must be done slowly, especially in heavy sections or large welded assemblies, in order to avoid introducing new thermal stresses and possible cracking during the stress-relief treatment itself (Ref 5.22).

The objective of stress relieving is not to produce major changes of mechanical properties by recrystallization as do the subcritical annealing treatments discussed in the previous section. Rather the stress relief is accomplished by recovery mechanisms that precede recrystallization, a situation that is aided by the difference in the kinetics between the two mechanisms (Ref 5.8). Recovery starts almost immediately on heating and reaching temperature. The rate of recovery is very high initially and decreases with increasing time at temperature. Recrystallization, on the other hand, requires an incubation period and starts very slowly. Therefore, it is possible to relieve residual stresses without changing mechanical properties significantly. For example, a recent investigation (Ref 5.23) of stress relief in cold extruded mild steel bars shows that residual stresses are almost completely relieved without any hardness decrease after heating at 500 °C (932 °F) for 1 h. The latter result shows that the strengthening produced by cold working can be used without the harmful effects of residual surface tensile stresses present in the as-deformed condition. Other aspects of stress relieving are discussed in Ref 5.24.

Mechanical Properties of Ferrite-Pearlite Microstructures

As discussed earlier, the objective of normalizing and full-annealing heat treatments of carbon steels is to produce microstructures consisting of ferrite and pearlite. Some relationships that show the effect of various parameters on the mechanical properties of steels containing ferrite and pearlite are described here.

Figure 5.21 shows a set of mechanical properties for ferrite-pearlite microstructures as a function of steel carbon content. Yield and tensile strengths increase and reduction of area, a measure of ductility, decreases with increasing carbon content because of the increase in pearlite content. The microstructures vary from essentially 100% ferrite to 100% pearlite as carbon is increased to 0.8%, the eutectoid carbon content. The divergence of the yield and ultimate strength curves with increasing carbon content indicates that pearlite increases the work hardening rate.

In addition to the mechanical properties that characterize the strength and ductility of steels, toughness or the energy absorbed during fracture is also of considerable engineering importance (Ref 5.26). Ferritic steels are unique in that they show a transition from ductile to brittle fracture when broken at successively lower temperatures. Generally the transition for a given steel and microstructure is determined by breaking a series of V-notched bars by impact loading at temperatures above and below room temperature. The specimens and machine for the testing are standardized in what is known as the Charpy impact test (Ref 5.27). The ductile fracture typical of higher temperatures proceeds by the growth of microvoids around carbides and/or inclusion particles, a fracture process that requires large amounts of shear or plastic deformation and, therefore, absorbs considerable energy. In contrast, the low-temperature brittle fracture of ferrite proceeds by cleavage between {100} planes of the ferrite grains. Little plastic deformation accompanies cleavage, and therefore little energy is absorbed during this type of fracture.

Figure 5.22(a) shows a mixture of cleavage and ductile fracture on the overload fracture surface of a mild steel. Particles are associated with some of the large conical holes in the fracture surface, and clusters of fine dimples characteristic of ductile fracture are visible in the upper portions of the micrograph. Figure 5.22(b) shows primarily cleavage fracture and is characterized by very flat fracture facets. Cleavage steps (arrows) from one set of cleavage planes to another in a given grain are also a characteristic feature of cleavage fracture.

When energy absorbed during impact fracture is plotted as a function of testing temperature, a transition curve results. Figure 5.23 shows a family of transition curves for steels containing from 0.11% to 0.80% carbon. The energy absorbed by ductile fracture is known as the "shelf energy" because it reaches a plateau or is essentially constant as a function of temperature. Figure 5.23 shows that increasing carbon content lowers shelf energy, and that, therefore, increasing amounts of pearlite adversely affect the ductile fracture toughness. The transition temperature marking the transition between ductile and brittle fracture is also adversely affected by increasing carbon content. Figure 5.23 shows that high-carbon steels with large amounts of pearlite have

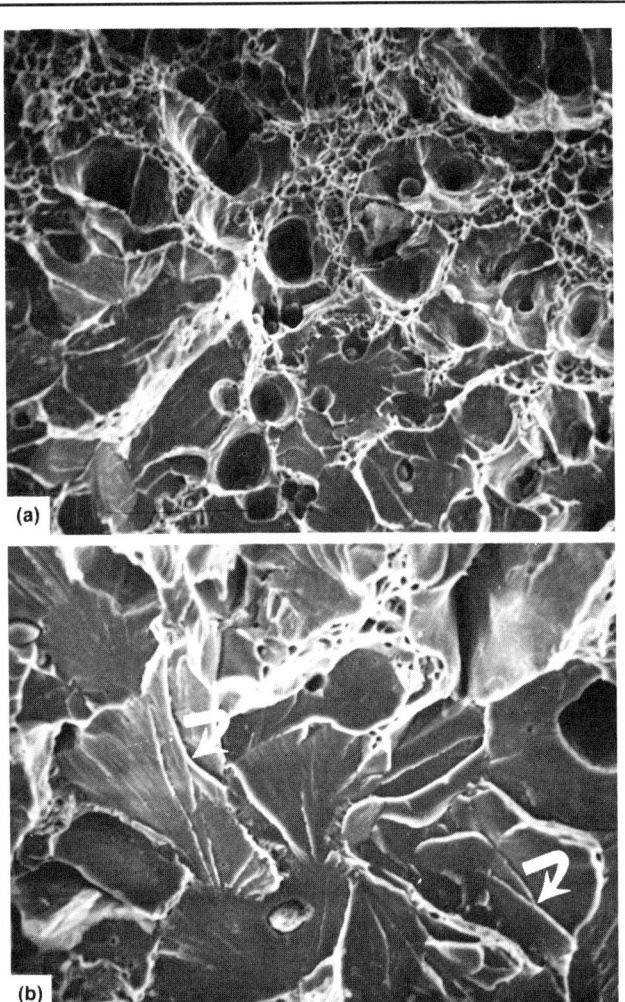

Fig. 5.22. (a) Mixture of cleavage and ductile fracture. Note fine dimples characteristic of ductile overload fracture in upper part of micrograph. Magnification, 500×; shown here at 75%. (b) Cleavage fracture. Arrows point to steps between cleavage planes of different elevation in a given grain. Magnification, 1000×; shown here at 75%. Scanning electron micrographs. Courtesy of D. Yaney, Colorado School of Mines, Golden

high transition temperatures and therefore will fail in a brittle manner even well above room temperature. Low-carbon steels, on the other hand, have subzero transition temperatures and are quite tough

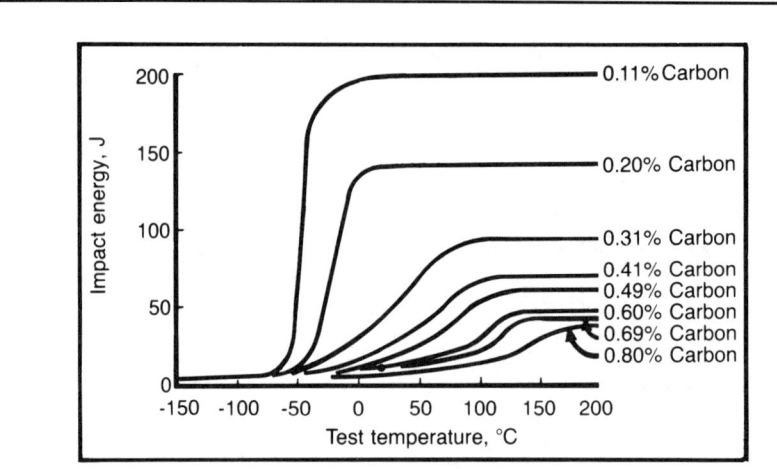

Fig. 5.23. Change in impact transition curves with increasing pearlite content in normalized carbon steels. (Ref 5.28)

at room temperature. Changes in shelf energy and transition temperature (taken as the testing temperature where fracture is 50% cleavage and 50% shear) as a function of carbon content are summarized in the upper part of Fig. 5.21.

At any given carbon content, the mechanical properties and toughness of a steel may be significantly affected not only by pearlite content but also by ferrite grain size and chemical composition. The refinement of ferritic grain size, for example, increases both strength and toughness. Good steel mill and heat treatment practice is therefore directed to producing as fine a ferrite grain size as possible for critical applications. Alloying, low finishing temperatures for hot rolling, and low austenitizing temperatures for normalizing are all techniques used to keep grain size small. Recently, a new class of ferritic steels has been developed to take advantage of the high strength and toughness of very fine-grained steels. The steels are referred to as high-strength low-alloy (HSLA) steels (Ref 5.29), and usually contain less than 0.2% carbon. Grain size control is achieved by microalloying with small amounts of vanadium or niobium that produce very fine carbides. The carbides limit austenite recrystallization and/or grain growth during hot rolling at low finishing temperatures (see Chapter 7) and as a result the ferrite that forms from that austenite on cooling is remarkably fine.

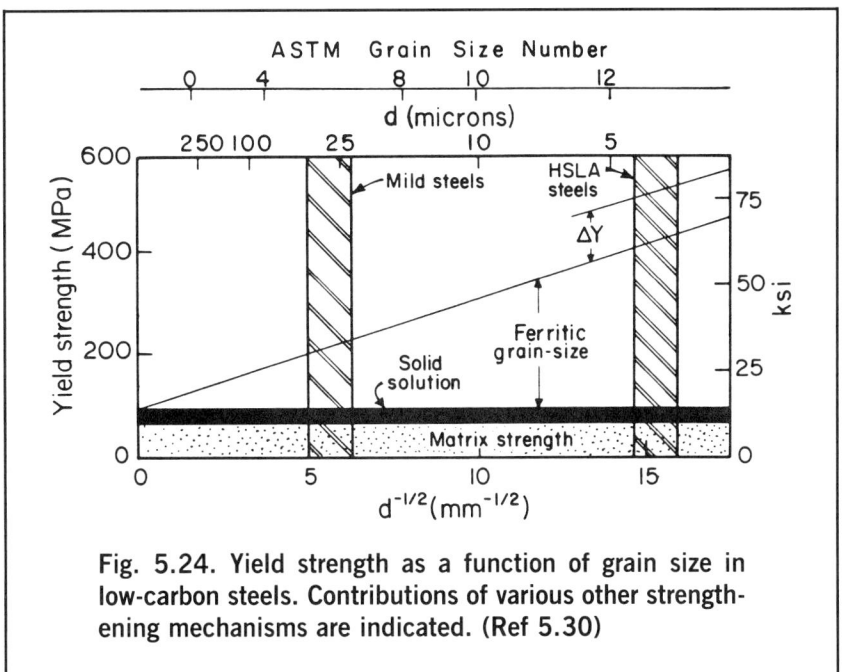

Fig. 5.24. Yield strength as a function of grain size in low-carbon steels. Contributions of various other strengthening mechanisms are indicated. (Ref 5.30)

Figure 5.24 shows yield strength as a function of grain size plotted as $d^{1/2}$ (Ref 5.30). Bands for mild steels and HSLA steels are indicated, and the very strong strengthening effect of grain size refinement is shown. HSLA steels also receive a small increment in strength, ΔY, by precipitation or dispersion strengthening associated with microalloying.

The effects of the various microstructural and composition parameters on the mechanical properties of steels with ferrite-pearlite microstructures have been statistically analyzed by multiple linear regression analysis (Ref 5.31, 5.32). The resulting empirical equations are limited to steels containing less than 0.25% carbon and therefore to microstructures that are largely ferritic. The following selected equations for yield strength and impact transition temperature illustrate the effects of the various parameters:

Yield strength (MPa) (± 31 MPa) =
$K + 37(\% \text{ Mn}) + 83(\% \text{ Si}) + 2918(N_f) + 15.1(d^{-1/2})$ (Eq 5.2)

Impact transition temperature (°C) (± 30 °C) =
$-19 + 44(\%\text{Si}) + 700 N_f - 11.5(d^{-1/2}) + 2.2(\% \text{ pearlite})$
(Eq 5.3)

In these equations, K is 88 MPa for air-cooled steel and 62 MPa for furnace-cooled steel (Ref 5.33); N_f is the free nitrogen dissolved in the ferrite lattice (i.e., not combined as a stable nitride); and d is the mean linear intercept in polygonal ferrite (mm). The beneficial effect of the ferrite grain size on both strength and toughness is apparent from the equations. The effect of manganese and silicon is to increase both yield and tensile strength by solid solution strengthening of the ferrite. The manganese and silicon replace iron on the bcc lattice of ferrite, and are said to dissolve substitutionally. Nitrogen, however, dissolves interstitially and the equations show that it not only is a very potent strengthener of mild steels but also significantly promotes brittle cleavage fracture. The equations also show that the colonies of pearlite in low-carbon steels have no statistically important effect on yield strength. However, other equations show that pearlite does increase tensile strength, an effect attributed to the decrease in yield extension produced by larger pearlite content (Ref 5.31). Figure 5.11 shows a similar greater effect of pearlite on ultimate strength than on yield strength, especially beyond the 0.25% carbon for which Eq 5.2 and Eq 5.3 hold.

In addition to the largely ferritic microstructures of mild steels with carbon contents less than 0.25%, ferrite-pearlite microstructures in medium-carbon steels (Ref 5.34) and fully pearlitic microstructures of eutectoid steels (Ref 5.35) have also been analyzed to correlate various microstructural parameters with mechanical properties. Whereas low-carbon steels are used for structural applications where relatively low strength but good ductility and toughness are required for formability and service conditions, the eutectoid steels are used where high hardness and wear resistance are of major concern. Railroad rails are an example of an important application where the latter properties are important, and the need for improved rail steels has led to the development of the following equations for yield strength and toughness of pearlitic steels (Ref 5.35):

$$\text{Yield strength (MPa)} = 2.18(S^{-1/2}) - 0.40(P^{-1/2}) - 2.88(d^{-1/2}) + 52.30 \quad \text{(Eq 5.4)}$$

$$\text{Transition temperature (°C)} = -0.83(P^{-1/2}) - 2.98(d^{-1/2}) + 217.84 \quad \text{(Eq 5.5)}$$

where S is the pearlite interlamellar spacing, P is the pearlite colony size, and d is the austenite grain size.

Equations 5.4 and 5.5 show that the most important parameter affecting yield strength is the interlamellar spacing while the transition temperature is affected most strongly by the austenite grain size from which the pearlite develops. The transition temperature for fully pearlitic steels is invariably above room temperature, and therefore cleavage on the {100} planes of the ferrite in pearlite is the characteristic fracture mode. The size of the cleavage facets is a strong function of but always smaller than the austenite grain size and appears to be related to common ferrite orientations in several adjoining pearlite colonies (Ref 5.36). Air cooling or normalizing of eutectoid rail steels containing only manganese and silicon produces fine interlamellar pearlite spacings, generally on the order of 2000 Å (200 nm) (Ref 5.37). Alloying of rail steels with vanadium, chromium, and molybdenum produces even finer pearlite interlamellar spacings (Ref 5.38-5.40).

Microalloyed Bar and Forging Steels

Forgings for moderate- and high-strength applications have commonly been made from hardenable steels which are quenched and tempered after forging. Another group of steels now available for moderate-strength forgings are medium-carbon microalloyed steels with ferrite-pearlite microstructures (Ref 5.41). Carbon contents are 0.3 to 0.5% and the steels contain small amounts of vanadium, niobium, and titanium. These steels harden to ultimate tensile strengths up to 145 ksi (1000 MPa) directly on cooling from forging temperatures, and therefore do not require separate quench and tempering heat treatments. At hardness between 25 and 30 HRC, fatigue strengths of the microalloyed steels are comparable but impact strengths are lower than those of quenched and tempered steels of the same hardness.

Gladman *et al.* (Ref 5.34) evaluated the strength of medium-carbon ferrite-pearlite steels relative to microstructural and chemical factors and developed the following equation:

$$\sigma_{ys} \text{ (MPa)} = 15.4 \, \{f_\alpha^{1/3} [2.3 + 3.8 \, (\%Mn) + 1.13 d^{-1/2}] \\ + (1 - f_\alpha^{1/3}) [11.6 + 0.25 S_p^{-1/2}] \\ + 4.1 \, (\%Si) + 27.6 \, (\%N)\} \quad \text{(Eq 5.6)}$$

where f_α is the volume fraction of ferrite; S_p is the interlamellar spacing of pearlite, the distance (in mm) from the center of one cementite lamella to the center of the next, measured normal to the plane of the lamellae; and d is the mean linear intercept grain diameter (in mm). All ferrite-pearlite steels but those microalloyed with vanadium developed strengths which correlated well with those

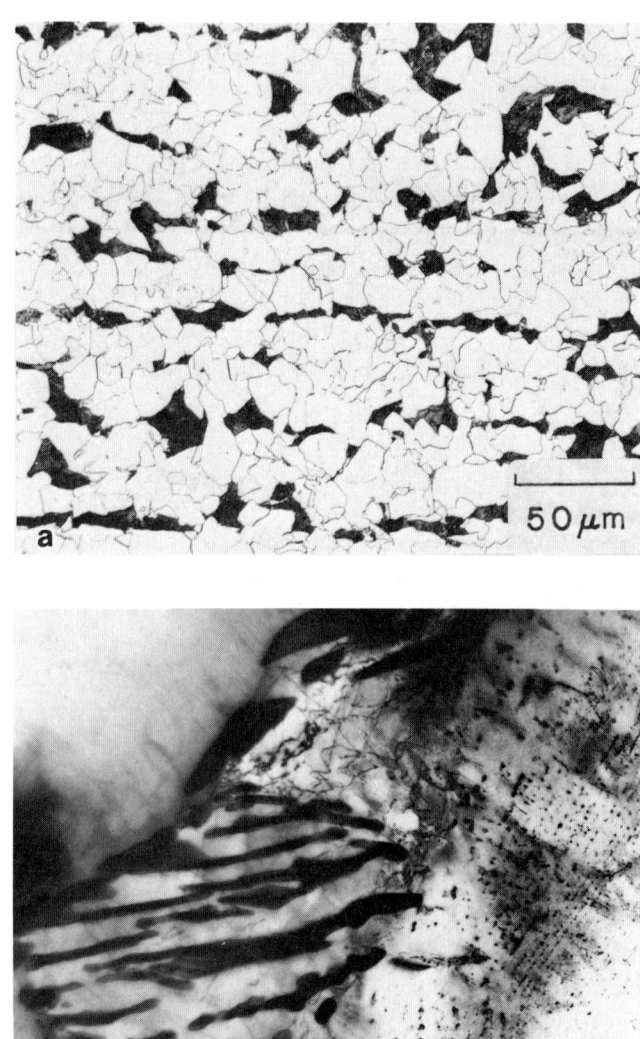

Fig. 5.25. Microstructure of 0.2C-0.15V steel. (a) Ferrite and pearlite. Light micrograph. (b) Ferrite, pearlite, and V(C,N) precipitates. Transmission electron micrograph. Courtesy of S.W. Thompson, Colorado School of Mines, Golden. (Ref 5.42)

HEAT TREATMENTS TO PRODUCE FERRITE AND PEARLITE / 141

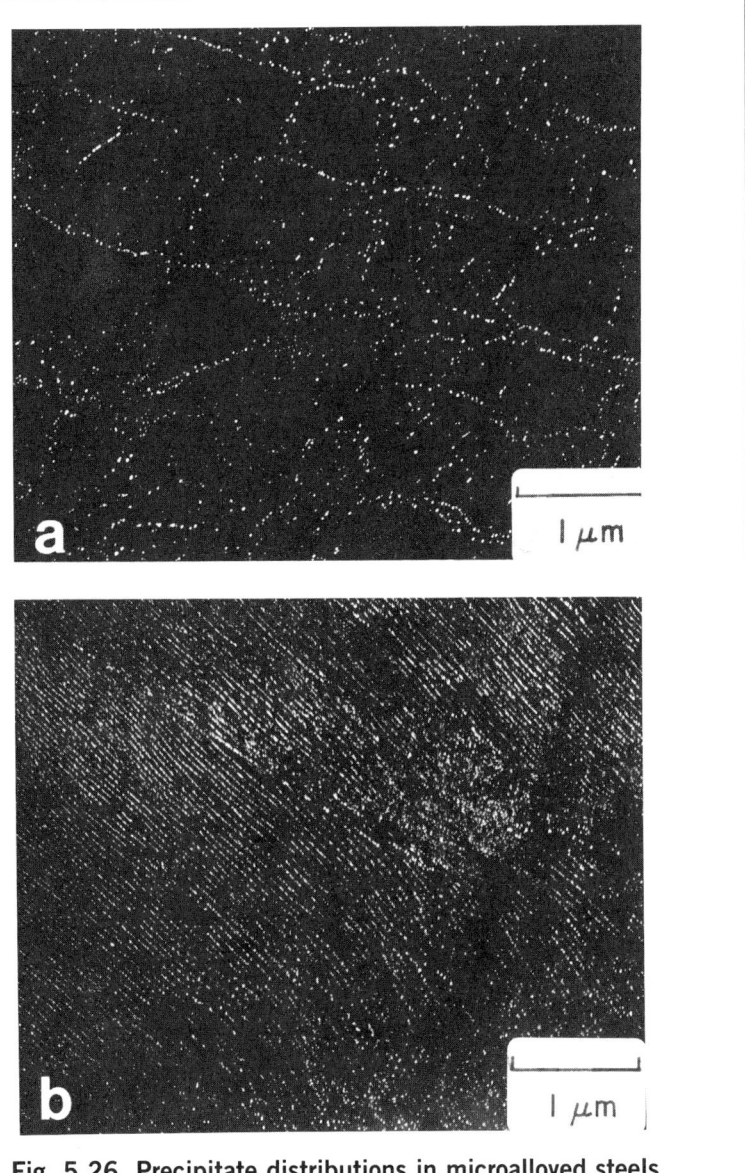

Fig. 5.26. Precipitate distributions in microalloyed steels containing vanadium and niobium. (a) Nb-rich precipitates on deformed austenite substructure. (b) V-rich interphase precipitates. Dark-field transmission electron micrographs. Courtesy of S.W. Thompson, Colorado School of Mines, Golden. (Ref 5.42)

calculated from Eq 5.6. The vanadium-containing steels showed higher than expected yield strengths, and this discovery became the basis for developing the microalloyed medium-carbon forging steels.

Figure 5.25 shows the ferrite-pearlite microstructure of a 0.2% carbon steel containing 0.15% vanadium (Ref 5.42). No microalloy precipitation is visible in the light micrograph, but when the same structure is examined in the electron microscope, many very fine precipitates are revealed in the ferrite. This fine-scale precipitation of vanadium carbonitride, V(C,N), plus large volume fractions of pearlite, are the major strengthening components of medium-carbon microalloyed ferrite-pearlite steels. The V(C,N) particles tend to form aligned arrays produced by interphase decomposition of austenite.

Sometimes niobium is added as a microalloying element, and in structures finished hot worked at low temperatures, Nb(C,N) particles precipitate on austenitic deformation substructure, as shown in Fig. 5.26(a). In alloys with both niobium and vanadium, interphase precipitation as well as Nb(C,N) precipitates on austenite substructure may form (Fig. 5.26b). Such substructure precipitation effectively retards austenite recrystallization, and when the unrecrystallized austenite transforms, very fine ferrite grain sizes result. This approach is effectively used in flat rolled products where low finish hot working temperatures can be achieved, but forgings are generally worked at temperatures above those at which Nb(C,N) precipitation occurs.

A number of approaches are being used to refine microstructure and increase toughness of direct-cooled forging steels. These include the use of lower forging temperatures, lower carbon alloys, sulfur additions to nucleate ferrite and thereby break up massive pearlite colonies, and bainitic microstructures (Ref 5.43).

References

5.1 Heat Treating of Carbon and Low-Alloy Steels, in *Metals Handbook*, Vol 2, 8th ed., American Society for Metals, Metals Park, OH, 1964, p 1-10

5.2 K. Nakazawa and G. Krauss, Martensite and Fracture in 52100 Steel, *Met Trans A*, Vol 9A, 1978, p 681-689

5.3 M. Atkins, *Atlas of Continuous Cooling Transformation Diagrams for Engineering Steels*, British Steel Corp., Sheffield, 1977

5.4 S. Chattopadhyay and C.M. Sellars, Quantitative Measurements of Pearlite Spheroidization, *Metallography*, Vol 10, 1977, p 89-105

5.5 E.C. Rollason, *Fundamental Aspects of Molybdenum on Transformation of Steel*, Climax Molybdenum Co., London

5.6 P. Payson, W.L. Hodapp, and J. Leeder, The Spheroidizing of Steel by Isothermal Transformation, *Trans ASM*, Vol 28, 1940, p 306-332

5.7 R.L. Fullman, Measurement of Particle Sizes in Opaque Bodies, *Trans AIME*, Vol 197, 1953, p 447-452
5.8 P.G. Shewmon, *Transformations in Metals*, McGraw-Hill, New York, 1969
5.9 D. Witmer and G. Krauss, Effect of Thermal History on the Recrystallization Behavior of Low Carbon 0.305 Mn Steels Containing Oxygen and Sulfur, *Trans ASM*, Vol 62, 1969, p 447-456
5.10 W.C. Leslie, *The Physical Metallurgy of Steels*, McGraw-Hill, New York, 1981
5.11 J. Pan, Research in progress, Colorado School of Mines, Golden, 1989
5.12 G. Langford and M. Cohen, Strain Hardening of Iron by Severe Plastic Deformation, *Trans ASM*, Vol 62, 1969, p 623-638
5.13 D.Z. Yang, E.L. Brown, D.K. Matlock, and G. Krauss, Ferrite Recrystallization and Austenite Formation in Cold-Rolled Intercritically Annealed Steel, *Met Trans A*, Vol 16A, 1985, p 1385-1392
5.14 P.R. Mould, An Overview of Continuous-Annealing Technology for Steel-Sheet Products, in *Metallurgy of Continuous-Annealed Sheet Steel*, B.L. Bramfitt and P.L. Mangonon, Jr. (Eds.), TMS-AIME, Warrendale, PA, 1982
5.15 A.W. Cottrell, *Dislocations and Plastic Flow in Crystals*, Oxford University Press, London, 1953
5.16 R.A. Kot and B.L. Bramfitt (Eds.), *Fundamentals of Dual-Phase Steels*, TMS-AIME, Warrendale, PA, 1981
5.17 D.K. Matlock, F. Zia-Ebrahimi, and G. Krauss, Structure, Properties, and Strain Hardening of Dual-Phase Steels, in *Deformation, Processing and Structure*, G. Krauss (Ed.), American Society for Metals, Metals Park, OH, 1984, p 47-87
5.18 D.Z. Yang, D.K. Matlock, and G. Krauss, The Effect of Cold-Rolling on Aging of an Intercritically Annealed Mn-Si-C Steel, in *Technology of Continuously Annealed Cold-Rolled Sheet Steel*, R. Pradhan (Ed.), TMS-AIME, Warrendale, PA, 1985, p 319-339
5.19 J.E. Indacochea, Dual Phase Behavior and Aging of a Renitrogenized Steel, M.S. thesis, Colorado School of Mines, Golden, 1979
5.20 W.C. Leslie, R.M. Fisher, and N. Sen, Morphology and Crystal Structure of Carbides Precipitated from Solid Solution in Alpha Iron, *Acta Metall*, Vol 7, 1969, p 632-644
5.21 W.C. Leslie, The Quench-Aging of Low-Carbon Iron and Iron-Manganese Alloys, *Acta Metall*, Vol 9, 1961, p 1004-1021
5.22 A. Vinckier and A. Dhooge, Reheat Cracking in Welded Structures During Stress Relief Heat Treatments, *J Heat Treating*, Vol 1, 1979, p 72-80
5.23 M.B. Adeyemi, R.A. Stark, and G.F. Modlen, Isothermal Stress Relief of Cold Extruded Mild Steel Rods, *Proceedings of the Metals Society, Heat Treatment '79*, Birmingham, England, 22-24 May 1979
5.24 K.E. Thelning, *Steel and Its Heat Treatment*, Bofors Handbook, 1075, Butterworths, London, 1975
5.25 A.R. Rosenfield, G.T. Hahn, and J.D. Embury, Fracture of Steels Containing Pearlite, *Met Trans*, Vol 3, 1972, p 2797-2804
5.26 K.W. Burns and F.B. Pickering, Deformation and Fracture of Ferrite-Pearlite Structures, *JISI*, Vol 202, 1964, p 899-906
5.27 G.E. Dieter, Jr., *Mechanical Metallurgy*, McGraw-Hill, New York, 1961, p 371-375

5.28 F.B. Pickering, The Optimization of Microstructures in Steel and Their Relationship to Mechanical Properties, in *Hardenability Concepts with Applications to Steel*, D.V. Doane and J.S. Kirkaldy (Eds.), AIME, Warrendale, PA, 1978, p 179-228

5.29 *Microalloying 75*, Union Carbide Corp., 1977, distributed by American Society for Metals, Metals Park, OH

5.30 M. Cohen and S.S. Hansen, Microstructural Control in Microalloyed Steels, in *MiCon 78: Optimization of Processing, Properties and Service Performance Through Microstructural Control*, ASTM STP 672, H. Abrams, G.N. Maniar, D.A. Nail, and H.D. Solomon (Eds.), ASTM, 1979, p 34-52

5.31 F.B. Pickering, The Effect of Composition and Microstructure on Ductility and Toughness, in *Toward Improved Ductility and Toughness*, Climax Molybdenum Development Co., Japan, 1971, p 9-31

5.32 R.W.K. Honeycombe and F.B. Pickering, Ferrite and Bainite in Alloy Steels, *Met Trans*, Vol 3, 1972, p 1099-1112

5.33 T. Gladman, D. Dulieu, and I.D. McIvor, Structure-Property Relationships in High-Strength Microalloyed Steels, in *Microalloying 75*, Union Carbide Corp., 1977, distributed by American Society for Metals, Metals Park, OH, p 32-54

5.34 T. Gladman, I.D. McIvor, and F.B. Pickering, Some Aspects of the Structure-Property Relationships in High-Carbon Ferrite-Pearlite Steels, *JISI*, Vol 210, 1972, p 916-930

5.35 J.M. Hyzak and I.M. Bernstein, The Role of Microstructure on the Strength and Toughness of Fully Pearlitic Steels, *Met Trans*, Vol 74, 1976, p 1217-1224

5.36 Y-J. Park and I.M. Bernstein, Mechanism of Cleavage Fracture in Fully Pearlitic 1080 Rail Steel, in *Rail Steels—Developments, Processing and Use*, STP 644, ASTM, 1978, p 287-302

5.37 D.E. Sonon, J.V. Pellegrino, and J.M. Wandrisco, A Metallurgical Examination of Control-Cooled, Carbon-Steel Rails with Service-Developed Defects, in *Rail Steels—Developments, Processing and Use*, STP 644, ASTM, 1978, p 99-117

5.38 G.K. Bouse, I.M. Bernstein, and D.H. Stone, in *Rail Steels—Developments, Processing and Use*, STP 644, ASTM, 1978, p 145-166

5.39 S. Marich and P. Curcio, in *Rail Steels—Developments, Processing and Use*, STP 644, ASTM, 1978, p 167-210

5.40 Y.E. Smith and F.B. Fletcher, in *Rail Steels—Developments, Processing and Use*, STP 644, ASTM, 1978, p 212-232

5.41 G. Krauss and S.K. Banerji (Eds.), *Fundamentals of Microalloying Forging Steels*, TMS-AIME, Warrendale, PA, 1987

5.42 S.W. Thompson and G. Krauss, Precipitation and Fine Structure in Medium-Carbon Vanadium and Vanadium/Niobium Microalloyed Steels, *Met Trans A*, to be published

5.43 K. Grassl, S.W. Thompson, and G. Krauss, New Options for Steel Selection for Automotive Applications, SAE Technical Paper No. 890508, Society of Automotive Engineers, Warrendale, PA, 1989

CHAPTER 6

Hardness and Hardenability

A martensitic microstructure is the hardest microstructure that can be produced in any carbon steel, but it can be produced only if the transformation of austenite to mixtures of ferrite and cementite is avoided. This chapter first describes the relationship of martensite hardness to carbon content, and then discusses other factors that determine whether or not that hardness can be achieved throughout a given part fabricated from a given steel. The term hardenability is used to describe both the ease of martensite formation and the technology that relates section size, cooling rates, and composition to the hardening of steel. This chapter emphasizes principles and the classical approach to hardenability as developed by Grossmann, Bain, and their contemporaries and then describes more recent approaches to the continually developing technology of hardenability.

Hardness and Carbon Content

The maximum hardness that can be produced in any given carbon steel is that associated with a fully martensitic microstructure. Figure 6.1 shows the much higher hardness of martensite relative to that of ferrite-pearlite or spheroidized microstructures for the entire range of carbon content usually found in steels. The high hardness and associated high strength, fatigue resistance, and wear resistance are the prime reasons for the quenching heat treatments that produce martensite. Almost all martensite is tempered, and depending on the amount of tempering, hardness in a given quench and tempered steel may vary from close to the maximum shown for martensite to the minimum associated with the spheroidized carbide structure. Heat

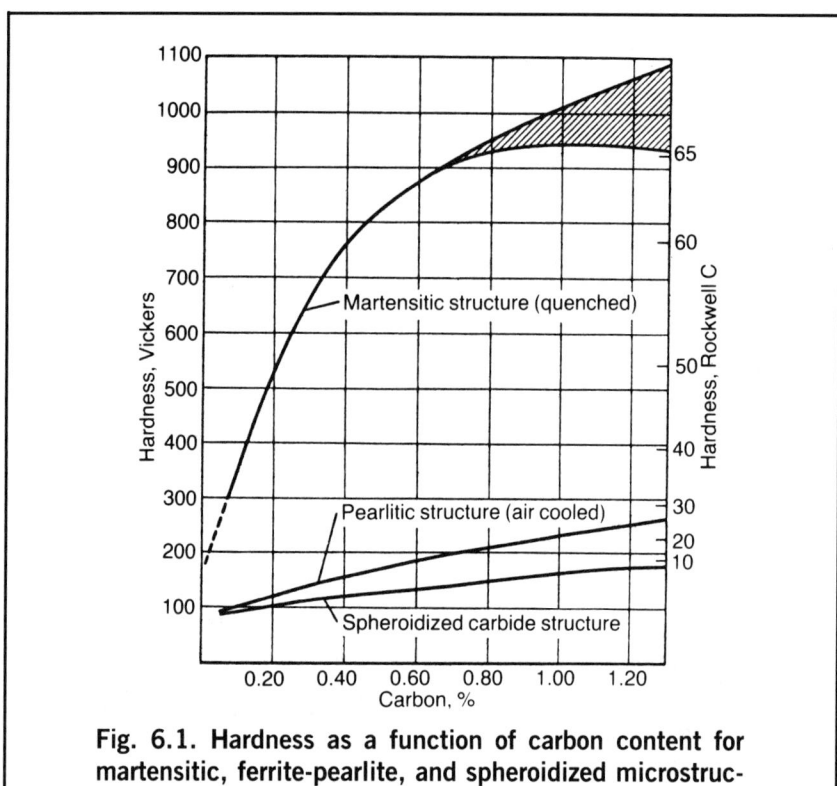

Fig. 6.1. Hardness as a function of carbon content for martensitic, ferrite-pearlite, and spheroidized microstructures in steels. Cross-hatched area shows effect of retained austenite. (Ref 6.1)

treatments to form martensite are generally applied to steels containing more than 0.3% carbon. In these steels the gains in hardness are most substantial. Also, steels containing less than 0.3% carbon tend to be difficult to harden and are usually used with the ferrite-pearlite microstructures produced by the heat treatments described in Chapter 5. Rockwell C readings below 20 are not considered valid and are included in Fig. 6.1 only for comparative purposes.

Figure 6.2 is a summary plot of many investigations of the hardness of martensitic microstructures as a function of the carbon content of steels and Fe-C alloys, and shows the range of hardness that may develop in largely martensitic microstructures in steels of a given carbon content. Special care was taken in all of the investigations to insure that no proeutectoid phases or mixtures of ferrite and cementite formed. However, the martensitic microstructures may have contained various amounts of retained austenite because M_f drops below room temperature even in low-carbon steels. For

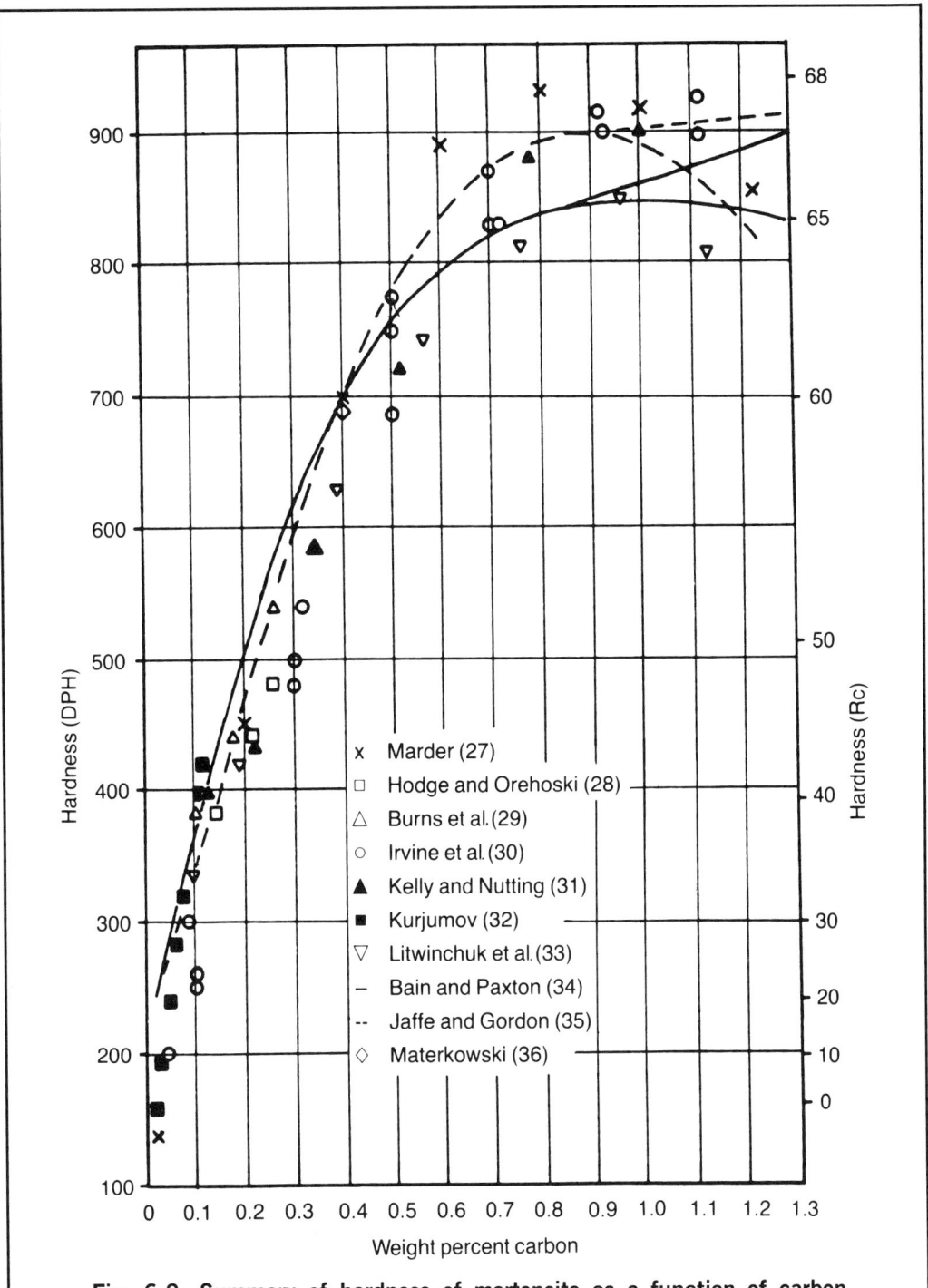

Fig. 6.2. Summary of hardness of martensite as a function of carbon content in Fe-C alloys and steels. (Ref 6.2)

example, Fig. 3.8 showed that small amounts of retained austenite are present at room temperature in steels with carbon content as low as 0.3%. The most significant effect of retained austenite on hardness occurs in steels containing more than 0.7% carbon; Fig. 6.1 and several of the sets of data in Fig. 6.2 show the decrease in hardness that develops with increasing amounts of retained austenite in high-carbon steels.

Some of the investigators whose data is shown in Fig. 6.2 quenched specimens in liquid nitrogen (−196 °C) in order to reduce retained austenite and thereby increase hardness. For example, the continuous curve after Bain and Paxton (Ref 6.1), based on as-quenched hardness at room temperature, is lower than the dashed curve after Jaffee and Gordon (Ref 6.3), who cooled their specimens in liquid nitrogen. The data points marked by x's were taken from specimens cooled in liquid helium (−269 °C) (Ref 6.4) and tend to be higher than the hardness of steels not as deeply cooled. The effectiveness of the subzero treatments is of course greatest in steels containing more than 0.4% carbon, where significant amounts of retained austenite (see Fig. 3.8) may be present at room temperature.

Apart from differences in retained austenite content, some of the variation in the maximum hardness of various carbon levels might also be due to aging or differences in austenitic grain size. Figure 6.3 shows that room temperature aging significantly increases the hardness of martensitic Fe-Ni-C alloys (Ref 6.5, 6.6). Similar hardness changes with time have been observed in Fe-C martensites (Ref 6.4); thus, if attention is not paid to the time after quenching at which hardness measurements are made, some variation contributing to scatter in reported hardness values may occur.

Austenitic grain size has also been observed to affect the strength of martensite in low-carbon steels (Ref 6.7, 6.8). When the austenite grain size is reduced, significant increases in strength occur. The relationship between austenite grain size and martensite structure is a result of the unique structure of martensite in low- and medium-carbon steels. The martensite laths, as described in Chapter 3, are arranged in packets whose size is directly related to austenite grain size (see Chapter 7). Thus either martensite packet size or austenite grain size may be used to correlate with mechanical properties. Figure 6.4 shows the increase in yield strength with decreasing martensite packet size in an Fe-0.2C alloy. Packet size (D) is plotted as $D^{-1/2}$ in what is referred to as a Hall-Petch plot. An interesting observation is that the slope of the Fe-0.2C martensite curve is steeper than that of lath martensite in an Fe-Mn alloy without carbon. This observation

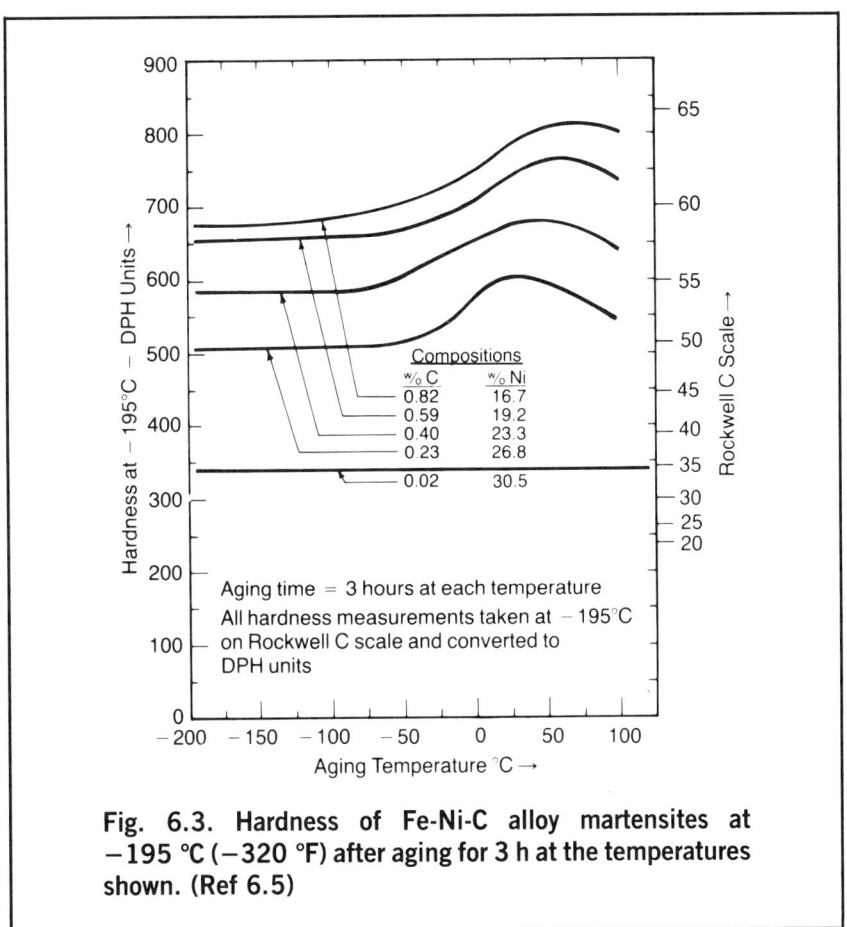

Fig. 6.3. Hardness of Fe-Ni-C alloy martensites at −195 °C (−320 °F) after aging for 3 h at the temperatures shown. (Ref 6.5)

was explained by the segregation of carbon atoms to packet boundaries where they make the initial yielding process more difficult; the more so the finer the packet size (Ref 6.8).

Martensite Strength

The reason for the very high hardness of carbon-containing martensites has long intrigued metallurgists. Cohen in the 1962 Howe Memorial Lecture (Ref 6.5) followed the historical development of the theories of martensite strength in steels, and emphasized the important role that carbon atoms trapped in the octahedral interstitial sites play in the strengthening of martensite. Figure 6.5 shows a schematic representation of the displacement of the iron atoms due to carbon atoms in the body-centered tetragonal lattice of martensite. This

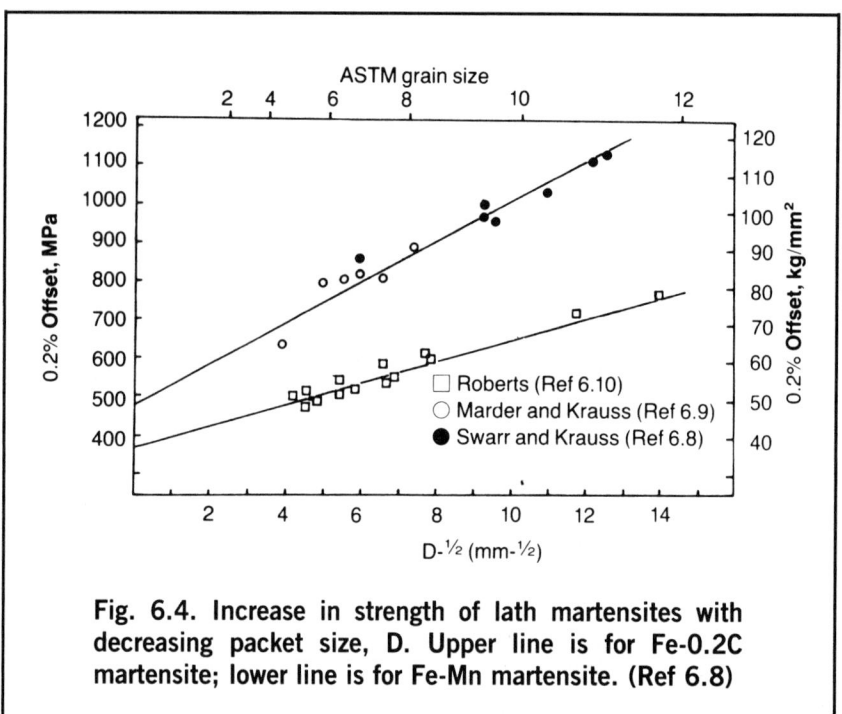

Fig. 6.4. Increase in strength of lath martensites with decreasing packet size, D. Upper line is for Fe-0.2C martensite; lower line is for Fe-Mn martensite. (Ref 6.8)

distortion of the iron lattice makes the movement of dislocations very difficult and is considered to be a major cause of the high strength of martensite.

In addition to the solid solution strengthening by carbon, the substructure of martensite also contributes to strength. Chapter 3 showed that martensitic transformation was unique in that it introduced a high density of dislocations and/or fine twins into a martensite lath or plate. The contribution to the strength of martensite by the substructure is relatively constant as a function of carbon content and, except at low carbon concentrations, does not make nearly as great a contribution as does the carbon solid solution strengthening. The following equation (Ref 6.11, 6.12) for the 0.2% offset yield strength of martensite ($\sigma_{0.2}$), determined from a series of Fe-Ni-C alloys with subzero M_s temperatures, permits a quantitative assessment of the carbon and substructure contributions:

$$\sigma_{0.2} \text{ (MPa)} = 461 + 1.31 \times 10^3 \text{ (w/o C)}^{1/2} \qquad \text{(Eq 6.1)}$$

The second term shows the strong effect of carbon and that the strengthening of martensite follows a square-root dependency with

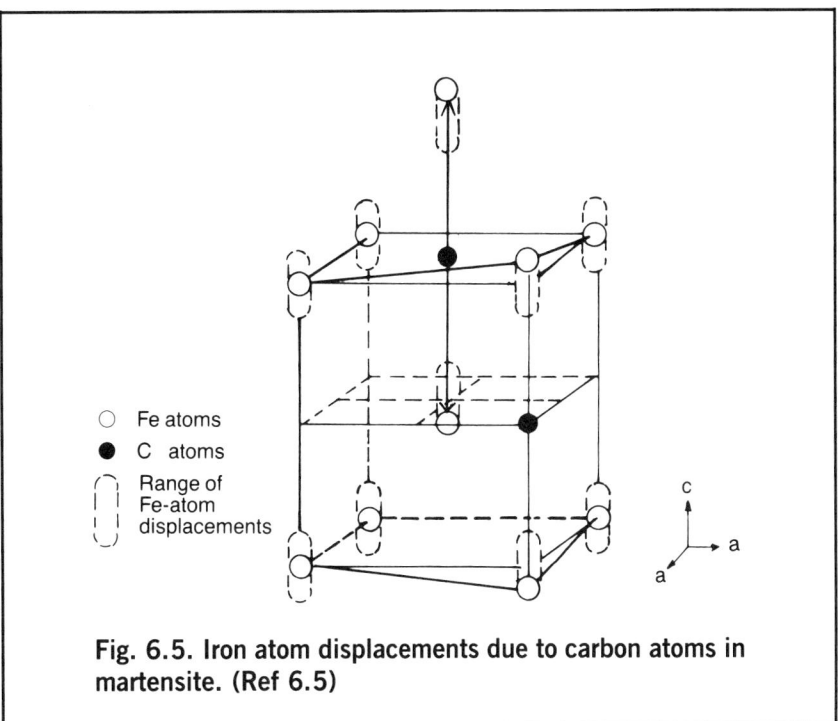

Fig. 6.5. Iron atom displacements due to carbon atoms in martensite. (Ref 6.5)

carbon content, a functional relationship that correlates well with the initial rapid increase in strength with carbon content and the more gradual strength increases at higher carbon contents. The first term includes the strengthening contribution of 20% nickel (20 000 psi or 138 MPa), the friction stress or stress to move dislocations in pure bcc iron (10 000 psi or 69 MPa), and the contribution of the martensitic substructure (37 000 psi or 255 MPa). Equation 6.1 holds for unaged martensite, a result possible because of the low M_s temperatures of the Fe-Ni-C alloys. The martensite, therefore, was formed and mechanically tested at low temperatures where aging was minimal.

Carbon steels, especially those of low carbon content, have high M_s temperatures and undergo considerable carbon atom rearrangement during quenching before reaching room temperature, a process referred to as autotempering. The carbon atoms segregate to the dislocation fine structure and/or lath and packet boundaries (Ref 6.13). One result of the segregation, a very high dependency of strength on packet size, has already been mentioned (Ref 6.8). Despite the effects of carbon atom segregation, the yield strength of low-carbon martensite still follows a square root dependency on carbon content, as shown in the following equation (Ref 6.14):

$$\sigma_{0.2} \text{ (MPa)} = 413 + 1.72 \times 10^3 \text{ (w/o C)}^{1/2} \qquad \text{(Eq 6.2)}$$

This equation was determined from low-carbon Fe-C alloys containing up to 0.2% carbon and also fits data for martensitic steels containing 0.08 to 0.24% carbon and 0.4 to 0.5% manganese (Ref 6.15). Again, the first term includes all of the structural contributions to strength, including the austenitic grain size or packet size (in this case austenite grain size was roughly constant, between ASTM 7 and 9), lathsize, and dislocation fine structure.

Definitions of Hardenability

The above discussion shows that the maximum hardness of any steel is associated with a fully martensitic structure. This microstructure, however, can only be produced if the diffusion-dependent transformation of austenite can be suppressed by sufficiently rapid cooling. There are a number of factors that affect cooling rates throughout a given part and the response of a given steel to those cooling rates. Thus, the formation of martensite and high hardness may vary considerably throughout a given cross section or between identical cross sections fabricated from different steels. The subject of hardenability deals with the latter variations.

Hardenability is defined as the "susceptibility to hardening by rapid cooling" (Ref 6.16), or as "the property, in ferrous alloys, that determines the depth and distribution of hardness produced by quenching" (Ref 6.17). Both of these definitions emphasize hardness. As discussed above, the source of hardening is the formation and presence of martensite, and therefore a third definition of hardenability, "the capacity of a steel to transform partially or completely from austenite to some percentage of martensite at a given depth when cooled under some given conditions," more accurately describes the physical process underlying hardening. Siebert, Doane, and Breen in their comprehensive book on hardenability prefer the latter structural definition (Ref 6.18).

Hardness Distribution

An experimental approach that demonstrates the striking effect of various factors on hardenability is the quenching of series of round bars of various diameters. The bars are completely austenitized, quenched, and tempered. Hardness readings are then taken along diameters of the bar cross sections in order to show the distribution of hardness as a function of distance from the surface to the center of the bar.

Figures 6.6 and 6.7 show the results of water quenching bars of SAE 1045 steel, a plain carbon steel, and SAE 6140 steel, an alloy steel, respectively (Ref 6.16). The chemical compositions of the two steels are given in Table 6.1.

Plain carbon and alloy steels are classified by the Society of Automotive Engineers (SAE) and the American Iron and Steel Institute (AISI) and are manufactured to various ranges of compositions (Ref 6.19). For example, the AISI-SAE specifications for steel designated as 1045 permit carbon in the range of 0.42 to 0.50% and manganese in the range of 0.60 to 0.90%. It is therefore important to state the exact composition of a heat of steel (as in Table 6.1) for the most accurate interpretation of the response to hardening.

Figure 6.6 shows that the maximum hardness in the SAE 1045 steel can be achieved only on the surface of bars with small diameters. Even in a 0.5-in. (12.7-mm) diameter bar, the hardness in the interior drops significantly. With increasing bar diameter, the surface hardness of the SAE 1045 steel drops significantly and the center hardness continues to decrease. The alloy steel, SAE 6140, on the other hand, develops higher hardness than the SAE 1045 steel at all bar diameters (see Fig. 6.7) but nevertheless still shows large variations in hardness from the surface to the center of the bars, especially in the larger sizes.

Figures 6.6 and 6.7 show the effects of bar diameter and alloy content on hardness distribution of water-quenched rounds. A third factor that influences hardness distribution is the rate of quenching. Figures 6.8 and 6.9 show the results of oil quenching on the hardness distribution in round bars of various diameters for the SAE 1045 and 6140 steels, respectively. Oil is a much less severe quenching medium than water, and so the cooling rates of oil-quenched bars are appreciably lower than those of water-quenched bars. Figure 6.8 shows that the hardening response of the SAE 1045 steel to oil quenching is very low. Even in the 0.5-in. (12.7-mm) diameter bar the surface hardness is well below the hardness expected from a fully martensitic structure of a 0.48% carbon steel (see Fig. 6.1 and 6.2). It is apparent, therefore, that the slower cooling associated with oil quenching has not been able to prevent the diffusion-controlled transformation to ferrite and/or pearlite in the SAE 1045 steel. The SAE 6140 steel, however, hardens well in the same bar sizes (see Fig. 6.9) and only in the larger sizes does the hardness distribution fall off significantly.

Comparison of Fig. 6.6 through 6.9 shows that the alloy steel, SAE 6140, is much more hardenable than the plain carbon SAE 1045

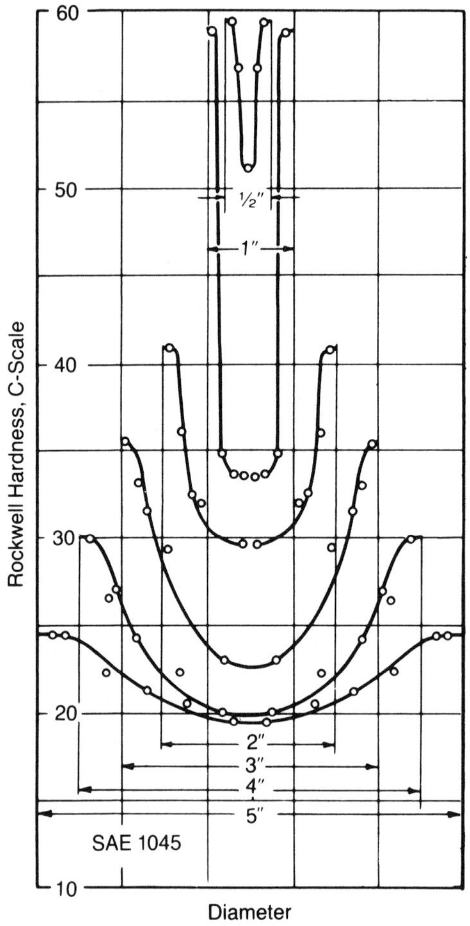

Fig. 6.6. Hardness distributions in water-quenched bars of SAE 1045 steel. The various bar diameters are indicated. (Ref 6.16)

steel. SAE 6140 is therefore said to have a higher hardenability than the SAE 1045 steel. The plain carbon steel can be hardened but only in small sections and/or with very severe quenches. Fundamentally, the alloying elements in the SAE 6140 steel increase the time required for austenite to decompose to ferrite and/or ferrite-cementite mixtures, and thereby make it possible to form martensite at lower cooling rates. The effects of alloying elements on the diffusion-controlled decomposition of austenite in many steels are summarized

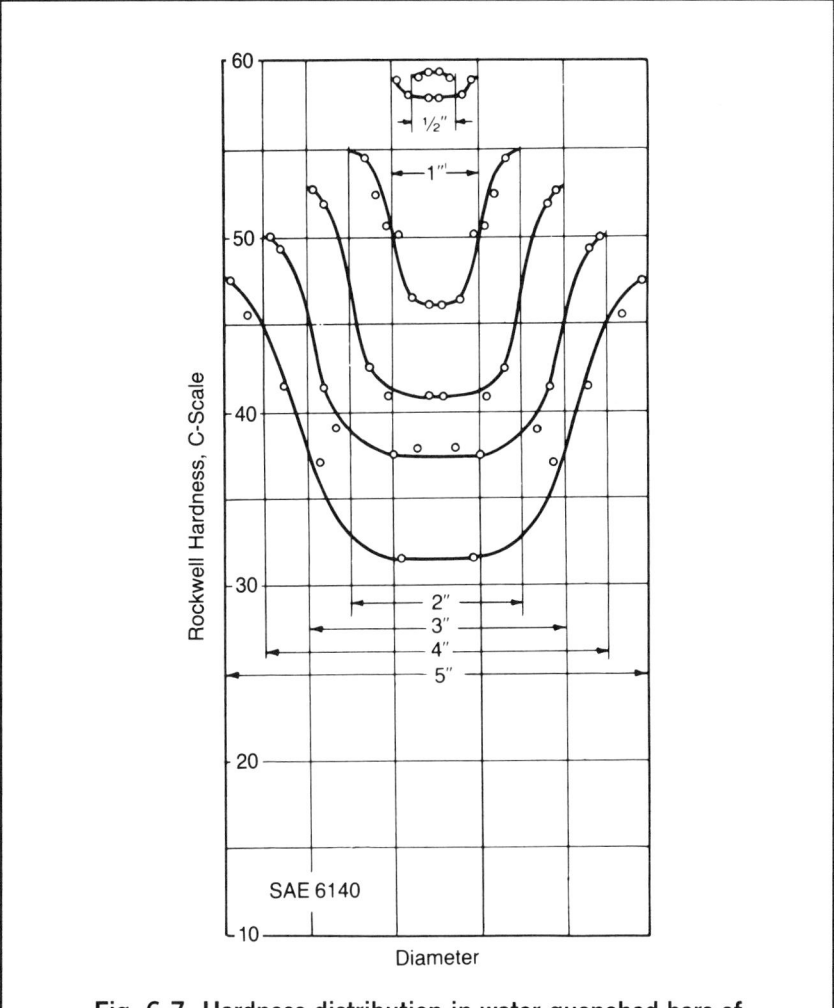

Fig. 6.7. Hardness distribution in water-quenched bars of SAE 6140 steel. The various bar diameters are indicated. (Ref 6.16)

Table 6.1. Compositions of Steels Used in Bar Quenching Experiments

Steel	Composition, %						
	C	Mn	P	S	Si	Cr	V
SAE 1045	0.48	0.60	0.022	0.016	0.17	..	..
SAE 6140	0.42	0.73	0.027	0.023	0.25	0.94	0.17

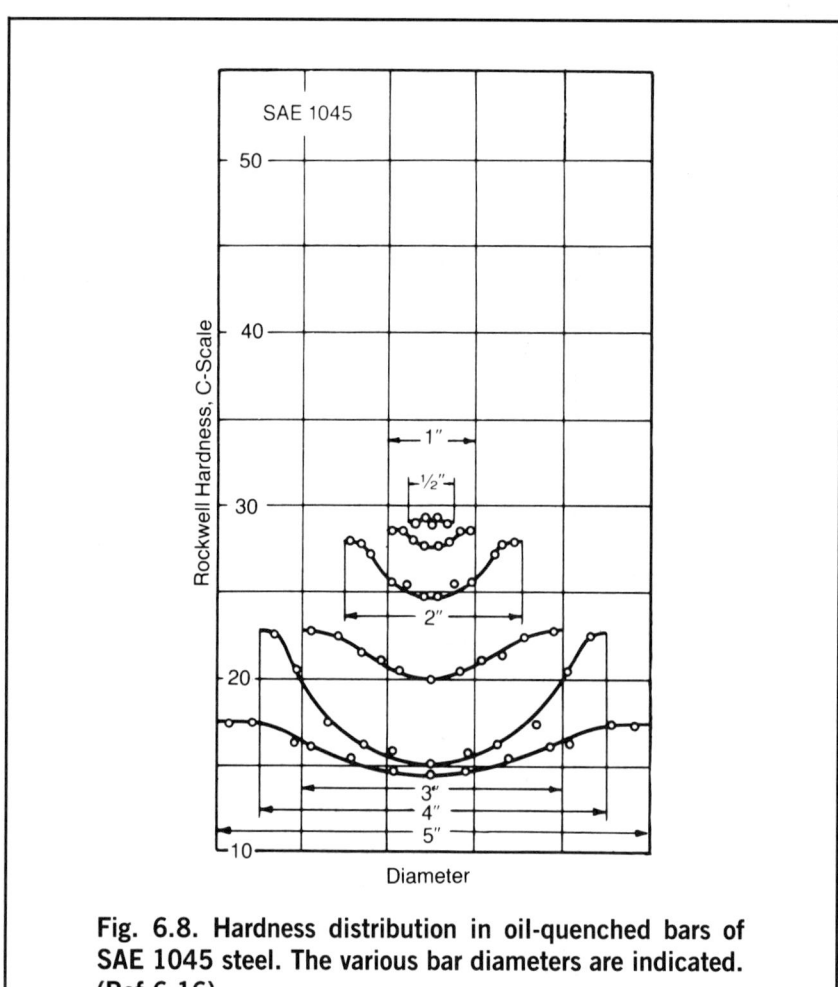

Fig. 6.8. Hardness distribution in oil-quenched bars of SAE 1045 steel. The various bar diameters are indicated. (Ref 6.16)

in the IT and CT diagrams contained in the atlases described in Chapter 4.

Factors Affecting Cooling Rates

Two important factors influence cooling rates or the rates at which heat can be removed from a steel part. One is the ability of the heat to diffuse from the interior to the surface of the steel specimen, and the other is the ability of the quenching medium to remove heat from the surface of the part.

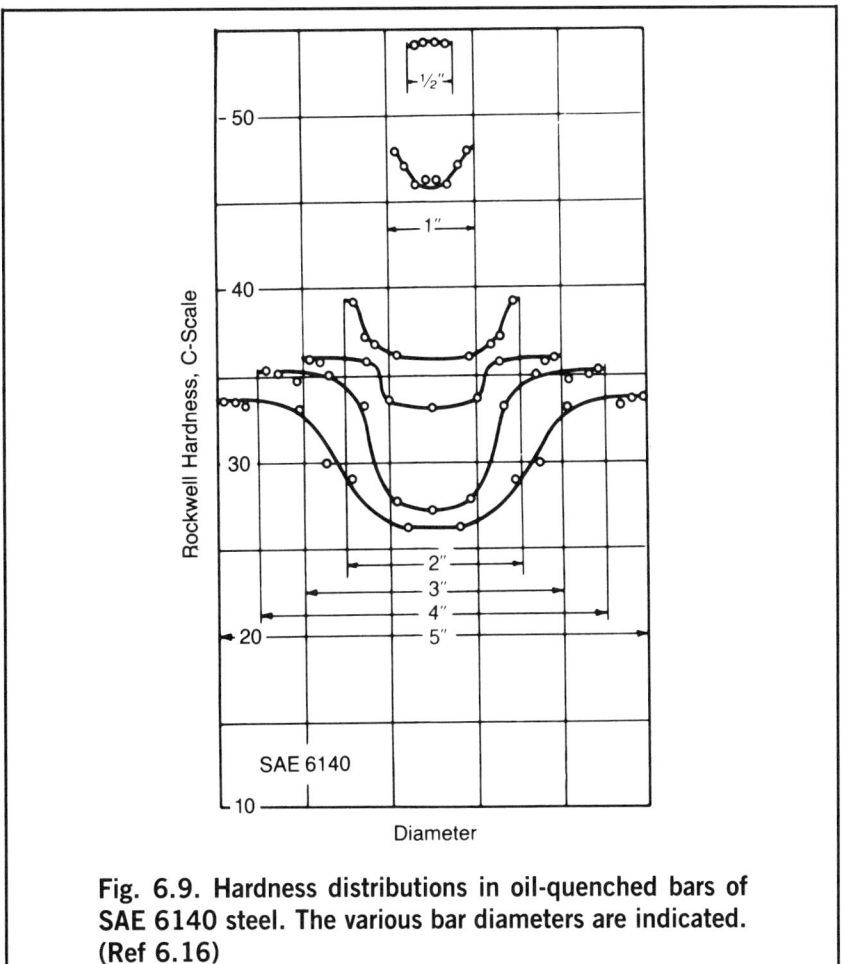

Fig. 6.9. Hardness distributions in oil-quenched bars of SAE 6140 steel. The various bar diameters are indicated. (Ref 6.16)

The ability of a steel to transfer heat is characterized by its thermal diffusivity (units of area per unit time) or the ratio of its thermal conductivity to the volume specific heat. The thermal diffusivity of austenitic transformation products increases with decreasing temperature, and plots of thermal diffusivity and conductivity for various structures as a function of temperature are reproduced in Ref 6.18. For a given quenching medium, the thermal diffusivity determines the temperature distribution as a function of position at any given time in the quenching process. For example, Fig. 6.10 shows cooling rates as a function of position in a quenched 1-in. diameter bar. The slower cooling rates at positions removed from the surface of the bar permit more time for diffusion-controlled transformations, and it is

158 / STEELS: HEAT TREATMENT AND PROCESSING PRINCIPLES

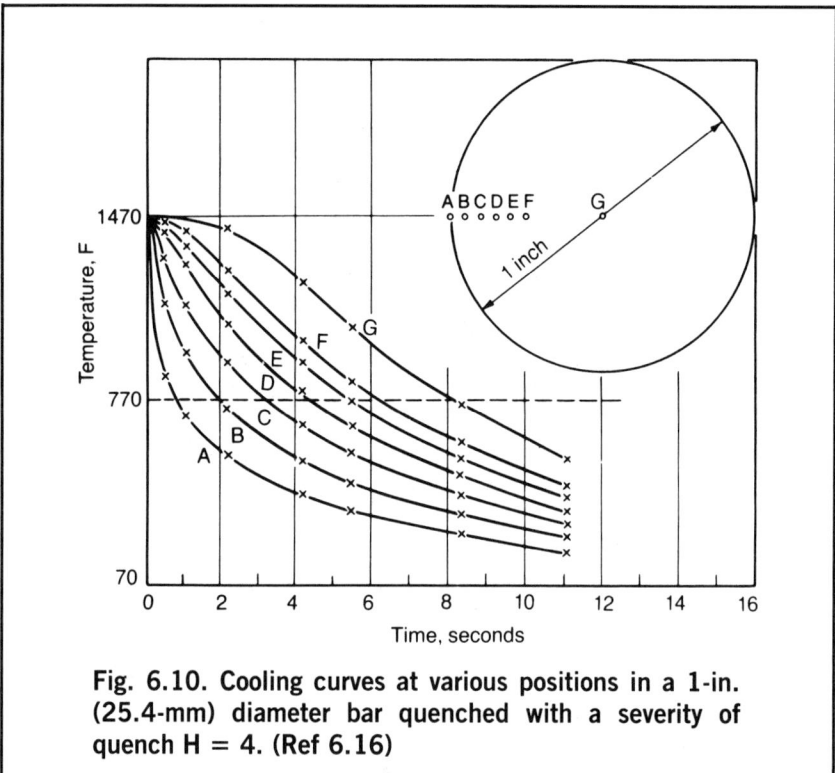

Fig. 6.10. Cooling curves at various positions in a 1-in. (25.4-mm) diameter bar quenched with a severity of quench H = 4. (Ref 6.16)

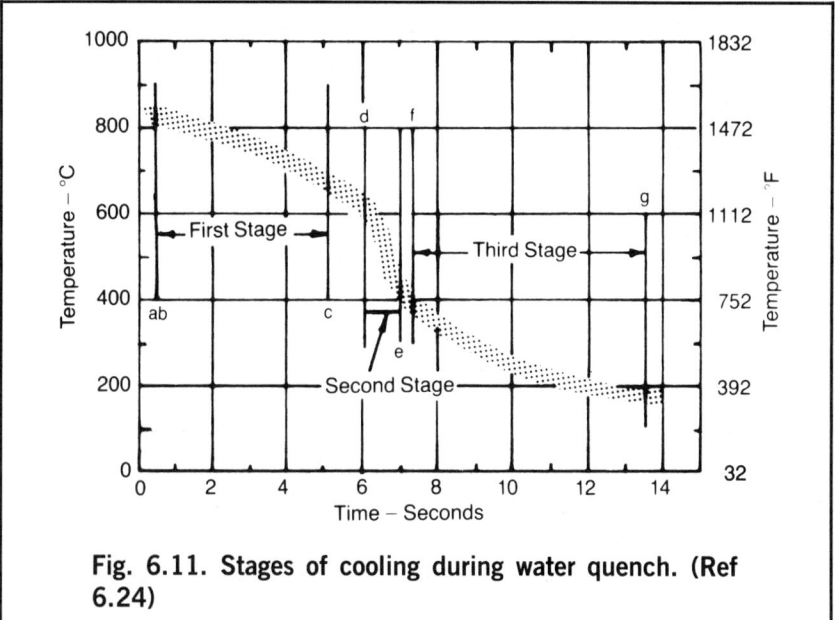

Fig. 6.11. Stages of cooling during water quench. (Ref 6.24)

this type of cooling behavior that results in the low center hardness of the bars shown in Fig. 6.6 through 6.9, especially in the larger sizes. Practically, however, there is little control of thermal properties possible in steels, and the most important control of cooling rates is performed by proper selection of quenching media.

The transfer of heat at the interface of a steel part and a quenching medium is a complex process that depends primarily on the emissivity of the steel (or the rate at which the surface of the steel radiates heat) and convection currents within the quenching medium that remove heat from the interface. The complexity of the process is illustrated in Fig. 6.11, a curve obtained by measuring temperature as a function of time in the center of a 0.5-in. (12.7-mm) diameter bar of steel during water quenching (Ref 6.20). Three stages of cooling are shown. The first stage is associated with the development of a layer of water vapor or steam immediately adjacent to the surface of the steel. The steam insulates the surface and produces a low cooling rate. In the second stage the vapor blanket breaks down and water comes in contact with the steel. The water vaporizes, but bubbles away, thereby continually bringing more water in contact with the surface. Cooling is quite rapid in this stage. When the surface temperature of the steel drops below the boiling point, vaporization stops and cooling is controlled by convection and conduction at the fluid-metal interface. The latter or third stage is characterized by relatively slow rates of cooling.

Understanding the cooling process has important practical consequences. For example, if the low cooling rate of the first stage results in ferrite or pearlite, efforts should be made to increase the cooling rate in this stage. Agitation of the part or the quenchant or the use of brine solutions for quenching are effective in reducing the duration of the first stage of cooling.

Severity of Quench

The effectiveness of a given quenching medium is ranked by a parameter referred to as its "severity of quench". This measure of cooling or quenching power is identified by the letter "H", and is determined experimentally by quenching a series of round bars of a given steel. Figure 6.12 shows schematically the results of oil and water quenching bars of SAE 3140 steel (Ref 6.16). SAE 3140 is a nickel-chromium alloy steel containing nominally 0.40% carbon. The crosshatched areas represent the unhardened areas of the various bars, assuming that less than 50% martensite represents an unhardened microstructure. The larger the bar diameter (D), the greater the

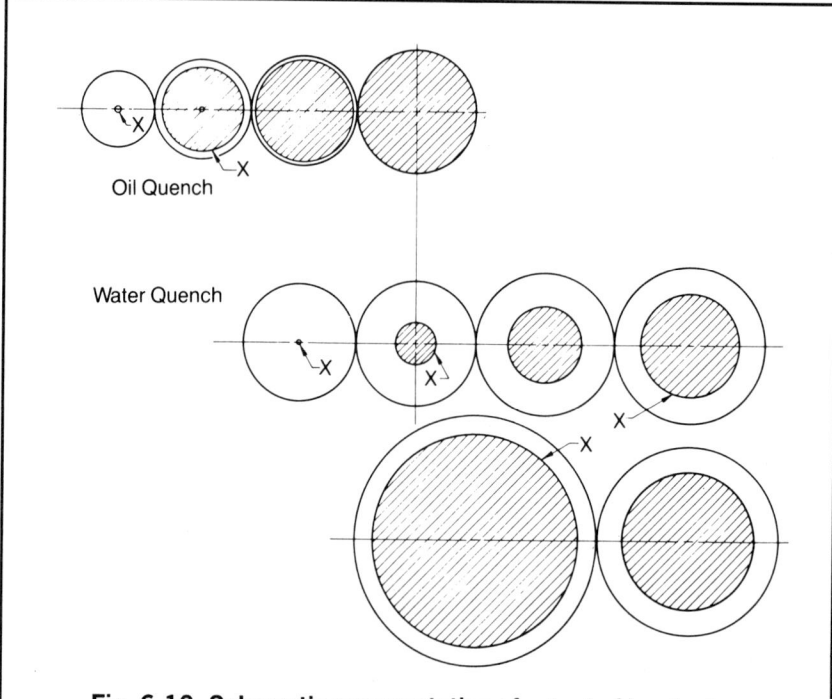

Fig. 6.12. Schematic representation of extent of hardening in oil-quenched and water-quenched bars of SAE 3140 steel of various diameters. The cross-hatched areas represent the unhardened core. (Ref 6.16)

unhardened diameter (D_u). Figure 6.13 plots the results of Fig. 6.12 as D_u/D versus D for both the oil- and water-quenched series. The steeper curve is associated with the oil quench, a result of the reduced ability of oil quenching to produce hardening in heavier sections. When the curves of Fig. 6.13 are matched to one of the large number of calculated curves that are characteristic of a wide range of quench severities (see Fig. 6.14), H can be determined. The matching can be performed by plotting D_u/D versus D as in Fig. 6.13 on transparent paper and finding the best correspondence to a D_u/D versus HD curve in Fig. 6.14. When the HD values are divided by corresponding D values, the H value is obtained. For example, point A in Fig. 6.13, corresponding to a bar diameter of 1.83 in. (4.65 cm) would fall on an HD value of 2.6 in Fig. 6.14 when the curves are matched. Then H = 2.6/1.83 = 1.4 for the water quench of this example.

Table 6.2 lists H values for a number of commonly used quenches. The increase in severity of quench from air, H = 0.02, through brine

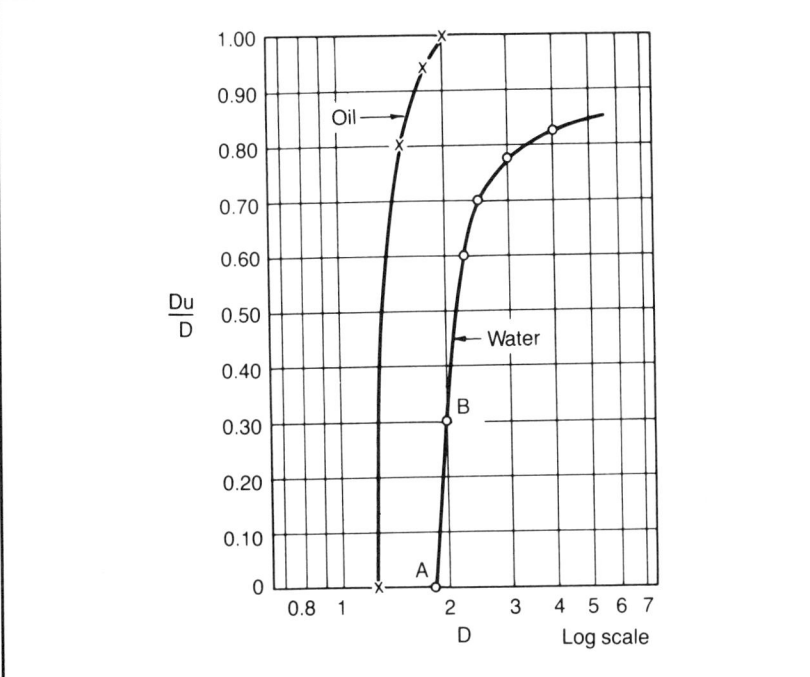

Fig. 6.13. Ratio of unhardened to hardened diameters as a function of bar diameter for oil- and water-quenched bars of 3140 steel. (Ref 6.16)

quenching, H = 2, is shown. Also, the very strong effect of agitation or circulation on increasing the severity of quench in any given quenching medium is apparent. Another useful ranking of quenching media relative to water is shown in Table 6.3. This table not only ranks the

Table 6.2. Severity of Quench (H) for Various Quenching Media (Ref 6.16, 6.21)

	Air	Oil	Water	Brine
No circulation of fluid or agitation of piece	0.02	0.25 to 0.30	0.9 to 1.0	2
Mild circulation (or agitation)	...	0.30 to 0.35	1.0 to 1.1	2 to 2.2
Moderate circulation	...	0.35 to 0.40	1.2 to 1.3	...
Good circulation	...	0.4 to 0.5	1.4 to 1.5	...
Strong circulation	0.05	0.5 to 0.8	1.6 to 2.0	...
Violent circulation	...	0.8 to 1.1	4	5

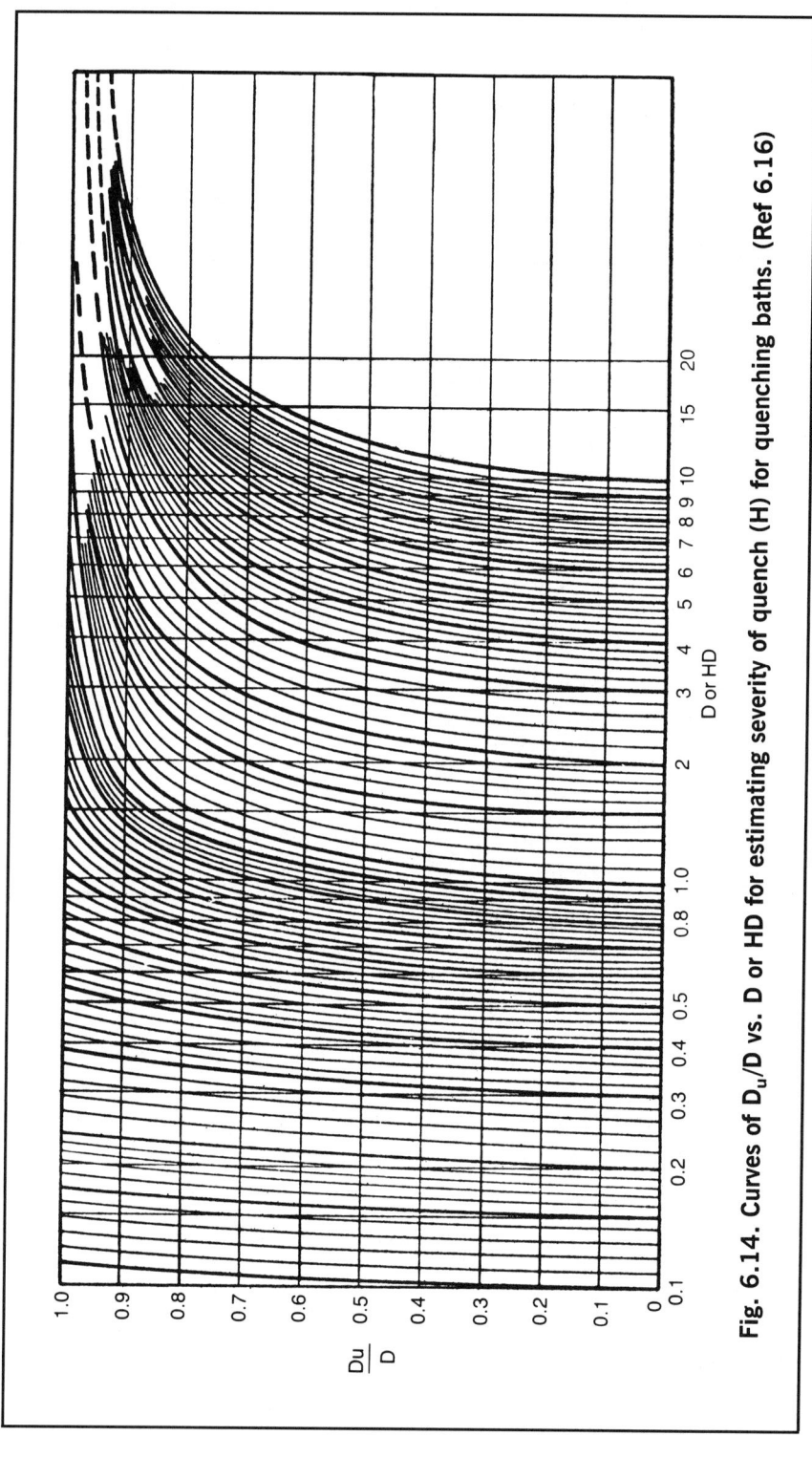

Fig. 6.14. Curves of D_u/D vs. D or HD for estimating severity of quench (H) for quenching baths. (Ref 6.16)

Table 6.3. Relative Cooling Rates in Different Quenching Media (Ref 6.16, 6.22)

Quenching medium	Cooling rate (a) from 717 to 550 °C (1328 to 1022 °F) relative to that for water at 18 °C (65 °F)	Quenching medium	Cooling rate (a) from 717 to 550 °C (1328 to 1022 °F) relative to that for water at 18 °C (65 °F)
Aqueous solution, 10% LiCl	2.07	Oil 20204	0.20
Aqueous solution, 10% NaOH	2.06	Oil, Lupex light	0.18
Aqueous solution, 10% NaCl	1.96	Water at 122 F	0.17
Aqueous solution, 10% Na_2CO_3	1.38	Oil 25441	0.16
Aqueous solution, 10% H_2SO_4	1.22	Oil 14530	0.14
Water at 32 F	1.06	Emulsion of 10% oil in water	0.11
Water at 65 F	1.00	Copper plates	0.10
Aqueous solution, 10% H_3PO_4	0.99	Soap water	0.077
Mercury	0.78	Iron plates	0.061
$Sn_{30}Cd_{70}$ at 356 F	0.77	Carbon tetrachloride	0.055
Water at 77 F	0.72	Hydrogen	0.050
Rape seed oil	0.30	Water at 166 F	0.047
Trial oil No. 6	0.27	Water at 212 F	0.044
Oil P20	0.23	Liquid air	0.039
Oil 12455	0.22	Air	0.028
Glycerin	0.20	Vacuum	0.011

(a) Determined by quenching a 4-mm nichrome ball, which when quenched from 860 °C (1580 °F) into water at 18 °C (65 °F) cooled at the rate of 1810 °C (3260 °F) per second over the range 717 to 550 °C (1328 to 1022 °F). This cooling rate in water at 18 °C (65 °F) is rated as 1.00 in the table, and the rates in the other media are compared with it. (Ref 6.22)

cooling media, but also shows the large number of media available for cooling at various rates.

Quantitative Hardenability

Up to this point a number of important aspects of hardenability have been described. High hardness is related to martensite formation, which in turn is dependent upon cooling rate. Cooling rate is affected by both the specimen size and severity of quench. However, there still remain the questions as to how hardenability is evaluated as a function of steel composition and how the effect of the large number of quenching media on hardness distribution can be evaluated without the time-consuming approach of quenching a series of round bars in the various quenching media. The first and now classical approach to these questions, described below, was developed in the 1930s and 1940s by Grossmann and Bain (Ref 6.16) and their many colleagues and contemporaries.

The Grossmann and Bain approach to hardenability is based on the definition of two parameters: the critical size and the ideal size. The critical size is the largest size of a bar quenched in a given medium which contains no unhardened core after quenching. An important aspect of this definition is that the hardness that separates the hardened from the unhardened core of a bar is associated with a microstructure assumed to contain only 50% martensite. This assumption underlies all of the graphical information associated with the Grossmann-Bain approach to hardenability. The reason for selecting the 50% martensite criterion for the critical diameter is shown in Fig. 6.15. Etching differences between the hardened surface of a bar and the unhardened center are most clearly developed close to the 50% martensite-50% pearlite zone in a bar. Likewise, when a quenched bar is broken, the same 50% martensite zone correlates well with a transition from very smooth or faceted intergranular fracture (now known to be related to austenitizing and the presence of impurities such as phosphorus) associated with a predominantly martensitic structure to a rough, transgranular surface associated with ductile fracture of the softer nonmartensitic transformation products of austenite. Therefore, both etching and fracture observations, frequently on a macroscopic scale, could be readily used to evaluate depth of hardening at the 50% martensite level. Detection of martensite at levels above 50% in the microstructure would be much more difficult.

Not only does the etching and fracture response of a quenched bar change abruptly at the 50% martensite level, but hardness also changes rapidly as bar diameter increases through those associated with 50% martensite. Figure 6.16 shows center hardness as function of bar diameter for the chromium-nickel SAE 3140 steel quenched in oil and water. Each quenching medium produces a different critical diameter associated with the rapid changes in hardness with bar diameter close to Rockwell C 50. Judging the position of 50% martensite from hardness changes with bar diameter can be difficult, as for example in the water quenched data of Fig. 6.16. Therefore, the probable hardness associated with a 50% martensite structure, similar to those given for fully martensitic microstructures in Fig. 6.1 and 6.2, was determined as a function of carbon content. Figure 6.17 shows such a plot based on data from plain carbon steels. Alloy steels are expected to show somewhat higher hardness values, as is the case for the SAE 3140 steel. From Fig. 6.17 the hardness at 50% martensite for a 0.40% carbon steel would be expected to be Rockwell C 40, but Fig.

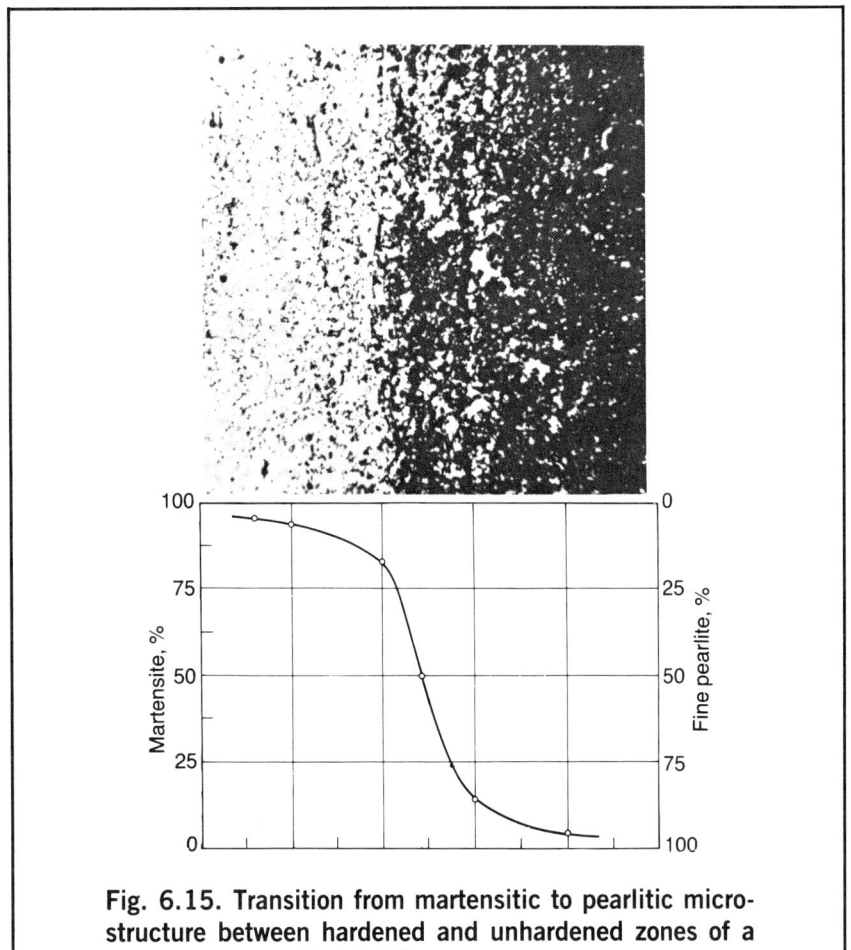

Fig. 6.15. Transition from martensitic to pearlitic microstructure between hardened and unhardened zones of a quenched steel. (Ref 6.16)

6.16 shows that the critical diameter was selected at Rockwell C 50. A possible explanation for this discrepancy may be the presence of large amounts of bainite having relatively high hardness together with 50% martensite in alloy steels, whereas ferrite and pearlite of relatively lower hardness might coexist with 50% martensite in plain carbon steels.

In summary of the above discussion, the critical size or diameter of a steel of given composition is directly related to a given quenching medium. The higher the quench severity the greater the critical size. The ideal size, on the other hand, is defined as the size of bar hardened to 50% martensite by a theoretically perfect quench in which it is assumed the surface of the bar cools instantly to the temperature of the quenching medium. The ideal size is a true measure of the hardenabil-

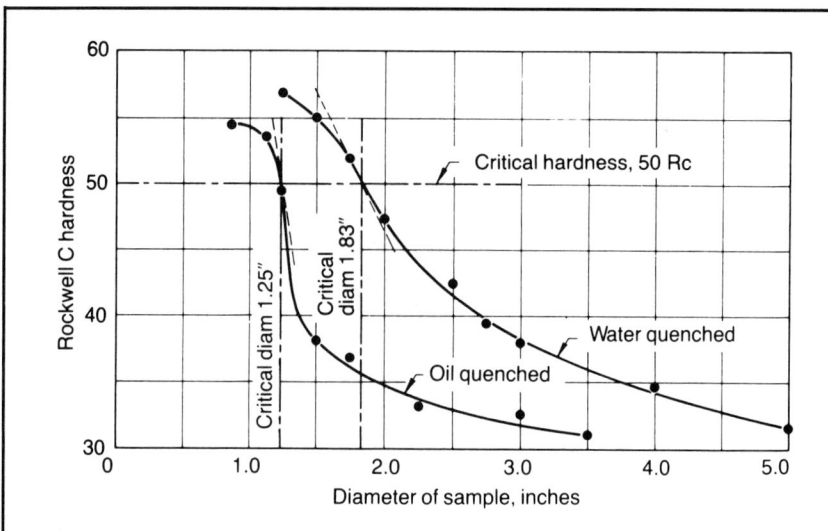

Fig. 6.16. Hardness at the center of water- and oil-quenched bars of SAE 3140 steel of various diameters. (Ref 6.16)

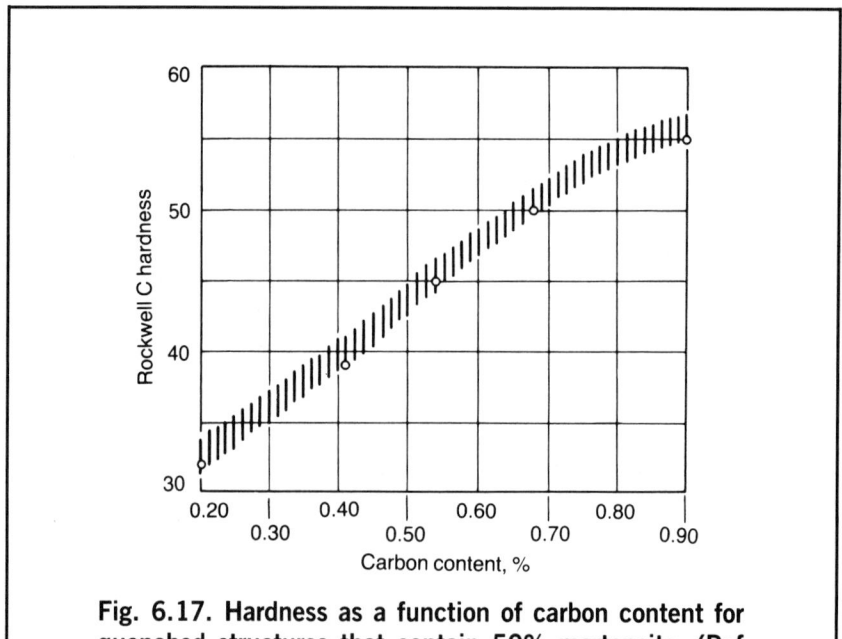

Fig. 6.17. Hardness as a function of carbon content for quenched structures that contain 50% martensite. (Ref 6.16)

Table 6.4. Composition and Multipliers for a Ni-Cr Steel

	Carbon	Manganese	Element Silicon	Nickel	Chromium
Concentration....	0.50%	0.80%	0.25%	1.00%	0.28%
Multiplier	0.24	3.7	1.2	1.4	1.6

ity associated with a given steel composition, and it can also be used to determine the critical size of the steels quenched in media of different quench severities.

Figures 6.18 and 6.19 show plots of critical size (D) versus ideal size (D_I) for various quench severities (H). The straight line identified by a quench severity of infinity shows that the critical size equals the ideal size for a theoretically perfect quench. However, as quench severity decreases, Fig. 6.18 and 6.19 show that the critical size for a given D_I decreases. Thus the concept of the ideal size permits a rapid estimate of the bar size that will harden to the 50% martensite level in quenches over the entire range of severities. Similar curves between critical plate thickness, ideal plate thickness, and quench severity have also been developed (Ref 6.16).

Determination of Ideal Size

As noted in the above section, the ideal diameter is a true measure of the hardenability of a steel and can be used to compare the hardening response of different steels to the same quenching medium. Three factors, austenitic grain size, carbon content, and alloy content, affect the ideal diameter. Fundamentally, an increase in any of these factors reduces the rate at which the diffusion-controlled transformations of austenite occur and thereby makes martensite formation more likely at a given cooling rate.

Figure 6.20 shows the relationship of ideal diameter to carbon content and austenite grain size. This plot is used to establish a base hardenability, D_I, for a steel based on its carbon content and grain size. The base hardenability is then multiplied by factors as given in Fig. 6.21 for the various concentrations of alloying elements. As an example, Table 6.4 shows multiplying factors for concentrations of elements in a nickel-chromium steel containing 0.5% carbon (Ref 6.16). If the steel has an austenitic grain size of No. 7, then the base ideal diameter from Fig. 6.20 is 0.24 in. (6.1 mm). After multiplying by the factors in Table 6.4, an ideal diameter of 2.4 in. (61 mm) is obtained for the steel. The multiplying

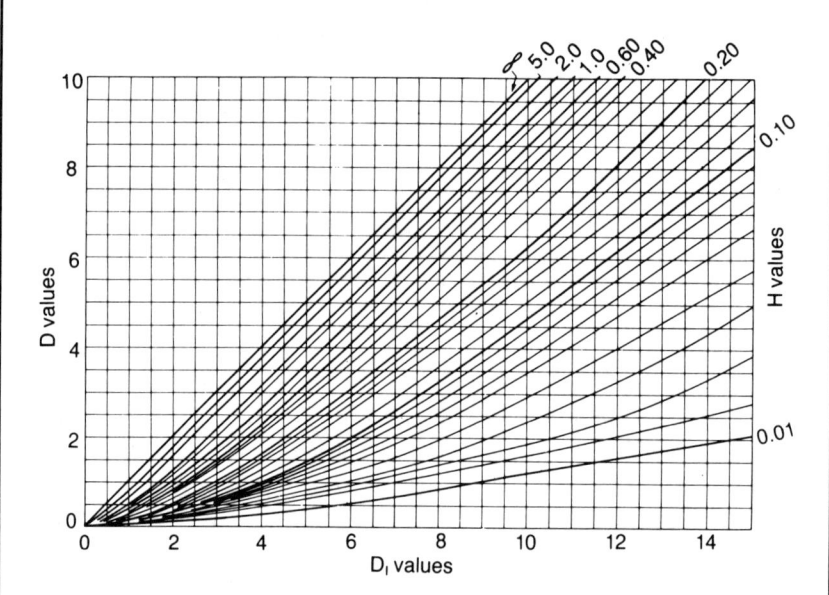

Fig. 6.18. Relationship between actual critical size (D), ideal critical size (D_I), and severity of quench (H). (Ref 6.16)

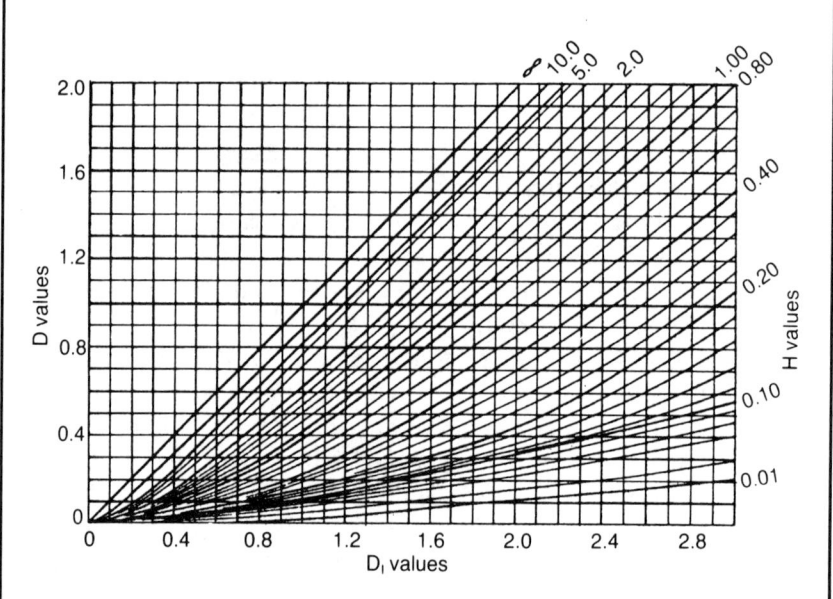

Fig. 6.19. Relationships similar to those shown in Fig. 6.18 but at a larger scale. (Ref 6.16)

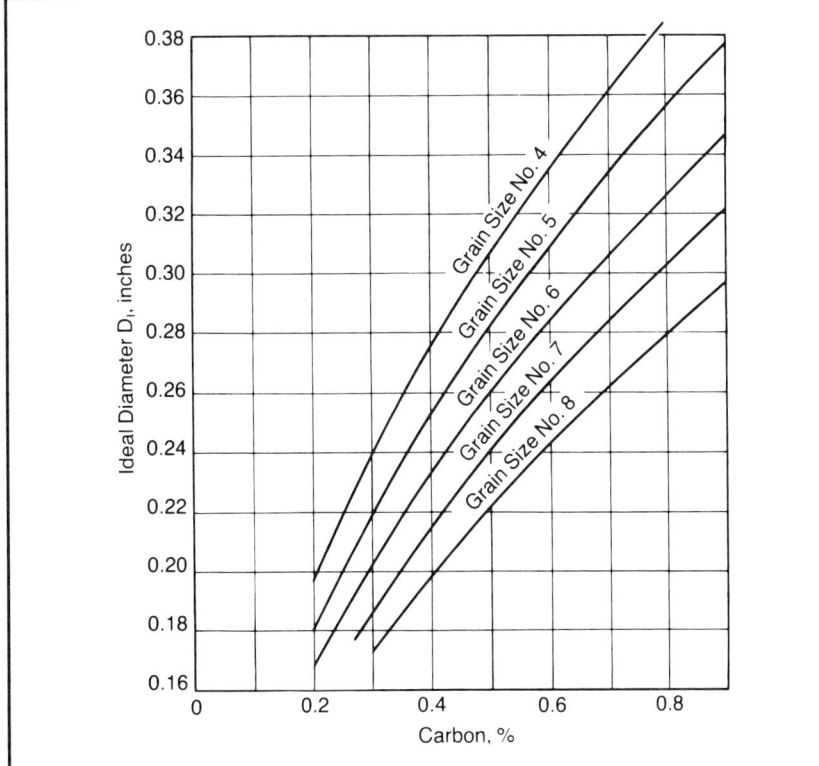

Fig. 6.20. Hardenability, expressed as ideal critical size, as a function of austenite grain size and carbon content of iron-carbon alloys. (Ref 6.16)

factors have been reviewed and revised over the years, and the reader is referred to Ref 6.18 for a recent complete compilation of multiplying factors for the common alloying elements.

Table 6.5 lists ranges of D_I for a number of commercial steels. Compositions of these steels are given in Ref 6.19. The letter H at the end of the SAE-AISI designation indicates that the steels are produced to specified hardenability limits. The range in D_I for a given steel is a result of the acceptable ranges of composition for that grade and other factors such as grain size and the concentrations of residual elements.

Jominy Test for Hardenability

Another important approach to the evaluation of hardenability is the use of the end-quench test developed by Jominy and Boegehold (Ref

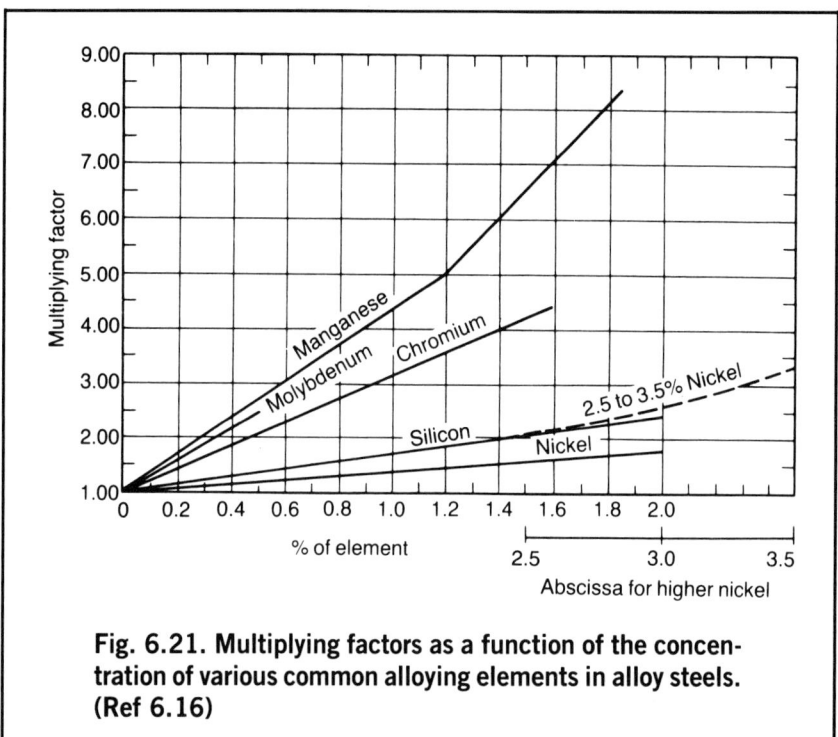

Fig. 6.21. Multiplying factors as a function of the concentration of various common alloying elements in alloy steels. (Ref 6.16)

6.23). The test is now commonly referred to as the Jominy test, and has the great advantage of characterizing the hardenability of a given steel from a single specimen rather than from a series of round bars.

Figure 6.22 shows the shape and dimensions of a Jominy specimen and the fixture for supporting the specimen in a quenching system. The specimen is cooled at one end by a column of water; thus the entire specimen experiences a range of cooling rates between those associated with water and air cooling. After quenching, parallel flats are ground on opposite sides of the specimen, and hardness readings are taken every 1/16 in. from the quenched end, and plotted as shown in Fig. 6.23. Hardenability differences between different grades of steels can be readily compared if Jominy curves are available. For example, Fig. 6.24 shows hardenability differences between different grades of alloy steels containing 0.5% carbon. Higher hardness persists to greater distances from the quenched end in the more hardenable steels.

The Jominy test method is now standardized in specifications of the American Society for Testing and Materials (ASTM Method A 255) and the Society of Automotive Engineers (SAE Standard J406). Figure 6.25 shows the method of presentation of the end-quench data for a single heat of AISI 8650 steel (Ref 6.25). For any grade of steel a

HARDNESS AND HARDENABILITY / 171

Table 6.5. Hardenabilities (Stated as a Range of D_I Values) for Various Steels (Ref 6.16)

Steel	D_I	Steel	D_I	Steel	D_I
1045	0.9 to 1.3	4135 H	2.5 to 3.3	8625 H	1.6 to 2.4
1090	1.2 to 1.6	4140 H	3.1 to 4.7	8627 H	1.7 to 2.7
1320 H	1.4 to 2.5	4317 H	1.7 to 2.4	8630 H	2.1 to 2.8
1330 H	1.9 to 2.7	4320 H	1.8 to 2.6	8632 H	2.2 to 2.9
1335 H	2.0 to 2.8	4340 H	4.6 to 6.0	8635 H	2.4 to 3.4
1340 H	2.3 to 3.2	X4620 H	1.4 to 2.2	8637 H	2.6 to 3.6
2330 H	2.3 to 3.2	4620 H	1.5 to 2.2	8640 H	2.7 to 3.7
2345	2.5 to 3.2	4621 H	1.9 to 2.6	8641 H	2.7 to 3.7
2512 H	1.5 to 2.5	4640 H	2.6 to 3.4	8642 H	2.8 to 3.9
2515 H	1.8 to 2.9	4812 H	1.7 to 2.7	8645 H	3.1 to 4.1
2517 H	2.0 to 3.0	4815 H	1.8 to 2.8	8647 H	3.0 to 4.1
3120 H	1.5 to 2.3	4817 H	2.2 to 2.9	8650 H	3.3 to 4.5
3130 H	2.0 to 2.8	4820 H	2.2 to 3.2	8720 H	1.8 to 2.4
3135 H	2.2 to 3.1	5120 H	1.2 to 1.9	8735 H	2.7 to 3.6
3140 H	2.6 to 3.4	5130 H	2.1 to 2.9	8740 H	2.7 to 3.7
3340	8.0 to 10.0	5132 H	2.2 to 2.9	8742 H	3.0 to 4.0
4032 H	1.6 to 2.2	5135 H	2.2 to 2.9	8745 H	3.2 to 4.3
4037 H	1.7 to 2.4	5140 H	2.2 to 3.1	8747 H	3.5 to 4.6
4042 H	1.7 to 2.4	5145 H	2.3 to 3.5	8750 H	3.8 to 4.9
4047 H	1.8 to 2.7	5150 H	2.5 to 3.7	9260 H	2.0 to 3.3
4047 H	1.7 to 2.4	5152 H	3.3 to 4.7	9261 H	2.6 to 3.7
4053 H	2.1 to 2.9	5160 H	2.8 to 4.0	9262 H	2.8 to 4.2
4063 H	2.2 to 3.5	6150 H	2.8 to 3.9	9437 H	2.4 to 3.7
4068 H	2.3 to 3.6	8617 H	1.3 to 2.3	9440 H	2.4 to 3.8
4130 H	1.8 to 2.6	8620 H	1.6 to 2.3	9442 H	2.8 to 4.2
4132 H	1.8 to 2.5	8622 H	1.6 to 2.3	9445 H	2.8 to 4.4

hardenability band (see Fig. 6.26) develops because of the small variations in composition allowable in the grade. The SAE/AISI steels designated by the letter H (H-steels) are guaranteed to meet established hardenabilities.

A very important feature of the Jominy test is that each position of the specimen corresponds to a well-known cooling rate. The top scale of Fig. 6.25 shows approximate cooling rates corresponding to positions on the Jominy specimen. As developed previously, it is the cooling rate that determines the amount of martensite, and therefore the degree of hardness, that develops at a given point in a steel specimen. Therefore, if cooling rates as a function of position in parts of various geometries are known, it is possible to use Jominy curves to plot hardness profiles in the parts. Such correlations of cooling rate as a function of position in various sizes of bars and plates quenched in various media are available (Ref 6.25). Figure 6.27 shows equivalent cooling rates for

172 / STEELS: HEAT TREATMENT AND PROCESSING PRINCIPLES

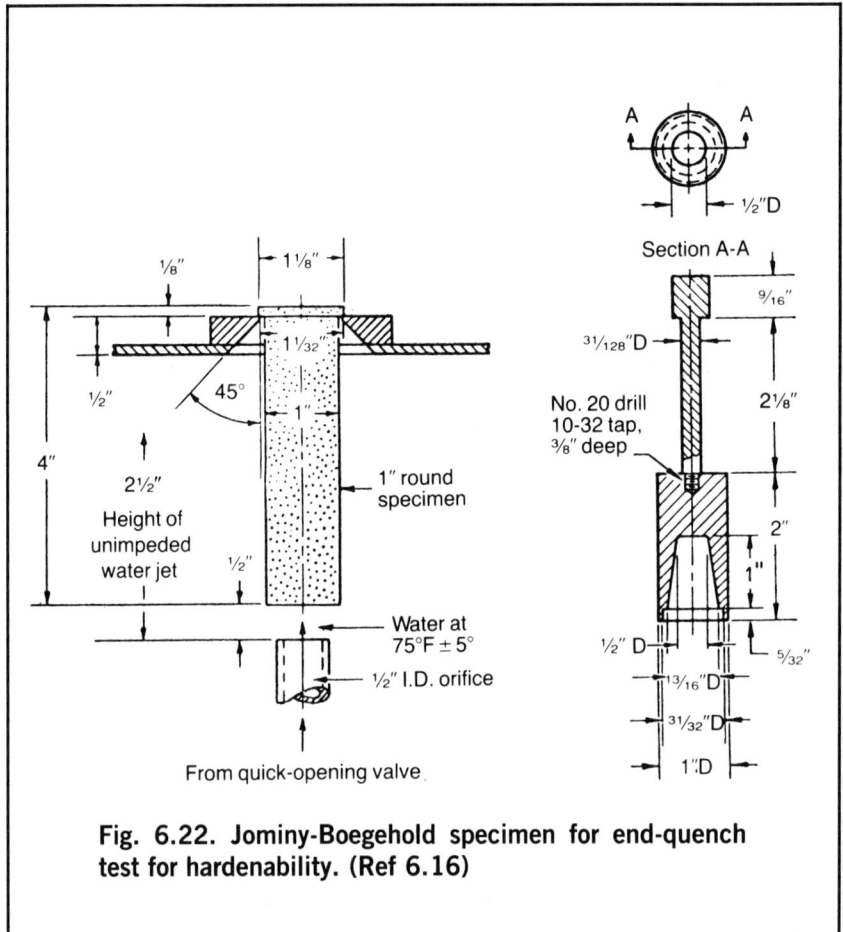

Fig. 6.22. Jominy-Boegehold specimen for end-quench test for hardenability. (Ref 6.16)

four positions in round bars quenched in water and oil. As bar diameter increases, the cooling rates at the surface and interior points decrease (see top scale of Fig. 6.27). The cooling rates correspond to equivalent distances from the quenched end (see bottom scale of Fig. 6.27), and those distances can be used to determine the hardness distribution in the rounds from appropriate Jominy curves.

The use of the Jominy data as described above is a highly accurate method of selecting steels of just the right hardenability for a given required hardness distribution. A steel can be selected that will not only satisfy the hardness requirements but also have just the right alloy content, therefore permitting selection at minimum cost from the many steels that might have sufficient or even excess hardenability for the application. On the other hand, alloy steels that can be hardened

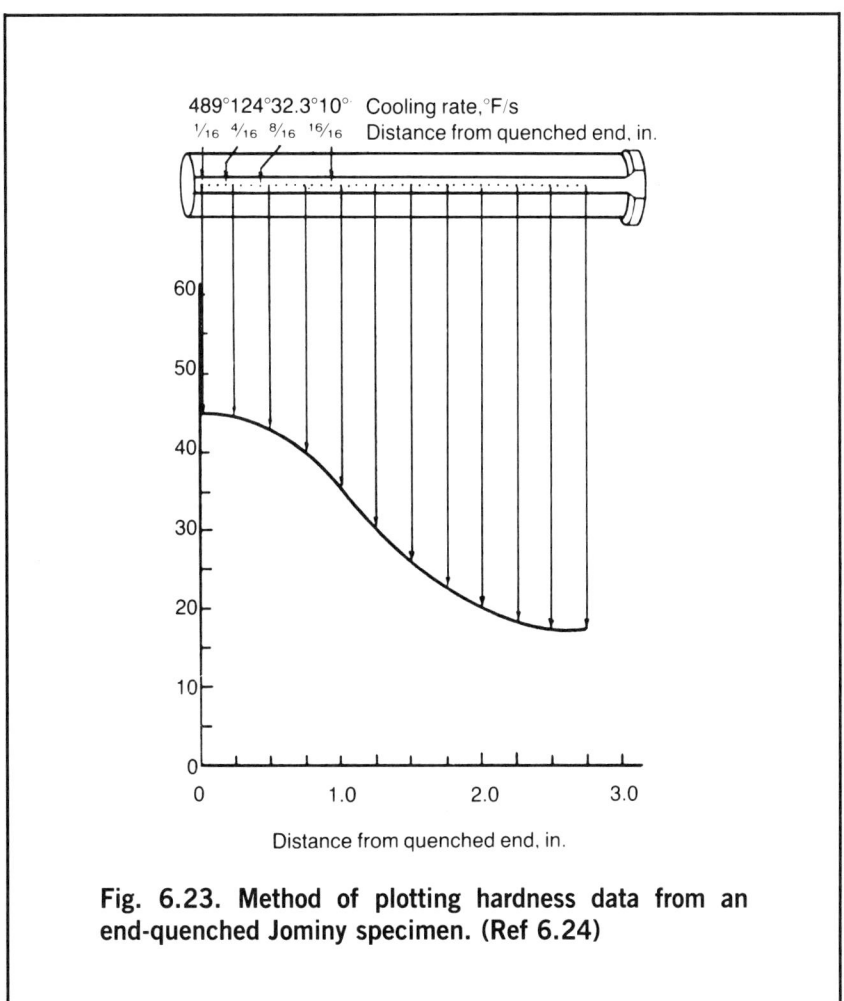

Fig. 6.23. Method of plotting hardness data from an end-quenched Jominy specimen. (Ref 6.24)

by moderate quenching may be selected to replace leaner steels in which the severe quenching required to obtain high hardness causes quench cracking.

Recent Developments

The technology associated with hardenability is continually developing. A measure of this activity is the recent publication of two volumes, *Hardenability Concepts with Applications to Steel* (Ref 6.26) and *The Hardenability of Steels—Concepts, Metallurgical Influences and Industrial Applications* (Ref 6.18). A detailed review of these two

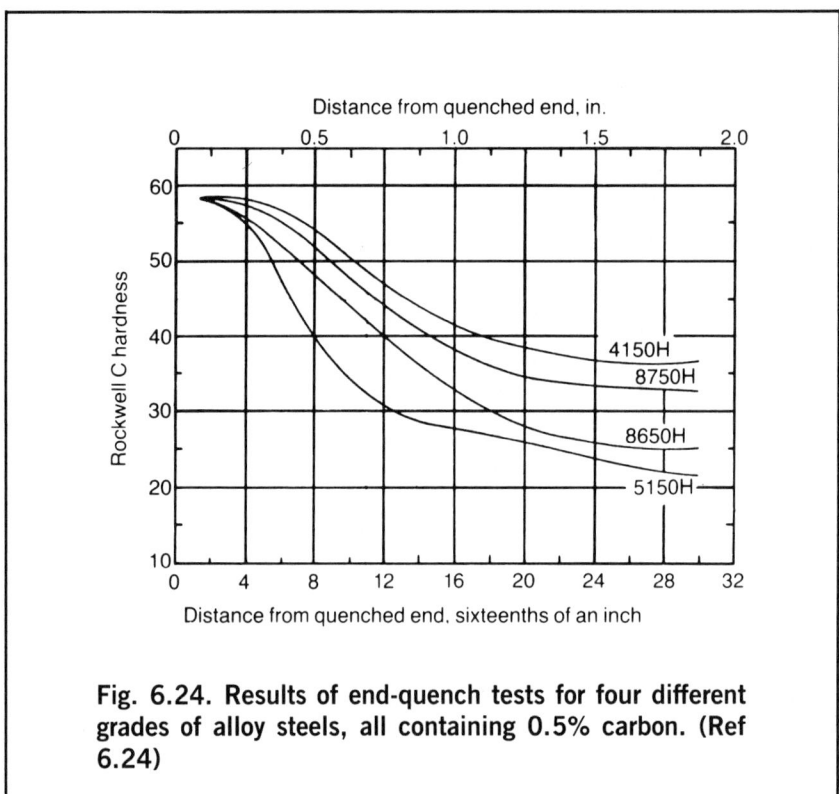

Fig. 6.24. Results of end-quench tests for four different grades of alloy steels, all containing 0.5% carbon. (Ref 6.24)

volumes is published in Ref 6.27. The principles of hardenability and much of the hardenability testing as described in the preceding sections of this chapter remain the same, but the present emphasis is on developing more reliable and systematic hardenability data that can be used in rapid computer prediction of hardenability and the selection of hardenable steels for given applications. Emphasis has also been placed on evaluating the hardenability of shallow hardening low-carbon steels (Ref 6.28), high-carbon steels (Ref 6.29), and the boron steels (Ref 6.30, 6.31), all of which did not receive a great deal of attention in the early days when the hardenability of medium-carbon steels was of greatest importance.

Several computerized on-line systems for hardenability calculations have been developed and are presently in use. The International Harvester Company has developed a system identified as CHAT, an acronym for Computer Harmonized Application Tailored process. This system is a two-part system in which the first part determines the hardenability requirements in terms of D_I for a given application, and

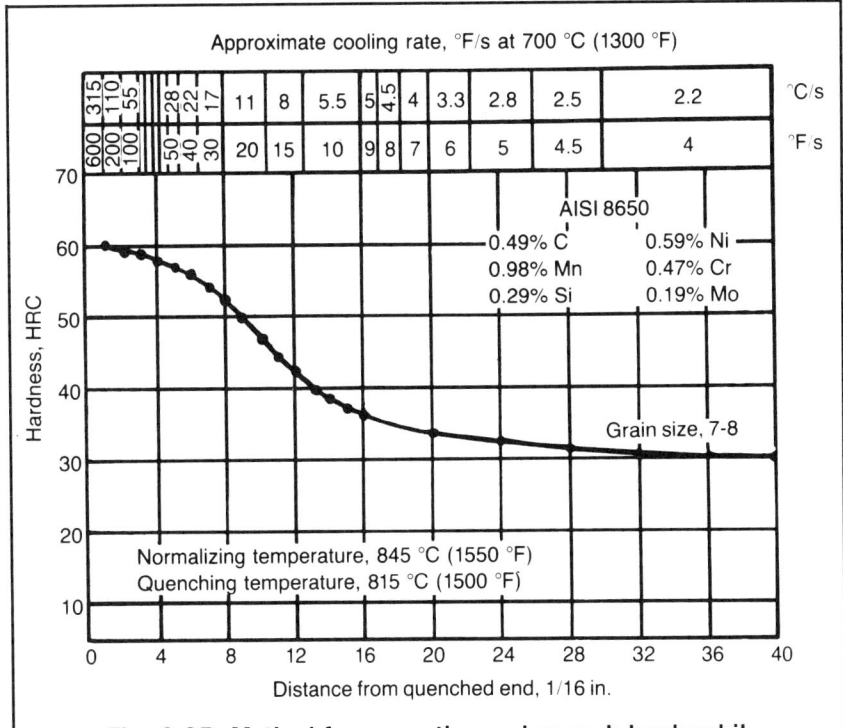

Fig. 6.25. Method for presenting end-quench hardenability data. Data presented here are for AISI 8650 steel. Note relationship of cooling rate (top) to distance from the quenched end. (Ref 6.25)

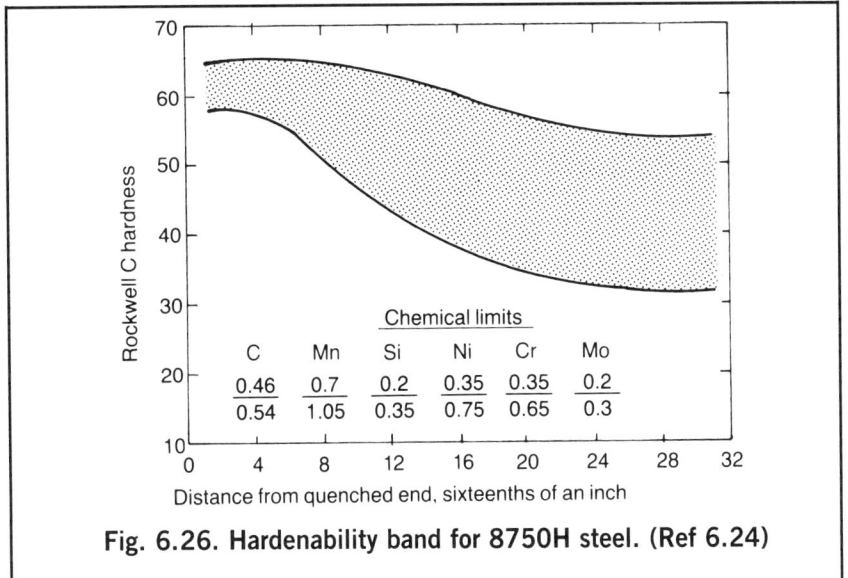

Fig. 6.26. Hardenability band for 8750H steel. (Ref 6.24)

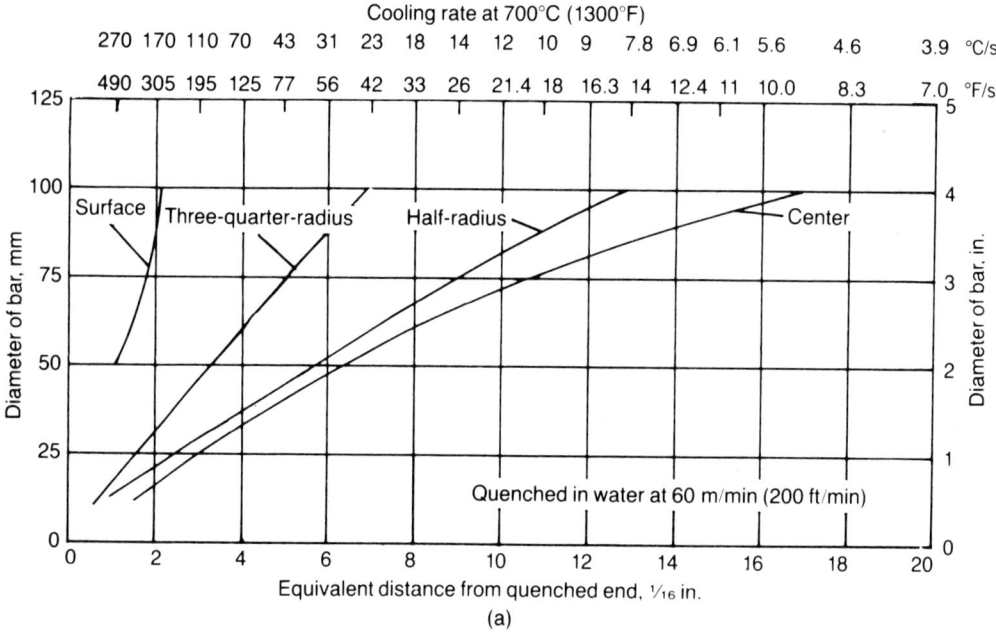

(a)

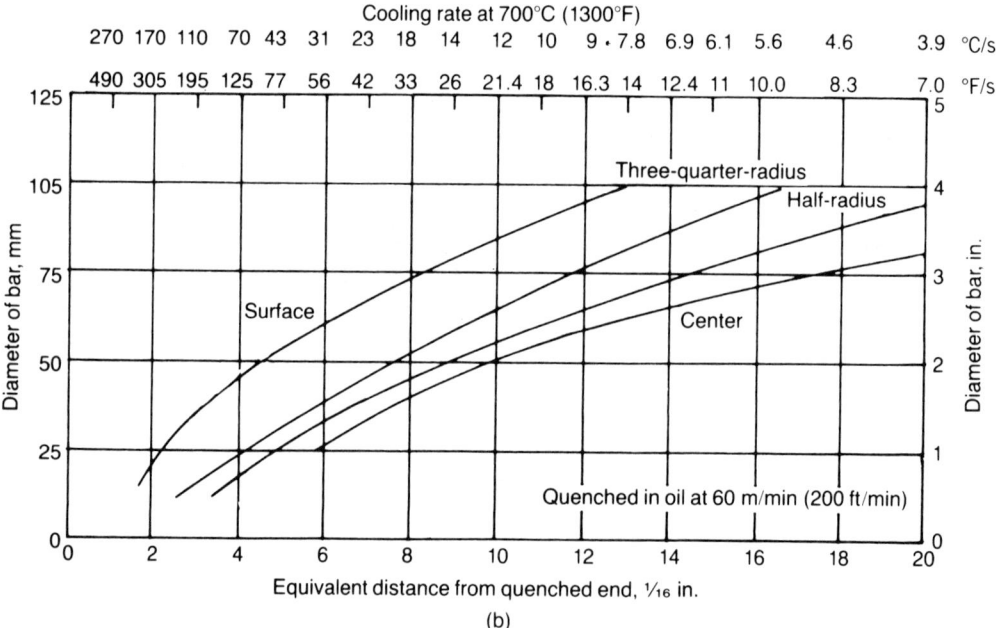

(b)

Correlation of equivalent cooling rates in the end-quenched hardenability specimen and quenched round bars free from scale. Data for surface hardness are for "mild agitation"; other data are for 60 m/min (200 ft/min).

Fig. 6.27. Equivalent cooling rates for round bars quenched in water and oil. (Ref 6.25)

the second part is used to design or select the most economical steel with composition adequate for the requirements. Another system, developed by the Creusot-Loire Corporation in France, predicts structures produced at various cooling rates by the use of CT diagrams and then calculates hardness, yield and tensile strengths expected from martensitic, bainitic, and/or ferrite-pearlite microstructures in the as-quenched and as-tempered conditions. Another system, the Minitech Alloy Steel Information System, calculates the inflection point of Jominy curves as well as entire Jominy curves for many types of steels. The reader is referred to Ref 6.26 for detailed articles about the development and applications of these and other systems for hardenability calculations.

References

6.1 E.C. Bain and H.W. Paxton, *Alloying Elements in Steel*, 2nd ed., American Society for Metals, Metals Park, OH, 1961

6.2 G. Krauss, Martensitic Transformation, Structure and Properties in Hardenable Steels, in *Hardenability Concepts with Applications to Steel*, D.V. Doane and J.S. Kirkaldy (Eds.), AIME, Warrendale, PA, 1978, p 229-248

6.3 L.D. Jaffee and E. Gordon, Temperability of Steels, *Trans ASM*, Vol 49, 1957, p 359-369

6.4 A.R. Marder, The Morphology and Strength of Iron-Carbon Martensite, Ph.D. Dissertation, Lehigh University, Bethlehem, PA, 1968

6.5 M. Cohen, The Strengthening of Steel, *Trans TMS-AIME*, Vol 224, 1962, p 638-657

6.6 P.G. Winchell and M. Cohen, The Strength of Martensite, *Trans ASM*, Vol 55, 1962, p 347-361

6.7 T.E. Swarr and G. Krauss, Boundaries and the Strength of Low Carbon Ferrous Martensites, in *Grain Boundaries in Engineering Materials*, Claitor's Publishing Division, Baton Rouge, LA, 1975, p 127-138

6.8 T.E. Swarr and G. Krauss, The Effect of Structure on the Deformation of As-Quenched and Tempered Martensite in an Fe-0.2% C Alloy, *Met Trans A*, Vol 7A, 1976, p 41-48

6.9 A.R. Marder and G. Krauss, The Effect of Morphology on the Strength of Lath Martensite, *Proceedings of Second International Conference on the Strength of Metals and Alloys*, Vol III, American Society for Metals, Metals Park, OH, 1970, p 822-823

6.10 M.J. Roberts, Effect of Transformation Substructure on the Strength and Toughness of Fe-Mn Alloys, *Met Trans*, Vol 1, 1970, p 3287-3294

6.11 M. Cohen, Strengthening Mechanisms in Steel, *Trans J.I.M.*, Vol 9, 1968, Supplement

6.12 M.J. Roberts and W.S. Owen, Solid Solution Hardening and Thermally Activated Deformation in Iron-Nickel-Carbon Martensites, *J Iron Steel Inst*, Vol 206, 1968, p 375-384

6.13 G.R. Speich, Tempering of Low-Carbon Martensite, *Trans TMS-AIME*, Vol 245, 1969, p 2553-2564

6.14 G.R. Speich and H. Warlimont, Yield Strength and Transformation Substructure of Low-Carbon Martensite, *J Iron Steel Inst*, Vol 206, 1968, p 385-392

6.15 W.H. McFarland, Mechanical Properties of Low-Carbon Alloy-Free Martensite, *Trans TMS-AIME*, Vol 233, 1965, p 2028-2035

6.16 M.A. Grossmann and E.C. Bain, *Principles of Heat Treatment*, 5th ed., American Society for Metals, Metals Park, OH, 1964

6.17 Definition Relating to Metals and Metalworking, *Metals Handbook*, Vol 1, 8th ed., American Society for Metals, Metals Park, OH, 1961, p 20

6.18 C.A. Siebert, D.V. Doane, and D.H. Breen, *The Hardenability of Steels—Concepts, Metallurgical Influences, and Industrial Applications*, American Society for Metals, Metals Park, OH, 1977

6.19 Classification and Designation of Carbon and Alloy Steels, in *Metals Handbook*, Vol 1, 9th ed., American Society for Metals, Metals Park, OH, 1978, p 117-143

6.20 N.B. Pilling and T.D. Lynch, Cooling Properties of Technical Quenching Liquids, *Trans AIME*, Vol 62, 1920, p 665

6.21 M.A. Grossmann and M. Asimov, Hardenability and Quenching, *Iron Age*, Vol 145, 1940, p 25-29, 39-45

6.22 F. Wever, *Archiv für das Eisenhüttenwesen*, Vol 5, 1936-37, p 367

6.23 W.E. Jominy and A.L. Boegehold, *Trans ASM*, Vol 26, 1938, p 574

6.24 G.F. Melloy, *Hardness and Hardenability*, P.D. Harvey (Ed.), Metals Engineering Institute, Metals Park, OH, 1977

6.25 C.F. Jatczak, Hardenability of Carbon and Alloy Steels, in *Metals Handbook*, Vol 1, 9th ed., American Society for Metals, Metals Park, OH, 1978, p 471-526

6.26 D.V. Doane and J.S. Kirkaldy (Eds.), *Hardenability Concepts with Applications to Steel*, AIME, Warrendale, PA, 1978

6.27 D.V. Doane, Application of Hardenability Concepts in Heat Treatment of Steel, *J Heat Treating*, Vol 1, 1979, p 5-30

6.28 R.A. Grange, Estimating the Hardenability of Carbon Steels, *Met Trans*, Vol 4, 1973, p 2231-2244

6.29 C.F. Jatczak, Hardenability in High Carbon Steels, *Met Trans*, Vol 4, 1973, p 2267-2277

6.30 P. Maitrepierre, D. Thivellier, J. Rofes-Vernis, D. Rousseau, and R. Tricot, Microstructure and Hardenability of Low-Alloy Boron-Containing Steels, in *Hardenability Concepts with Applications to Steel*, AIME, Warrendale, PA, 1978, p 421-447

6.31 B.M. Kapadia, Prediction of the Boron Hardenability Effect in Steel—A Comprehensive Review, in *Hardenability Concepts with Applications to Steel*, AIME, Warrendale, PA, 1978, p 448-482

CHAPTER 7

Austenite in Steels

The formation of austenite and the control of austenite grain size are vital aspects of many of the annealing and hardening heat treatments described in earlier chapters. Both the transformation behavior of austenite and the mechanical properties of the microstructures formed from austenite are influenced by austenite grain size. This chapter discusses the formation of austenite from various types of starting microstructures and shows how austenite grain size may be revealed in various types of transformed structures. Finally, the control of austenite grain size by particle dispersions and hot rolling practice is described.

Austenite and Properties

The austenite in carbon steels, although it is stable only above the Ae_1 temperature, strongly influences the transformation and deformation behavior of heat treated steels. With respect to transformation, austenite grain boundaries are preferred sites for the nucleation of proeutectoid phases and pearlite. Therefore, if the austenite grain size in a steel is coarse, fewer nucleation sites are available and diffusion-controlled transformation of the austenite is retarded. As a result, hardenability is increased. The effect of austenite grain size on hardenability has already been discussed in Chapter 6, and the effect of austenite grain size on the base hardenability (D_I) is shown in Fig. 6.20. Austenite grain size also affects martensite transformation kinetics by its effect on M_s temperatures. In Fe-Ni and Fe-Ni-C alloys decreasing austenite grain size significantly lowers M_s (Ref 7.1). The latter result is attributed to the higher strength of fine-grained austenite which in turn increases the shear resistance of the austenite-to-martensite transformation. Similar effects of grain size on M_s may occur in carbon steels.

The preferred nucleation and growth of proeutectoid phases and pearlite at austenitic grain boundaries establishes a direct relationship between austenite grain size and the grain size of the transformation products: the finer the austenite grain size, the finer the grain size of the ferrite-cementite products. In normalized or annealed low-carbon steels, where the microstructure is predominantly ferritic, not only the strength but also the toughness of the steel is increased with decreasing grain size. Equations showing the effect of ferrite grain size on the mechanical properties of ferritic steels are given in Chapter 5. The reduction of austenite grain size in low-carbon steels, therefore, offers the possibility of significant improvement in the properties of ferritic microstructures and, in fact, plays a very important part in the recent development of low-carbon high-strength low-alloy (HSLA) steels as discussed later in this chapter.

In hardened steels, formation of martensite in fine-grained austenite is preferred because of the improved mechanical properties that result. The increase in yield or flow strength of low-carbon lath martensite with decreasing austenite grain size has already been described in the discussion of martensite hardness in Chapter 6. Lath martensite is unique in that the martensite units form in parallel arrays called packets (see Chapter 3) which subdivide the parent austenite grain. Each packet is effectively a grain because most of the laths have the same orientation. Figure 7.1 shows that packet size is directly related to austenitic grain size, and helps explain why the strength of lath martensite may be related to either prior austenite grain size or packet size (Ref 7.2-7.5).

The impact toughness of hardened steels, as determined by Charpy impact testing, improves with decreasing austenitic grain size. The reasons for this effect of austenite grain size are complex and may be in part related to the segregation of impurity atoms to austenite grain boundaries during austenitizing (Ref 7.6). In steels austenitized at high temperatures, and therefore with coarse austenitic grain size, fracture of Charpy impact specimens frequently occurs along prior austenite grain boundaries. Such intergranular fracture is quite brittle and often reflects the effects of grain boundary segregation. Analytical techniques such as Auger electron spectroscopy verify the presence of elements such as phosphorus in very thin layers on the grain boundary fracture surfaces of quench and tempered steels (Ref 7.6). Since tempering in certain temperature ranges aggravates intergranular fracture, a more complete discussion of mechanisms of embrittlement in hardened steel is deferred to the chapter on tempering (see Chapter 8).

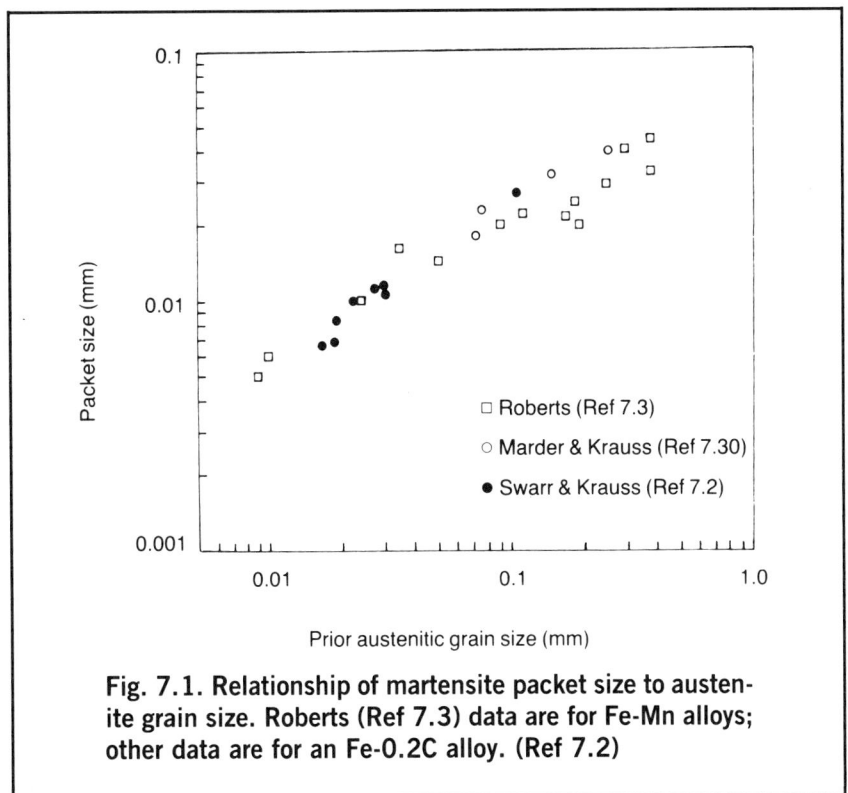

Fig. 7.1. Relationship of martensite packet size to austenite grain size. Roberts (Ref 7.3) data are for Fe-Mn alloys; other data are for an Fe-0.2C alloy. (Ref 7.2)

However, when toughness of martensitic steels is evaluated by the propagation of a very sharp crack that has been fatigued into a specimen, a testing technique referred to as fracture toughness testing, coarse-grained specimens appear to show better fracture toughness than fine-grained specimens (Ref 7.7). This apparent anomaly is resolved by understanding the differences between fracture toughness and Charpy V-notch impact toughness testing and their relationship to the microstructure of hardened steel (Ref 7.8). The probability that the sharp crack used to obtain fracture toughness lies away from grain boundaries and within the martensite and retained austenite of a hardened steel is greater for a coarse-grained specimen than for a fine-grained specimen. Therefore, in coarse-grained specimens, the detrimental effects of grain boundary embrittlement are initially minimized when the sharp crack begins to propagate, while the beneficial matrix effects of homogenization, retained austenite, and the solution of second-phase particles due to high-temperature austenitizing are enhanced. In Charpy V-notch impact testing, however, the large extent of the stress field ahead of the notch immediately causes stress to be applied to embrittled grain boundaries, and even during fracture toughness testing, the propagating crack will soon find and follow any embrittled grain boundaries (Ref 7.9). The best metallurgical and heat treatment practice, then, is to avoid high-temperature austenitizing that produces coarse grains and the attendant high susceptibility to intergranular fracture of hardened steels.

Austenite Formation

The manner in which austenite forms in a given steel depends very much on the microstructure present prior to heating for austenitizing. This section describes austenite formation from microstructures consisting of pearlite, ferrite, spheroidized cementite, and martensite.

Figure 7.2 shows the development of austenite in the pearlitic microstructure of a eutectoid steel, from the work of Vilella as published in Ref 7.10. A series of specimens were heated into the austenite field, held for the times shown, and quenched. Areas of austenite formation are visible as white patches within the lamellar pearlitic structure. The austenite, of course, transformed to martensite on quenching, but the etching differences between the newly formed martensite and the preexisting pearlite allow clear delineation of the extent of the austenite formation. In addition to the gradual development of the austenite, Fig. 7.2 shows that not all of the cementite is dissolved as the austenite grows into the pearlite. Some of the cementite persists in the form of spheroidized particles (the small dark spots in the white areas), and only dissolves with longer holding times at temperature. More recent work (Ref 7.11) confirms the persistence of cementite in the austenite after initial austenite formation.

When the amount of austenite formed in Fig. 7.2 is plotted as a function of austenitizing time, the curve shown in Fig. 7.3 results. The austenite formation requires some incubation or time for the first nuclei to form and then proceeds at a more rapid rate as more nuclei develop and grow. The time dependence is expected because of the carbon diffusion required to produce an austenite containing 0.8% carbon from the low-carbon (0.02%) ferrite and high-carbon (6.67%) cementite from the pearlite. At higher temperatures the diffusion rate increases, and austenite forms more rapidly. Figure 7.4 shows the acceleration of austenite formation in a pearlitic 0.80% carbon steel when the austenitizing temperature was raised from 730 to 751 °C (1346 to 1385 °F) (Ref 7.12).

In microstructures consisting initially of ferrite and spheroidized cementite particles, austenite forms first at the interface between the carbides and the ferrite. Ferrite and cementite, therefore, combine to

AUSTENITE IN STEELS / 183

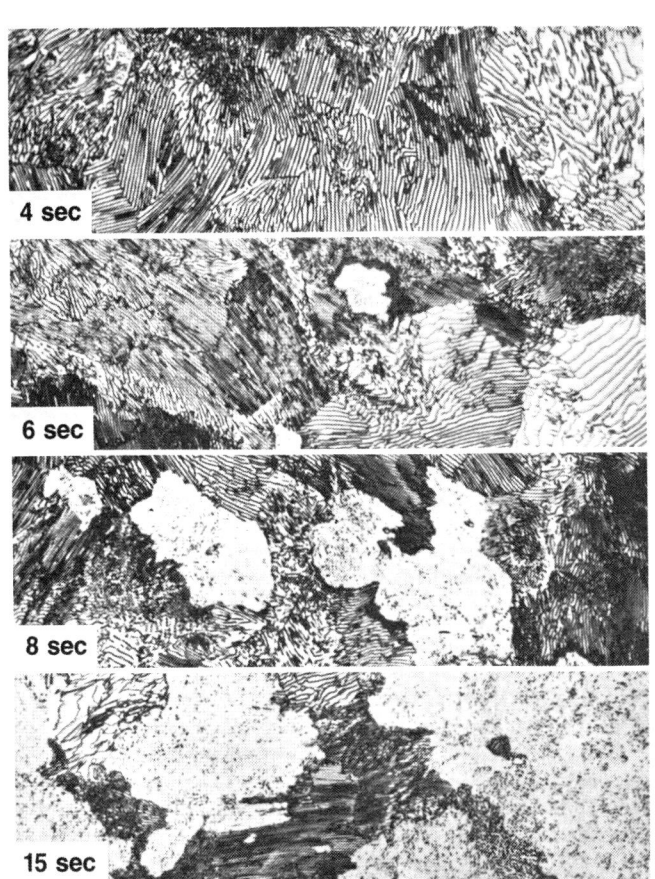

Fig. 7.2. Formation of austenite (light patches) from pearlite as a function of time. (Ref 7.10)

form austenite, exactly the reverse of the reaction that forms pearlite (see Eq 2.1). Figure 7.5 shows the development of austenite at carbide particles in a spheroidized low-carbon steel (Ref 7.10). The cementite particles are soon enveloped by the austenite, and subsequent austenite formation depends on carbon diffusion through the austenite as the carbides dissolve. Figure 7.6 shows that the latter process leads to a slower rate of austenite formation compared to that in a pearlitic steel (see Fig. 7.3) where the closely spaced ferrite and cementite lamellae reduce the diffusion distances for austenite formation. Judd and Paxton (Ref 7.13) and Speich and Szirmae (Ref 7.11) show similar effects associated with austenite formation in spheroidite.

184 / STEELS: HEAT TREATMENT AND PROCESSING PRINCIPLES

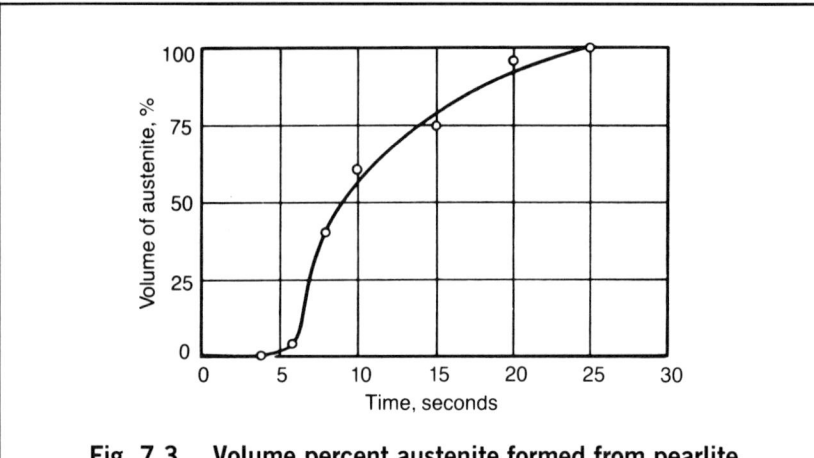

Fig. 7.3. Volume percent austenite formed from pearlite in a eutectoid steel as a function of time at a constant austenitizing temperature. (Ref 7.10)

Figure 7.7 shows schematically the preferred nucleation sites for austenite formation in three different microstructures (Ref 7.11). In ferritic microstructures, nucleation of austenite occurs primarily at ferrite grain boundary faces and edges. In spheroidized microstructures, austenite nucleates at cementite particles associated with fer-

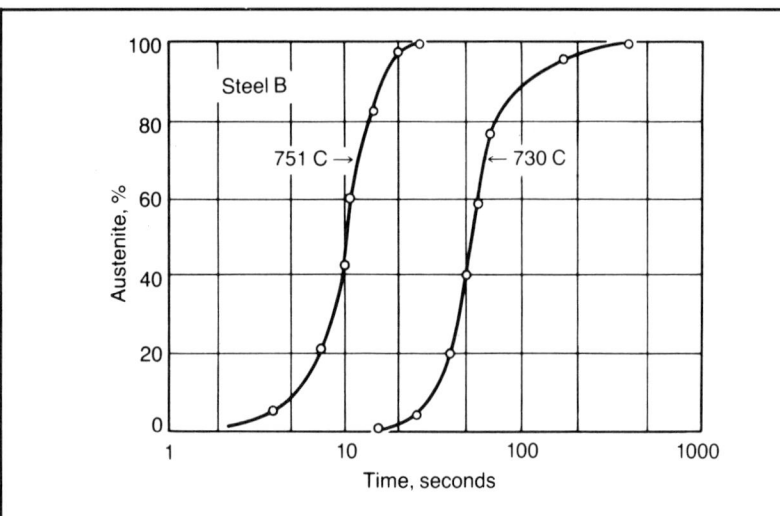

Fig. 7.4. Effect of austenitizing temperature on the rate of austenite formation from pearlite in eutectoid steel. (Ref 7.10)

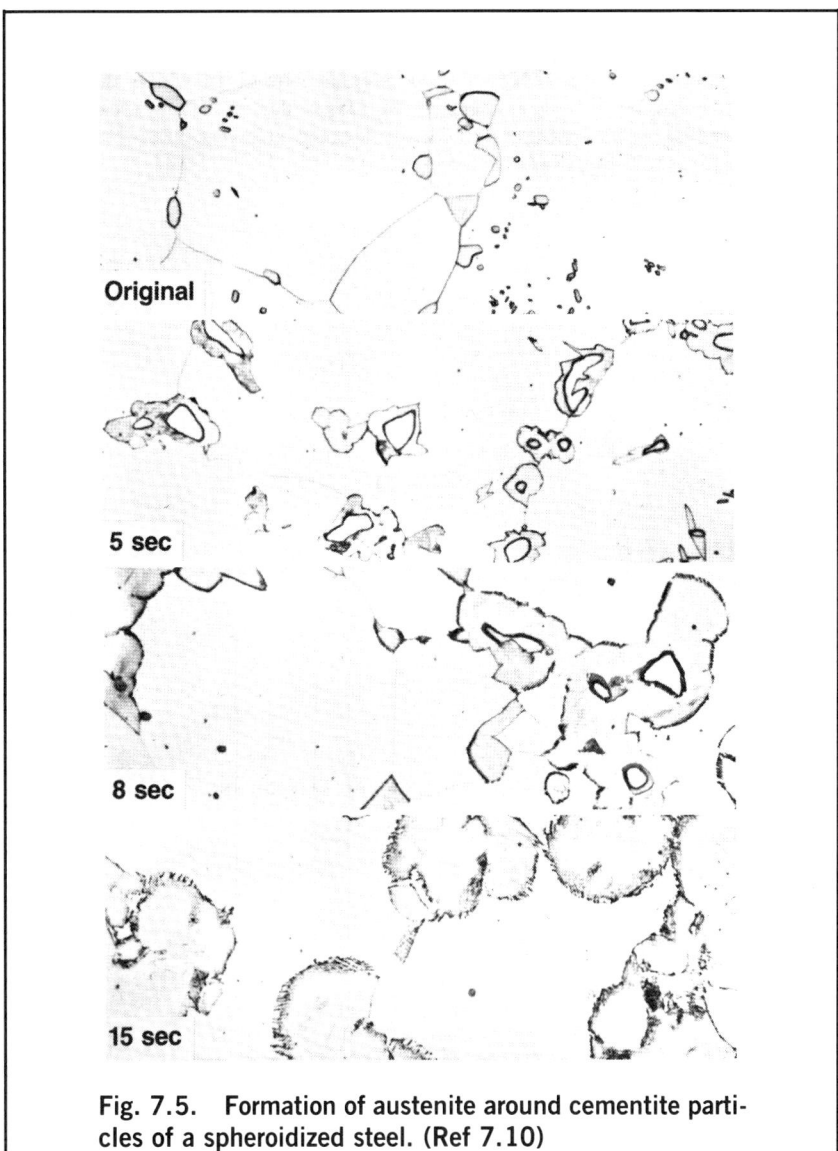

Fig. 7.5. Formation of austenite around cementite particles of a spheroidized steel. (Ref 7.10)

rite grain boundaries. In pearlite, austenite nucleates primarily at intersections of pearlite colonies but also at the interfaces of ferrite and cementite lamellae within a colony.

The formation of austenite in martensitic microstructures of carbon steels produces two morphologies of austenite crystals (Ref 7.13-7.15). One morphology is equiaxed and tends to form at prior austenite grain boundaries. Figure 7.8(a) shows an example of equi-

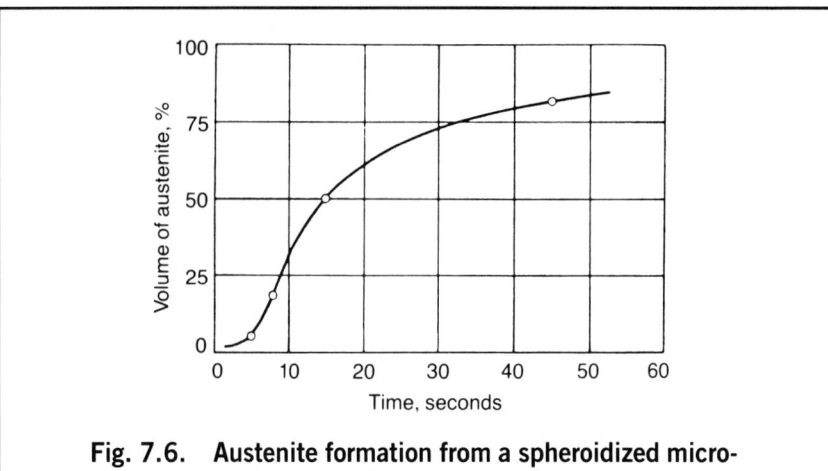

Fig. 7.6. Austenite formation from a spheroidized microstructure as a function of time. (Ref 7.10)

axed austenite grains that have formed in martensitic 4340 steel heated to 715 °C (1319 °F). The dark etching component of the microstructure in Fig. 7.8(a) is fine tempered martensite with the structure shown in Fig. 7.8(b). The other austenite morphology is acicular or lath-like and forms by nucleation between the laths of martensite. In martensitic carbon steels which readily temper even on heating to the austenitizing temperature, the mechanism of the austenite formation appears to be diffusion controlled and associated with fine carbide

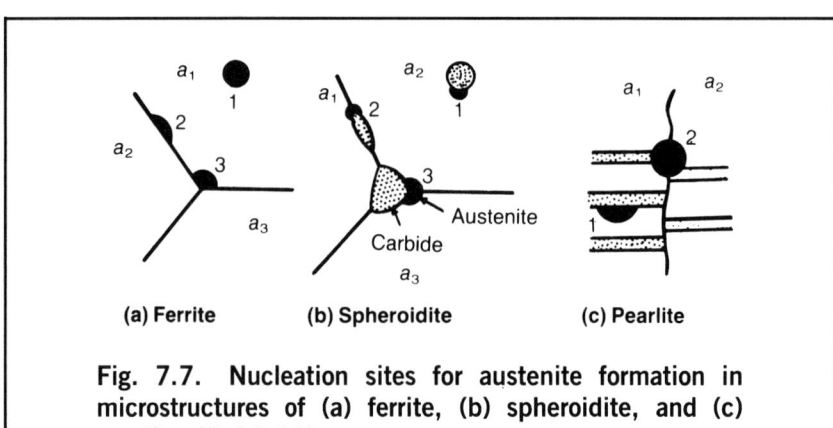

Fig. 7.7. Nucleation sites for austenite formation in microstructures of (a) ferrite, (b) spheroidite, and (c) pearlite. (Ref 7.11)

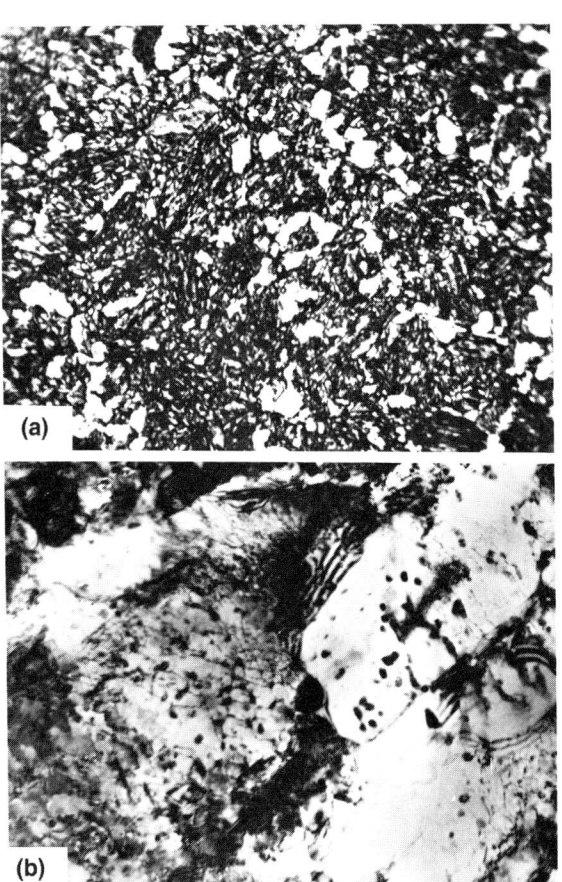

Fig. 7.8. (a) Formation of equiaxed austenite grains (white patches) in tempered martensite of a hardened 4340 steel specimen rapidly heated to 715 °C (1319 °F), held for 6 s, and oil quenched. Light micrograph. Nital etch. Magnification, 1300×; shown here at 75%. (b) Tempered martensite in same specimen. Very fine equiaxed grains of ferrite with spheroidized carbides formed. Transmission electron micrograph. Magnification, 20 000×; shown here at 75%. (T. Scoonover and G. Krauss, unpublished research)

particles (Ref 7.14, 7.15). However, in carbon steels with high nickel content, it has been suggested (Ref 7.16) that the lath-like austenite may be formed by a shear process similar to that described below for Fe-Ni alloys. At any rate, the formation of austenite from martensite or tempered martensite is extremely rapid. In a study of AISI 4340 steel (Ref 7.14), austenite was formed in as little as 2 s at temperatures between 790 and 870 °C (1450 and 1600 °F) by rapidly heating thin specimens in a lead bath. The martensite that formed from that austenite was harder than normally obtained, an indication that the rapid austenite formation introduced imperfections that subsequently affected martensite strength. The higher than normal hardness observed after high-rate short-time austenitizing, as sometimes produced by induction heating, has been referred to as superhardness (Ref 7.17).

In Fe-30Ni alloys, the formation of austenite from martensite is clearly accomplished by a martensitic or shear mechanism. Figure 7.9(a) shows tilting surface relief produced when many plates of austenite formed in a large plate of martensite heated to 340 °C (644 °F). When the surface relief was polished away, platelets of austenite were revealed by etching (see Fig. 7.9b). Figure 7.10 presents two transmission electron micrographs of the same area in an Fe-33Ni alloy, one taken before and one after the transformation of the martensite to austenite. Figure 7.10(a) shows a portion of a martensite plate with very fine twins and the adjacent retained austenite. The circle shows the area from which a diffraction pattern was taken to confirm the presence of the martensite and the austenite. When the martensite was heated, austenite formed (see Fig. 7.10b). The new austenite is characterized by a very high density of tangled dislocations which substantially raises the strength of the shear-formed austenite relative to annealed austenite (Ref 7.18, 7.19), and provides the strain energy for recrystallization of the austenite during annealing (Ref 7.19).

The shear transformation of austenite from martensite in Fe-Ni alloys occurs even at very slow heating rates in the absence of carbon. In Fe-Ni-C alloys (22 to 32% nickel, 0.004 to 0.6% carbon), however, the shear transformation occurs only at high heating rates, and is replaced by a diffusion-controlled formation of austenite at slower heating rates (Ref 7.20). The latter type of austenite formation is associated with an equiaxed morphology, relatively low dislocation density, and a lack of tilting surface relief, and is probably characteristic of equiaxed austenite formation from martensite in carbon steels.

Austenite Grain Size

Austenite grain size is usually reported in terms of ASTM grain size numbers, where n, the grain size number, is obtained from the

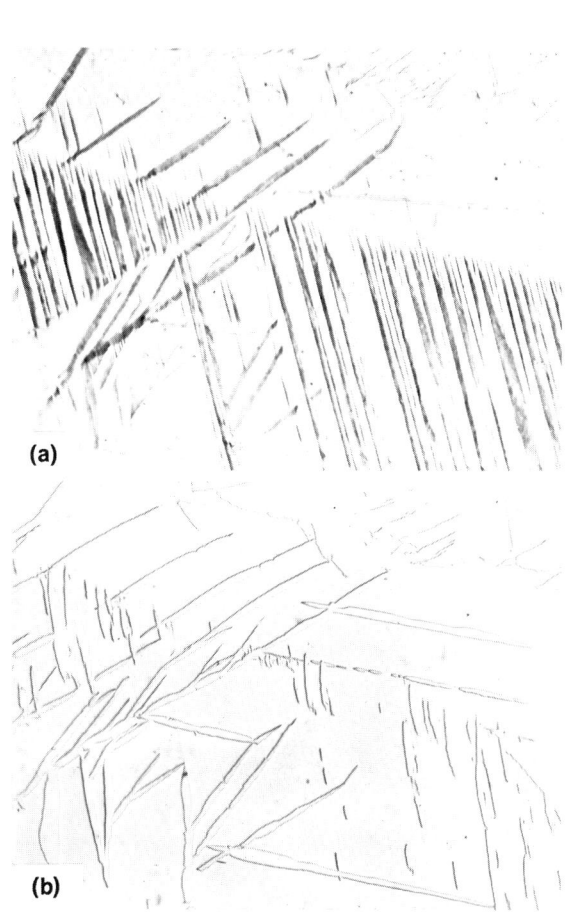

Fig. 7.9. (a) Tilting surface relief produced by formation of plates of austenite in a large plate of martensite in an Fe-32Ni alloy heated to 340 °C (644 °F). (b) Same area after polishing and etching. Both light micrographs. Magnification, 500×; shown here at 75%. (G. Krauss and W. Pitsch, unpublished research at Max-Planck Institut für Eisenforschung, Düsseldorf, Germany)

expression $2^{(n-1)}$ which gives the number of grains per square inch in a microstructure examined at a magnification of 100×. Figure 7.11 shows prior austenite grain sizes ranging from ASTM No. 1 to ASTM No. 9 in a hardened medium-carbon steel. In this series of micrographs

190 / STEELS: HEAT TREATMENT AND PROCESSING PRINCIPLES

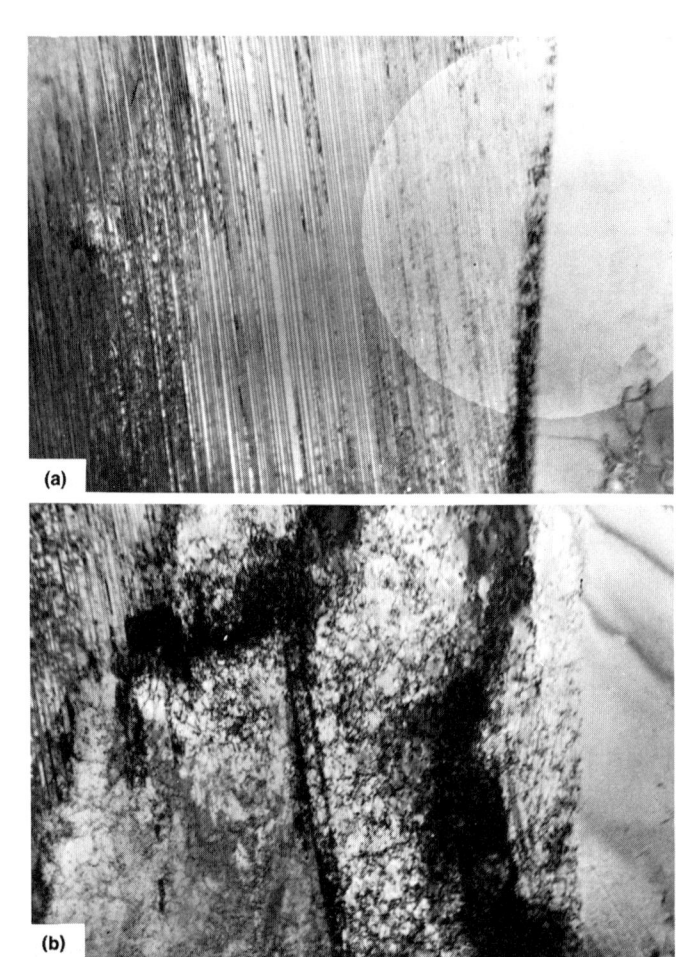

Fig. 7.10. (a) Portion of a martensite plate with fine transformation twins and adjacent retained austenite in an Fe-33Ni alloy. (b) Same area after transformation of martensite to austenite. Both transmission electron micrographs. Magnification, 40 000×; shown here at 60%. (G. Krauss and W. Pitsch, unpublished research at Max-Planck Institut für Eisenforschung, Düsseldorf, Germany)

the austenite grains are revealed rather clearly by the etching differences associated with various orientations of martensite in the different austenite grains.

AUSTENITE IN STEELS / 191

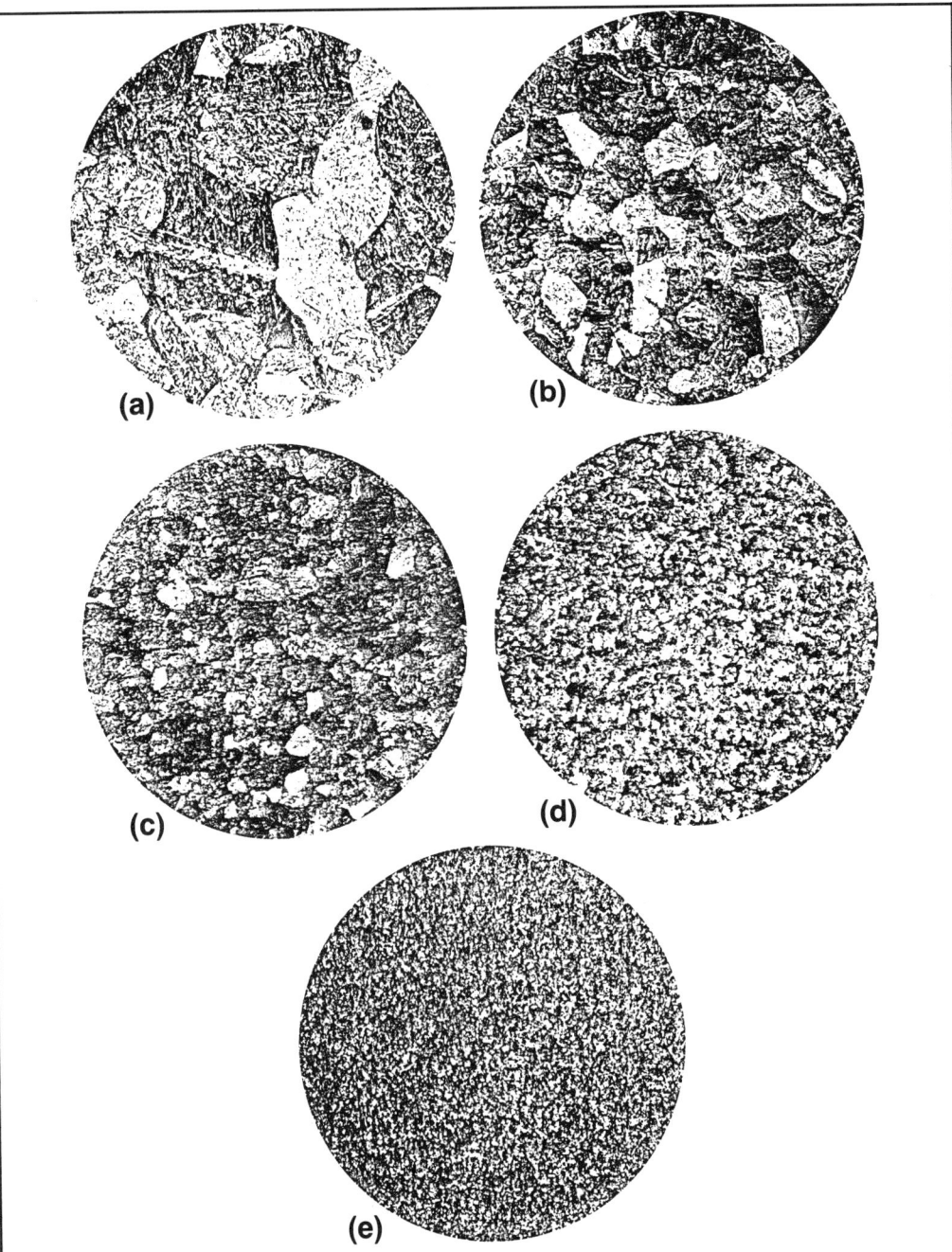

Fig. 7.11. Martensitic microstructures with prior austenite grain sizes of (a) ASTM No. 1; (b) ASTM No. 3; (c) ASTM No. 5; (d) ASTM No. 7; and (e) ASTM No. 9. These microstructures were prepared by lightly tempering and etching in a hydrochloric-picric acid solution in alcohol. Magnification, 100×; shown here at 50%. (Ref 7.10)

192 / STEELS: HEAT TREATMENT AND PROCESSING PRINCIPLES

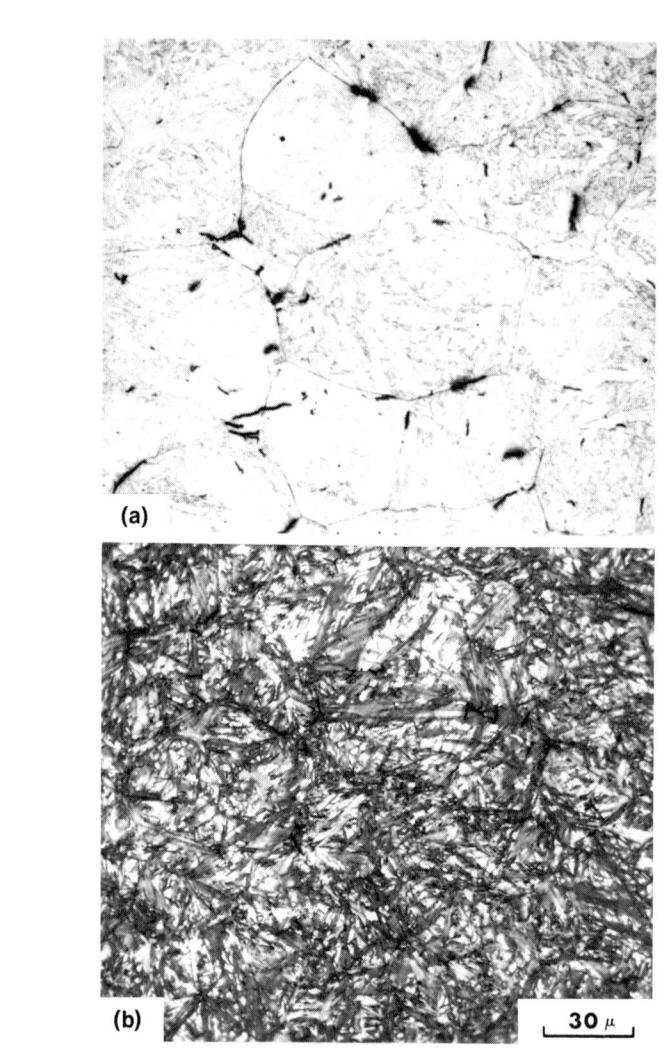

Fig. 7.12. Microstructure of Fe-1.22C alloy austenitized at 890 °C (1740 °F) for 2 min and water quenched. (a) Etched to show austenite grain boundaries. (b) Etched to show martensite and retained austenite. Etchants given in text. (Ref 7.21)

Frequently, however, austenite grain size is difficult to reveal, especially in quench and tempered steels, and special etching or heat treatments must be used to develop the austenite structure. For example (Ref 7.21), Fig. 7.12 shows the microstructure of a quenched

Fe-1.22C alloy etched to show the austenite grain boundaries (a), and to show the plate martensite (b). The austenitic grain boundaries were revealed by etching untempered specimens in a boiling solution of 25 g NaOH, 2 g picric acid, and 100 ml H_2O for 15 min, followed by etching lightly in nital (Ref 7.22). When nital alone was used to etch the structure, as in Fig. 7.12(b), the austenite grain boundaries were not at all obvious. A similar set of micrographs for a martensitic Fe-0.2C alloy is shown in Fig. 7.13 (Ref 7.2). In this case, the prior austenite grain boundaries were revealed after tempering by etching in a solution of 80 ml H_2O, 28 ml oxalic acid (10%), and 4 ml H_2O_2, and the martensite was best developed by etching in 2% nital. The effectiveness of many of the special etchants used to reveal austenite grain boundaries may be due to impurity atom segregation at austenite grain boundaries during austenitizing or tempering.

The various techniques available to reveal austenitic grain size, including special etching techniques, have recently been reviewed by Millsop (Ref 7.23). Table 7.1, taken from his work, lists the various methods, the steels or structures most responsive to each technique, and some of the limitations. The McQuaid-Ehn test, based on carburizing for 8 h at 925 °C (1690 °F) and slow cooling to allow a proeutectoid cementite network to mark the grain boundaries, sometimes results in coarsening of the original austenite grain size. The oxidation and vacuum grooving methods are based on preferential oxidation and preferential evaporation of the austenite grain boundaries, respectively, and are useful as indicated. Slow cooling, to reveal austenite grain boundaries by the formation of proeutectoid ferrite or cementite networks, is very effective provided the carbon content of the steel is not too low. In the latter case, too much ferrite forms and the outlining effect of a thin network is lost.

A useful table for interrelationships between ASTM grain size number and other measurements of grain size is included in ASTM specification E 112–63 (Ref 7.24) and included as Table 7.2 here. Frequently grain size is measured by an intercept technique, and Table 7.2 makes possible the conversion of intercept sizes to ASTM grain size number.

Austenite Grain Size Control

Steels may be coarse grained or fine grained. Fine-grained steels are deoxidized with aluminum and contain fine aluminum nitride particles that restrain austenite grain growth. Coarse-grained steels are generally deoxidized with silicon, a practice which does not

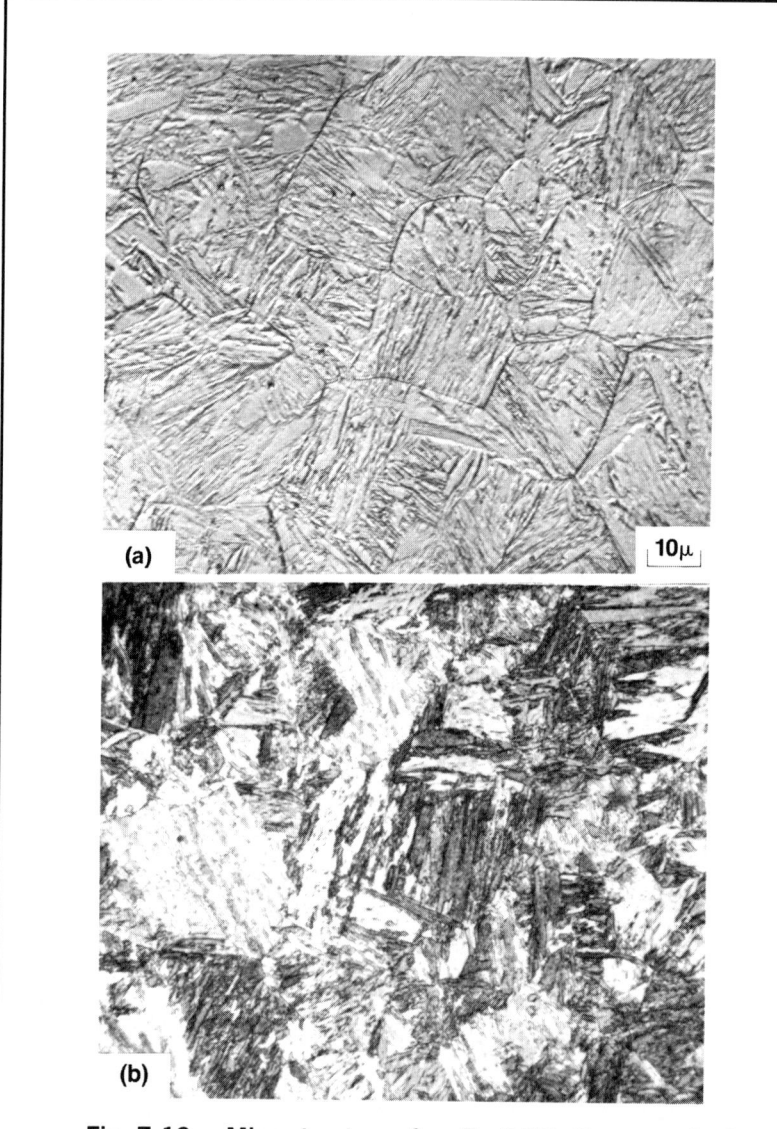

Fig. 7.13. Microstructure of an Fe-0.2C alloy quenched to form martensite. (a) Etched to show austenite grain boundaries. (b) Etched to show martensite. Etchants given in text. (Ref 7.2)

produce particle dispersions effective in inhibiting austenite grain growth. Two types of austenite grain coarsening behavior occur in the two types of steels. Figure 7.14 shows that coarse-grained steels coarsen gradually with increasing austenitizing temperature, but

AUSTENITE IN STEELS / 195

Table 7.1. Methods to Reveal Austenitic Grain Size (Ref 7.23)

Method	Comments
Picric acid solutions	(a) Used for a wide range of steels having martensitic or bainitic structures (b) Room temperature etch (c) Unpredictable, does not always work (d) May give information on unrecrystallized grain shape
McQuaid-Ehn carburization	(a) Used for a limited range of steels (mainly hypo-eutectoid) (b) Lengthy 8 h treatment at 925 °C (1690 °F) (c) May not reflect true grain size of as-received steels
Oxidation	(a) Used for a limited range of steels (mainly hypo-eutectoid) (b) Heat treatment for 1 h at 855 °C (1575 °F) (c) May not reflect true grain size of as-received steels
Vacuum grooving	(a) Used for a wide range of steels (b) Heat treatment for 1 h or less at 900 °C (1652 °F) (c) Full potential not known (d) May not reflect true grain size of as-received steels
Delineation by ferrite or cementite	(a) Used for a range of hypo- and hyper-eutectoid steels (b) Heating above Ac_3, followed by controlled cooling (c) Used for some as-received steels (carbon composition range limited)

fine-grained steels show no grain growth until a relatively high austenitizing temperature is reached. When coarsening finally occurs in a fine-grained steel, it occurs very rapidly. The temperature of the abrupt coarsening is referred to as the "grain coarsening temperature". Sometimes the abrupt coarsening is referred to as "discontinuous grain growth" or "secondary recrystallization". The latter term is used because the kinetics of discontinuous grain growth are very similar to recrystallization, which has a slow initial or incubation stage, a more rapid second stage, and finally another slow stage.

In many finishing heat treatments, the austenitizing temperature never exceeds 980 °C (1800 °F). Fine-grained steels, therefore, retain a fine austenite grain size, but the austenite of coarse-grained steels might coarsen significantly, especially during carburizing treatments that are frequently performed over many hours at 930 to 955 °C (1700 to 1750 °F). Figure 7.15 shows an example of a great difference in grain

Table 7.2. Micrograin Size Relationships (Ref 7.24)

ASTM micrograin size No.	Calculated "diameter" of average grain		Average intercept distance(a)		Calculated area of average grain section		Average number of grains per mm³	Nominal grains per mm² at 1×	Nominal grains per in.² at 100×
	mm	in. ×10⁻³	mm	in. ×10⁻³	mm ×10⁻³	in. ×10⁻⁶			
00(b)....	0.508	20.0	0.451	17.8	258	400	7.63	3.88	0.250
0........	0.359	14.1	0.319	12.6	129	200	21.6	7.75	0.50
0.5......	0.302	11.9	0.268	10.6	91.2	141	36.3	11.0	0.707
1.0......	0.254	10.0	0.226	8.88	64.5	100	61.0	15.5	1.0
....	0.250	9.84	0.222	8.74	62.5	96.9	64.0	16.0	1.03
1.5......	0.214	8.41	0.190	7.47	45.6	70.7	103	21.9	1.41
....	0.200	7.87	0.178	6.99	40.0	62.0	125	25.0	1.61
....	0.180	7.09	0.160	6.29	32.4	50.2	171	30.9	1.99
2.0......	0.180	7.07	0.160	6.28	32.3	50.0	172.3	31.0	2.0
2.5......	0.151	5.95	0.134	5.30	22.8	35.4	290	43.8	2.83
....	0.150	5.91	0.133	5.24	22.5	34.9	296	44.4	2.87
3.0......	0.127	5.00	0.113	4.44	16.1	25.0	488	62.0	4.0
....	0.120	4.72	0.107	4.20	14.4	22.3	578.9	69.4	4.48
3.5......	0.107	4.20	0.0948	3.73	11.4	17.7	821	87.7	5.66
....	0.090	3.54	0.0799	3.15	8.10	12.6	1 370	123	7.97
4.0......	0.0898	3.54	0.0797	3.14	8.06	12.5	1 380	124	8.0
4.5......	0.076	2.97	0.0671	2.64	5.70	8.84	2 320	175	11.3
....	0.070	2.76	0.0622	2.45	4.90	7.59	2 920	204	13.2
5.0......	0.064	2.50	0.0564	2.22	4.03	6.25	3 910	248	16.0
....	0.060	2.36	0.0533	2.10	3.60	5.58	4 630	278	17.9
5.5......	0.0534	2.10	0.0474	1.87	2.85	4.42	6 570	351	22.6
....	0.050	1.97	0.0444	1.75	2.50	3.88	8 000	400	25.8
6.0......	0.045	1.77	0.0399	1.57	2.02	3.13	11 000	496	32.0
....	0.040	1.58	0.0355	1.40	1.60	2.48	15 600	625	40.3
6.5......	0.038	1.49	0.0335	1.32	1.43	2.21	18 600	701	45.3
....	0.035	1.38	0.0311	1.22	1.23	1.90	23 000	816	52.7
7.0......	0.032	1.25	0.0282	1.11	1.01	1.56	31 000	992	64.0
....	0.030	1.18	0.0267	1.05	0.90	1.40	37 000	1 110	71.7
7.5......	0.027	1.05	0.0237	0.933	0.713	1.10	52 500	1 400	90.5
....	0.025	0.984	0.0222	0.874	0.625	0.969	64 000	1 600	103
8.0......	0.0224	0.884	0.0199	0.785	0.504	0.781	88 400	1 980	128
....	0.0200	0.787	0.0178	0.699	0.40	0.620	125 000	2 500	161
8.5......	0.0189	0.743	0.0168	0.660	0.356	0.552	149 000	2 810	181
9.0......	0.0159	0.625	0.0141	0.555	0.252	0.391	250 000	3 970	256
....	0.0150	0.591	0.0133	0.524	0.225	0.349	296 000	4 440	287
9.5......	0.0134	0.526	0.0119	0.467	0.178	0.276	420 000	5 610	362
10.0.....	0.0112	0.442	0.00997	0.392	0.126	0.195	707 000	7 940	512
....	0.0100	0.394	0.00888	0.350	0.10	0.155	1.00×10^6	10 000	645
10.5.....	0.00944	0.372	0.00838	0.330	0.089	0.138	1.19×10^6	11 200	724
....	0.00900	0.354	0.00799	0.315	0.081	0.126	1.37×10^6	12 300	797
....	0.00800	0.315	0.00710	0.280	0.064	0.0992	1.95×10^6	15 600	1 010
11.0.....	0.00794	0.313	0.00705	0.278	0.063	0.0977	2.00×10^6	15 900	1 020
....	0.00700	0.276	0.00622	0.245	0.049	0.0760	2.92×10^6	20 400	1 320
11.5.....	0.00667	0.263	0.00593	0.233	0.045	0.0691	3.36×10^6	22 400	1 450
....	0.00600	0.236	0.00533	0.210	0.036	0.0558	4.63×10^6	27 800	1 790
12.0.....	0.00561	0.221	0.00498	0.196	0.031	0.0488	5.66×10^6	31 700	2 050
....	0.00500	0.197	0.00444	0.175	0.025	0.0388	8.00×10^6	40 000	2 580
12.5.....	0.00472	0.186	0.00419	0.165	0.022	0.0345	9.51×10^6	44 900	2 900
....	0.00400	0.158	0.00355	0.140	0.0160	0.0248	15.62×10^6	62 500	4 030
13.0.....	0.00397	0.156	0.00352	0.139	0.0158	0.0244	16.0×10^6	63 500	4 100
13.5.....	0.00334	0.131	0.00296	0.117	0.011	0.0173	26.9×10^6	89 800	5 800
....	0.00300	0.118	0.00266	0.105	0.009	0.0140	37.0×10^6	111 000	7 170
14.0.....	0.00281	0.111	0.00249	0.0981	0.0079	0.0122	45.2×10^6	127 000	8 200
....	0.00250	0.098	0.00222	0.0874	0.00625	0.00969	64.0×10^6	160 000	10 300

(a) Value of Heyn intercept for equiaxed grains. (b) The use of 00 is recommended instead of "−1" or "minus 1" to avoid confusion.

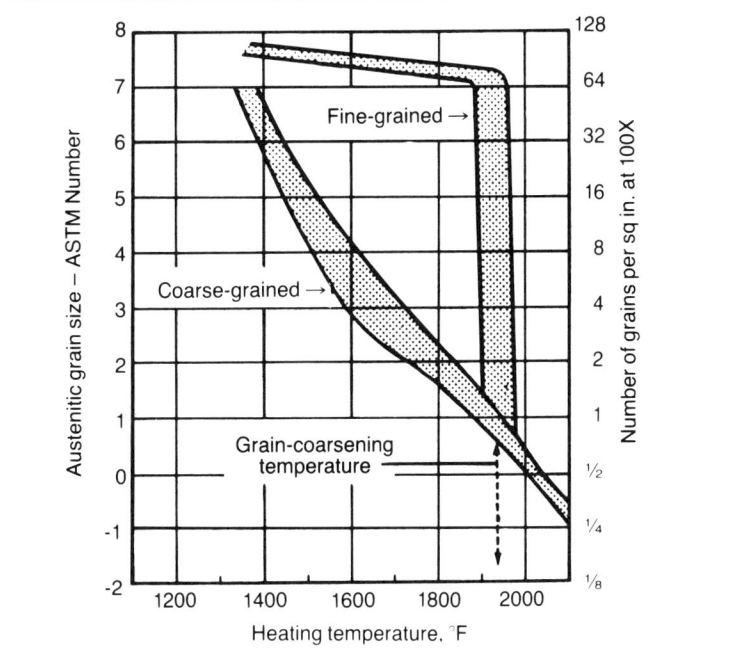

Fig. 7.14. Austenite grain size as a function of austenitizing temperature for coarse-grained and fine-grained steels. (Ref 7.25)

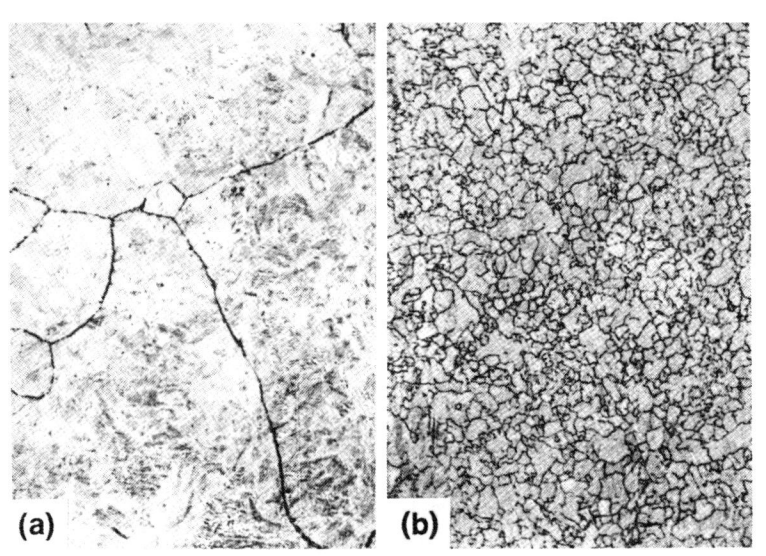

Fig. 7.15. Comparison of austenitic grain size in (a) coarse-grained SAE 1015 steel and (b) fine-grained SAE 4615 steel after carburizing. (Ref 7.10)

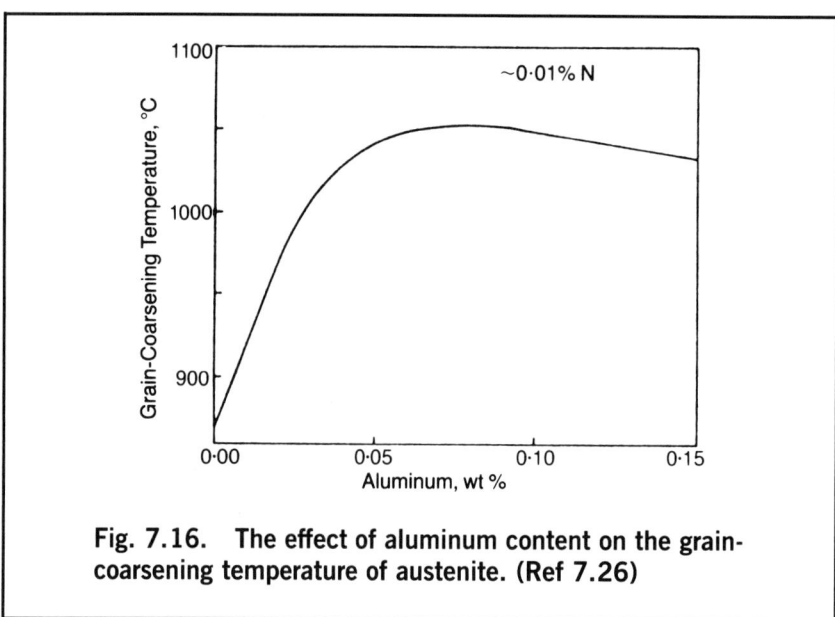

Fig. 7.16. The effect of aluminum content on the grain-coarsening temperature of austenite. (Ref 7.26)

size that developed when steels of different coarsening behavior were carburized.

Recognition of the two different types of austenite grain coarsening behavior has led to an almost universal use of steel-making practices that produce fine-grained steels for critical heat treated parts. The mechanism of grain size control is based on the fact that aluminum that does not combine with oxygen during deoxidation combines with nitrogen and forms a dispersion of fine aluminum nitride particles. The fine nitride particles in turn pin the austenite grain boundaries, and thereby inhibit grain growth. However, with increasing temperature the aluminum particles coarsen and/or dissolve, and very rapid grain growth occurs when the particle distribution becomes too coarse to restrain grain growth effectively.

Figure 7.16 shows the effect of aluminum content on the grain coarsening temperature in mild steel (Ref 7.26). Aluminum contents up to 0.08% increase the coarsening temperature, but higher additions cause a slight lowering of the coarsening temperature. The reaction Al + N = AlN is reversible, and at high temperatures some of the AlN particles go back in solution to permit grain coarsening, especially in steels with lower aluminum contents. Figure 7.17 shows the solubility of aluminum nitride at three different aluminum contents. The grain coarsening temperatures (arrows) are indicated on the curves (Ref 7.26). At high aluminum contents grain coarsening precedes substantial solution of aluminum nitride, and therefore the reduced coarsen-

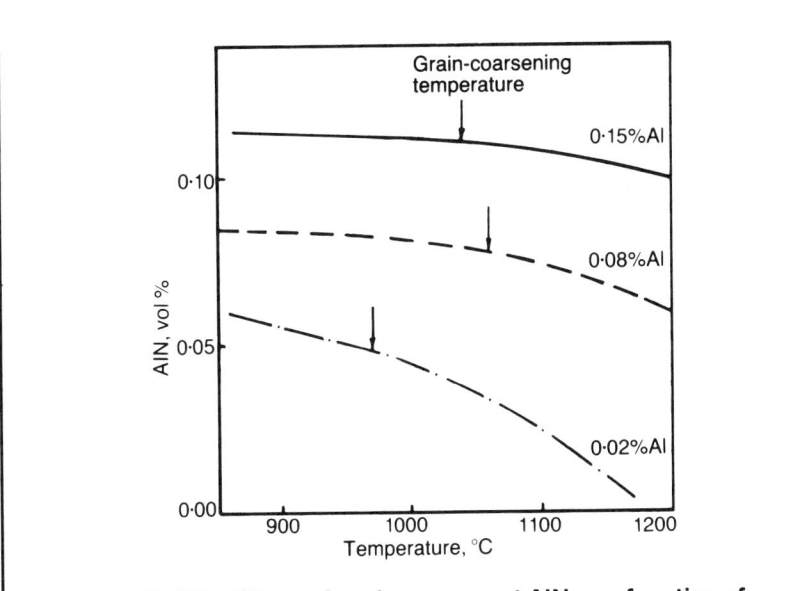

Fig. 7.17. Change in volume percent AlN as a function of temperature in mild steel containing 0.01% nitrogen and aluminum as shown. Grain-coarsening temperatures are indicated by arrows. (Ref 7.26)

ing temperature at high aluminum content is believed to be the result of coarse particles produced during solidification and hot working rather than coarsening and/or solution of the aluminum nitride particles during austenitizing.

Austenite grain size may be controlled by elements other than aluminum. In particular, the transition metals such as titanium, vanadium, and niobium that are strong carbide and nitride formers behave similarly to aluminum with respect to producing particle dispersions that inhibit austenite grain growth. Figure 7.18 shows the solubility products for aluminum nitride and various transition metal carbides and nitrides in austenite as a function of temperature, as collected by Aronsson from a number of sources (Ref 7.27). The behavior of the titanium, niobium, and vanadium compounds is similar to that of aluminum nitride: solubility in austenite decreases with decreasing temperature. Thus, precipitation of alloy carbides or nitrides will occur at low austenitizing temperatures if sufficient quantities of the alloying elements are present.

Small amounts, frequently less than 0.05%, of niobium, vanadium, and titanium are used in high-strength low-alloy (HSLA) steel (Ref 7.28).

200 / STEELS: HEAT TREATMENT AND PROCESSING PRINCIPLES

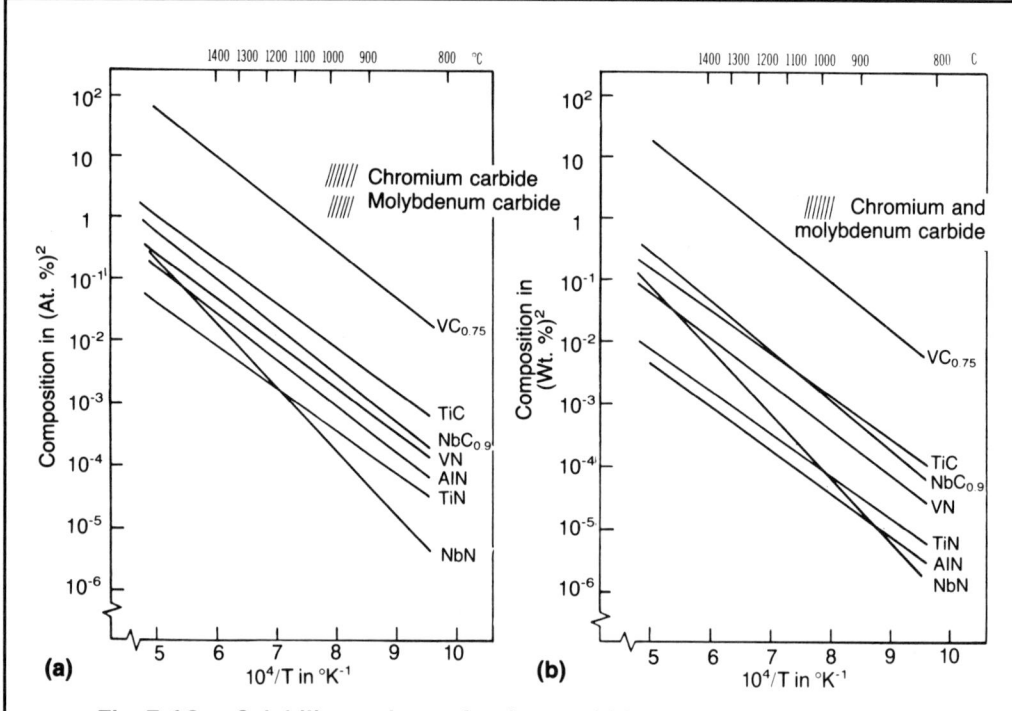

Fig. 7.18. Solubility products of various carbides and nitrides in austenite (Ref 7.27). Ordinate in (a) is content in atomic percent; ordinate in (b) is content in weight percent.

The very small additions of the alloying elements have led to the description of HSLA steels as microalloyed steels. HSLA steels have low carbon contents and therefore have largely ferritic microstructures. However, very fine ferrite grain sizes together with precipitation of alloy carbonitride particles raise the strengths of the HSLA steels well above those of normally processed mild steels. Mild steels with ferrite-pearlite microstructures usually have yield strengths on the order of 30 ksi (207 MPa), while the HSLA steels may have yield strengths ranging from 50 to 80 ksi (345 MPa to 550 MPa).

The very fine ferrite grain sizes in HSLA steels are obtained through control of austenite grain size by the interaction of alloy carbonitride precipitation and hot rolling. Figure 7.19 shows schematically the effects of processing and temperature on the austenite and ferrite grain size that develops in a microalloyed steel. At high hot working temperatures, where heavy sections are first reduced in size and maximum hot workability is required, the alloy carbides, nitrides, and/or carbonitrides are in solution as would be expected from Fig.

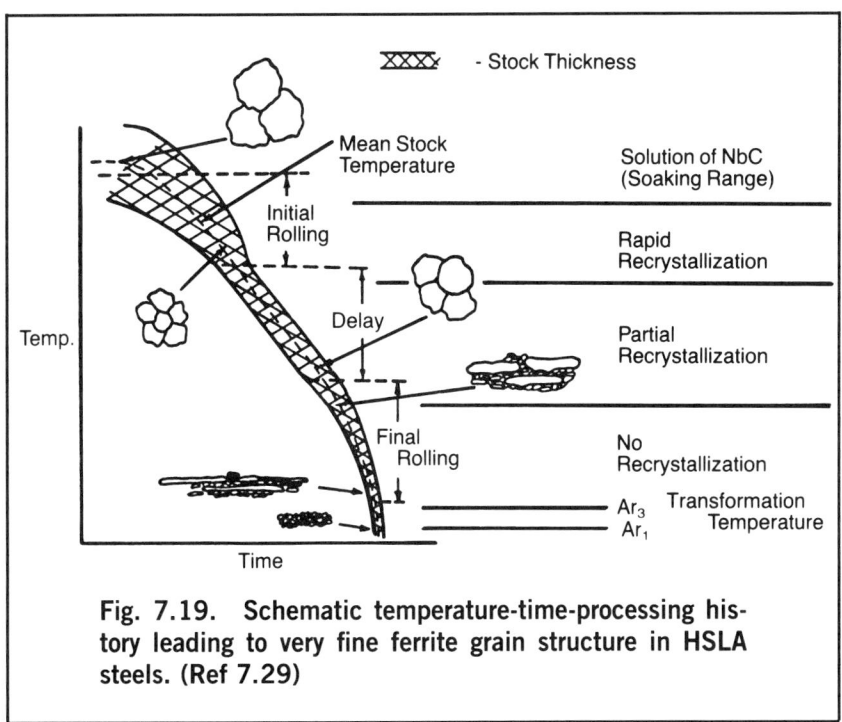

Fig. 7.19. Schematic temperature-time-processing history leading to very fine ferrite grain structure in HSLA steels. (Ref 7.29)

7.18. Austenite grain size is quite coarse at this stage of processing. Without the restraining effect of fine particles, recrystallization and grain growth occur almost simultaneously with deformation at this stage of hot working. However, if finish rolling is performed just above or close to the Ar_3 temperature, fine alloy carbonitrides precipitate from the austenite because of their reduced solubility (see Fig. 7.18) at low austenitizing temperatures. The fine particles prevent austenite grain growth and even recrystallization, the amount of retardation depending on alloy concentration, amount of deformation, and the rolling temperature. In the extreme case the austenite grains become highly deformed and elongated, as shown in Fig. 7.19. On cooling below Ar_3, and through Ar_1, then, ferrite grains form on the closely spaced grain boundaries of the unrecrystallized austenite. As a result, very fine ferrite grain sizes that give the HSLA steels their high strength and toughness are produced.

References

7.1 P.J. Brofman, G.S. Ansell, T.J. Nichol, and G. Judd, The Effect of Fine Grain Size on the Martensitic Transformation in Fe-Ni and Fe-Ni-C

Alloys, *Mechanical Behavior of Metals and Alloys Associated with Displacive Phase Transformations*, Joint U.S.-Japan Conference, Rensselaer Polytechnic Institute, Troy, NY, June 1979

7.2 T.E. Swarr and G. Krauss, Boundaries and the Strength of Low Carbon Ferrous Martensites, in *Grain Boundaries in Engineering Materials*, J.L. Walter, J.H. Westbrook, and D.A. Woodford (Eds.), Claitors Publishing Division, Baton Rouge, LA, 1975, p 127-138

7.3 M.J. Roberts, Effects of Transformation Substructure on the Strength and Toughness of Fe-Mn Alloys, *Met Trans*, Vol 1, 1970, p 3287-3294

7.4 R.A. Grange, Strengthening Steel by Austenite Grain Refinement, *Trans ASM*, Vol 59, 1966, p 26-48

7.5 T.E. Swarr and G. Krauss, The Effect of Structure on the Deformation of As-Quenched and Tempered Martensite in an Fe-0.2% C Alloy, *Met Trans A*, Vol 7A, 1976, p 41-48

7.6 W.C. Johnson and J.M. Blakely (Eds.), *Interfacial Segregation*, American Society for Metals, Metals Park, OH, 1979

7.7 G.Y. Lai, W.E. Wood, R.A. Clark, V.F. Zackay, and E.R. Parker, The Effect of Austenitizing Temperature on the Microstructure and Mechanical Properties of As-Quenched 4340 Steel, *Met Trans*, Vol 5, 1974, p 1663-1670

7.8 R.O. Ritchie, B. Francis, and W.L. Server, Evaluation of Toughness in AISI 4340 Alloy Steel Austenitized at Low and High Temperatures, *Met Trans A*, Vol 7A, 1976, p 831-838

7.9 K. Nakazawa and G. Krauss, Microstructure and Fracture of 52100 Steel, *Met Trans A*, Vol 9A, 1978, p 681-689

7.10 M.A. Grossmann and E.C. Bain, *Principles of Heat Treatment*, 5th ed., American Society for Metals, Metals Park, OH, 1964

7.11 G.R. Speich and A. Szirmae, Formation of Austenite from Ferrite and Ferrite-Carbide Aggregates, *Trans TMS-AIME*, Vol 245, 1969, p 1063-1074

7.12 G.A. Roberts and R.F. Mehl, The Mechanism and Rate of Formation of Austenite from Ferrite-Cementite Aggregates, *Trans ASM*, Vol 31, 1943, p 613

7.13 R.R. Judd and H.W. Paxton, Kinetics of Austenite Formation from a Spheroidized Ferrite-Carbide Aggregate, *Trans TMS-AIME*, Vol 242, 1968, p 206-215

7.14 T.M. Scoonover and G. Krauss, High-rate Short-time Austenitizing of 4340 Steel, *Met Eng Q,* Vol 12, 1972, p 41-48

7.15 S. Matsuda, Microstructural and Kinetic Studies of Reverse Transformation in a Low Carbon Alloy Steel, in *New Aspects of Martensitic Transformation*, Japan Institute of Metals, 1976, p 363-367

7.16 S. Watanabe, Y. Ohmori, and T. Kunitaki, Formation of Austenite from Lath-like Martensite, in *New Aspects of Martensitic Transformation,* Japan Institute of Metals, 1976, p 368-374

7.17 *Induction Hardening and Tempering*, ASM Committee on Induction Hardening, American Society for Metals, Metals Park, OH, 1964, p 133

7.18 G. Krauss, Fine Structure of Austenite Produced by the Reverse Martensitic Transformation, *Acta Met,* Vol 11, 1963, p 499-509

7.19 G. Krauss and M. Cohen, Strengthening and Annealing of Austenite Formed by Reverse Martensitic Transformation, *Trans AIME*, Vol 224, 1962, p 1212-1221

7.20 C.A. Apple and G. Krauss, The Effect of Heating Rate on the Martensite to Austenite Transformation in Fe-Ni-C Alloys, *Acta Met*, Vol 20, 1972, p 849-856

7.21 R.P. Brobst and G. Krauss, The Effect of Austenite Grain Size on Microcracking in Martensite of an Fe-1.22C Alloy, *Met Trans*, Vol 5, 1975, p 457-462

7.22 A. Benscoter of Bethlehem Steel Corporation, Bethlehem, PA, private communication 1975

7.23 R. Millsop, A Survey of Austenite Grain Size Measurements, in *Hardenability Concepts with Applications to Steels*, TMS-AIME, Warrendale, PA, 1978, p 316-333

7.24 Estimating the Average Grain Size of Metals, ASTM Designation: E 112-63, American Society for Testing and Materials, Philadelphia, 1963

7.25 G.F. Melloy, *Austenite Grain Size—Its Control and Effects*, Metals Engineering Institute, American Society for Metals, Metals Park, OH, 1968

7.26 T. Gladman, The Effect of Aluminum Nitride on the Grain Coarsening Behavior of Austenite, in *Metallurgical Developments in Carbon Steels*, Special Report 81, The Iron and Steel Institute, London, 1963, p 68-70

7.27 B. Aronsson, Gefügeaufbau und Mechanische Eigenschaften einiger martensitischer Stähle unter besonderer Berücksichtigung des Einflusses von Niob und Molybdän, in *Steel Strengthening Mechanisms*, Climax Molybdenum Co., Greenwich, CT, 1969, p 77-87

7.28 *Microalloying 75*, Union Carbide Corporation, 1977, distributed by American Society for Metals, Metals Park, OH

7.29 J.D. Baird and R.R. Preston, Relationships Between Processing, Structure and Properties in Low Carbon Steels, in *Processing and Properties of Low Carbon Steel*, J.M. Gray (Ed.), TMS-AIME, Warrendale, PA, 1973, p 1-46

7.30 A.R. Marder and G. Krauss, The Effect of Morphology on the Strength of Lath Martensite, in *Proceedings of Second International Conference on the Strength of Metals and Alloys*, American Society for Metals, Metals Park, OH, 1970, p 822-823

CHAPTER 8

Tempering of Steel

Virtually all steels that are hardened are subjected to a subcritical heat treatment referred to as tempering. Tempering improves the toughness of as-quenched martensitic microstructures but lowers strength and hardness. This chapter describes the mechanical property and microstructural changes that develop during tempering. The most important structural change is the formation of various distributions of iron and alloy carbides as the supersaturation of the as-quenched martensite is relieved and equilibrium mixtures of phases are approached with increased tempering. Finally, several types of embrittlement that may occur on tempering are described.

Mechanical Property Changes

Martensite, the object of the quenching treatments described in Chapter 6, is quite hard but it is also very brittle. The brittleness of martensitic microstructures is due to a number of factors that may include the lattice distortion caused by carbon atoms trapped in the octahedral sites of the martensite (see Fig. 6.5), impurity atom segregation at austenite grain boundaries, carbide formation during quenching, and residual stresses produced during quenching. Tempering is the heat treatment of hardened steels that has reduction of brittleness or increased toughness as its major objective. Any temperature up to the lower critical may be used for tempering; thus an extremely wide variation in properties and microstructure ranging from those of as-quenched martensite to spheroidized carbides in ferrite can be produced by tempering. Ultimately it is the balance of hardness (or strength) and toughness required in service that determines the conditions of tempering for a given application.

Figure 8.1 shows impact toughness as a function of tempering temperature for hardened 0.4 and 0.5% carbon steels (Ref 8.1). There

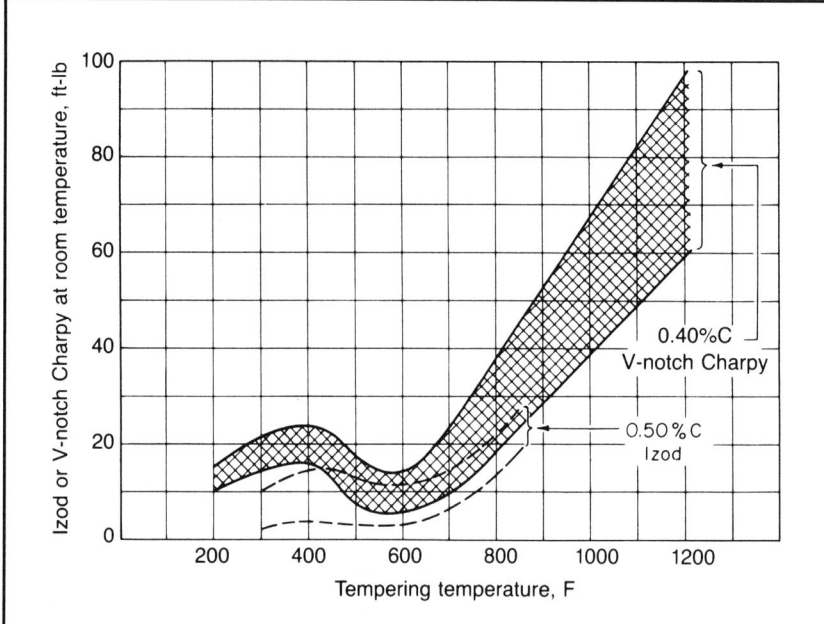

Fig. 8.1. Impact toughness as a function of tempering temperature of hardened, low-alloy, medium-carbon steels. (Ref 8.1)

are two tempering temperature ranges that produce significant improvement in toughness from that of the as-quenched state. Tempering in the range of 150 to 200 °C (300 to 400 °F) produces a modest increase in toughness that is adequate for applications that require high strength and fatigue resistance (medium-carbon steels) or where loading is primarily compressive as in bearings and gears (high-carbon steels). The latter applications require the high hardness and associated good wear resistance that high-carbon martensite and light tempers provide. Tempering above 425 °C (800 °F) is the other important tempering temperature range. Figure 8.1 shows that toughness improves significantly after tempering in this range, but as noted below hardness and strength also decrease significantly. Therefore, tempering above 425 °C (800 °F) is used where high toughness is of major concern, and strength and hardness are important but of secondary concern.

Figure 8.1 shows that toughness may actually decrease if steels are tempered in the range of 260 to 370 °C (500 to 700 °F). This decrease in toughness is referred to as tempered martensite embrittlement, 350 °C embrittlement, or 500 °F embrittlement and is discussed in more detail later in this chapter. As a result of this embrittlement, the tempering range between 260 to 370 °C (500 to 700 °F) is generally avoided in commercial practice. Another type of embrittlement, temper embrittlement, may develop in martensitic steels tempered above 425 °C (800 °F). Temper embrittlement occurs in certain alloy steels as a result of holding in or slow cooling through certain tempering temperature ranges, and is also discussed in more detail later in this chapter.

Finally, Figure 8.1 also shows the substantial effect that increasing carbon content has on impact toughness by comparing the results of tempering 0.5% carbon steels to those of 0.4% carbon steels. Steels with carbon contents of 0.5% or greater have very low impact toughness and are used only where high hardness, wear resistance, and/or edge retention are of prime importance. For example, hand tools, such as screw driver blades and cutting blades of all sorts, are made from quenched and low-temperature tempered medium- and high-carbon steels. Wear resistance and cutting edge retention are excellent in these applications, but the higher the carbon content of the steel, the more susceptible the tool becomes to fracture under bending or tensile stresses. The reasons for decreasing toughness with increasing carbon content of quenched and tempered steels are discussed in the last section of this chapter.

Figure 8.2, taken from a variety of sources by Grossmann and Bain (Ref 8.1), shows how hardness decreases from the maximum associated with as-quenched martensite with increasing tempering temperature. The effect of carbon content is also shown. The lower hardness of low-carbon steels in the as-quenched condition and throughout tempering is emphasized in the curves. Therefore, if maximum hardness is desired, a high-carbon steel should be selected and tempering should be restricted to the 150 to 200 °C (300 to 400 °F) temperature as noted above.

Figure 8.2 indicates a slight hardness increase on low-temperature tempering of the highest carbon steels. Figure 8.3, from an investigation of the early stages of tempering of martensite in an Fe-1.22C alloy (Ref 8.2), shows tempering times and temperatures that produce an increase in hardness above that associated with the as-quenched state. This increase in hardness is a result of the precipitation of a dense distribution of very fine transition carbide particles within the martensite plates.

Generally, the interplay of hardness and toughness is of major concern in the heat treatment and application of quench and tempered steels. However, the changes in other mechanical properties with increasing tempering are also tabulated for common grades of

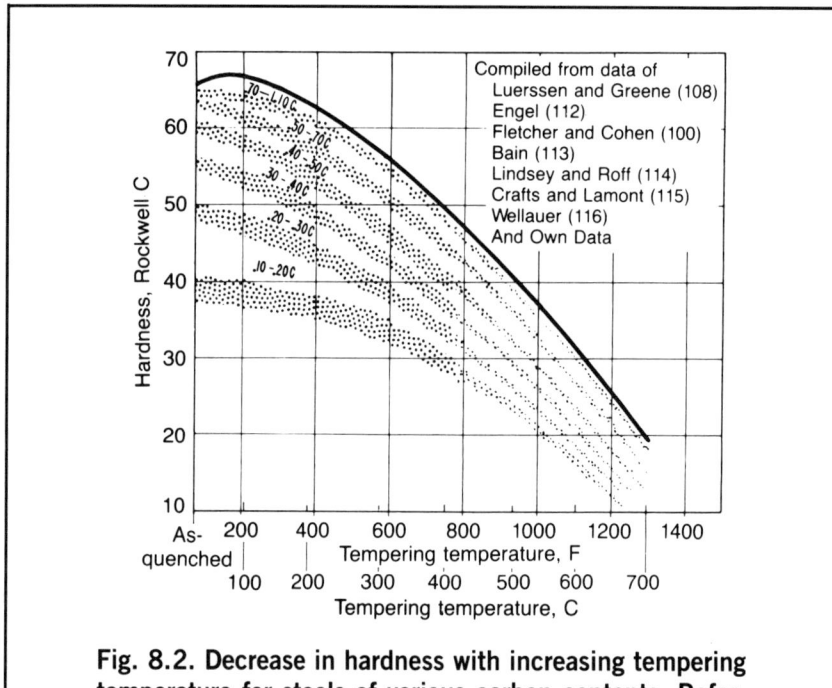

Fig. 8.2. Decrease in hardness with increasing tempering temperature for steels of various carbon contents. Reference numbers after investigators are from list in Grossmann and Bain. (Ref 8.1)

carbon and alloy steel bars (Ref 8.3) and are quite important for the selection of steels and design of heat treatments for some applications. Figure 8.4 shows the changes in mechanical properties that occur when an oil-quenched AISI 4340 steel is tempered at temperatures above 200 °C (400 °F). Both yield strength and tensile strength decrease continuously and elongation and reduction of area increase with increasing tempering temperature. The as-quenched hardness and the mechanical properties of 4340 steel for selected tempering treatments as a function of bar size are listed in Table 8.1. The strength properties for a given treatment decrease with increasing bar diameter (see Chapter 6).

Figure 8.4 shows two other aspects of the mechanical behavior of tempered carbon steels. One is the fact that there is no decrease in ductility produced by tempering in the temperature range that produces tempered martensite embrittlement. Specimen design and testing account for this observation. The toughness data shown in Fig. 8.1 are based on impact toughness testing accomplished by loading notched specimens at a high strain rate. Figure 8.4, on the other hand,

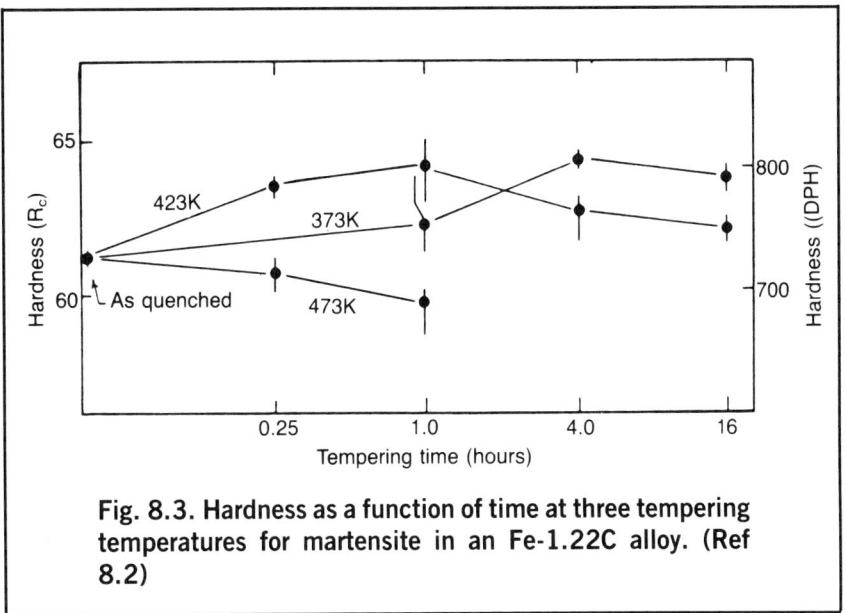

Fig. 8.3. Hardness as a function of time at three tempering temperatures for martensite in an Fe-1.22C alloy. (Ref 8.2)

is based on tensile testing of smooth round specimens at relatively slow strain rates. Thus at slow strain rates, without the stress concentrating effect of a notch, the microstructure of a steel tempered even in the range 260 to 370 °C (500 to 700 °F) can accommodate loading without undue embrittlement. On impact loading, however, the reverse is true and disregard of strain rate and notch effects may lead to unexpected failure in certain applications.

Figure 8.4 also shows that the yield and tensile strengths of the tempered 4340, at first well separated after tempering at low temperatures, tend to approach each other after tempering at high temperatures. This effect is a common characteristic of hardened carbon and low-alloy steels, and is related to differences in work-hardening behavior that develop on tempering. Figure 8.5 shows stress-strain curves that illustrate the changes in work hardening that develop with tempering of lath martensite in an Fe-0.2C alloy (Ref 8.4). In this case, the as-quenched martensite was obtained by quenching in a NaOH-NaCl solution and tempering was performed by heating in lead at 400 °C (750 °F) for 1 min. The work-hardening rate in the as-quenched specimen was quite high, as shown by the rapid increase in stress with increasing strain, while the stress-strain curve for the tempered specimen was almost flat, indicating a very low rate of work hardening. This difference in work hardening behavior is attributed to interaction of dislocations with relatively coarse particles of cementite

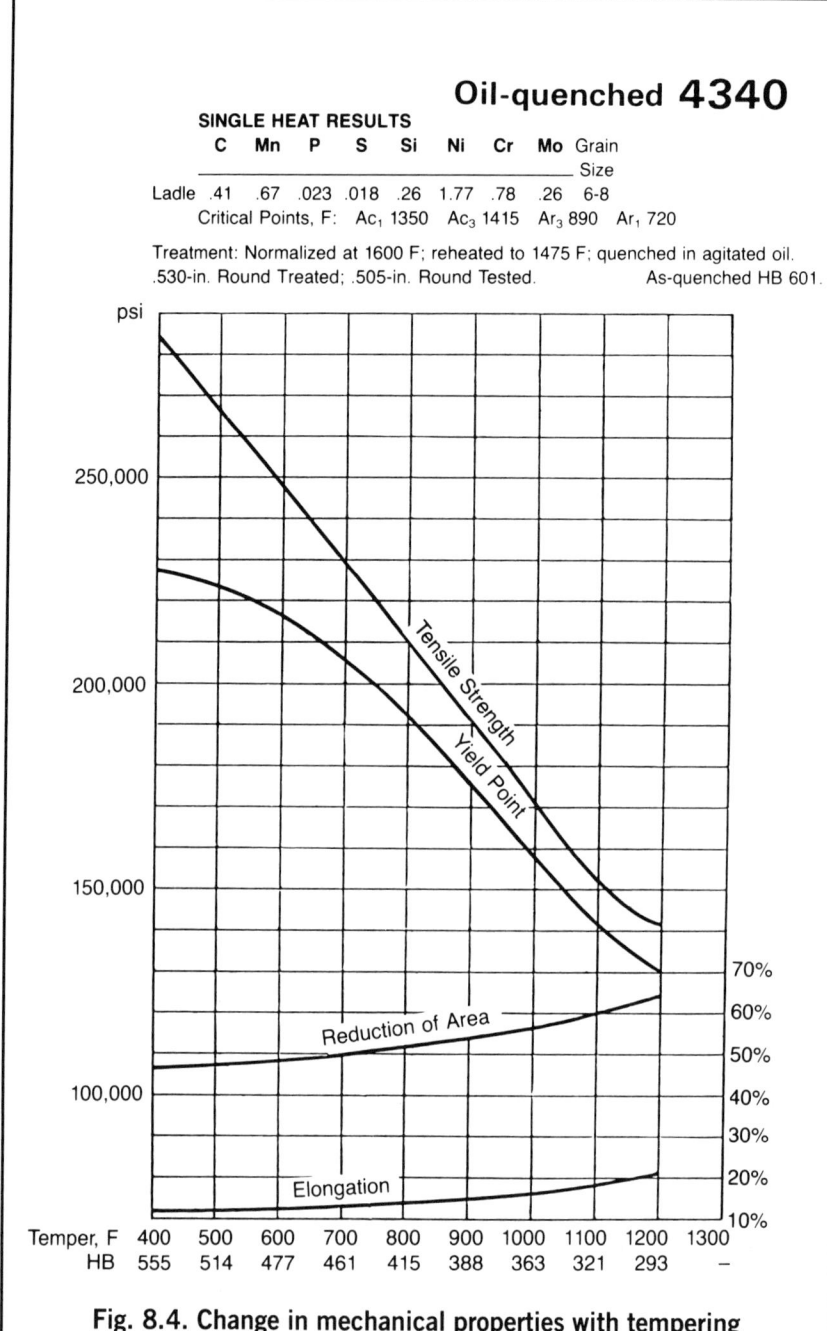

Fig. 8.4. Change in mechanical properties with tempering temperature for oil-quenched 4340 steel. (Ref 8.3)

Table 8.1. Mechanical Properties of Various Size Rounds of a Single Heat(a) of 4340 Steel After Various Heat Treatments Illustrating Mass Effects (Ref 8.3)

Size round mm	in.	Tensile strength MPa	ksi	Yield strength MPa	ksi	Elongation in 50 mm (2 in.), %	Reduction in area, %	Hardness, HB
Annealed: heated to 810 °C (1490 °F), furnace cooled 12 °C/h (20 °F/h) to 354 °C (670 °F), air cooled								
25.4	1	745	108	472	68.5	22.0	49.9	217
Normalized: heated to 871 °C (1600 °F), air cooled								
12.7	½	1448	210	972	141	12.1	35.3	388
25.4	1	1282	186	862	125	12.2	36.3	363
50.8	2	1220	177	793	115	13.5	37.3	341
101.6	4	1110	161	710	103	13.2	36.0	321
Oil quenched from 800 °C (1475 °F), tempered at 538 °C (1000 °F)								
12.7	½	1255	182	1165	169	13.7	45.0	363
25.4	1	1207	175	1145	166	14.2	45.9	352
50.8	2	1172	170	1103	160	16.0	54.8	341
101.6	4	1138	165	1000	145	15.5	53.4	331
Oil quenched from 800 °C (1475 °F), tempered at 593 °C (1110 °F)								
12.7	½	1145	166	1117	162	17.1	57.0	331
25.4	1	1138	165	1096	159	16.5	54.1	331
50.8	2	1014	147	958	139	19.0	60.4	293
101.6	4	924	134	793	115	19.7	60.7	269
Oil quenched from 800 °C (1475 °F), tempered at 650 °C (1200 °F)								
12.7	½	1000	145	938	136	20.0	59.3	285
25.4	1	958	139	883	128	20.0	59.7	277
50.8	2	931	135	834	121	20.5	62.5	269
101.6	4	855	124	730	106	21.7	63.0	255

As-quenched hardness (oil), HRC

Size round mm	in.	Surface	½ radius	Center
12.7	½	58	58	56
25.4	1	57	57	56
50.8	2	56	55	54
101.6	4	53	49	47

(a) Ladle composition: 0.40 C; 0.68 Mn; 0.020 P; 0.013 S; 0.28 Si; 1.87 Ni; 0.74 Cr; 0.25 Mo; grain size 7-8.

that form on tempering. In as-quenched specimens, dislocations tangle and form a tight substructure of fine cells with increasing deformation, but with large cementite particles present, the dislocations remain uniformly distributed and a well-defined dislocation cell structure

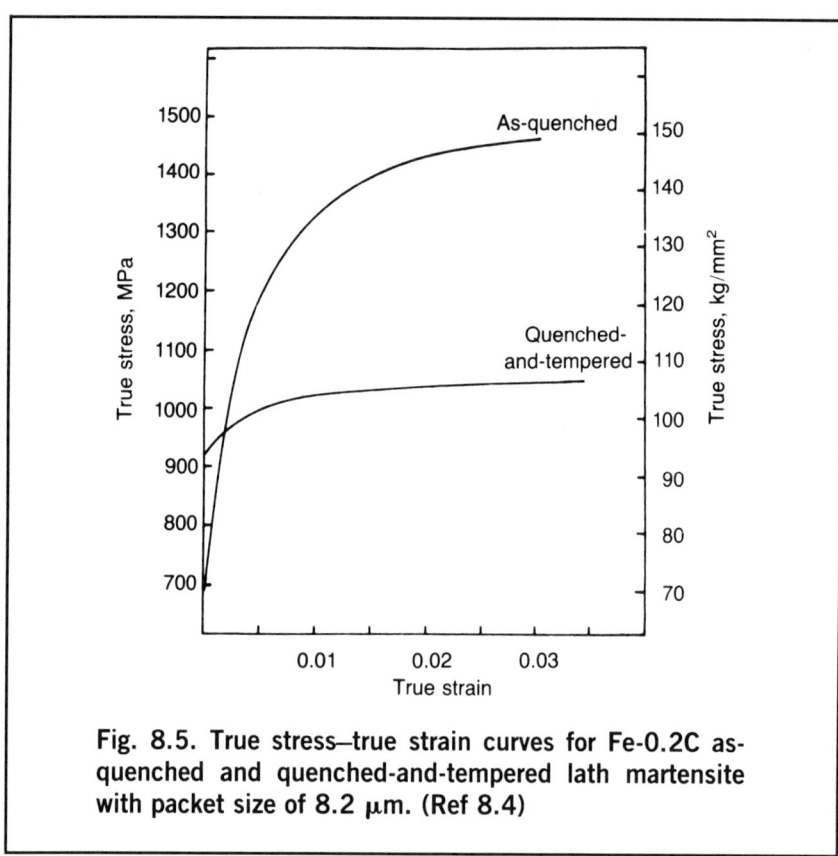

Fig. 8.5. True stress–true strain curves for Fe-0.2C as-quenched and quenched-and-tempered lath martensite with packet size of 8.2 μm. (Ref 8.4)

never develops. Figure 8.6 shows a uniform distribution of dislocations in the tempered Fe-0.2C martensite. This distribution did not change on deformation (Ref 8.4).

Alloying Elements and Tempering

In addition to increasing hardenability, certain alloying elements also help to retard the rate of softening during tempering. The most effective elements in this regard are strong carbide formers such as chromium, molybdenum, and vanadium. Without these elements, iron-carbon alloys and low-carbon steels soften rapidly with increasing tempering temperature as shown in Fig. 8.2. Figure 8.7 (Ref 8.5) similarly shows the softening as a function of tempering and carbon content in another form of diagram. This softening is largely due to the rapid coarsening of cementite with increasing tempering temperature, a process dependent on the diffusion of carbon and iron. If present in a

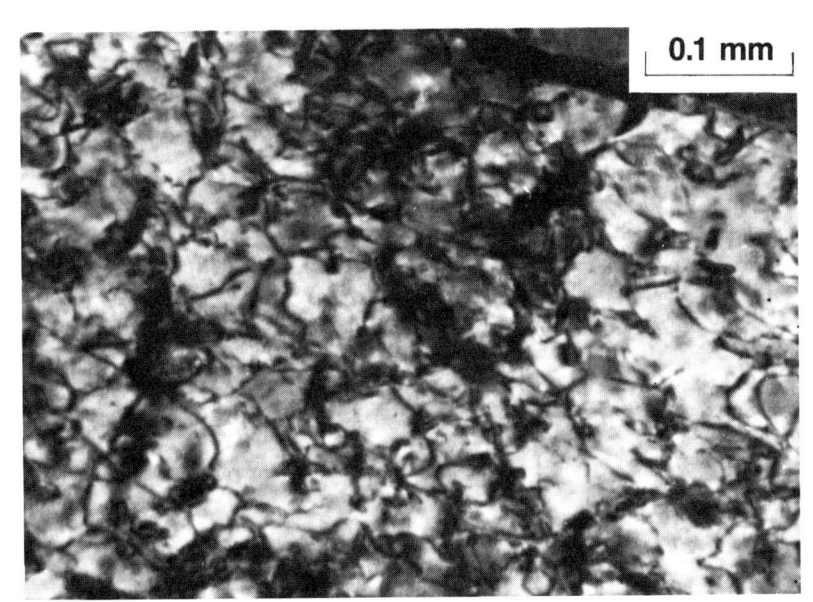

Fig. 8.6. Uniform distribution of dislocations and cementite particles in tempered Fe-0.2C martensite. Transmission electron micrograph. (Ref 8.4)

steel in sufficient quantity, however, the carbide-forming elements not only retard softening but also form fine alloy carbides that produce a hardness increase at higher tempering temperatures. This hardness increase is frequently referred to as secondary hardening.

Figure 8.8 shows secondary hardening in a series of steels containing molybdenum (Ref 8.6, 8.7). The higher the molybdenum content, the higher the hardness associated with the secondary hardening peak, and even at 0.47% molybdenum when no hardness peak is observed, a significant retardation of softening is apparent. The secondary hardening peaks develop only at high tempering temperatures because alloy carbide formation depends on the diffusion of the carbide-forming elements, a more sluggish process than that of carbon and iron diffusion. As a result, not only is a finer dispersion of particles produced, but also once formed, the alloy carbides are quite resistant to coarsening. The latter characteristic of the fine alloy carbides is used to advantage in tool steels that must not soften even though high temperatures are generated by their use in hot working dies or high-speed machining. Also, ferritic low-carbon steels containing chromium and molybdenum are used in pressure vessels and reactors

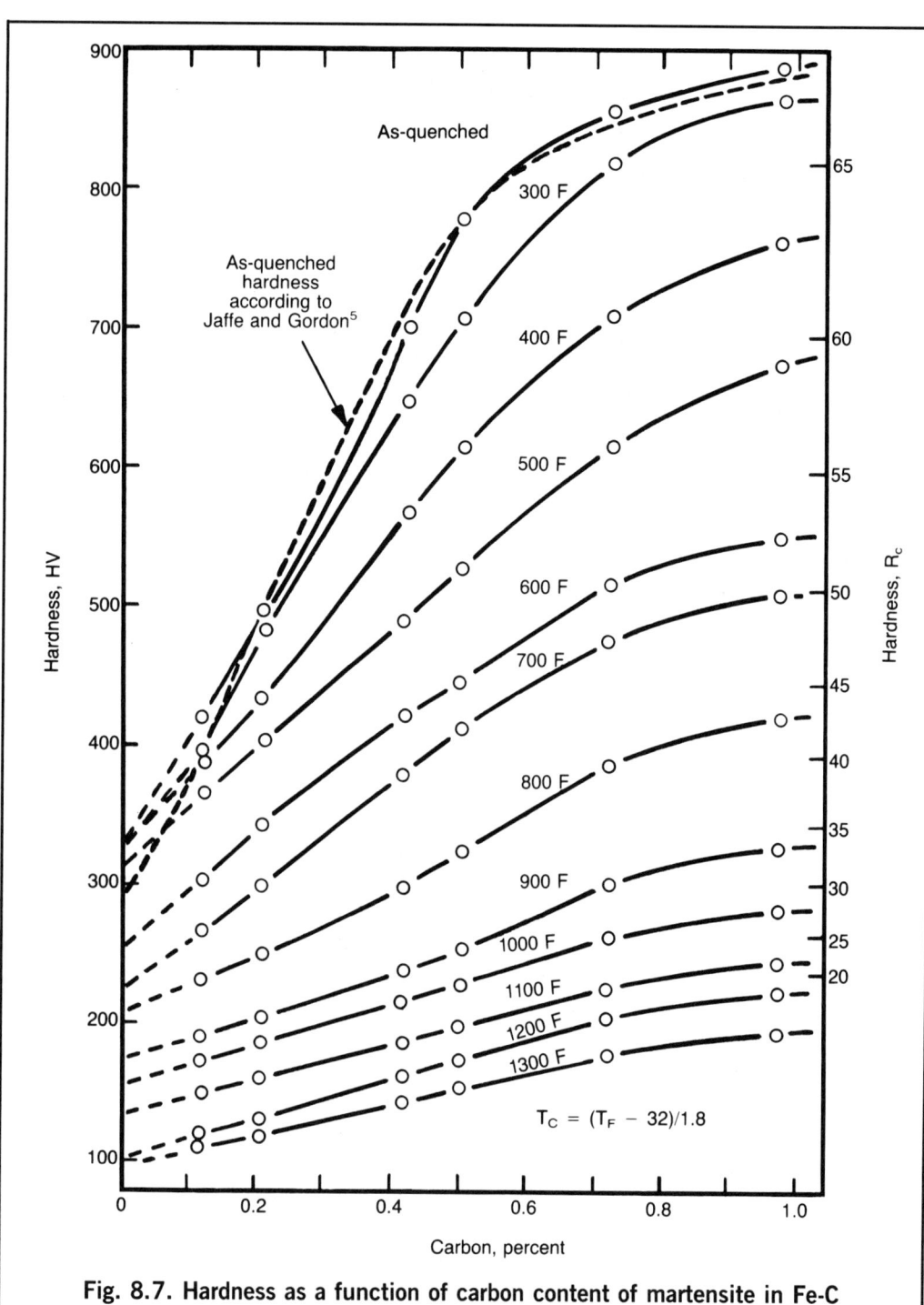

Fig. 8.7. Hardness as a function of carbon content of martensite in Fe-C alloys tempered at various temperatures. (Ref 8.5)

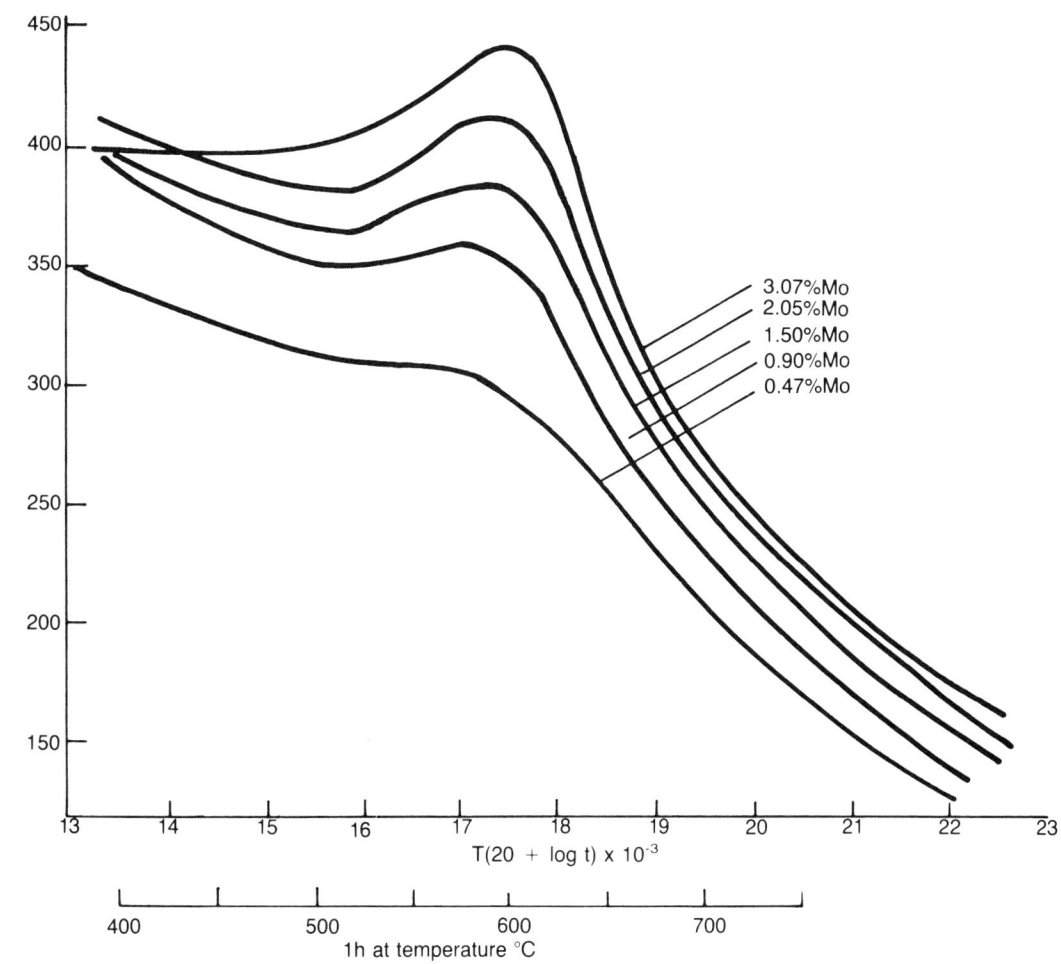

Fig. 8.8. Retardation of softening and secondary hardening during tempering of steels with various molybdenum content. (Ref 8.6)

operated at temperatures around 540 °C (1000 °F) because the alloy carbides resist coarsening at those temperatures and therefore provide good creep resistance.

Up to this point, tempering has been discussed with temperature as the major variable. The structural changes responsible for the property changes, however, are thermally activated and therefore dependent on both temperature and time. For example, if a single mechanism of structural change is operating during a stage of tempering, say the coarsening of cementite, a given hardness may be obtained by tempering at a high temperature for a short time or by

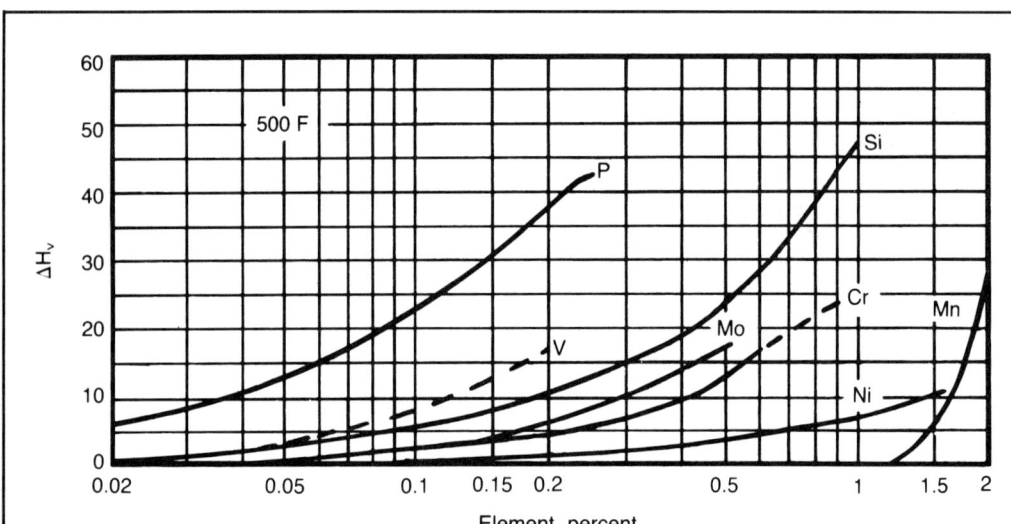

Fig. 8.9. Effect of alloying elements on the retardation of softening during tempering at 260 °C (500 °F) relative to Fe-C alloys. (Ref 8.5)

tempering at a lower temperature for a longer time. Generally if time is not mentioned, as is the case for most of the preceding figures, a constant tempering time of 1 h is assumed.

The interchangeability of time and temperature is accomplished by use of a tempering parameter, $T(20 + \log t) \times 10^{-3}$, where T is temperature in Kelvin, and t is time in hours. Figure 8.8 shows hardness changes plotted as a function of the tempering parameter as well as a function of tempering temperature where the time has been held constant at 1 h. Thus in a given alloy, tempering treatments with times other than 1 h may be selected to obtain a given hardness. While the tempering parameter is successfully applied to plain carbon steels, caution must be used in applying it to secondary hardening steels. During secondary hardening, the maximum hardness obtained on tempering is frequently a function of temperature (Ref 8.7, 8.8). For example, a higher maximum hardness may be obtained by holding at 600 °C (1112 °F) rather than at 700 °C (1292 °F), and it would be impossible to reproduce the 600 °C (1112 °F) hardness maximum even with very short-time tempering at 700 °C (1292 °F). This inability of different combinations of time and temperature to reproduce the same hardness is due to the somewhat coarser distribution of alloy carbides and/or their lower degree of coherency with the matrix at higher secondary hardening temperatures.

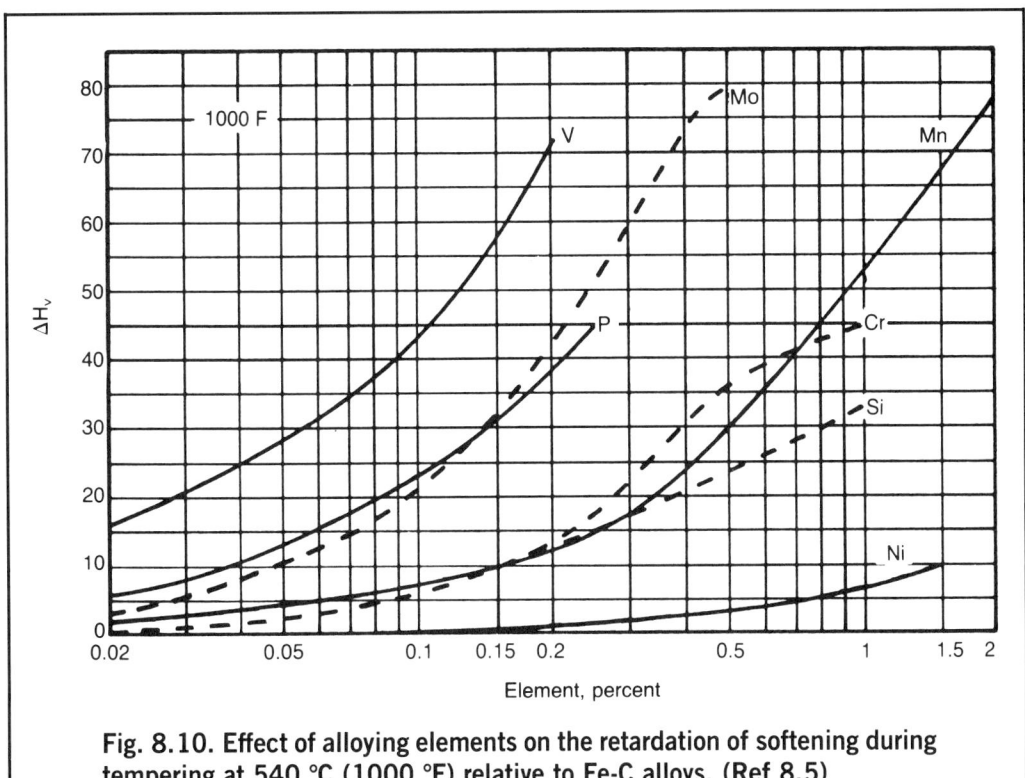

Fig. 8.10. Effect of alloying elements on the retardation of softening during tempering at 540 °C (1000 °F) relative to Fe-C alloys. (Ref 8.5)

The effect of alloying elements on hardness changes produced by tempering martensitic carbon and low-alloy steels has been summarized in the investigation by Grange, Hribal, and Porter (Ref 8.5). Steels with silicon, manganese, phosphorus, nickel, chromium, molybdenum, and vanadium additions up to 1.5% were examined. Graphs of hardness differences (ΔH_v) relative to tempered Fe-C alloys, as a function of alloying element content, were obtained for tempering temperatures between 200 and 700 °C (400 and 1300 °F) for constant tempering times of 1 h. Figures 8.9 and 8.10 show plots of ΔH_v versus alloying element content for martensite tempered at 260 °C (500 °F) and 540 °C (1000 °F), respectively. When the ΔH_v for each element at a given tempering temperature from plots similar to those of Fig. 8.9 and 8.10 is added to the base tempered hardness as determined by carbon content and tempering temperature (see Fig. 8.7), the final tempered hardness of a plain-carbon or low-alloy steel may be readily estimated.

Figures 8.9 and 8.10 reflect interesting differences due to the various alloying elements. The strong carbide formers, as discussed relative to secondary hardening, do not have a large effect until high temperatures are reached. Nickel has a very small and constant effect on tempered hardness at all temperatures, and since it is not a carbide-forming element, its influence is considered to be due to a weak solid solution strengthening effect. Silicon has a substantial retarding effect on softening around 316 °C (600 °F), an effect attributed (Ref 8.9) to its inhibition of the transformation of the low-temperature transition carbide to the more stable cementite. Manganese at low tempering temperatures has little effect on softening, but at higher temperatures has a strong effect, perhaps because of the incorporation of the manganese into the carbides at higher temperatures (Ref 8.10) and the attendant resistance to cementite coarsening that is associated with manganese diffusion.

Structural Changes on Tempering

The structure of a steel quenched to form martensite is highly unstable. Reasons for the instability include the supersaturation of carbon atoms in the body-centered tetragonal crystal lattice of martensite, the strain energy associated with the fine dislocation or twin structure of the martensite, the interfacial energy associated with the high density of lath or plate boundaries, and the retained austenite that is invariably present even in low-carbon steels. The supersaturation of carbon atoms provides the driving force for carbide formation; the high strain energy the driving force for recovery; the high interfacial energy the driving force for grain growth or coarsening of the ferrite matrix; and the unstable austenite the driving force for transformation to mixtures of ferrite and cementite on tempering. Thus, even without the alloying effects discussed in the preceding section, there are many factors at work to produce the microstructures responsible for the mechanical property changes that develop when martensitic carbon steel is tempered.

An important series of papers on tempering carbon steels was published by Cohen and his colleagues in the 1950s (Ref 8.11-8.14). As a result of systematic x-ray, dilatometric, and microstructural observations, three distinct stages of tempering were identified:

Stage I: The formation of a transition carbide, epsilon carbide (or eta carbide as discussed below), and the lowering of the carbon content of the matrix martensite to about 0.25% carbon.

Stage II: The transformation of retained austenite to ferrite and cementite.

Stage III: The replacement of the transition carbide and low-carbon martensite by cementite and ferrite.

The temperature ranges for the three stages overlap, depending on the tempering times used, but the temperature ranges of 100 to 250 °C (212 to 482 °F), 200 to 300 °C (392 to 572 °F), and 250 to 350 °C (482 to 662 °F) are generally accepted for the first, second, and beginning third stages, respectively (Ref 8.15). The formation of the alloy carbides responsible for secondary hardening is sometimes referred to as the fourth stage of tempering. Also, it is now recognized that carbon atom segregation to dislocations and various boundaries may occur during quenching and/or holding at room temperature (Ref 8.15-8.17) and carbon atom clustering in as-quenched martensite may precede carbide formation (Ref 8.18, 8.19) that occurs in the first stage of tempering. Even other structural changes due to carbon atom rearrangement have been found to precede the classical Stage I tempering of iron-carbon martensites (Ref 8.20, 8.21). Nagakura and his colleagues have identified a modulated structure associated with clustering of carbon atoms on (102) planes of martensite and a long-period ordered phase with an orthorhombic structure and composition of Fe_4C (Ref 8.20). The former structure forms on tempering between 0 and 90 °C (32 and 194 °F), while the latter structure forms between 60 and 80 °C (140 and 176 °F). Thus, tempering involves much more than three stages of tempering, but because of their central importance to understanding the behavior of tempered steels the three stages listed above will be discussed in more detail.

The transition carbide that forms in the first stage of tempering was first identified as having a hexagonal structure and designated epsilon (ϵ) carbide by Jack (Ref 8.22). More recently, Hirotsu and Nagakura (Ref 8.23, 8.24) have shown that the transition carbide has an orthorhombic structure isomorphous with transition metal carbides of the M_2C type. The transition carbide with the latter structure was designated as eta (η) carbide. The structures of the epsilon and eta carbides are very similar and are differentiated primarily by electron diffraction spots that come from a regular array of carbon atoms (or sublattice of carbon atoms) in the eta carbide (Ref 8.23, 8.24). Both the epsilon carbide, $Fe_{2.4}C$ (Ref 8.11), and the eta carbide, Fe_2C (Ref 8.23), have carbon contents substantially higher than that of the cementite, Fe_3C, that forms at higher temperatures. Kinetic studies show that the first stage of tempering is dependent on the diffusion of carbon through the martensite with an activation energy of 16 000 cal/mol (Ref 8.11).

Figures 8.11 through 8.13 are transmission electron micrographs that show various aspects of the transition carbide formation in the

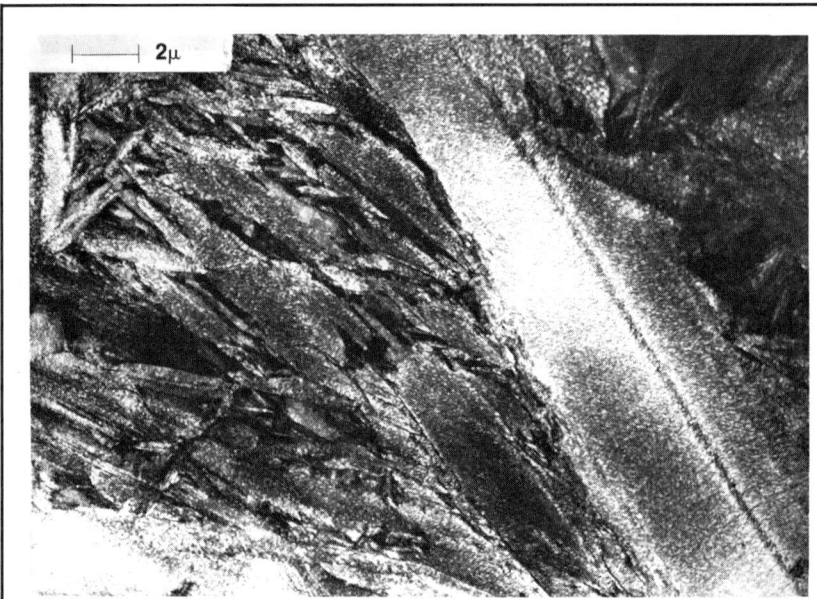

Fig. 8.11. Martensitic microstructure in an Fe-1.2C alloy tempered at 150 °C (302 °F). The microstructure consists of plates of various sizes containing uniform arrays of very fine carbides and retained austenite (black patches). Transmission electron micrograph. (Ref 8.25)

martensite of an Fe-1.22C alloy tempered at 150 °C (302 °F) for 16 h. Figure 8.11 shows a typical plate martensitic microstructure with plates of a variety of sizes and patches of retained austenite (black areas) between the plates. Each of the plates contains a highly uniform distribution of fine carbon particles. Figure 8.12 shows a typical array of transition carbides, identified as eta carbides (Ref 8.2), in a single plate of martensite. The carbides appear to be in the form of fine platelets, but Fig. 8.13, a dark-field micrograph taken with illumination from a carbide diffraction spot, shows that the eta carbide is actually present as rows of fine spherical particles about 2 nm in diameter (Ref 8.2, 8.23). The dark contrast in the plate-like morphology associated with the carbides in Fig. 8.12 is apparently due to strain effects between the martensitic matrix and the rows of particles.

The transformation of retained austenite during tempering occurs only after the transition carbide is well established. Figure 8.14 shows the rate of transformation of the retained austenite in an Fe-1.22C alloy at three different tempering temperatures. About 19% retained

Fig. 8.12. Distribution of eta carbide in martensite plate of an Fe-1.22C alloy tempered at 150 °C (302 °F) for 16 h. Transmission electron micrograph. Magnification, 80 000×; shown here at 75%. (Ref 8.2)

austenite, distributed as shown in Fig. 8.11, was initially present in the as-quenched structure. Even at 180 °C (356 °F) the retained austenite transformed completely to mixtures of ferrite and cementite if held for sufficiently long times. Analysis (Ref 8.25) of the austenite transformation kinetics in Fig. 8.14 yielded an activation energy of 1.15×10^5 J/mol (27 kcal/mol) in good agreement with the activation energies for the diffusion of carbon in austenite (Ref 8.26) and the activation energy for the second stage of tempering reported by Roberts, Averbach, and Cohen (Ref 8.11). Figure 8.15 shows that retained austenite is present in small amounts, about 2 and 4%, in as-quenched specimens of 4130 and 4340 steels, respectively, and that for tempering times of 1 h the transformation of retained austenite in these low-alloy medium-carbon steels begins only above 200 °C (392 °F). Transformation is complete at about 300 °C (572 °F) (Ref 8.27), and cementite becomes an important part of the microstructure, after tempering at 300 °C (572 °F) and higher temperatures.

Fig. 8.13. Rows of fine spherical eta carbide particles in a martensite plate of an Fe-1.22C alloy tempered at 150 °C (302 °F) for 16 h. Dark-field transmission electron micrograph. Magnification, 80 000×; shown here at 75%. (Ref 8.2)

The third stage of tempering consists of the formation of ferrite and cementite as required by the Fe-C diagram. However, there is some evidence, especially in high-carbon steels, that Hägg or chi (X)-carbide formation precedes cementite or theta (Θ)-carbide formation (Ref 8.28, 8.29). The chi carbide has a monoclinic structure, and the composition Fe_5C_2. However, despite the differences between cementite and chi carbide, the relatively complex structures of the two carbide phases are similar and difficult to separate by x-ray or electron diffraction techniques. Therefore, in view of the experimental difficulty in separating the presence of chi carbide from that of cementite, the temperature and compositions of the steels in which chi carbide forms are not yet completely defined.

Figure 8.16 shows the dense carbide distribution that has formed in the martensite of an Fe-1.22C alloy tempered at 350 °C (662 °F). In this case, the carbides were best identified as chi carbide (Ref 8.30). Two carbide morphologies are present: those that have nucleated and grown within the martensite plates, and very long planar carbides that have formed along the plate interfaces, perhaps as a result of the

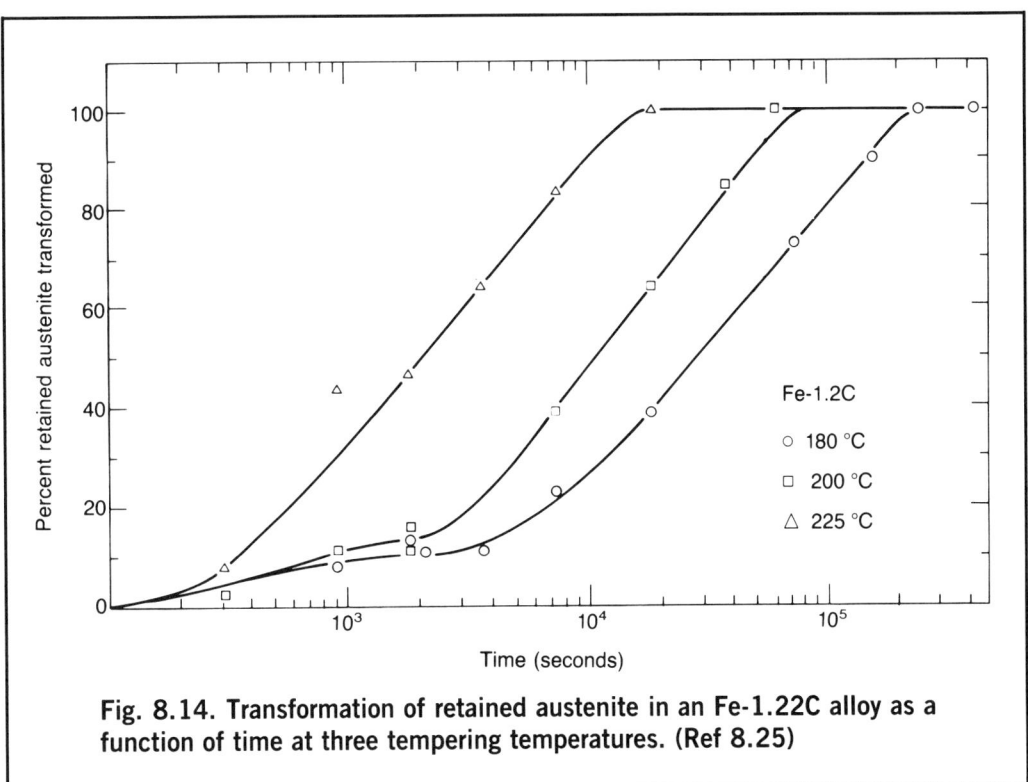

Fig. 8.14. Transformation of retained austenite in an Fe-1.22C alloy as a function of time at three tempering temperatures. (Ref 8.25)

transformation of retained austenite in the second stage of tempering. A third morphology of chi carbide and/or cementite in tempered high-carbon steels consists of parallel arrays of carbides formed on transformation twins sometimes present in high-carbon martensite, especially in the midrib portions of the plates (Ref 8.29). The carbides that have formed within the plates are coarser than the transition carbides and will eventually spheroidize if tempering is performed at higher temperatures. Nagakura *et al.* (Ref 8.20, 8.31) have shown that the transition from chi carbide, Fe_5C_2, to theta carbide (cementite), Fe_3C, takes place within single particles by development of sets of planes corresponding to higher order carbides of general composition $Fe_{2n+1}C_n$. The intergrowth of the various carbide structures is referred to as microsyntactic growth and requires only iron atom displacements and carbon diffusion.

The carbide structures and distributions that form in alloy steels and retard softening and/or produce secondary hardening during tempering are quite varied. Many of the alloy carbides and their formation on tempering have been characterized by Honeycombe and his colleagues. Much of this work, including descriptions of the carbide

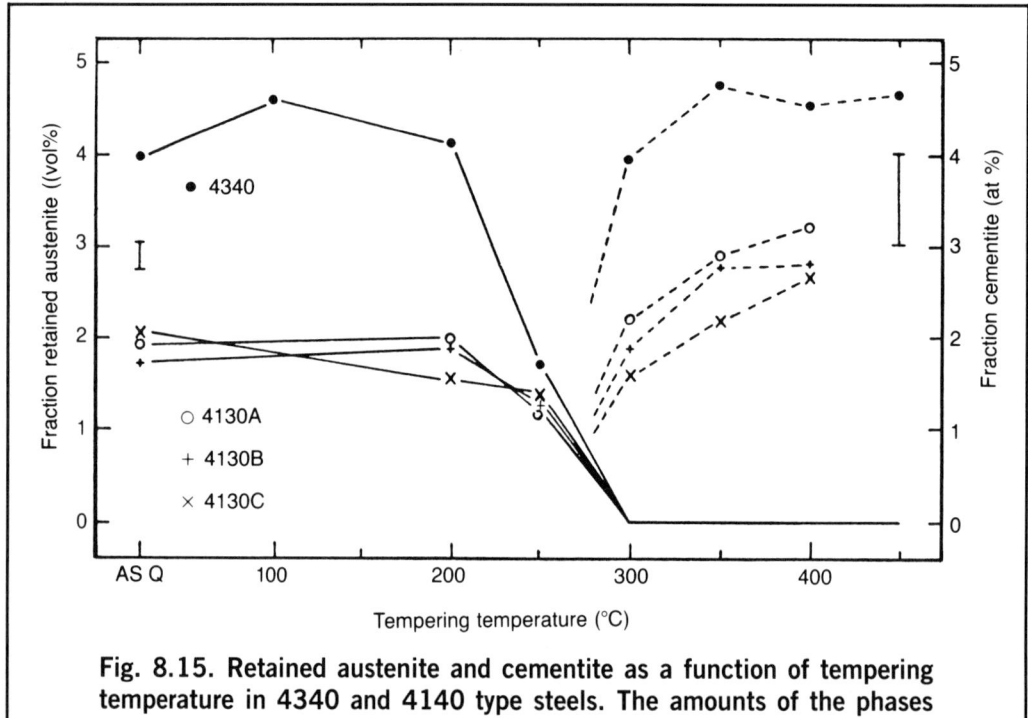

Fig. 8.15. Retained austenite and cementite as a function of tempering temperature in 4340 and 4140 type steels. The amounts of the phases were determined by Mössbauer spectroscopy. (Ref 8.27)

structures produced by tempering vanadium, molybdenum, tungsten, chromium, and titanium steels, is reviewed in Ref 8.32. The alloy carbide distributions formed in the secondary hardening range, 500 to 650 °C (932 to 1202 °F), depend on the nature of the cementite distribution formed at lower tempering temperatures, and the nature of the transformation of cementite to the alloy carbide. Honeycombe (Ref 8.32) presents evidence for two basic modes of alloy carbide formation on tempering. The carbides may form directly from the cementite, a mode referred to as *in situ* transformation, or the carbides may form by separate nucleation, after the cementite particles dissolve in the ferrite matrix. The independently nucleated alloy carbide particles are often nucleated on the dislocations residual from the as-quenched martensite, and tend to be much finer than the alloy carbides nucleated on the cementite particles.

Most of the structural changes discussed above have involved the formation of various types of carbides during tempering. There are also important changes in the martensitic matrix that accomplish the formation of fully tempered structures consisting of spheroidized carbides in a matrix of equiaxed ferrite grains. Figures 8.17

Fig. 8.16. Cementite and/or chi-carbide formation in martensitic structure of an Fe-1.22C alloy tempered at 350 °C (662 °F) for 1 h. Transmission electron micrograph. Magnification, 30 000×; shown here at 75%. (Ma, Ando, and Krauss, unpublished research, Colorado School of Mines, Golden)

through 8.20 show changes in the matrix structure that developed during the tempering of lath martensite in an Fe-0.2C alloy (Ref 8.33). Figure 8.17 shows that tempering at 400 °C (752 °F) for 15 min produces little change from the appearance of as-quenched lath martensite (see Chapter 3) on the scale resolvable with the light microscope. More pronounced changes are visible in a specimen tempered at 700 °C (1292 °F) (see Fig. 8.18) but even after this rather severe temper, the packet morphology with its parallel sub-units is still clearly visible. The major effects of tempering have been to eliminate many of the smaller laths and to produce coarse, spherical cementite particles at the prior austenite grain boundaries and within the packets. More severe tempering, 700 °C (1292 °F) for 12 h, begins to break up the remaining parallel blocks of crystals within the packets and more equiaxed ferrite grains begin to form (see Fig.

Fig. 8.17. Microstructure of lath martensite in an Fe-0.2C alloy after tempering at 400 °C (752 °F) for 15 min. Light micrograph. Nital etch. Magnification, 500×. (Ref 8.33)

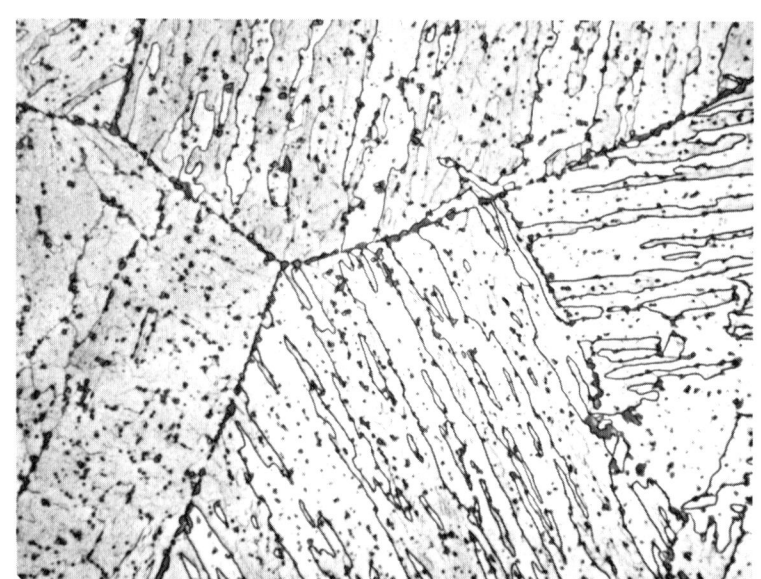

Fig. 8.18. Microstructure of lath martensite in an Fe-0.2C alloy after tempering at 700 °C (1292 °F) for 2 h. Light micrograph. Nital etch. Magnification, 500×. (Ref 8.33)

Fig. 8.19. Microstructures of lath martensite in an Fe-0.2C alloy after tempering at 700 °C (1292 °F) for 12 h. Light micrograph. Nital etch. Magnification, 500×. (Ref 8.33)

8.19). The equiaxed grains contain subboundaries made up of regular dislocation arrays as shown in the electron micrographs of Fig. 8.20.

Systematic measurement of the change in lath boundary per unit volume as a function of tempering of the Fe-0.2C martensite shows that the very high lath boundary area per unit volume of the fine laths in as-quenched martensite decreased very rapidly on tempering (Ref 8.33). This initial rapid decrease is primarily due to the elimination of the low-angle boundaries between laths of similar orientation. Simultaneously, fine carbides precipitate and help to stabilize the surviving lath boundaries to maintain their parallel orientation within the packets. All of these initial matrix changes occur as a result of recovery mechanisms. The dislocation density is effectively lowered not only by the reduction of dislocations within the laths but also by the elimination of the low-angle lath boundaries. Eventually, with coarsening of the carbide particles, the remaining large-angle boundaries rearrange themselves to produce more equilibrium junctions between grains as typical of the mechanisms associated with grain

228 / STEELS: HEAT TREATMENT AND PROCESSING PRINCIPLES

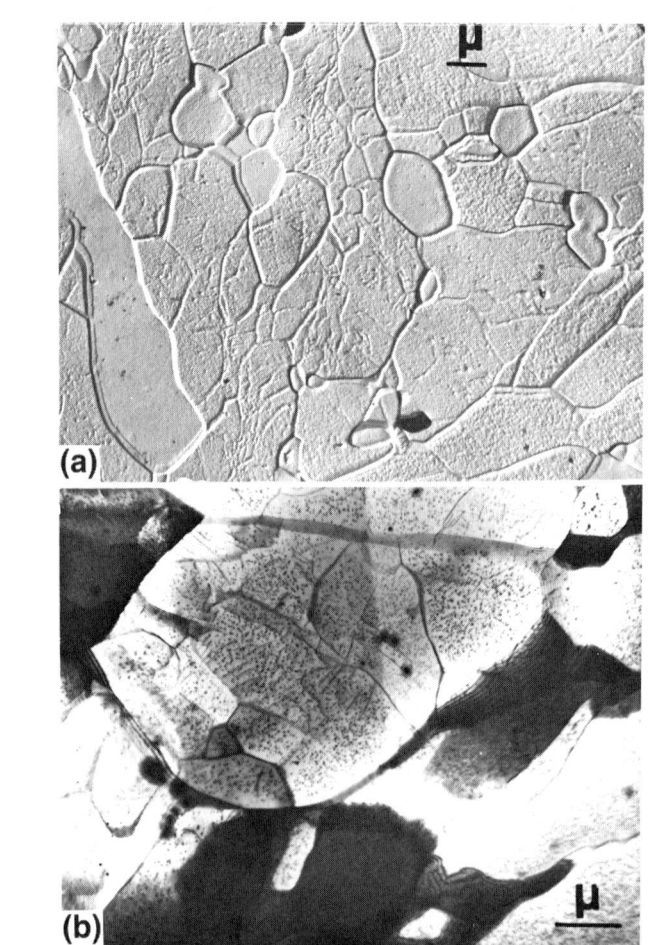

Fig. 8.20. Structure of lath martensite in an Fe-0.2C alloy after tempering at 700 °C (1292 °F) for 12 h. (a) Single-stage replica of polished and etched (nital) surface. (b) Transmission electron micrograph.

growth (Ref 8.34). Any residual dislocations within the laths then rearrange themselves into low-angle boundaries within the equiaxed grains as shown in Fig. 8.20. Such subdivisions of large grains by dislocation boundaries is referred to as polygonization. Thus, the formation of the equiaxed ferrite matrix that develops after long-time high-temperature tempering of a low-carbon lath martensite is accomplished by recovery and grain growth mechanisms (Ref 8.33, 8.35).

Apparently, the recovery mechanisms that operate early in tempering lower the strain energy of the as-quenched martensite to the point where there is no longer sufficient driving force for recrystallization.

Embrittlement Phenomena and Tempering

High-strength quench and tempered steels are susceptible to a number of different types of embrittlement. Some of the embrittlement mechanisms are due to structural changes introduced during processing and tempering, and some are due to the interaction of the environment with the quench and tempered microstructures. Examples of the first type of embrittlement are tempered martensite embrittlement, temper embrittlement, and embrittlement due to aluminum nitride formation. Examples of the second type are hydrogen embrittlement and liquid metal embrittlement. Often there is overlap between two types of embrittlement. For example, tempered martensite embrittlement may be severely aggravated by exposure to hydrogen. This section compares and describes general features of the two classic embrittlement phenomena, tempered martensite embrittlement and temper embrittlement. Each of the embrittlement phenomena is then described in more detail in separate sections.

The characteristics of tempered martensite embrittlement and temper embrittlement are well known, but the causes continue to receive intensive study. Tempered martensite embrittlement (TME) occurs after tempering between 260 to 370 °C (500 to 700 °F) and is also referred to as 350 °C embrittlement or 500 °F embrittlement. Temper embrittlement (TE) occurs after tempering in or cooling through the temperature range of 375 to 575 °C (707 to 1070 °F). In an effort to introduce a terminology more related to the steps necessary to induce the two types of embrittlement, McMahon has suggested the term "one-step embrittlement" for TME and the term "two-step embrittlement" for TE, the latter term because two tempering treatments or a heating step and a cooling step are sometimes required to induce temper embrittlement (Ref 8.36).

Both TME and TE are associated with a shift in the impact transition temperature, as shown schematically for two hypothetical steels in Fig. 8.21. Steel A represents an inherently tougher steel than Steel B in that its transition temperature in the unembrittled state is lower than that of Steel B. Thus, when an embrittlement phenomenon

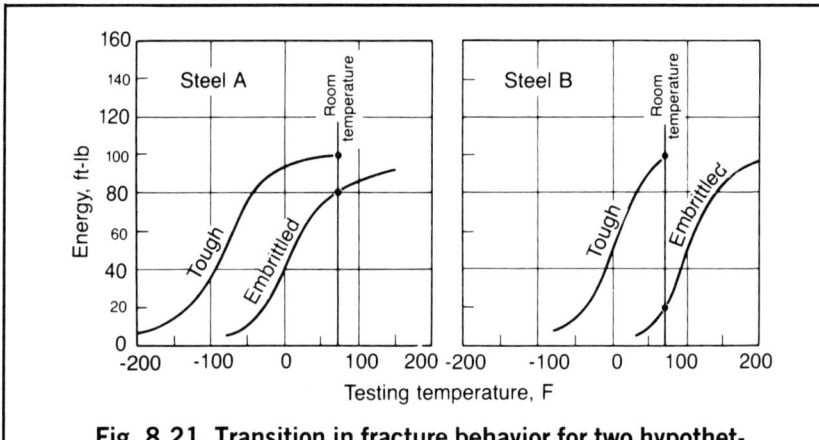

Fig. 8.21. Transition in fracture behavior for two hypothetical steels in the tough and embrittled conditions. (Ref 8.1)

shifts the transition temperature to higher temperatures by an equal amount for both steels, Fig. 8.21 shows that Steel A does not show as sharp a drop in room temperature toughness as does Steel B. Figure 8.21, therefore, emphasizes the importance of the testing temperature relative to the transition temperature, and shows that the results of room temperature testing may be misleading. For example, if Steel A is subjected to impact loading in service below room temperature, it will behave in a much more brittle fashion than indicated by the room temperature test.

Although there are similarities in the effects of the two types of embrittlement, from a practical standpoint TME and TE are separable into two different phenomena because they occur in two different temperature ranges and because TME is a much more rapid process than TE. TME develops within the normal 1-h time period for tempering, while TE takes many hours to accomplish. TE is therefore of major concern in heavy sections that are tempered at high temperatures to ensure a good balance of strength and toughness. In particular, large shafts and rotors for power generating equipment are susceptible to TE because even after tempering above the critical temperature range, the heavy sections cool very slowly over a period of many hours through the critical range for embrittlement. TME, on the other hand, develops during tempering for short times in the critical range and, therefore, is independent of section size and/or cooling rate after tempering.

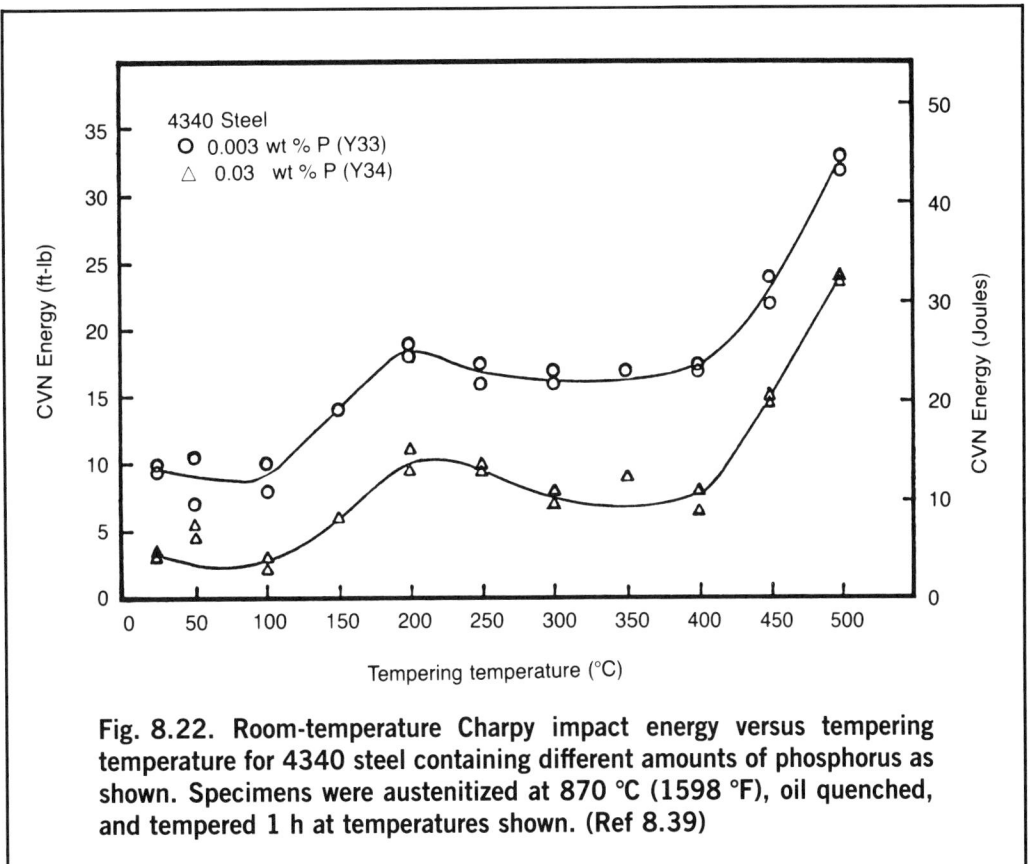

Fig. 8.22. Room-temperature Charpy impact energy versus tempering temperature for 4340 steel containing different amounts of phosphorus as shown. Specimens were austenitized at 870 °C (1598 °F), oil quenched, and tempered 1 h at temperatures shown. (Ref 8.39)

Tempered Martensite Embrittlement

Tempered martensite embrittlement may or may not be associated with impurity atom segregation to prior austenitic grain boundaries and may be associated with three different modes of fracture through the tempered martensite of specimens tempered between 260 to 370 °C (500 and 700 °F). The common factor for all the manifestations of TME, at least in medium-carbon hardenable steels, appears to be the formation of cementite during tempering. In steels with low impurity content and/or the effect of impurities minimized by the gettering action of an alloying element—for example, molybdenum interaction with phosphorus (Ref 8.37)—the source of the cementite that leads to brittle transgranular fracture is the decomposition of retained austenite in the second stage of tempering. This explanation for the cause of TME was first proposed by Thomas (Ref 8.38) after transmission electron microscopy revealed the presence of thin films of retained

232 / STEELS: HEAT TREATMENT AND PROCESSING PRINCIPLES

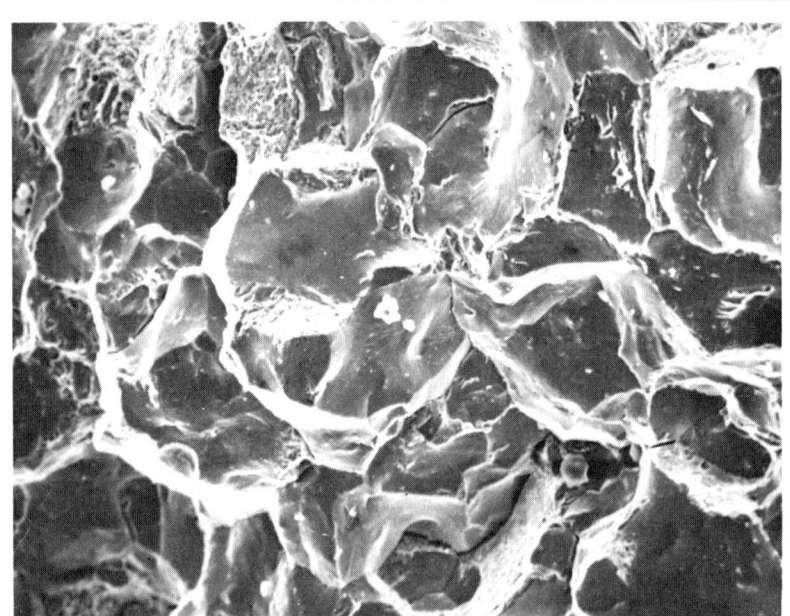

Fig. 8.23. Intergranular fracture of 4340 steel containing 0.03% phosphorus after tempering at 400 °C (752 °F). Specimen was broken by impact loading at room temperature. (Ref 8.39)

austenite between laths of martensite in as-quenched medium-carbon steels and the transformation of this retained austenite to thin plates of cementite on tempering. Figure 8.15 shows graphically that retained austenite is replaced by cementite in the critical temperature range for TME.

Figure 8.22 shows several characteristics of TME in two 4340 steels of the same composition except for phosphorus content (Ref 8.39). The impact toughness of the steel with the higher phosphorus content (0.03%) is inferior to that of the steel with the lower phosphorus content (0.003%) after tempering over the entire range of temperatures up to 500 °C (932 °F). Also, both steels show a trough or plateau in energy absorbed between 250 and 400 °C (482 and 752 °F). The lower toughness of the higher phosphorus steel was related to a substantial amount of intergranular fracture, about 20% after tempering up to 200 °C (392 °F), and over 80% after tempering between 300 and 400 °C (572 and 752 °F). Figure 8.23 shows the intergranular fracture along prior austenite grain boundaries of the high phosphorus 4340 steel broken at room temperature after tempering at 400 °C

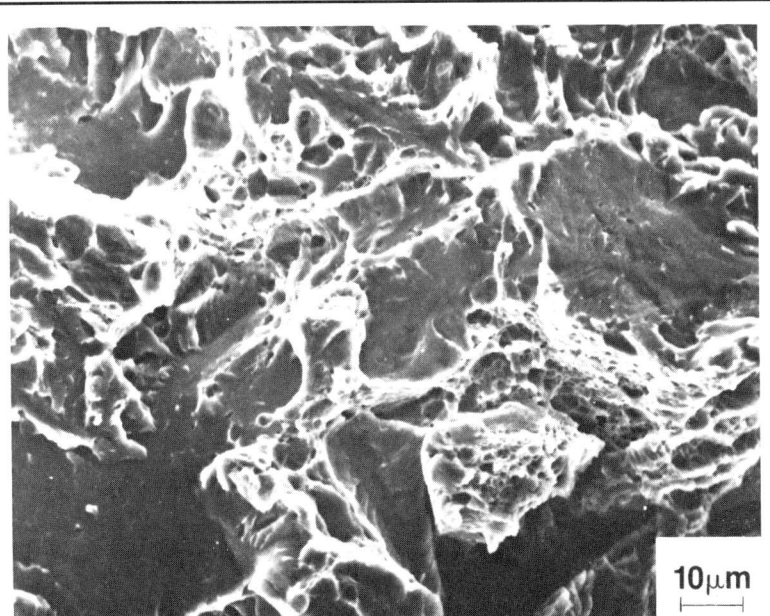

Fig. 8.24. Flat cleavage facets in 4340 steel containing 0.003% phosphorus after tempering at 350 °C (662 °F). Specimen was broken by impact loading at room temperature. (Ref 8.39)

(752 °F). The intergranular mode of fracture associated with TME is quite common and has been related to phosphorus segregation to the austenite grain boundaries during austenitizing (Ref 8.40-8.42). The phosphorus, therefore, is present at the grain boundaries in the as-quenched martensite and after tempering up to 200 °C (392 °F), but only after tempering in the range where cementite forms in tempered martensite does TME fully develop. The latter observation indicates that an interaction between phosphorus and cementite is necessary for the intergranular mode of TME.

Figure 8.24 shows the fracture surface of the low-phosphorus 4340 steel broken by impact testing at room temperature after tempering at 350 °C (662 °F). No intergranular fracture is apparent, but flat cleavage facets are interspersed among regions of ductile fracture. These cleavage facets were found to be oriented across the laths of a packet (Ref 8.39). The initiation of the translath cleavage was attributed to cracking of the relatively thick cementite particles as suggested by King, Smith, and Knott (Ref 8.43). Figure 8.25 shows interlath carbides in a specimen of low phosphorus content 4340 tempered at

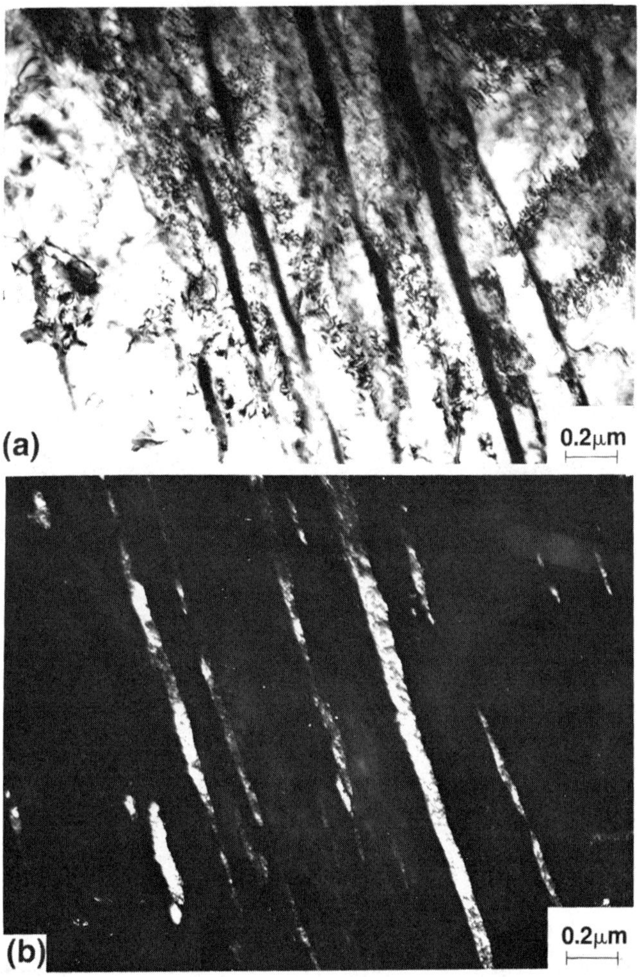

Fig. 8.25. Interlath carbides formed during tempering of 4340 steel containing 0.003% phosphorus at 350 °C (662 °F). (a) Bright-field image and (b) dark-field image using a cementite diffracted beam for illumination. Transmission electron micrographs. (Ref 8.39)

350 °C (662 °F). The carbides formed between laths as a result of the transformation of the 4% austenite retained in the as-quenched condition (see Fig. 8.15).

A ductile, transgranular mode of TME fracture is observed in lower carbon quenched and tempered steels such as 4130. Crack initiation and crack growth occur by microvoid coalescence around

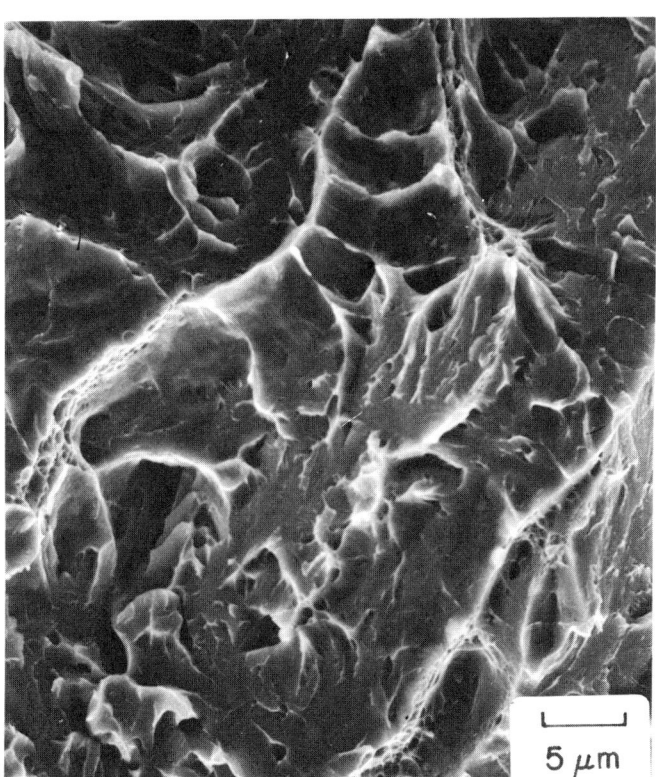

Fig. 8.26. Fracture surface of low-phosphorus 4130 steel tempered at 300 °C (572 °F). Scanning electron micrograph. (Ref 8.45)

carbide particles retained after austenitizing and produced during the second stage of tempering (Ref 8.44-8.46). Figure 8.26 shows the overload or unstable fracture from a CVN specimen of low-phosphorus 4130 steel tempered at 300 °C (572 °F). The fracture consists largely of microvoids, on the average larger than those observed in specimens tempered at 200 °C (392 °F), but some small, flat cleavage facets are also present. The overload fractures of the 4130 steel are preceded by shear initiation and stable crack growth, stages of crack growth accomplished by microvoid coalescence and ductile tearing (Ref 8.46).

Specimens of 4130 steel with low (0.002%) and high (0.017%) phosphorus exhibit almost identical sharp drops in room-temperature impact toughness after tempering at 300 °C (572 °F), leading to the conclusion that carbides produced during tempering, regardless of phosphorus content, are responsible for TME in the 4130 steel. Higher carbon steels, 4140 and 4150, by virtue of greater amounts of interlath and grain-boundary carbides, showed greater transgranular cleavage and intergranular fracture modes of TME (Ref 8.46).

Temper Embrittlement

Temper embrittlement is a long-standing metallurgical problem the causes of which are just now being understood because of new theoretical approaches and the recent availability of surface analysis equipment capable of detecting grain boundary segregation on an atomic scale (Ref 8.47). However, the many review articles (Ref 8.36, 8.47-8.51) on TE show that the tempering conditions and compositional factors that induce TE are well known even if the exact mechanisms still remain to be discovered.

Heat treatment factors include, as stated earlier, the fact that susceptible steels must be heated in or cooled through the critical temperature range of 375 to 575 °C (706 to 1070 °F) in order to develop TE. The embrittlement that results is detectable primarily by an increase in impact transition temperature as shown in Fig. 8.27 for a 3140 steel (containing nominally 1.15% nickel and 0.65% chromium) embrittled by both isothermal tempering and slow cooling through the critical temperature range. The embrittling kinetics follow C-curve behavior with tempering time and temperature, with a nose or minimum time for embrittlement at about 550 °C (1022 °F). One study (Ref 8.52) shows that it takes about an hour at 550 °C (1022 °F) for the first increase in transition temperature to be noticeable, and several hundred hours for the first signs of embrittlement at around 375 °C (706 °F), the lower temperature limit for TE. TE is reversible, and de-embrittlement may occur on heating to about 575 °C (1070 °F) after holding only a few minutes at temperature.

Compositional factors affecting TE include the requirement that specific impurities must be present for a steel to be susceptible. The impurities most detrimental are antimony, phosphorus, tin, and arsenic. Relatively small amounts of these elements, on the order of 100 ppm (0.01%) or less, have been shown to cause TE. Silicon and manganese in large amounts also appear to be detrimental. Plain carbon steels are not considered to be susceptible to TE, provided the manganese content is held below 0.5%. Alloy steels are most susceptible, especially the chromium-nickel steels that are frequently used for heavy rotors. Molybdenum, however, reduces susceptibility to TE and, in amounts of 0.5% or less, is an important alloying element added to steels to minimize TE.

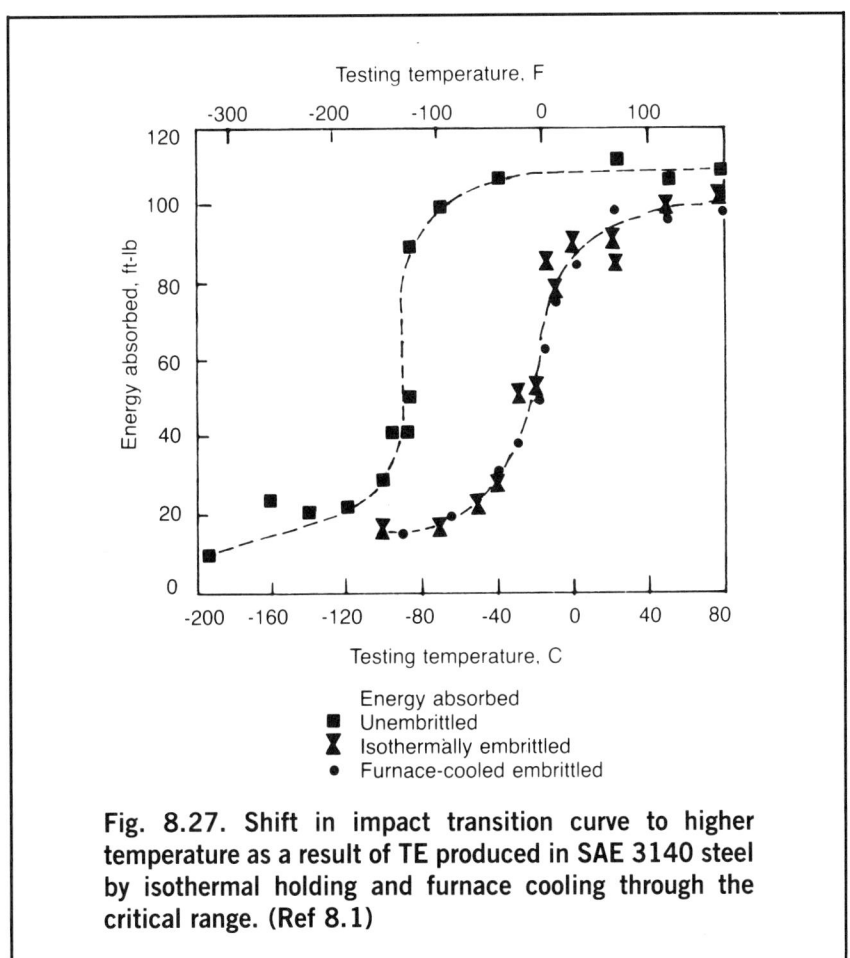

Fig. 8.27. Shift in impact transition curve to higher temperature as a result of TE produced in SAE 3140 steel by isothermal holding and furnace cooling through the critical range. (Ref 8.1)

The cause of TE has been difficult to identify because there is no readily resolvable microstructural feature identifiable with the characteristic intergranular fracture of embrittled specimens. The only metallographic evidence of embrittlement has been the ability of certain etchants to reveal prior austenite grain boundaries containing segregated phosphorus (Ref 8.47). Currently available surface analysis techniques, however, such as Auger electron spectroscopy (AES), which is capable of analyzing composition of atomic layers adjacent to the intergranular fracture surfaces, have contributed greatly to the understanding of TE. AES shows not only high concentrations of impurity atoms segregated to the fracture surfaces, but also gradients of the alloying elements such as nickel that may in fact stimulate the impurity element segregation to the prior austenite grain boundaries. For example, grain boundary carbides may reject the nickel as they grow and, therefore, produce nickel concentration gradients that in turn cause the concentration of impurity atoms (Ref 8.53). Further, the interactions of impurities and alloying elements that produce segregation have been treated in a thermodynamic model developed by Guttmann (Ref 8.54) with good success. Thus, research indicates that not just impurity elements but the interaction of impurities and alloying elements may be responsible for the segregation that leads to grain boundary decohesion in temper embrittled steels. For example, a quantitative assessment of the interactive cosegregation of phosphorus and common alloying elements shows that manganese weakly segregates on its own, and that the segregation of nickel, chromium, and molybdenum is driven by strong interactions with phosphorus (Ref 8.55). The grain-boundary interaction coefficients increase in the order nickel, manganese, chromium, and molybdenum. The very strong interaction between phosphorus and molybdenum correlates with the known beneficial effect of molybdenum on TE (Ref 8.37), and supports the formation of $(Mo,Fe)_3P$ or Mo-P atom clusters which prevent the segregation of phosphorus to grain boundaries. The Guttmann *et al.* study (Ref 8.55) also shows a strong repulsion between carbon and phosphorus, an interaction which is expected to strongly oppose phosphorus segregation.

As noted earlier, heavy forgings such as large pressure vessels and turbine rotors for electric power generation have historically been sensitive to temper embrittlement because of inherently slow cooling rates during processing or operation at temperatures in the critical TE range. Hydrogen flaking, the development of interior fissures due to decreasing hydrogen solubility during cooling, has also been a long-standing problem in the manufacture of heavy sections. The steels for rotor and pressure vessel applications are alloyed with chromium, nickel, manganese, and vanadium in order to provide sufficient hardenability in heavy sections for formation of bainitic microstructures with good creep resistance at temperatures around 400 °C (750 °F). Although the association of various elements with TE in these steels has been accumulating for many years (Ref 8.47-8.51), it is now largely by advances in steelmaking technology that steels can be manufactured to the level of cleanliness which is required to greatly reduce or eliminate TE (Ref 8.56).

Vacuum degassing has long been used to significantly reduce hydrogen flaking, and as steels have become cleaner, even greater efforts at hydrogen removal have been made to offset the loss of hydrogen gettering by sulfide inclusions (Ref 8.56). The most powerful attack on the removal of elements deleterious to toughness has been the development of ladle metallurgy, a steelmaking approach in which molten steel is refined in a vessel separate from the melting furnace. Desulfurization, dephosphorization, degassing, inclusion shape control, and alloying are all accomplished in a ladle to a degree not possible in a furnace (Ref 8.56). Thus phosphorus, a major contributor to TE, and sulfur, a major source of inclusions, can now be reduced to very low levels.

Low sulfur and the attendant reduction of sulfide inclusions greatly improve toughness, but increase sensitivity to overheating during forging, reduce machinability, and reduce trapping sites for hydrogen, secondary effects that must be weighed against the benefits of low sulfur (Ref 8.56).

Elements such as antimony, arsenic, and tin are not oxidizable during steelmaking and must be controlled by careful selection of steel scrap which is melted in electric furnaces. These elements have long been associated with TE, and more recently have been related to hot shortness in low-carbon manganese- and niobium-containing steels in the temperature range between 900 and 1100 °C (1652 and 2012 °F). Nachtrab and Chou (Ref 8.57) have shown that the reduction in hot ductility is related to intergranular cracking developed by austenite grain boundaries containing AlN and Nb(C,N) particles and segregated cobalt, tin, and antimony.

The most recent approach to eliminating TE is the reduction of manganese and silicon to very low levels, on the order of 0.01 to 0.03 wt.%, in rotor and nuclear reactor steels (Ref 8.58, 8.59). These elements have traditionally been used for alloying and deoxidation, but considerable information now directly ties manganese and silicon, even in moderate amounts, to TE by direct segregation or cosegregation with phosphorus or other alloying elements. For example, Weng and McMahon (Ref 8.60) show that 0.3% manganese greatly increases the susceptibility of a Ni-Cr-Mo-V rotor steel to TE relative to a steel without manganese, and that manganese and phosphorus strongly cosegregate in an Fe-Mn alloy. Other references linking manganese and silicon in TE are reviewed by Bodner *et al.* (Ref 8.59).

Aluminum Nitride Embrittlement

Another type of intergranular embrittlement sometimes encountered in hardened steels is that associated with aluminum nitride precipitation on prior austenite grain boundaries. Whereas a fine dispersion of aluminum nitride particles is desirable to control grain growth in steels (see Chapter 7), aluminum nitrides precipitated in the form of sheet-like particles on cooling from solidification, or reprecipitated on cooling after solution at high austenitizing temperatures,

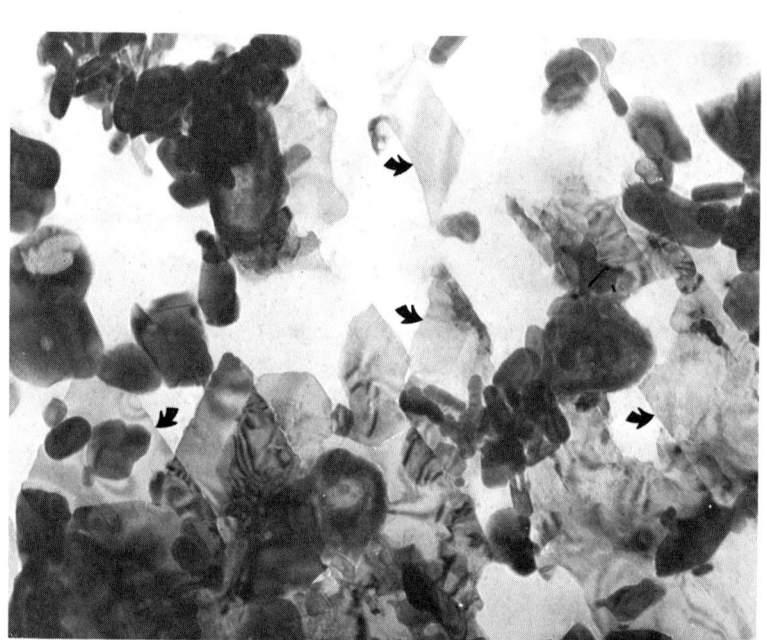

Fig. 8.28. Thin aluminum nitride particles (arrows) extracted from the intergranular fracture surface of a medium-carbon cast steel. Thick or dark particles are carbides. Extraction replica electron micrograph. Magnification, 82 500×; shown here at 75%. (G. Krauss, unpublished research, Colorado School of Mines, Golden)

may significantly reduce toughness. The intergranular fracture due to aluminum nitride particles is most frequently encountered in carbon steel castings (Ref 8.61, 8.62), and is sometimes referred to as "rock-candy" fracture because the coarse intergranular facets of the castings produce a macroscopically crystalline appearance. Figure 8.28 shows aluminum nitride and carbide particles extracted from the intergranular fracture surface of a medium-carbon steel casting. The aluminum nitride particles are characteristically very thin, with a plate or sheet-like morphology, and are readily identified by electron diffraction (Ref 8.61). Although the intergranular precipitation of aluminum nitrides is most closely tied to intergranular fracture in cast steels, plate-like aluminum nitride particles have also been observed in wrought medium-carbon steels (Ref 8.63) and, if present in high concentration because of high nitrogen and aluminum content, may contribute to the intergranular fracture associated with TME (Ref 8.40, 8.64). The solubility products, morphologies, and many effects of AlN in cast and wrought steels are comprehensively reviewed by Wilson and Gladman (Ref 8.65).

Liquid Metal Embrittlement

The exposure of steels to liquid metals may also cause brittle fracture by intergranular cracking (Ref 8.66). Plain carbon and low-alloy steels may be embrittled by exposure to liquid lead, cadmium, brass, aluminum bronze, copper, zinc, lead-tin solders, and lithium (Ref 8.67). The initiation of fracture by liquid metal embrittlement is not time dependent but begins immediately on wetting of the steels. Often very low stress is sufficient to cause fracture by this mechanism.

Of special interest with respect to heat treatment of carbon steels is the work of Breyer and his colleagues concerning embrittlement of steels by lead (Ref 8.68-8.70). Lead may cause embrittlement if externally applied or if present internally in a steel. The latter cause of embrittlement is magnified by the use of leaded steels for improved machinability. Figure 8.29 shows an extreme example of the effects of lead embrittlement in a leaded 4145 steel heat treated to strengths close to 200 ksi (1380 MPa). At testing temperatures between 200 and 480 °C (400 and 900 °F) the ductility is significantly reduced, with the most severe reduction to zero ductility occurring at and above the melting point of lead, 327 °C (621 °F). Generally the embrittlement is more severe the higher the strength level of the steel and therefore quench and tempered steels, if leaded, are especially susceptible. The fractures associated with the embrittlement are generally intergranular (Ref 8.70).

In summary, three conditions are necessary for lead embrittlement: (1) the presence of lead either externally or internally in the steel, (2) tensile loading, and (3) temperatures between 200 and 480 °C (400 and 900 °F). The absence of any one of these three conditions will prevent the brittle fracture associated with liquid lead embrittlement.

Hydrogen Embrittlement

There are many embrittling effects of hydrogen on steels: the ultimate strength of a steel may be reduced, ductility as measured by total elongation to fracture or reduction of area may be decreased, and crack growth may be significantly accelerated. The hydrogen respon-

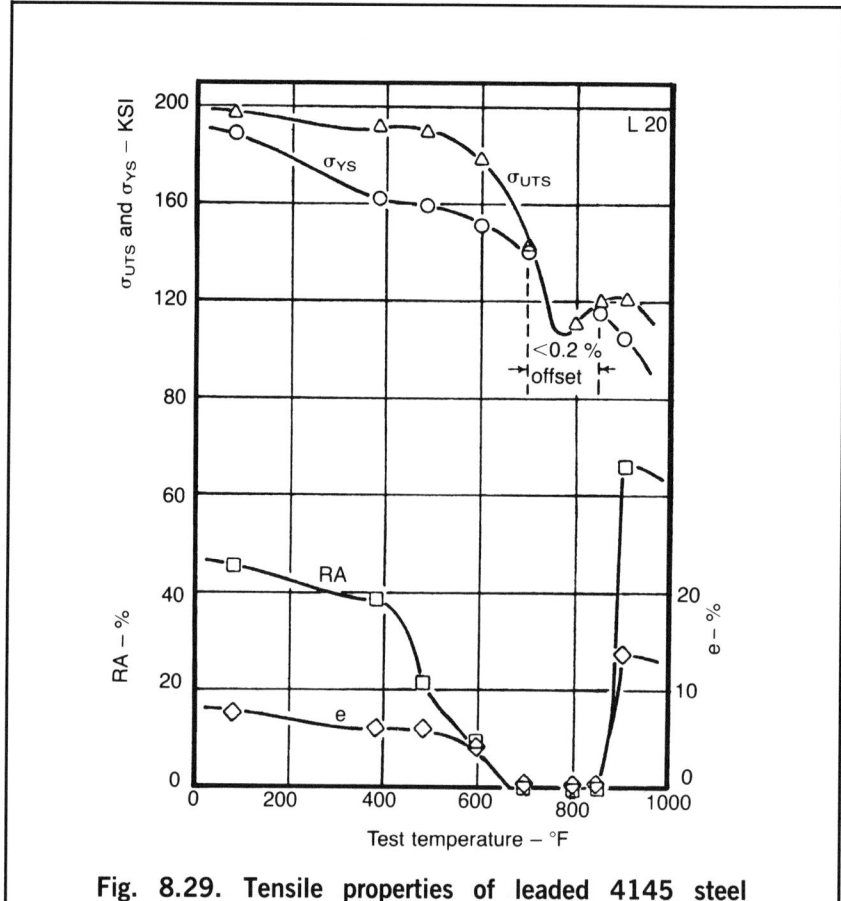

Fig. 8.29. Tensile properties of leaded 4145 steel quenched and tempered to strengths of 200 ksi (1380 MPa) as a function of tensile test temperature. (Ref 8.68)

sible for these effects may be present in the environment external to the steel or may be present internally as a result of steelmaking or processing operations such as pickling or electroplating. Hydrogen may promote a transition from a ductile to brittle fracture mode or it may reduce ductility without a change in fracture mode (Ref 8.71). Of considerable interest with regard to quench and tempered high-strength steels is the development of intergranular cracking in the presence of hydrogen. Some of the conditions affecting such cracking due to internal hydrogen have been presented in a paper by Johnson, Morlet, and Troiano (Ref 8.72) and are described briefly below. More recently, hydrogen-induced or hydrogen-assisted cracking has been related to the same type of impurity segregation effects responsible in

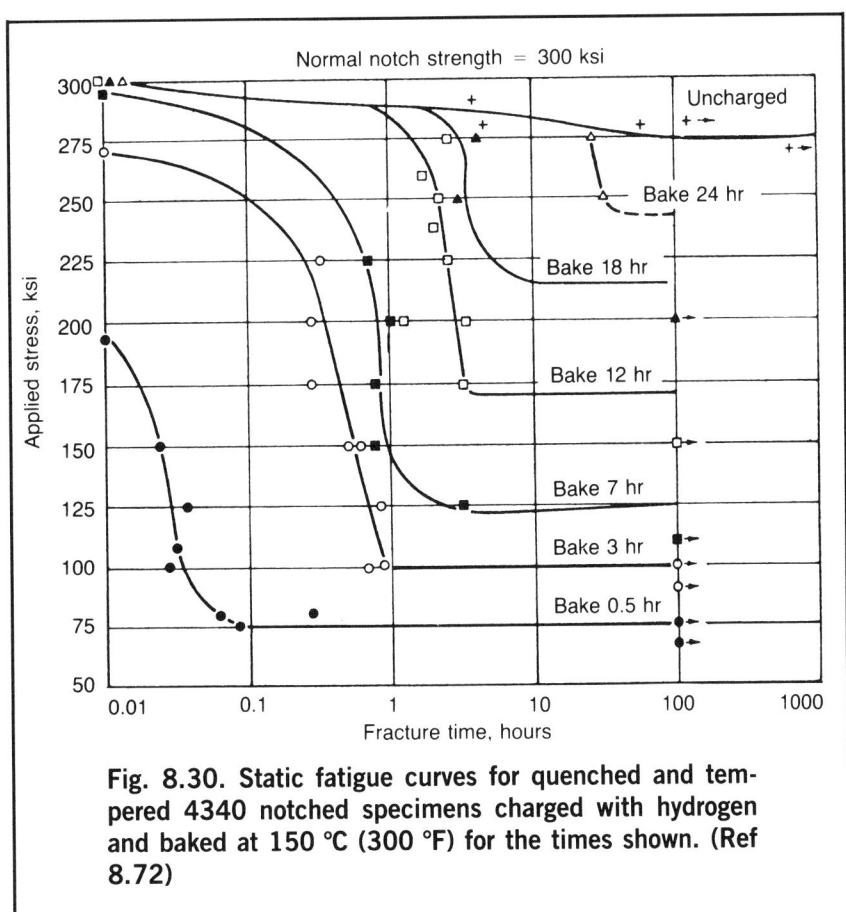

Fig. 8.30. Static fatigue curves for quenched and tempered 4340 notched specimens charged with hydrogen and baked at 150 °C (300 °F) for the times shown. (Ref 8.72)

part for TME (Ref 8.40) and TE (Ref 8.66). For more information on the manifold effects of hydrogen on materials the reader is referred to the books *Effect of Hydrogen on Behavior of Materials* and *Hydrogen Effects in Metals* (Ref 8.73, 8.74). A collection of important papers published in the literature between 1875 and 1977 concerning hydrogen effects is also available (Ref 8.75).

An important characteristic of hydrogen embrittlement in high-strength steel is the delay time required for failure. The higher the hydrogen content of a steel the lower the stress and the shorter the time for failure. Johnson, Morlet, and Troiano (Ref 8.72) studied this problem by cathodically charging with hydrogen a 4340 steel oil quenched and tempered to 230 ksi (1585 MPa) tensile strength. After charging, the specimens were immediately plated with cadmium, a procedure that delayed outgassing of the hydrogen and ensured a uniform distribution of hydrogen throughout the specimen section after baking. Figure 8.30

shows the effects of baking at 150 °C (300 °F) on the static fatigue (sustained loading) of the hydrogen-charged specimens. Increasing baking time effectively lowers hydrogen content even in the plated specimens, and sufficient baking eventually restores the strength of charged specimens to that of uncharged specimens. The horizontal portions of the curves in Fig. 8.30 are designated as static fatigue or endurance limits, i.e., the stress level below which failure would not occur no matter what the duration of stress application. As hydrogen content is decreased by baking, the static fatigue limit increases.

The specimens used to obtain Fig. 8.30 were notched and therefore the static fatigue limits hold for that particular notch geometry (Ref 8.72). In general the sharper the notch, the lower the static fatigue limits, an indication that a critical combination of hydrogen concentration and triaxial stress state is required for crack initiation. Another very important aspect of hydrogen cracking revealed by the Johnson study (Ref 8.72) was the observation of an incubation period for cracking. This incubation period is dependent on hydrogen concentration (see Fig. 8.30) and is related to the time for diffusion of hydrogen to the point of crack initiation within the triaxial stress field. The need for an incubation time for crack initiation means that high-strain rate tests might not detect hydrogen embrittlement, a situation just the reverse of TME and TE. Sustained loading is therefore the most sensitive means of detecting susceptibility to delayed hydrogen failure. The propagation of a crack initiated by internal hydrogen is discontinuous, again because of the requirement for hydrogen diffusion. Once the crack leaves behind the initial hydrogen concentration, it stops, and does not advance until sufficient hydrogen has diffused to the stress field at the crack tip (Ref 8.76).

Hydrogen is strongly attracted to dislocation cores and may be transported through a steel by dislocation motion (Ref 8.73, 8.74). Hydrogen also appears to increase the mobility of screw dislocations but reduces their ability to cross slip, thus causing slip to concentrate on relatively few slip planes (Ref 8.77). When the dislocations pile up at obstacles such as carbide or inclusion particles, also strong traps for hydrogen, the concentration of planar slip and hydrogen together lowers the cohesive strength on the slip planes, leading to the hydrogen fracture mode referred to as glide-plane decohesion (Ref 8.77).

Overheating of Forgings

Forging may be performed at very high temperatures, around 1400 °C (2552 °F), in the austenite phase field. Heating at such high temperatures causes MnS inclusions to dissolve and the manganese and sulfur to go into solution in the austenite. On cooling, fine MnS particles reprecipitate on the coarse austenite grain boundaries established during forging. After subsequent quench and tempering, on mechanical loading the steel may fracture by microvoid coalescence along the sulfide networks formed on the coarse prior austenite grain boundaries, a fracture morphology characteristic of overheated forgings (Ref 8.78-8.80).

The sulfide particle size and spacing are a function of manganese and sulfur contents, maximum forging temperature, and cooling conditions, and determine the severity of the reduction of toughness associated with overheating. Higher tempering temperatures increase the sensitivity to overheating-type fractures (Ref 8.78). The large plastic zones at notches or cracktips in stressed low-strength, well-tempered microstructures encompass large areas of the coarse intergranular sulfide networks. In higher strength microstructures, the plastic zones are smaller and may act only on small fractions of the sulfide networks.

Overheating can be reduced or eliminated in a number of ways (Ref 8.78-8.80). Forging temperatures can be reduced, but in some cases this might not be the most efficient manufacturing approach. Strong sulfide-forming elements such as rare earths, calcium, or zirconium could be added to stabilize sulfides and prevent their re-solution, but care must be taken not to introduce coarse particle dispersions which by themselves will reduce ductile fracture resistance. Increased manganese would also stabilize MnS particles, but is not recommended for heavy sections, because as discussed in the TE section, increased manganese would promote temper embrittlement. An attractive solution to overheating, now possible with advanced steelmaking techniques, is the reduction of manganese and sulfur to very low levels, as is being done in steels for very heavy forgings (Ref 8.56, 8.58, 8.59). Care must be taken to reduce both the manganese and sulfur to low enough levels (Ref 8.59). Reduction of sulfur alone results in distributions of fine MnS particles which rapidly dissolve and reprecipitate during forging (Ref 8.56).

Flow and Fracture of Tempered Steels

This section relates the microstructure of hardened medium-carbon steels to their strong carbon-dependent plastic deformation and fracture behavior. Hardenable steels are generally low alloy, containing at most a total of a few percent manganese, nickel, chromium, and molybdenum, primarily to ensure through hardening at quench rates which minimize distortion and quench cracking. Very high strengths, exceeding 200 ksi (1380 MPa), excellent fatigue resistance, and good toughness are attainable by low-temperature tempering, and more moderate strengths with very high toughness are attainable by tempering at higher temperatures.

Earlier chapters and sections have laid the groundwork for this discussion by describing the structure and morphology of martensite, the austenitizing process, and the structural changes and embrittlement phenomena which may occur during tempering. The picture of a quenched and tempered steel which emerges shows a microstructure with many components, as listed below:

- Inclusions
- Grain boundary segregation/precipitates
- Carbides/nitrides undissolved by austenitizing
- Coarse carbides produced by tempering

- Martensite laths or plates
- Martensite packet size (in low- and medium-carbon steels)
- Prior austenite grain structure
- Retained austenite
- Dislocation/twin substructure
- Fine intralath carbides (Stage I/secondary hardening)
- Interstitial (C,N) and substitutional atoms

Each component plays a role in the plastic flow and fracture process, but it is the composite structure which must respond to applied stresses and strains. Thus the mechanisms of plastic flow and fracture as influenced by all components determine strengthening and the toughness or fracture resistance of a given steel.

The above list of structural components in quenched and tempered steels is divided into two parts. The upper portion lists features at which cracking or microvoid formation initiate during fracture. The lower portion lists features which determine plastic flow or deformation behavior. The latter features establish the mechanical design properties of a steel, i.e., yield and ultimate tensile strengths, and the dynamic dislocation interactions which are responsible for strain hardening and the shape of the stress-strain curve.

The fracture-initiating particles or their effects are readily identified by examination of fracture surfaces by scanning electron microscopy. If the particles are separate and dispersed through the matrix, ductile fracture by microvoid nucleation, growth, and coalescence develops. The denser the distribution of particles, and generally the larger the particles, the lower will be the energy absorbed by the ductile fracture process. In CVN impact tests, the latter conditions are manifested by low upper-shelf energies.

The theoretical and empirical evolution of particle identification with ductile fracture has led to the acceptance of inclusions as an important microstructural component of steel. More important, the deleterious effect of inclusions on fracture has been a powerful driving force to reduce the inclusion content in structural steels. It should be noted, however, that inclusions in some types of steels are beneficial, as for example in grades with high machinability where sulfur is deliberately added to form sulfide inclusions. Inclusions are nonmetallic phases such as alumina, aluminates, sulfides, and silicates, in many combinations and morphologies, which are introduced during various stages of the steelmaking process. Kiessling and Lange (Ref 8.81), Ototani (Ref 8.82), and Leslie (Ref 8.83) present detailed reviews of the origins, nature, and effects of inclusions. Steelmaking innovations to reduce inclusion contents include improved deoxidation, shrouding of liquid streams to prevent reoxidation, vacuum degassing, argon blowing, and desulfurization, many of which are made possible by the good thermal, atmospheric, and chemical control associated with ladle metallurgy (Ref 8.56).

Some inclusions, such as manganese sulfides, are plastic during hot work and become elongated and flattened (Ref 8.83). Thus considerable anisotropy in properties and fracture can be introduced in hot rolled shapes, as documented in a study of high-strength quenched and tempered 4340 plate steels by Speich and Spitzig (Ref 8.84). This major source of anisotropy can be reduced by inclusion shape control with additions of such elements as titanium, zirconium, cobalt, and rare earth metals (REM). These elements combine with sulfur to produce sulfides which are less deformable during hot work.

Grain boundary segregation and precipitation phenomena, such as those which produce temper embrittlement, tempered martensite embrittlement, and hydrogen embrittlement, produce continuous fracture paths throughout the microstructure of high-strength steels. The associated intergranular fracture along prior austenite grain boundaries, because of reduced lattice or interface cohesive strength, is accompanied by very little plastic deformation, and therefore represents a brittle, stress-controlled mode of fracture in contrast to ductile, strain-controlled fracture with dispersed second-phase particles.

Given that inclusion contents can be significantly reduced by clean steel processing, and that the conditions which produce embrittlement can be avoided, there are still many particles at which ductile fracture of quenched and tempered steels may originate. The highest strengths are

248 / STEELS: HEAT TREATMENT AND PROCESSING PRINCIPLES

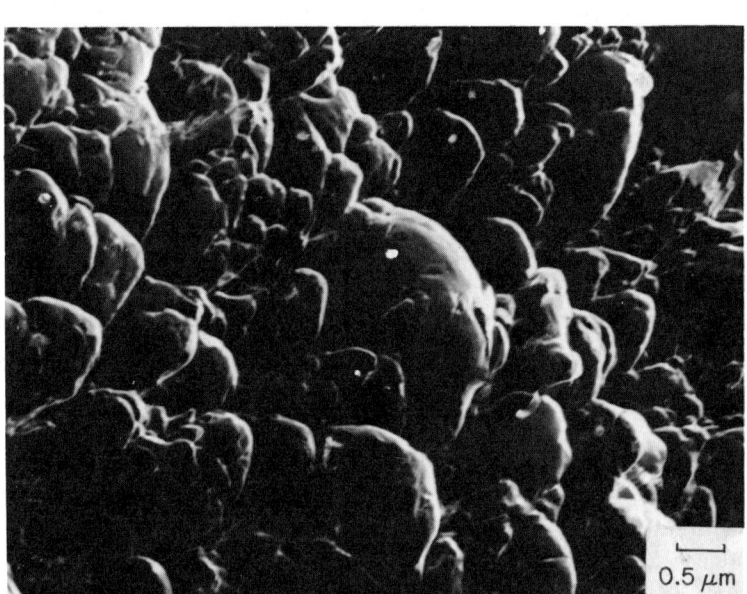

Fig. 8.31. Microvoids and extracted particles on fracture surface of 4130 steel tempered at 400 °C (752 °F). Extraction replica, dark-field transmission electron micrograph. (Ref 8.45)

produced by tempering between 150 and 200 °C (302 and 392 °F), below the temperature ranges which produce TME and TE. Quenched and tempered steels in this condition when overloaded invariably fail by shear or ductile fracture characterized by fine microvoid formation around fine particles. Although small, on the order of 0.05 μm in size (below the resolution of the light microscope), the particles are not produced by tempering. Tempering between 150 and 200 °C (302 and 392 °F) produces eta transition carbides which are much finer, on the order of 0.002 μm. These ultrafine carbides are apparently too small to initiate microvoids, and instead contribute to strain hardening of the tempered martensite. The void-initiating particles instead are carbides, perhaps stabilized by alloying elements, which are spheroidized and retained at the austenitizing temperatures typically used for hardening low-alloy, medium-carbon steels. Figure 8.31 shows an example of small microvoids and particles extracted from the shear initiation zone of a 4130 steel CVN specimen. The dimples or microvoids formed around small, spherical carbide particles, some of which are illuminated by a diffracted carbide beam in the dark-field micrograph. This specimen was tempered at 400 °C (752 °F), but similar particle distributions and fracture morphologies are found in as-quenched and low-temperature tempered specimens (Ref 8.44-8.46).

Higher austenitizing temperatures cause dissolution of retained carbides and improve the ductile fracture resistance of tempered martensite. However, higher austenitizing temperatures also cause excessive grain growth and exacerbate grain boundary segregation. These opposing effects of austenitizing temperature were demonstrated in a paper by Lai *et al.* (Ref 8.85), who evaluated the toughness of 4340 steel austenitized at temperatures as high as 1200 °C (2192 °F). They found that fracture toughness increased but CVN impact energy decreased with increasing austenitizing temperature. These apparently quite opposite results were explained by Ritchie *et al.* (Ref 8.86), who showed that the results were consistent with the two quite different tests used to evaluate toughness. Fracture toughness is obtained by loading specimens in which sharp cracks have been fatigued. The stress field or process zone ahead of the crack is quite small and acts on mostly matrix structure in a coarse-grained specimen. Thus improved matrix fracture resistance translates into higher fracture toughness. In contrast, CVN toughness is obtained by impact loading of a specimen with a machined notch of relatively large root radius. The stress field or process zone around the notch is quite large and encompasses even widely spaced features. As a result, intergranular fracture occurs along coarse, embrittled grain boundary networks and the tougher matrix microstructure of the high-temperature-austenitized specimens is by-passed.

Most hardened steels contain particles of several size distributions. Cox and Low (Ref 8.87), in a study of 4340 steel quenched from 843 °C (1550 °F) and tempered at around 430 °C (806 °F), showed that ductile fracture initiated at the largest inclusion particles. However, the growth of the microvoids around the large particles was limited by microvoid initiation at many finer particles, in this study the cementite particles produced during 400 °C (752 °F) tempering. The resulting dense planar distributions of fine, closely spaced microvoids between the large voids, referred to as void sheets, were considered to significantly limit the toughness of the 4340 steel. In related work, Garrison (Ref 8.88), in a study of 0.4% carbon steels quenched from 900 °C (1652 °F) and tempered at 200 °C (392 °F), found that specimens with distributions of coarse, widely spaced sulfide inclusions, other structural factors being equal, had better fracture toughness than did specimens with smaller, more closely spaced inclusions.

250 / STEELS: HEAT TREATMENT AND PROCESSING PRINCIPLES

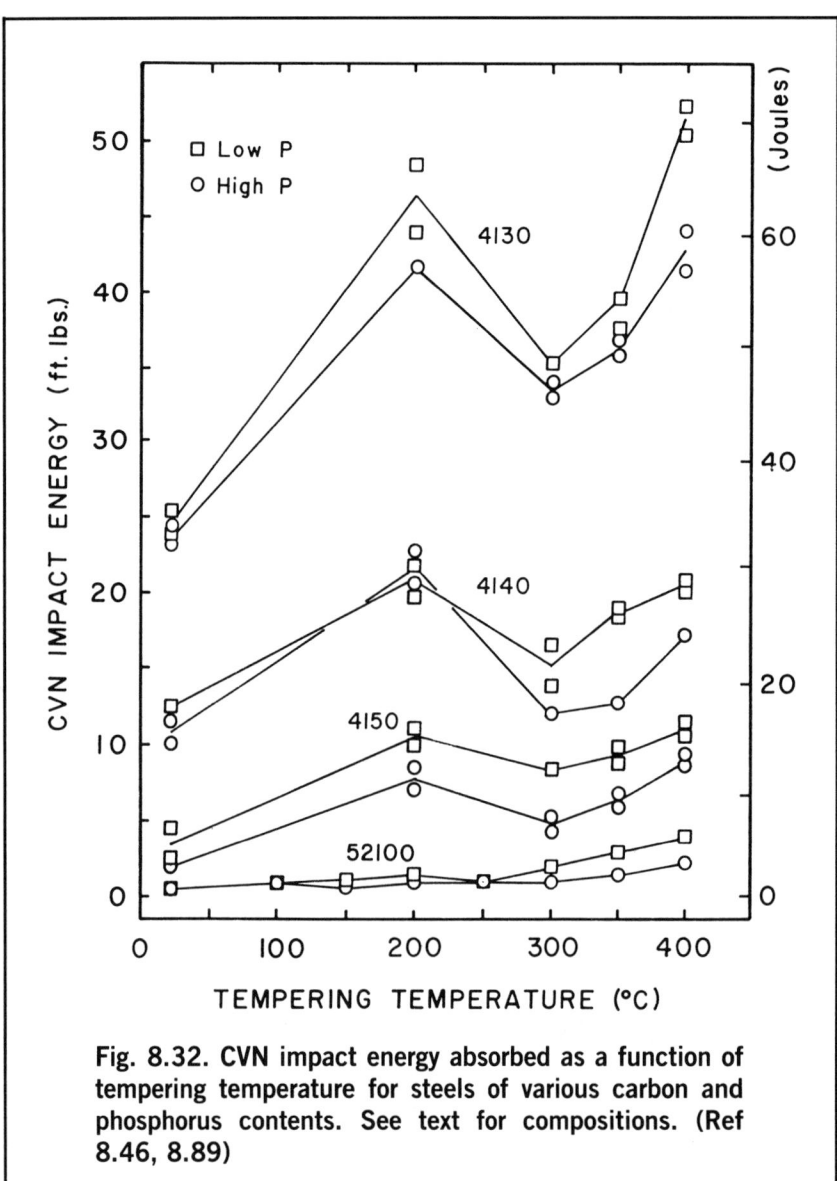

Fig. 8.32. CVN impact energy absorbed as a function of tempering temperature for steels of various carbon and phosphorus contents. See text for compositions. (Ref 8.46, 8.89)

Figure 8.32 shows CVN impact energy absorbed as a function of tempering temperature for a series of specimens with various carbon contents and two levels of phosphorus (Ref 8.46, 8.89). The 41xx steels contain nominally 1% chromium and 0.2% molybdenum and the 52100 steel contains nominally 1.5% chromium. High and low levels of phosphorus were respectively 0.02 and 0.002% in the 41xx steels and

Fig. 8.33. Shear along slip line field at notch root of CVN specimen. 4340 steel quenched and tempered at 200 °C (392 °F). (Ref 8.90)

0.023 and 0.009% in the 52100 steel. All of the ductile and brittle fracture modes, related to various particle arrays as described above, were shown by one or another of the specimens.

Typically, no matter what the appearance of the unstable fracture surface, fracture was initiated by some degree of shear along the slip line field at the root of the CVN specimen notches as shown in Fig. 8.33 (Ref 8.90). The load-time curves associated with CVN impact fracture are related to stages associated with the initiation, propagation, and shear lip formation (Ref 8.45, 8.46, 8.91). The lower the specimen toughness and its ability to plastically deform, the greater the fraction of the absorbed energy which is associated with crack initiation.

Figure 8.32 shows that the effect of phosphorus on lowering impact toughness is roughly constant no matter what the tempered conditions of the specimens. This is attributed to the segregation of phosphorus during austenitizing and its embrittling effect on critical carbide-matrix interfaces in as-quenched and all tempered specimens. The drop in CVN impact toughness at 300 °C (572 °F) is therefore attributed to carbide formation during the second stage of tempering as discussed earlier.

Figure 8.32 shows that the factor which has the greatest effect on the impact toughness of quenched and tempered steels is carbon content. For example, impact toughness drops dramatically in the 200 °C (392 °F) tempered specimens where fracture is exclusively ductile, from 60 J (45 ft · lbf) in the 0.3% carbon steel to less than 2 J (1.5 ft · lbf) in the 1% carbon steel. Certainly the density of void-initiating carbide particles increases with carbon, especially in the intercritically austenitized 52100 steel, but particle density alone cannot explain the strong effect of carbon on tempering. Another effect of increasing carbon is increased hardness and strength in a given tempered condition, and the way carbon modifies plastic deformation behavior must contribute strongly to the carbon-dependent impact toughness of quenched and tempered steels.

Figure 8.34 shows engineering stress-strain curves for the 4130, 4140, and 4150 steels austenitized at 900 °C (1652 °F), quenched, and tempered at 150 °C (302 °F), and Table 8.2 lists hardness and mechanical properties derived from the tensile tests (Ref 8.92). Figures 8.35 and 8.36 show flow stresses at various levels of strain derived from tension and compression tests, respectively. Elastic limits obtained from strain gage measurements are also included in Fig. 8.36.

The tempered steels show increasing rates of strain hardening, i.e., increases in stress per increment of strain, with increasing carbon content. Uniform elongation also increases with carbon content. This observation is consistent with the increase in strain hardening according to the criterion for tensile instability. This criterion equates the change in true strain hardening rate to the true strain at instability, i.e., at the maximum uniform true strain at which the load-carrying capacity due to strain hardening is not able to keep up with the decrease in load-bearing capacity resulting from decreasing specimen cross-sectional area (Ref 8.93). Although uniform elongation increases somewhat with carbon content, the post-uniform elongation decreases sharply in the higher carbon specimens. During post-uniform deformation, mechanisms of microvoid nucleation, growth, and coalescence dominate and eventually lead to final fracture. This last stage of ductile fracture is accelerated at the high post-instability stress levels attained by the 4150 steel specimens. Thus although higher stresses are required to initiate instability in the high-carbon steel, the ductile, shear fracture process occurs with significantly less strain and energy absorption than in the lower carbon steels.

The austenitic grain size and therefore the martensitic packet size of the specimens which yield the stress-strain curves of Fig. 8.34 were essentially constant. Martensite lath width distributions and of

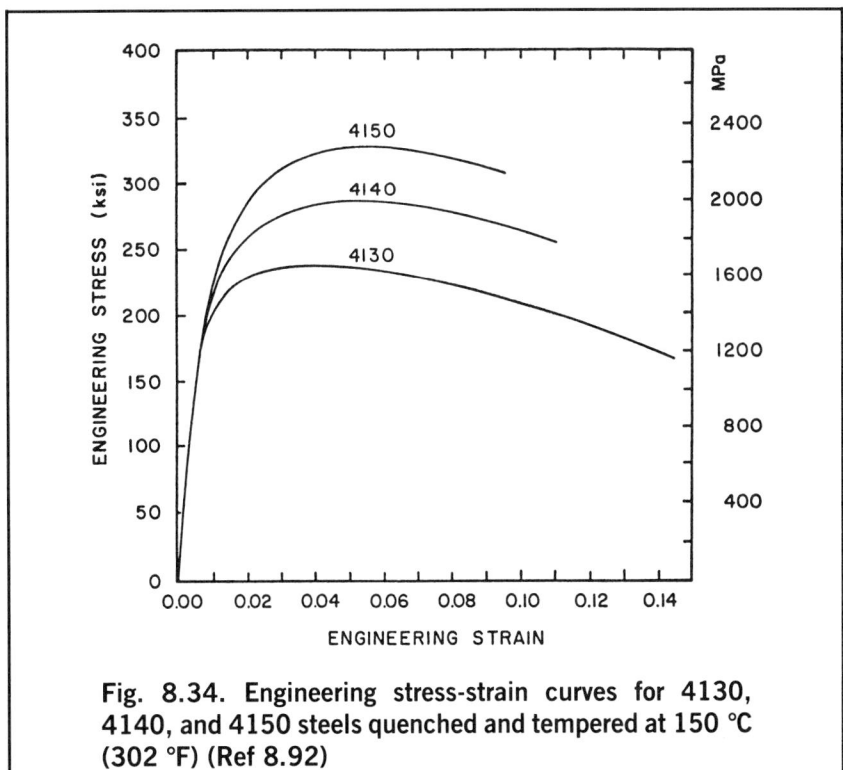

Fig. 8.34. Engineering stress-strain curves for 4130, 4140, and 4150 steels quenched and tempered at 150 °C (302 °F) (Ref 8.92)

course the substitutional alloy element content of the three steels were also constant. Three structural features varied with carbon content: the dislocation substructure, the eta transition carbide density, and retained austenite (Ref 8.92, 8.94). Figure 8.37 shows examples of the latter features. Dark-field transmission electron microscopy was required to reveal the transition carbides because they were obscured by strain fields in bright-field images. The dislocation substructure could not be directly observed, but is assumed to be reflected by the distribution of the transition carbides.

Table 8.2. Mechanical Properties of 150 °C (302 °F) Tempered Specimens of 41xx Steels (Ref 8.92)

Steel	0.2% Yield strength, MPa	Ultimate strength, MPa	Uniform elongation, %	Total elongation, %	Hardness, HRC
4130	1317	1661	4.2	14	50
4140	1462	1971	5.3	11	54
4150	1620	2261	5.6	9.5	58

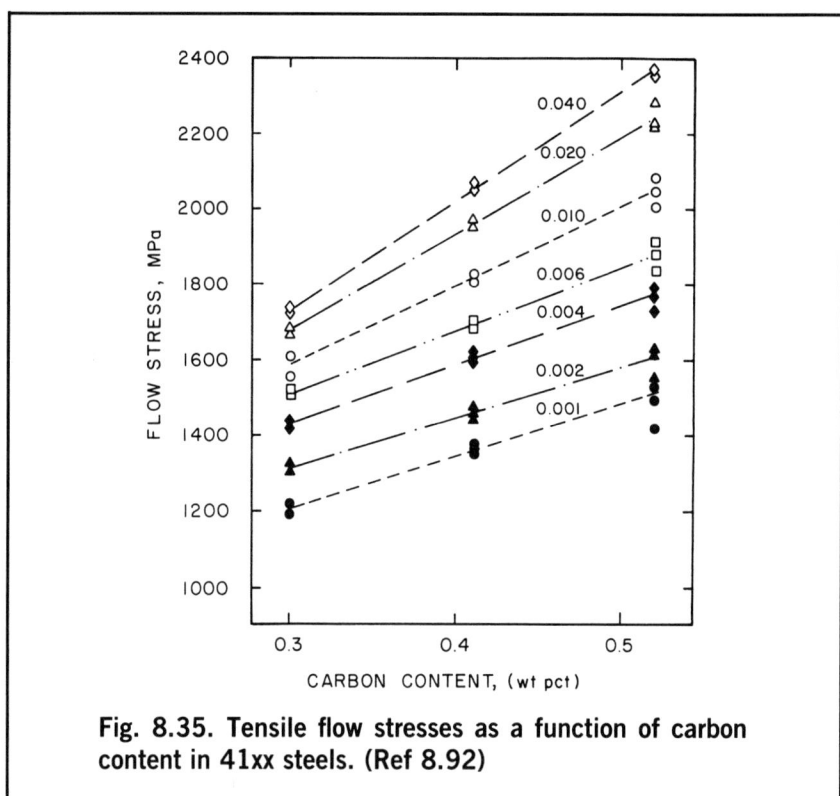

Fig. 8.35. Tensile flow stresses as a function of carbon content in 41xx steels. (Ref 8.92)

The transition carbides were identified as the orthorhombic eta carbides even in the 4130 steel, and were present in aligned rows of particles, each on the order of 2 nm (20 Å) in size. The density of the carbide clusters increased and their spacing decreased with increasing carbon content (Fig. 8.37b). The retained austenite was present between martensite laths (Fig. 8.37a) and increased with increasing carbon content. The 4130, 4140, and 4150 steels had 1.4, 3.8, and 5.0 vol.% retained austenite, respectively.

The above structural characterization of low-temperature tempered martensitic microstructures shows that the high strain hardening rates and high flow stresses of the higher carbon steels correlates primarily with refinement of the dislocation/transition carbide substructure with increasing carbon content. The retained austenite accounts for the decreasing elastic limits (Fig. 8.36) and very high initial strain hardening rates in the higher carbon steels. The retained austenite transforms to martensite by stress and strain-induced mechanisms depending on volume fraction, but these effects are exhausted at very low strains (Ref 8.95). Therefore, the major portion of strain

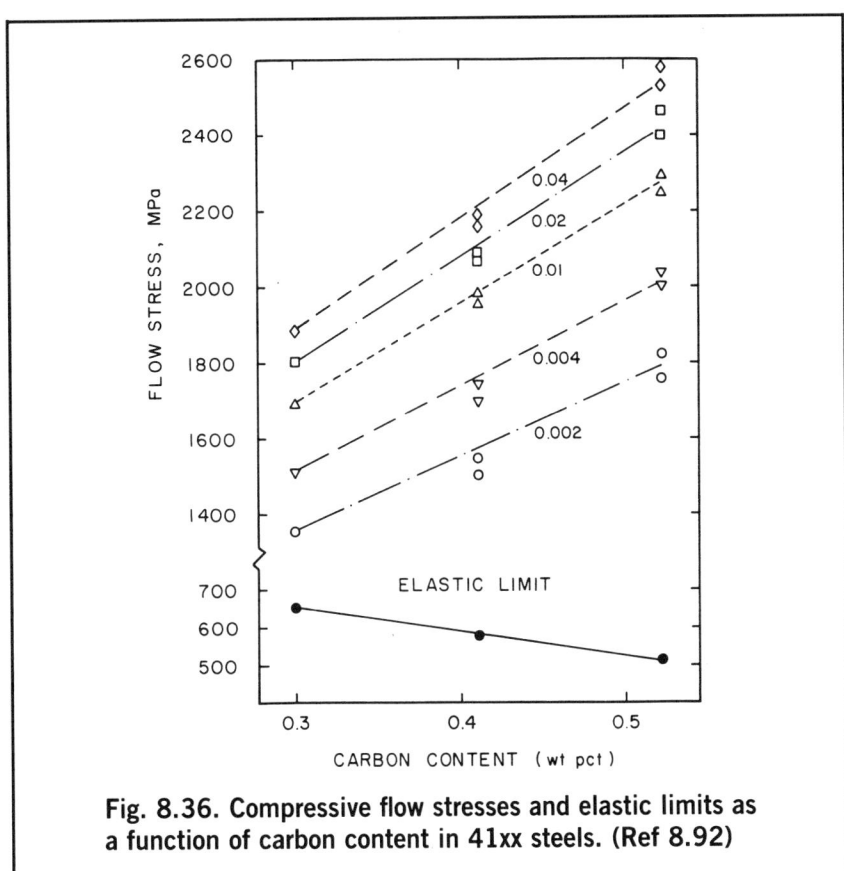

Fig. 8.36. Compressive flow stresses and elastic limits as a function of carbon content in 41xx steels. (Ref 8.92)

hardening and carbon-dependent strengthening is accomplished by dislocation interactions in the fine structure of the tempered lath martensite. Increased tempering coarsens both the dislocation substructure and carbide distributions, reduces strain hardening, and brings yield and ultimate strengths together, as shown earlier in Fig. 8.4 and 8.5.

References

8.1 M.A. Grossmann and E.C. Bain, *Principles of Heat Treatment*, 5th ed., American Society for Metals, Metals Park, OH, 1964

8.2 D.L. Williamson, K. Nakazawa, and G. Krauss, A Study of the Early Stages of Tempering in an Fe-1.2% C Alloy, *Met Trans A*, Vol 10A, 1979, p 1351-1363

8.3 *Modern Steels and Their Properties*, Handbook 2757, 7th ed., Bethlehem Steel Corp., Bethlehem, PA, 1972

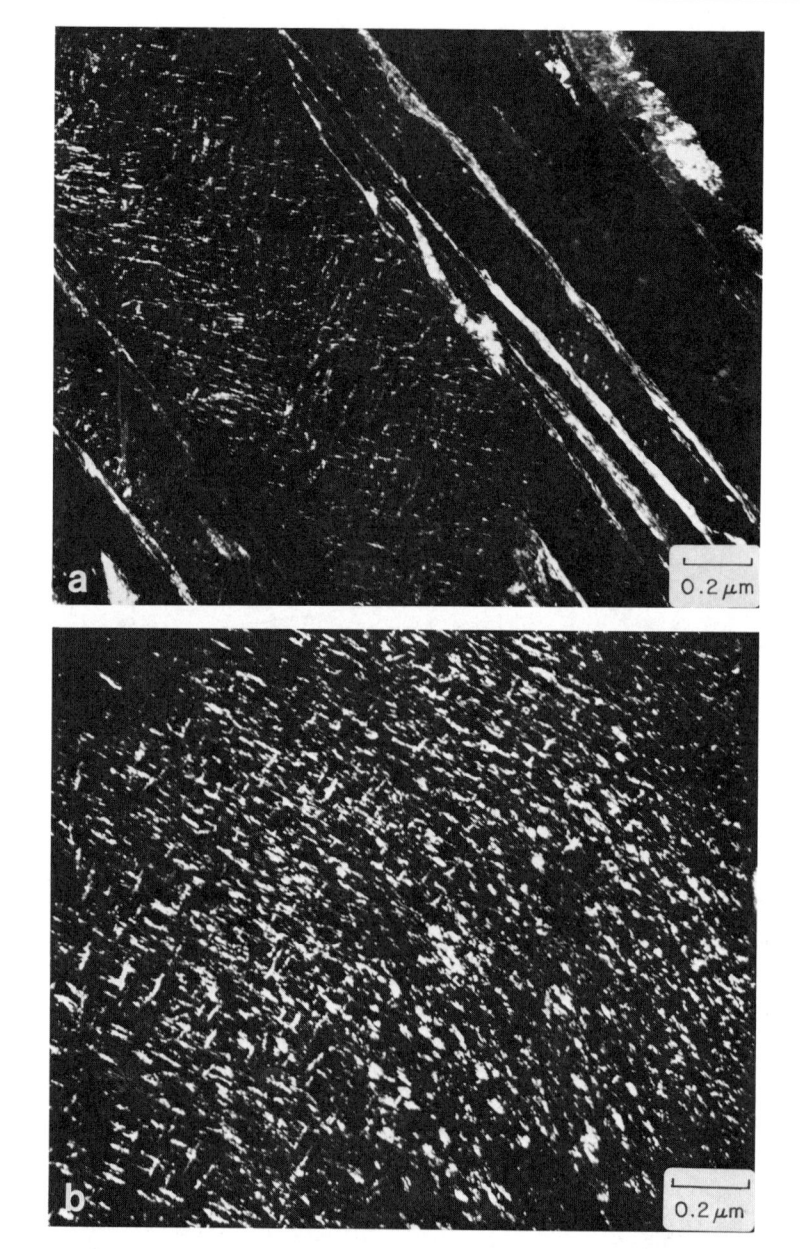

Fig. 8.37. (a) Interlath retained austenite (white diagonal bands) and eta transition carbides in 4130 steel tempered at 150 °C (302 °F). (b) Dense eta transition carbide precipitation in a martensite lath in 4150 steel tempered at 150 °C (302 °F). Dark-field transmission electron micrographs. Courtesy of J.M.B. Losz, Colorado School of Mines, Golden

8.4 T. Swarr and G. Krauss, The Effect of Structure on the Deformation of As-Quenched and Tempered Martensite in an Fe-0.2% C Alloy, *Met Trans A*, Vol 7A, 1976, p 41-48

8.5 R.A. Grange, C.R. Hribal, and L.F. Porter, Hardness of Tempered Martensite in Carbon and Low-Alloy Steels, *Met Trans A*, Vol 8A, 1977, p 1775-1785

8.6 E.C. Rollason, Fundamental Aspects of Molybdenum in Transformation of Steel, Climax Molybdenum Co., London

8.7 K.J. Irvine and F.B. Pickering, The Tempering Characteristics of Low-Carbon Low-Alloy Steels, *JISI*, Vol 194, 1960, p 137-153

8.8 E. Smith and J. Nutting, The Tempering of Low-Alloy Creep-Resistant Steels Containing Chromium, Molybdenum, and Vanadium, *JISI*, Vol 187, 1957, p 314-329

8.9 A.G. Allten and P. Payson, The Effect of Silicon on the Tempering of Martensite, *Trans ASM*, Vol 45, 1953, p 498-532

8.10 W. Crafts and J.L. Lamont, Effects of Alloys in Steel as Resistance to Tempering, *Trans AIME*, Vol 172, 1947, p 222-243

8.11 C.S. Roberts, B.L. Averbach, and M. Cohen, The Mechanism and Kinetics of the First Stage of Tempering, *Trans ASM*, Vol 45, 1953, p 576-604

8.12 B.S. Lement, B.L. Averbach, and M. Cohen, Microstructural Changes on Tempering Iron-Carbon Alloys, *Trans ASM*, Vol 46, 1954, p 851-881

8.13 F.E. Werner, B.L. Averbach, and M. Cohen, The Tempering of Iron-Carbon Martensite Crystals, *Trans ASM*, Vol 49, 1957, p 823-841

8.14 B.S. Lement, B.L. Averbach, and M. Cohen, Further Study of Microstructural Changes on Tempering Iron-Carbon Alloys, *Trans ASM*, Vol 47, 1955, p 291-319

8.15 G.R. Speich, Tempered Ferrous Martensitic Structures, in *Metals Handbook*, Vol 8, 8th ed., American Society for Metals, Metals Park, OH, p 202-204

8.16 G.R. Speich and W.C. Leslie, Tempering of Steel, *Met Trans*, Vol 3, 1972, p 1043-1054

8.17 G.R. Speich, Tempering of Low-Carbon Martensite, *Trans TMS-AIME*, Vol 245, 1969, p 2553-2564

8.18 J. Genin and P.A. Flinn, Mössbauer Effect Study of the Clustering of Carbon Atoms during the Room-Temperature Aging of Iron-Carbon Martensite, *Trans TMS-AIME*, Vol 242, 1968, p 1419-1430

8.19 G.V. Kurdjumov and A.G. Khachaturyan, Phenomena of Carbon Atom Redistribution in Martensite, *Met Trans*, Vol 3, 1972, p 1069-1076

8.20 S. Nagakura, Y. Hirotsu, M. Kusunoki, T. Suzuki, and Y. Nakamura, Crystallographic Study of the Tempering of Martensitic Carbon Steel by Electron Microscopy and Diffraction, *Met Trans A*, Vol 14A, 1983, p 1025-1031

8.21 G. Krauss, Tempering and Structural Change in Ferrous Martensitic Structures, in *Phase Transformations in Ferrous Alloys*, A.R. Marder and J.I. Goldstein (Eds.), TMS-AIME, Warrendale, PA, 1984, p 101-123

8.22 K.H. Jack, Structural Transformations in the Tempering of High Carbon Martensitic Steel, *JISI*, Vol 169, 1951, p 26-36

8.23 Y. Hirotsu and S. Nagakura, Crystal Structure and Morphology of the Carbide Precipitated from Martensite High Carbon Steel during the First Stage of Tempering, *Acta Met*, Vol 20, 1972, p 645-655

8.24 Y. Hirotsu and S. Nagakura, Electron Microscopy and Diffraction Study of the Carbide Precipitated at the First Stage of Tempering of Martensitic Medium Carbon Steel, *Trans Jpn Inst Met*, Vol 15, 1974, p 129-134

8.25 T.A. Balliett and G. Krauss, The Effect of the First and Second Stages of Tempering on Microcracking in Martensite of an Fe-1.22% C Alloy, *Met Trans A*, Vol 7A, 1976, p 81-86

8.26 C. Wells, W. Batz, and R.F. Mehl, Diffusion Coefficient of Carbon in Austenite, *Trans AIME*, Vol 188, 1950, p 553-560

8.27 D.L. Williamson, R.G. Shupmann, J.P. Materkowski, and G. Krauss, Determination of Small Amounts of Austenite and Carbide in a Hardened Medium Carbon Steel by Mössbauer Spectroscopy, *Met Trans A*, Vol 10A, 1979, p 379-382

8.28 Y. Imai, Phases in Quenched and Tempered Steels, *Trans Jpn Inst Met*, Vol 16, 1975, p 721-734

8.29 Y. Ohmori, Hägg Carbide Formation and Its Transformation into Cementite During the Tempering of Martensite, *Trans Jpn Inst Met*, Vol 13, 1972, p 119-127

8.30 C-B. Ma, T. Ando, D.L. Williamson, and G. Krauss, Chi-Carbide in Tempered High Carbon Martensite, *Met Trans A*, Vol 14A, 1983, p 1033-1045

8.31 S. Nagakura, T. Suzuki, and M. Kusunoki, Structure of the Precipitated Particles at the Third Stage of Tempering of Martensitic Iron-Carbon Steel Studied by High Resolution Electron Microscopy, *Trans Japan Inst Metals*, Vol 22, 1981, p 699-709

8.32 R.W.K. Honeycombe, *Steels, Microstructure and Properties*, Edward Arnold Ltd and American Society for Metals, Metals Park, OH, 1981

8.33 R.W. Caron and G. Krauss, The Tempering of Fe-C Lath Martensite, *Met Trans*, Vol 3, 1972, p 2381-2389

8.34 P.G. Shewmon, *Transformations in Metals*, McGraw-Hill, New York, 1969, p 220

8.35 R.M. Hobbs, G.W. Lorimer, and N. Ridley, Effect of Silicon on the Microstructure of Quenched and Tempered Medium-Carbon Steels, *JISI*, Vol 210, 1972, p 757-764

8.36 C.J. McMahon, Jr., Strength of Grain Boundaries in Iron-base Alloys, in *Grain Boundaries in Engineering Materials*, Claitors Publishing Division, Baton Rouge, LA, 1975, p 525-552

8.37 C.J. McMahon, Jr., A.K. Cianelli, and H.C. Feng, The Influence of Mo on P-induced Temper Embrittlement in Ni-Cr Steel, *Met Trans A*, Vol 8A, 1977, p 1055-1057

8.38 G. Thomas, Retained Austenite and Tempered Martensite Embrittlement, *Met Trans A*, Vol 9A, 1978, p 439-450

8.39 J.P. Materkowski and G. Krauss, Tempered Martensite Embrittlement in SAE 4340 Steel, *Met Trans A*, Vol 10A, 1979, p 1643-1651

8.40 S.K. Banerji, C.J. McMahon, Jr., and H.C. Feng, Intergranular Fracture in 4340-type Steels: Effects of Impurities and Hydrogen, *Met Trans A*, Vol 9A, 1978, p 237-247

8.41 B.J. Schultz and C.J. McMahon, Jr., STP-499, ASTM, 1972, p 104

8.42 H. Ohtani and C.J. McMahon, Jr., Modes of Fracture in Temper Embrittled Steels, *Acta Met*, Vol 23, p 377-386

8.43 J.E. King, R.F. Smith, and J.F. Knott, Toughness Variations During Tempering of a Plain Carbon Martensitic Steel, *Fracture 1977*, Vol 2, ICF4, Waterloo, Canada

8.44 R.G. Schupmann, "A Study of Tempered Martensite Embrittlement in 4130 Type Steels," M.S. Thesis, Colorado School of Mines, Golden, 1978

8.45 F. Zia-Ebrahimi and G. Krauss, The Evaluation of Tempered Martensite Embrittlement in 4130 Steel by Instrumented Charpy V-Notch Testing, *Met Trans A*, 14A, 1983, p 1109-1119

8.46 F. Zia-Ebrahimi and G. Krauss, Mechanisms of Tempered Martensite Embrittlement in Medium-Carbon Steels, *Acta Met*, Vol 32, 1984, p 1767-1777

8.47 I. Olefjord, Temper Embrittlement, Review 231, *Inter Met Rev*, Vol 23, 1978, p 149-163

8.48 J.H. Hollomon, Temper Brittleness, *Trans ASM*, Vol 36, 1946, p 473-540

8.49 B.C. Woodfine, Temper Brittleness: A Critical Review of the Literature, *JISI*, Vol 173, 1953, p 229-240

8.50 J.M. Capus, The Mechanism of Temper Brittleness, in *Temper Embrittlement in Steel*, STP 407, ASTM, Philadelphia, 1968, p 3-19

8.51 C.J. McMahon, Jr., Temper Brittleness—An Interpretive Review, in *Temper Embrittlement in Steel*, STP 407, ASTM, 1968, p 127-167

8.52 F.L. Carr, M. Goldman, L.D. Jaffee, and D.C. Buffum, Isothermal Temper Embrittlement of SAE 3140 Steel, *Trans TMS-AIME*, Vol 197, 1953, p 998

8.53 C.J. McMahon, Jr., E. Furubayashi, H. Ohtani, and H.C. Feng, A Study of Grain Boundaries During Temper Embrittlement of a Low Carbon Ni-Cr Steel Doped with Antimony, *Acta Met*, Vol 24, 1976, p 695-704

8.54 M. Guttmann, The Link between Equilibrium Segregation and Precipitation in Ternary Solutions Exhibiting Temper Embrittlement, *Met Science*, Vol 10, 1976, p 337-341

8.55 M. Guttmann, Ph. Dumoulin, and M. Wayman, The Thermodynamics of Interactive Co-segregation of Phosphorus and Alloying Elements in Iron and Temper-Brittle Steels, *Met Trans A*, Vol 13A, 1982, p 1693-1711

8.56 R.L. Bodner and R.F. Cappellini, Effect of Residual Elements in Heavy Forgings: Past, Present, and Future, *MiCon 86: Optimization of Processing, Properties, and Service Performance Through Microstructural Control*, B.L. Bramfitt, R.C. Benn, C.R. Brinkman, and G.F. Vander Voort (Eds.), ASTM STP 979, American Society for Testing and Materials, Philadelphia, 1988, p 47-82

8.57 W.T. Nachtrab and Y.T. Chou, High Temperature Ductility Loss in Carbon-Manganese and Niobium-Treated Steels, *Met Trans A*, Vol 17A, 1986, p 1995-2006

8.58 R.L. Bodner, J.R. Michael, S.S. Hansen, and R.I. Jaffee, Progress in the Design of an Improved High-Temperature 1 pct CrMoV Rotor Steel, *Proceedings 30th Mechanical Working and Steel Processing Conference*, ISS/AIME, Warrendale, PA, 1988

8.59 R.L. Bodner, T. Ohhashi, and R.I. Jaffee, Effects of Mn, Si, and Purity on the Design of 3½ NiCrMoV, 1 CrMoV and 2½ Cr-1 Mo Bainitic Alloy Steels, *Met Trans A*, Vol 20, 1989, p 1445-1460

8.60 Y. Weng and C.J. McMahon, Jr., The Effect of Manganese on Intergranular Embrittlement in Iron and Steel, *Grain Boundary Structure and*

Related Phenomena, Proceedings of JIMS-4 (1986), supplement to *Trans Japan Inst Metals*, p 579-585

8.61 J.A. Wright and A.G. Quarrell, Effect of Chemical Composition on the Occurrence of Intergranular Fracture in Plain Carbon Steel Castings Containing Aluminum and Nitrogen, *JISI*, Vol 197, 1962, p 299-307

8.62 N.H. Croft, A.R. Entwisle, and G.J. Davies, Intergranular Fracture of Steel Castings, in *Advances in Physical Metallurgy and Applications of Steels*, The Metals Society, London, 1982, p 286-295

8.63 J.P. Materkowski, "Tempered Martensite Embrittlement in 4340 Steel as Related to Phosphorus Content and Carbide Morphology," M.S. Thesis, Colorado School of Mines, Golden, 1978

8.64 J.M. Capus and G. Mayer, The Influence of Trace Elements on Embrittlement Phenomena in Low-Alloy Steels, *Metallurgia*, Vol 62, 1960, p 133-138

8.65 F.G. Wilson and T. Gladman, Aluminum Nitride in Steel, *Int Mater Rev*, Vol 33 (No. 5), 1988, p 221-286

8.66 C.L. Briant and S.K. Banerji, Intergranular Failure in Steel: the Role of Grain Boundary Composition, Review 232, *Inter Met Rev*, Vol 23, 1978, p 164-199

8.67 Liquid Metal Embrittlement, in *Metals Handbook*, Vol 10, 8th ed., American Society for Metals, Metals Park, OH, p 228-229

8.68 S. Mostovoy and N.N. Breyer, The Effect of Lead on the Mechanical Properties of 4145 Steel, *Trans ASM*, Vol 61, 1968, p 219-232

8.69 W.R. Warke and N.N. Breyer, Effect of Steel Composition on Lead Embrittlement, *JISI*, Vol 209, 1971, p 779-784

8.70 R.D. Zipp, W.R. Warke, and N.N. Breyer, A Comparison on Elevated Temperature Tensile Fractures in Nonleaded and Leaded 4145 Steel, STP 453, ASTM, 1969, p 111-133

8.71 I.M. Bernstein, R. Garber, and G.M. Pressouyre, Effect of Dissolved Hydrogen on Mechanical Behavior of Metals in *Effect of Hydrogen on Behavior of Materials*, A.W. Thompson and I.M. Bernstein (Eds.), TMS-AIME, Warrendale, PA, 1976, p 37-58

8.72 H.H. Johnson, J.G. Morlet, and A.R. Troiano, Hydrogen, Crack Initiation and Delayed Failure in Steel, *Trans TMS-AIME*, Vol 212, 1958, p 528-536

8.73 A.W. Thompson and I.M. Bernstein (Eds.), *Effect of Hydrogen on Behavior of Material*, TMS-AIME, Warrendale, PA, 1976

8.74 I.M. Bernstein and A.W. Thompson (Eds.), *Hydrogen Effects on Metals*, TMS-AIME, Warrendale, PA, 1981

8.75 *Hydrogen Damage*, American Society for Metals, Metals Park, OH, 1977

8.76 A.R. Troiano, The Role of Hydrogen and Other Interstitials in the Mechanical Behavior of Metals, *Trans ASM*, Vol 52, 1960, p 54-80

8.77 C.J. McMahon, Jr., Effects of Hydrogen on Plastic Flow and Fracture in Iron and Steel, in *Hydrogen Effects on Metals*, I.M. Bernstein and A.W. Thompson (Eds.), TMS-AIME, Warrendale, PA, 1981, p 219-233

8.78 T.J. Baker, Use of Scanning Electron Microscopy in Studying Sulfide Morphology on Fracture Surfaces, in *Sulfide Inclusions in Steel*, J.J. de Barbadillo and E. Snape (Eds.), American Society for Metals, Metals Park, OH, 1975, p 135-158

8.79 N.P. McLeod and J. Nutting, Influence of Manganese on Susceptibility of Low-alloy Steel to Overheating, *Metals Technol*, Vol 9, 1982, p 399-404
8.80 G.E. Hale and J. Nutting, "Overheating of Low-Alloy Steels," *Int Metal Rev*, Vol 29, 1984, p 273-298
8.81 R. Kiessling and N. Lange, *Non-metallic Inclusions in Steel*, 2nd ed., The Metals Society, London, 1978
8.82 T. Ototani, *Calcium Clean Steel*, B. Ilschner and N.J. Grant (Eds.), Springer-Verlag, Berlin and New York, 1986
8.83 W.C. Leslie, Inclusions and Mechanical Properties, *ISS Trans*, Vol 2, 1983, p 1-24
8.84 G.R. Speich and W.A. Spitzig, Effect of Volume Fraction and Shape of Sulfide Inclusions on Through-Thickness Ductility and Impact Energy of High-Strength 4340 Plate Steels, *Met Trans A*, Vol 13A, 1982, p 2239-2257
8.85 G.Y. Lai, W.E. Wood, R.A. Clark, V.F. Zackey, and E.R. Parker, The Effect of Austenitizing Temperature on the Microstructure and Mechanical Properties of As-Quenched 4340 Steel, *Met Trans*, Vol 5, 1974, p 1663-1690
8.86 R.O. Ritchie, B. Francis, and W.L. Server, Evaluation of Toughness in AISI 4340 Alloy Steel Austenitized at Low and High Temperatures, *Met Trans A*, Vol 7A, 1976, p 831-838
8.87 T.B. Cox and J.R. Low, Jr., An Investigation of the Plastic Fracture of AISI 4340 and 18 Nickel-200 Grade Maraging Steels, *Met Trans*, Vol 5, 1974, p 1457-1470
8.88 W.M. Garrison, Jr., The Effects of Silicon and Nickel Additions on the Sulfide Spacing and Fracture Toughness of a 0.4 Carbon Low Alloy Steel, *Met Trans A*, Vol 17A, 1986, p 669-678
8.89 D.L. Yaney, The Effects of Phosphorus and Tempering on the Fracture of AISI 52100 Steel, M.S. thesis, Colorado School of Mines, Golden, 1987
8.90 Gu Baozhu and G. Krauss, The Effect of Low-Temperature Isothermal Heat Treatments on the Fracture of 4340 Steel, *J Heat Treat*, Vol 4, 1986, p 365-372
8.91 M. Leap, D.K. Matlock, and G. Krauss, Correlation of the Charpy Test to Fracture Mechanics in a Vanadium Modified 1045 Steel, in *Fundamentals of Microalloying Forging Steels*, G. Krauss and S.K. Banerji (Eds.), TMS-AIME, Warrendale, PA, 1987, p 113-152
8.92 Gu Baozhu, J.M.B. Losz, and G. Krauss, Substructure and Flow Strength of Low-Temperature Tempered Medium Carbon Martensite, in *Proceedings of the International Conference on Martensitic Transformations*, The Japan Institute of Metals, 1986, p 367-374
8.93 D.K. Matlock, F. Zia-Ebrahimi, and G. Krauss, Structure, Properties and Strain Hardening of Dual-Phase Steels, in *Deformation, Processing and Structure*, G. Krauss (Ed.), American Society for Metals, Metals Park, OH, 1984, p 47-87
8.94 M.B. Losz and G. Krauss, unpublished research, Colorado School of Mines, Golden
8.95 M.A. Zaccone and G. Krauss, Elastic Limits and Microplastic Response in Ultrahigh Strength Carbon Steels, *Met Trans A*, Vol 20A, 1989, p 188-191

CHAPTER 9

Special Heat Treatments

An important consideration in many of the heat treatments described up to this point is the variation in cooling rate between the surface and center of heat treated parts. The differential cooling produces stresses that may cause distortion and even cracking. A number of precautions may be taken to reduce the residual stress, but two special heat treatments, martempering and austempering, have been developed specifically to minimize residual stress formation and/or to increase the toughness of heat treated steel parts. This chapter describes the origin of residual stress and various approaches to minimizing it. Martempering and austempering are then discussed. Also described briefly are thermomechanical treatments that use both mechanical and thermal processing to produce desirable combinations of properties, and a new approach to heat treating low-carbon steels based on intercritical annealing.

Residual Stress, Distortion, and Quench Cracking

Two processes that occur during the cooling of heat treated steel parts are responsible for the dimensional changes that ultimately result in residual stresses and distortion. One is the volume expansion that occurs when the close-packed, fcc structure of austenite transforms to the more open crystal structures of ferrite, cementite, and martensite. The other is the normal thermal contraction that occurs during cooling of any single phase or combination of phases in the absence of a phase transformation. The volume expansion due to austenite transformation is the dominant factor in any heat treatment that involves cooling from the austenite phase field, while the thermal contraction is the dominant factor in subcritical heat treatments. Residual stresses and distortion arise because cooling rate is a function of section size or position in a part (as discussed in Chapter 6) and therefore the volume changes occur at different times in different locations during the cooling process. Heat treatments that involve austenitizing produce residual stress patterns quite different from those produced by subcritical annealing.

When a steel part is cooled from the austenite phase field, the surface cools most rapidly and austenite transformation with the attendant volume expansion occurs there first. Generally, the still hot, ductile austenite of the untransformed interior accommodates readily to the changes in surface dimensions. However, when the interior transforms at some later time in the cooling process, its expansion is restrained by the hardened, transformed surface layer. This restraint places the interior of the part in compression while at the same time the interior expansion places the surface in tension. Just the reverse is true for subcritically heat treated parts. The surface cools and contracts first, and again the still hot and ductile interior accommodates readily. However, when the interior eventually cools, its contraction is opposed by the higher strength surface. The restraint on the contraction places the interior in tension while the surface is, in turn, placed in compression by the contracting interior.

The stresses produced by the different rates of cooling throughout a part produce distortion if the stresses generated are high enough to produce nonuniform yielding or plastic deformation on cooling. Even without yielding, stresses up to the yield point may be present on reaching room temperature, and such residual stresses will be superimposed on the applied stresses in service unless tempering or stress-relief treatments are performed. Surface compressive stresses are desirable, and oppose tensile stresses applied in service. Residual tensile stresses, however, add to the applied tensile stresses and effectively lower strength and fatigue resistance in service.

The most critical situation with respect to cooling stresses is in steels austenitized and quenched to form martensite. There the surface tensile stresses, especially on severe quenching, may become high enough to cause quench cracking rather than yielding. High-carbon and tool steels that are inherently hard and brittle are especially susceptible to quench cracking. The quench cracks are invariably intergranular, and their formation may be related to some of the same factors that cause intergranular fracture as a result of tempered martensite embrittlement (discussed in Chapter 8).

Practically, the tendency to distortion, quench cracking, and/or high residual stress formation during heat treatment may be reduced by any change in processing that reduces the differences in the rates of cooling between the surface and interior of a part. Quite effective in this regard is a change to more moderate quenching for hardening, even to the point of air cooling for certain tool steels, if problems are encountered with more severe quenching. The effectiveness of a less severe quench in hardening may require the use of a more hardenable steel, and the matching of steel composition, section size, and cooling rates is an important application of hardenability. Reducing the temperature difference between austenitizing and the quenchant is also sometimes effective in reducing quench cracking or distortion. For example, steels carburized at 950 °C (1750 °F) are often cooled to 840 °C (1550 °F) prior to quenching and may be quenched in oil heated to 70 °C (170 °F) instead of oil held at room temperature. The austempering and martempering treatments described below are also used to minimize differential cooling effects throughout a heat treated part.

The discussion to this point has described the effects of variations in cooling rate through the cross section of a heat treated steel part. Exactly the same considerations apply during heating with the direction of the thermal and transformation volume changes reversed. Thus distortion or cracking may develop in the heating portion of a heat treatment cycle if a steel part cannot accommodate the stresses generated by uneven heating through its cross section. Reheat cracking during stress relief heat treatments of welded low-alloy heat resisting steels is one example of difficulties caused by heating for a subcritical heat treatment (Ref 9.1). Apparently the coarse grained structures of the heat-affected zone have such poor ductility that the deformation required for stress relief cannot be accommodated.

Another situation where difficulty on heating may be encountered is in the austenitizing of high-carbon and tool steels (Ref 9.2). In order to minimize distortion or cracking problems on heating, these steels are preheated prior to heating to the final austenitizing temperature. The work is heated slowly to the preheat temperature and held long enough to ensure that the temperature is uniform throughout the part. Hot work die steels and high-speed steels, especially those containing large amounts of tungsten and molybdenum that lower thermal conductivity, are susceptible to distortion and/or cracking on heating, and preheating of these steels prior to high-temperature austenitizing is essential. Preheating temperatures range between 600 to 800 °C (1110 to 1470 °F) and are tabulated together with complete hardening cycles for the various grades of tool steels by Wilson (Ref 9.2).

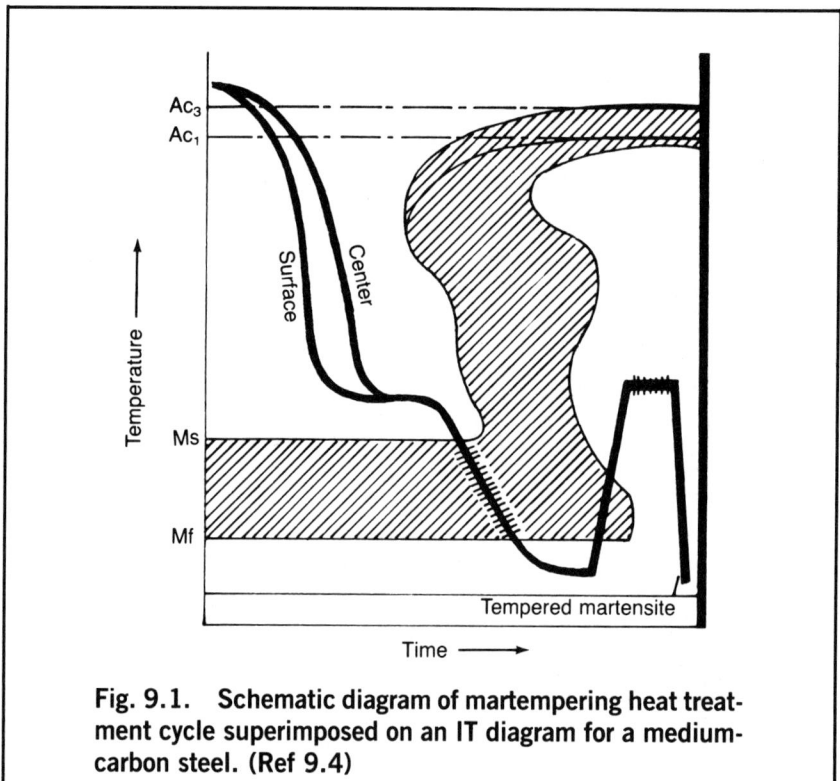

Fig. 9.1. Schematic diagram of martempering heat treatment cycle superimposed on an IT diagram for a medium-carbon steel. (Ref 9.4)

Martempering

Martempering or interrupted quenching is a hardening treatment that consists of quenching to a temperature above the M_s, usually by quenching into a salt bath, holding for a time sufficient for the temperature to become uniform, and then air cooling through the M_s to room temperature (Ref 9.3). Tempering is then performed as required. Figure 9.1 shows schematically the temperature-time sequence for martempering superimposed on a transformation diagram. As noted, the surface and the center of a part cool at different rates to the hold temperature above M_s, and then cool uniformly as martensite is formed on air cooling. The equalization of temperature throughout a part prior to martensite formation ensures that the transformation stresses across the part will be minimal and therefore that cracking and distortion will also be minimized.

An important aspect of martempering is that no transformation product other than martensite should form. Therefore, steels that are suitable for martempering must have sufficient hardenability not only

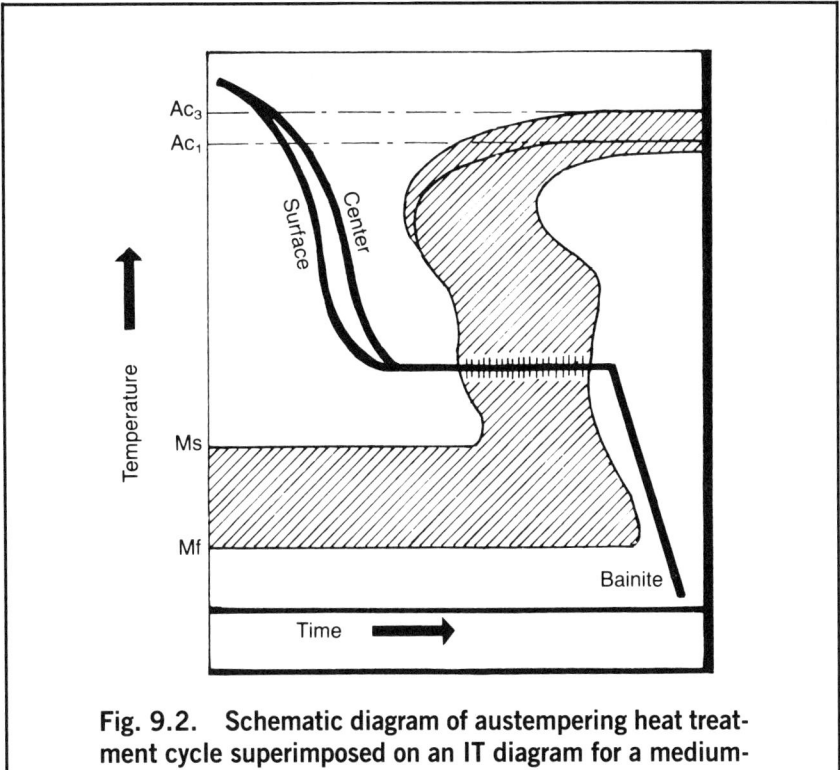

Fig. 9.2. Schematic diagram of austempering heat treatment cycle superimposed on an IT diagram for a medium-carbon steel. (Ref 9.4)

with respect to higher temperature transformation products such as ferrite and pearlite, but also with respect to bainite that might form just above the M_s. Another consideration in this regard is that hot salt has a quenching severity slightly lower than oil quenching. A steel for martempering, therefore, must also have sufficient hardenability to compensate for the reduced rate of cooling. The air cooling through the martensite transformation range is also very important. Water quenching, even though the temperature is uniform throughout the part just above M_s, will almost invariably lead to cracking (Ref 9.3).

Austempering

Austempering is another hardening treatment designed to reduce distortion and cracking in higher carbon steels. The object of austempering, however, is to form bainite rather than martensite. Figure 9.2 shows schematically the heat treatment cycle for austempering. The

Table 9.1. Comparison of Mechanical Properties Produced by Austempering and Quench and Tempering (Ref 9.3)

Steel composition: 0.74C, 0.37 Mn, 0.145Si, 0.039S, 0.044P	
New Method, Direct from Austenite	**Quench-and-Temper Method**
Heat 5 min at 790 °C (1450 °F)	Heat 5 min at 790 °C (1450 °F)
Quench into lead alloy bath at 305 °C (580 °F)	Quench into oil at 21 °C (70 °F)
Let specimens remain in bath for 15 min	Temper immediately in lead alloy bath, 30 min at 315 °C (600 °F)
Quench into water	Quench into water
Mechanical Properties (Average of 6 tests)	**Mechanical Properties** (Average of 6 tests)
Rockwell C hardness 50.4	Rockwell C hardness 50.2
Ultimate strength, ksi 282.7	Ultimate strength, ksi 246.7
Yield point, ksi 151.3	Yield point, ksi 121.7
Elongation, % in 6 inches 1.9	Elongation, % in 6 inches 0.3
Reduction of area, % 34.5	Reduction in area, % 0.7
Impact, ft-lb(a) 35.3	Impact, ft-lb(a) 2.9

(a) Foot-pounds absorbed in breaking 0.180-inch round, unnotched specimens.

steel is austenitized, quenched in molten salt held at a temperature above M_s, and then allowed to transform to bainite at that temperature. No tempering is required. As shown, the temperatures of the center and surface of an austempered part come together at the holding temperature, and the absence of thermal gradients in the subsequent transformation to bainite minimizes the stresses generated during austempering.

As in martempering, a steel suitable for austempering must have sufficient hardenability to avoid the higher temperature austenite transformation products when quenched into heated molten salt with a relatively low quench severity. Carbon steel parts for austempering are therefore limited in size to obtain sufficiently high cooling rates to avoid pearlite. If an alloy steel is selected to compensate for the reduced quenching efficiency of the hot molten salt, the bainite hardenability may also be increased to the point where very long times are required for complete transformation. Thus a major processing advantage of austempering, the fact that no tempering is required, may be offset by the increased holding time for bainite formation.

A major advantage of austempering, apart from significantly reducing distortion or cracking of high-carbon steel parts, is that the toughness is greatly improved relative to tempered martensite at the same hardness level. Table 9.1 shows that at a hardness of Rockwell C 50, austempered 0.74% carbon steel has much better ductility and

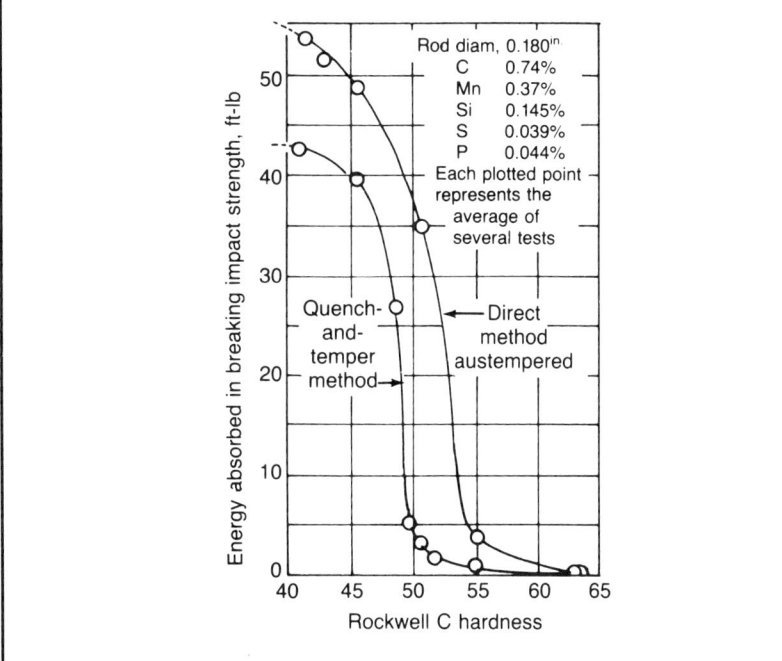

Fig. 9.3. Comparison of impact toughness of austempered and quenched-and-tempered carbon steel as a function of hardness. (Ref 9.3)

impact toughness than the same steel in the quench and tempered condition. It should be noted that the quenched steel has a high phosphorus content and was tempered at 315 °C (600 °F), conditions that indicate the low impact toughness was most probably due to tempered martensite embrittlement as discussed in Chapter 8. Figure 9.3 is another comparison of the impact toughness produced by austempering and conventional quench and tempering of the 0.74% carbon steel (Ref 9.5). The comparison is made at different hardness levels. Austempering is clearly most beneficial in the hardness range between Rockwell C 50 and 55.

Thermomechanical Treatments

Thermomechanical treatments are processing treatments that combine plastic deformation with thermal processing or heat treatment in order to produce microstructures and improved properties not obtained by independently applied conventional heat treatment or

Table 9.2. U.S. Classification of Thermomechanical Treatments (Ref 9.8)

CLASS I:	Deformation occurs before the austenite transformation. Austenite is deformed in the stable austenite range above the critical temperature (A_1) or in the unstable region above the pearlite nose or in the bay region between the pearlite and bainite noses.
CLASS II:	Deformation during the austenite transformation. Depending on the deformation temperature, as well as the M_S and M_D temperatures, the transformation products can be either pearlite, bainite or martensite. The martensite transformation can be due to a strain-induced or stress-assisted transformation.
CLASS III:	Deformation after austenite transformation to martensite or other transformation products.

(a) M_D is temperature above which martensite cannot be formed by plastic deformation.

working operations. Generally, increased strength with improvement in ductility and/or toughness are the objectives of thermomechanical treatments. Intensive research in the 1950s and 1960s showed that these objectives could be achieved, and resulted in a U.S. classification system that recognized three types of thermomechanical treatments (Ref 9.6, 9.7). Table 9.2, after Azrin (Ref 9.8), describes the three classes of

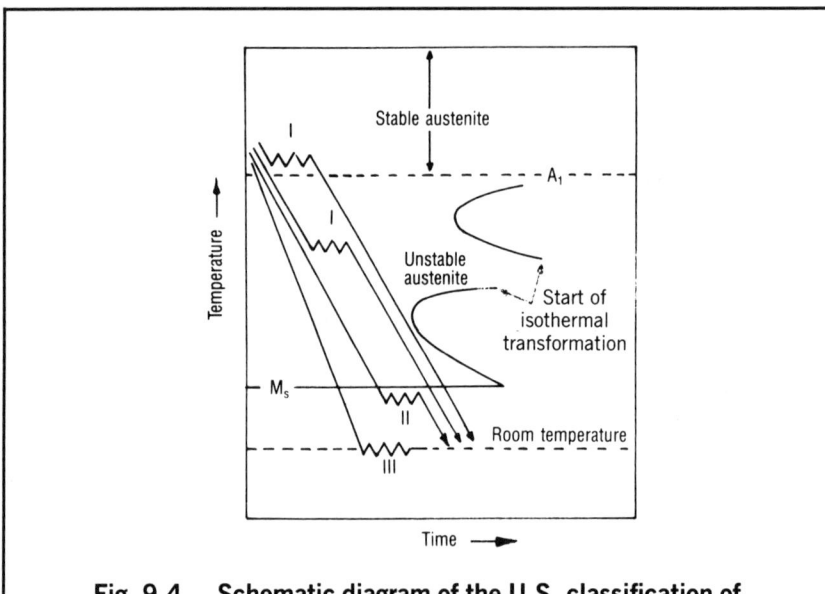

Fig. 9.4. Schematic diagram of the U.S. classification of thermomechanical treatments superimposed on a time-temperature-transformation diagram. (Ref 9.8)

Table 9.3. Soviet Classification of Thermomechanical Treatments (Ref 9.8)

SHT:	Standard Heat Treatment — Conventional heat treatment without deformation.
TMT:	Thermomechanical Treatment — A combined thermal and mechanical treatment generally involving a phase transformation.
HTTMT:	High Temperature Thermomechanical Treatment — Deformation above the recrystallization temperature.
LTTMT:	Low Temperature Thermomechanical Treatment — Deformation below the recrystallization temperature.
CTMT:	Combined Thermomechanical Treatment — HTTMT followed by LTTMT.
PTMT:	Preliminary Thermomechanical Treatment — Deformation by HTTMT or LTTMT or cold working followed by rapid reaustenitizing and quenching.
MTT:	Mechanico-thermal Treatment — Deformation at room or elevated temperature with or without subsequent annealing or aging applied to a material which does not undergo a phase transformation. As with TMT, deformation can be below (LTMTT) or above (HTMTT) the recrystallization temperature.

thermomechanical treatment and Fig. 9.4 shows the treatments schematically superimposed on an idealized transformation diagram. The Soviets have also actively pursued the development of thermomechanical treatments and their classification system is shown in Table 9.3 and Fig. 9.5, also after Azrin (Ref 9.8). As Fig. 9.4 shows, the thermomechanical treatments, recognized as such, have the modification of martensite (either by the deformation of the austenite preceding transformation or by the deformation of the martensite after transformation) as the major approach to improved properties. Very high strengths therefore result from this type of thermomechanical process.

Ausforming is a special term for the class I type of thermomechanical treatment in which austenite is deformed in the bay of the time-temperature-transformation diagram prior to martensite formation. A special variant of this treatment involves the use of highly alloyed steels with an initial structure containing metastable austenite (Ref 9.9). These steels are referred to as transformation induced plasticity (TRIP) steels because they transform to martensite on straining. The strain induced martensite in turn resists plastic instability or necking and thereby produces good combinations of very high strength and ductility. Figure 9.6, after Zackay et al. (Ref 9.9), compares strengths and ductilities produced by various thermomechanical techniques. The properties of the low-alloy high-strength steels indicated are those developed by conventional quench and tempering of carbon steels. The maraging steels are highly alloyed

272 / STEELS: HEAT TREATMENT AND PROCESSING PRINCIPLES

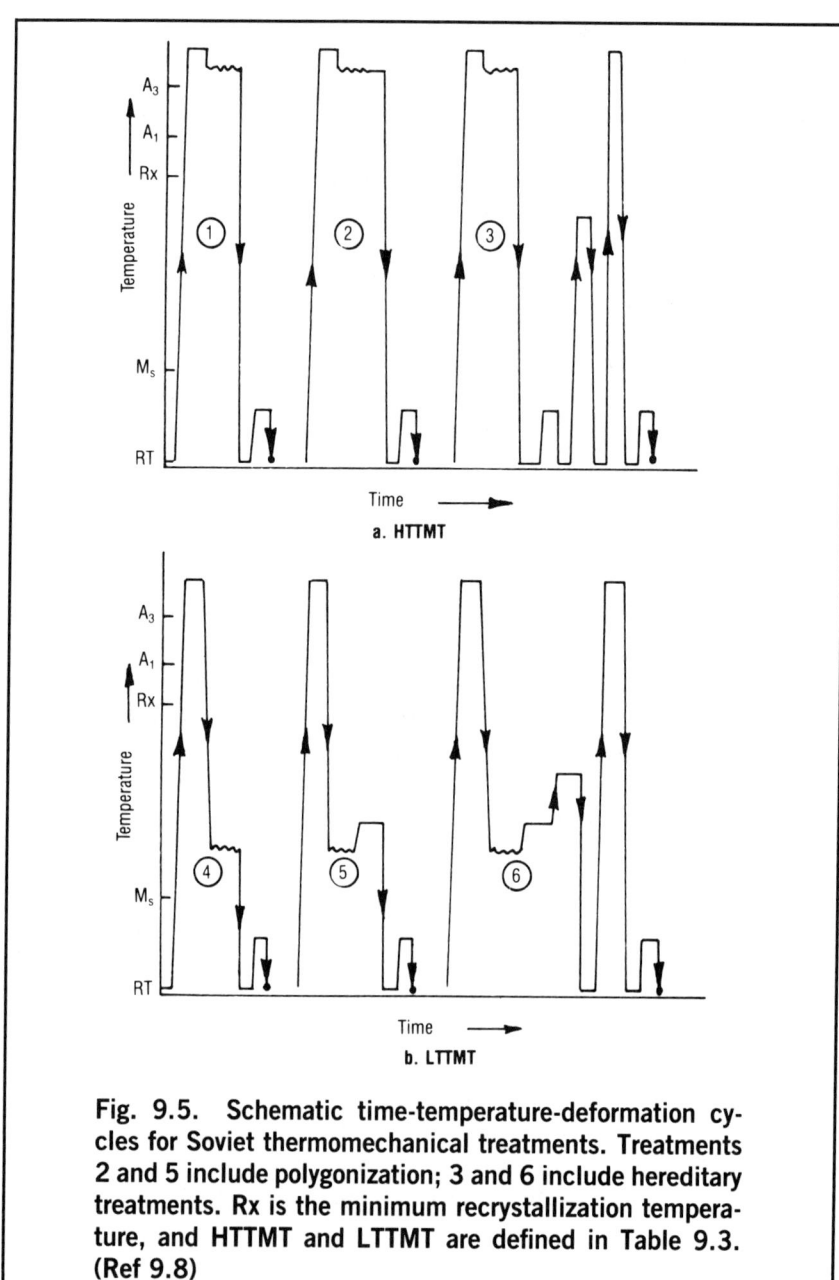

Fig. 9.5. Schematic time-temperature-deformation cycles for Soviet thermomechanical treatments. Treatments 2 and 5 include polygonization; 3 and 6 include hereditary treatments. Rx is the minimum recrystallization temperature, and HTTMT and LTTMT are defined in Table 9.3. (Ref 9.8)

steels, containing about 18% nickel, 8% cobalt, and small amounts of aluminum and titanium. High strength and good toughness are obtained by precipitation of compounds such as Ni_3Mo and Ni_3Ti in a matrix of carbon-free lath martensite (Ref 9.10).

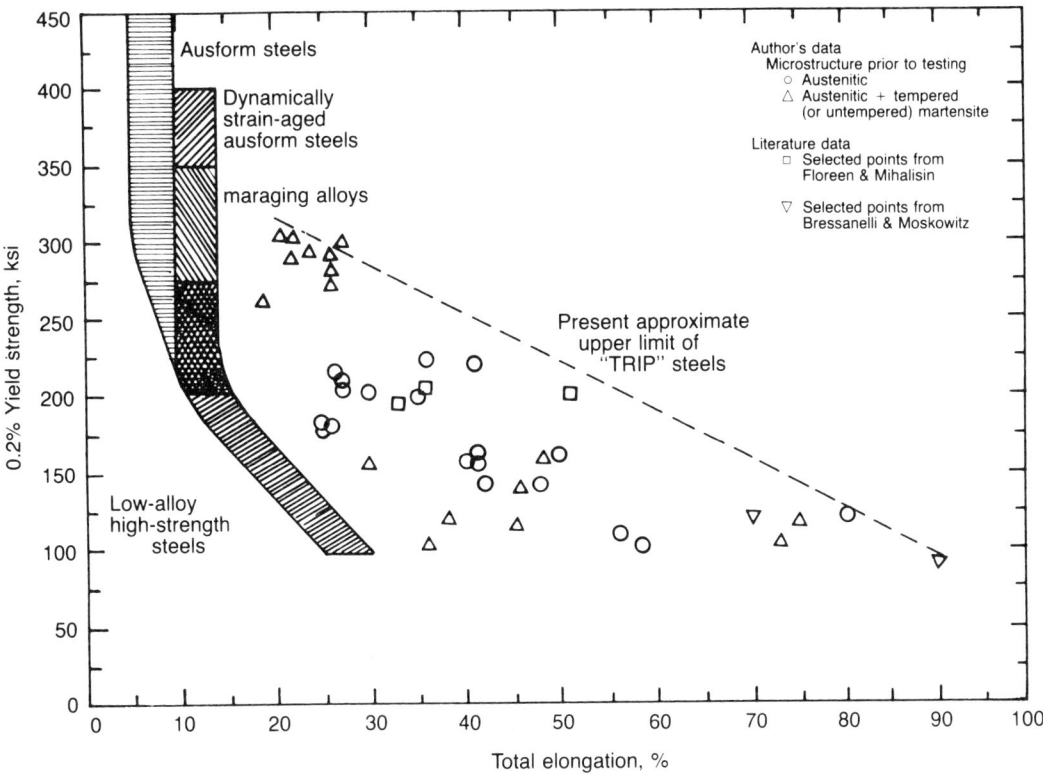

Fig. 9.6. Yield strength versus total elongation for various classes of high-strength steels and thermomechanical treatments. (Ref 9.9)

While the physical metallurgy of thermomechanical treatments such as those shown in Fig. 9.4 and 9.5 is well established, very few industrial applications of thermomechanical treatments have been developed (Ref 9.8, 9.11). Reasons for this situation include technical factors such as the difficulty in machining the very hard thermomechanically treated products, the deterioration in mechanical properties when thermomechanically processed parts are joined by welding, the high alloy content of steels that respond best to thermomechanical treatment, and the difficulty in achieving the uniform deformation required for improved properties in parts with complex shapes. The status of thermomechanical treatments in the U.S. and the U.S.S.R. is reviewed in the reports by Henning (Ref 9.11) and Azrin (Ref 9.8), respectively, and the reader is referred to these publications for the practical aspects and limitations of thermomechanical treatments as well as extensive bibliographies of the literature on thermomechanical treatments.

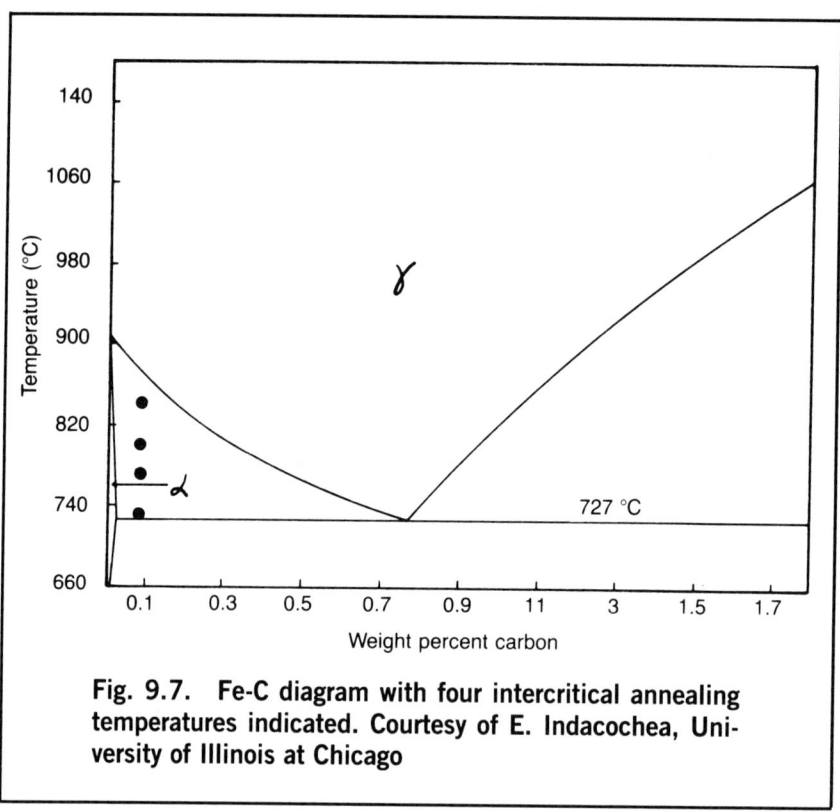

Fig. 9.7. Fe-C diagram with four intercritical annealing temperatures indicated. Courtesy of E. Indacochea, University of Illinois at Chicago

Perhaps the most important application of thermomechanical processing actually in use is the controlled rolling of microalloyed HSLA steels (see Chapter 7) in regard to grain size control. The processing of HSLA low-carbon steels, however, depends on very fine ferrite grain sizes and precipitation for strengthening and not martensite formation, and therefore falls outside the classification systems shown in Table 9.2.

Intercritical Heat Treatment

A new approach to developing good combinations of strength and ductility in low-carbon steels consists of intercritical annealing of hot or cold rolled sheet steels with microstructures consisting initially of ferrite and pearlite. Figure 9.7 shows the Fe-C diagram with four temperatures indicated between the A_1 and A_3 temperatures: the higher the intercritical annealing temperature, the more austenite

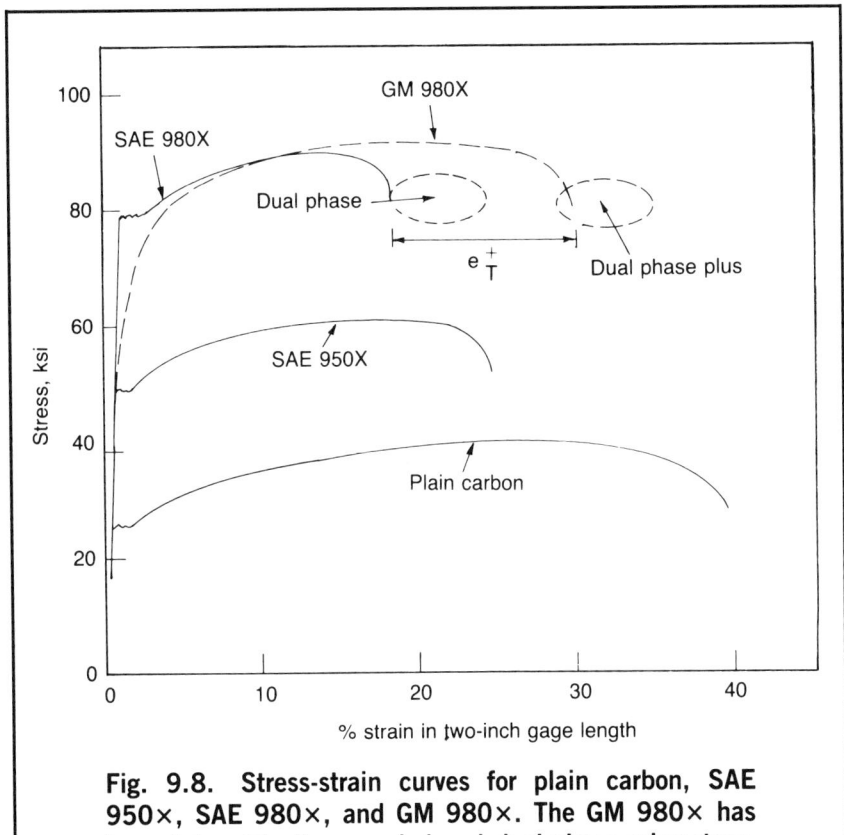

Fig. 9.8. Stress-strain curves for plain carbon, SAE 950×, SAE 980×, and GM 980×. The GM 980× has been intercritically annealed and dual-phase microstructures produced. The two dashed ellipses indicate reported ranges of elongation for dual-phase steels. (Ref 9.12)

formed during annealing. As discussed below, on cooling the austenite may transform into various microstructures whose formation is dependent on the cooling rate and the hardenability of the austenite.

Intercritical annealing for producing high strength, formable sheet steels was first applied by Rashid (Ref 9.12) and Hayami and Furukawa (Ref 9.13). Figure 9.8 shows the basis for three stages in the development of ferritic low-carbon steels. The lower stress-strain curve represents the deformation behavior of mild steel with ferrite-pearlite microstructures. The yielding is discontinuous and yield strengths are typically 30 ksi (207 MPa). SAE 950X and SAE 980X are HSLA steels with yield strengths of 50 ksi (345 MPa) and 80 ksi (562 MPa), respectively. The microstructures still consist of ferrite and pearlite, but the ferrite grain size is highly refined because of controlled rolling and microalloying with vanadium. GM 980X is similar to SAE 980X but has been intercritically annealed to convert the pearlite to martensite. Rashid termed the resulting microstructure "dual-phase" to distinguish the ferrite-martensite microstructure from the ferrite-pearlite microstructure of conventionally treated mild steels or HSLA steels.

Figure 9.8 shows that the intercritical annealing treatment has eliminated discontinuous yielding and significantly increased ductility relative to SAE 980X. These changes in deformation behavior are not yet fully explained, but the formation of martensite during cooling from the intercritical annealing temperature is important in that it introduces a high density of dislocations into the ferrite. These dislocations move immediately on application of stress, and thereby produce the continuous yielding associated with intercritically annealed steels.

The microstructures produced by cooling the austenite formed during intercritical annealing are very much dependent on alloy composition and cooling rate. Severe quenching is required to form martensite in dilute alloys, but air cooling is frequently sufficient to form considerable martensite in steels with higher hardenability. Even though the steels are low carbon, the intercritical annealing concentrates the carbon in the austenite, another helpful factor for improved hardenability.

Figure 9.9 shows the variety of microstructures that may form in an intercritically annealed steel when cooling rate is varied (Ref 9.14). In this case the steel contained 0.08% carbon, 1.47% manganese, and 0.053% niobium. The austenite of the specimen cooled at the highest rate has transformed primarily to martensite, the uniformly dark structure (see Fig. 9.9a). Surrounding the martensite is a white layer where some of the austenite has transformed during cooling to ferrite ("new" ferrite) prior to the martensite formation. This new ferrite is considered to grow epitaxially (with the same crystal orientation) from the ferrite retained at the intercritical annealing temperature (Ref 9.15). Figure 9.9(b) and (c) show that the amount of new ferrite increases with decreasing cooling rate. Also, less martensite and more ferrite-carbide mixtures form with decreasing cooling rate. These microstructural changes are summarized schematically in the microstructure map shown in Fig. 9.10 (Ref 9.14). Also shown in the top part of Fig. 9.10 is a schematic plot of the mechanical property changes with microstructure or cooling rate. The highest strengths and lowest ductilities are associated with microstructures that are most rapidly cooled and contain the most martensite. Good combinations of strength and ductility result from intermediate cooling rates that produce more complex microstructures. The slowest cooling rates produce structures

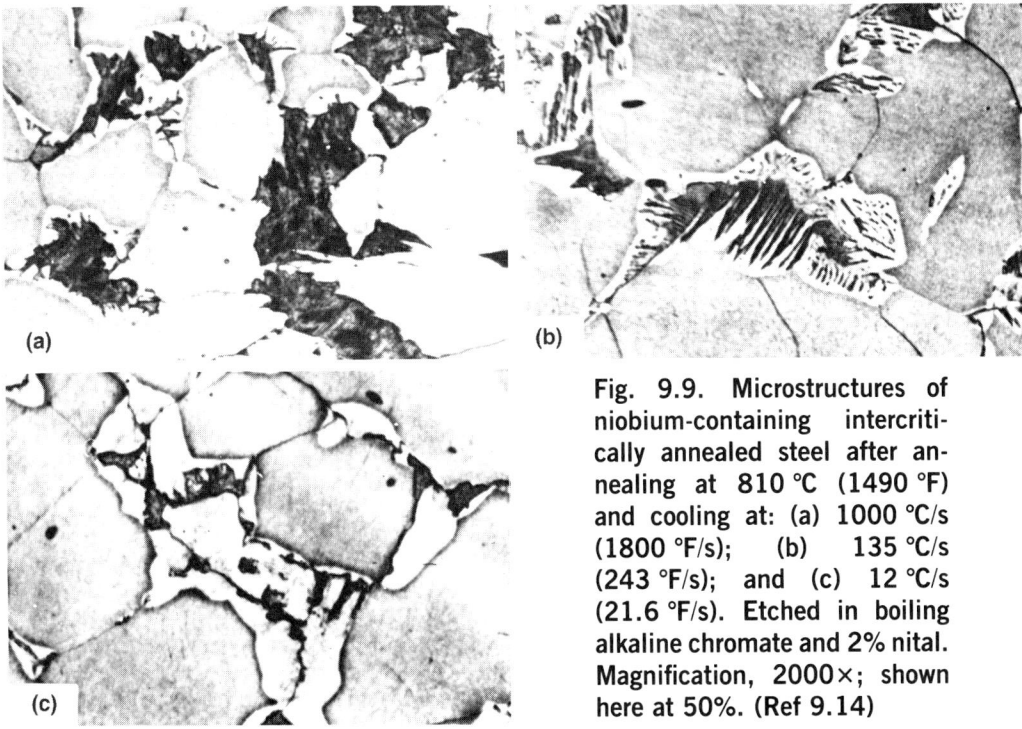

Fig. 9.9. Microstructures of niobium-containing intercritically annealed steel after annealing at 810 °C (1490 °F) and cooling at: (a) 1000 °C/s (1800 °F/s); (b) 135 °C/s (243 °F/s); and (c) 12 °C/s (21.6 °F/s). Etched in boiling alkaline chromate and 2% nital. Magnification, 2000×; shown here at 50%. (Ref 9.14)

typical of mild steels and have substantially lower strengths and higher ductilities. Discontinuous yielding also returns for the latter microstructures.

The old ferrite or ferrite retained during intercritical annealing is stained gray by the etchant (Ref 9.15) used to reveal the microstructures in Fig. 9.9. Figure 9.11 is a transmission electron micrograph taken from an area adjacent to the austenite transformation product in a specimen treated similarly to that shown in Fig. 9.9(b). The new ferrite is free of precipitate particles that have formed in the old ferrite matrix and no interfaces or boundaries separate the two types of ferrite. The dislocation density in the ferrite is also rather low because of the reduced amount of martensite that has formed during cooling at the intermediate rate.

The above discussion shows that intercritically annealed steels offer good combinations of strength and formability, a characteristic that makes this type of heat treatment attractive for vehicle applications where weight saving through increased strength is viable. The reader is referred to the proceedings of the symposium identified in Ref 9.14 for a number of papers on all aspects of intercritical annealing and dual-phase steel processing.

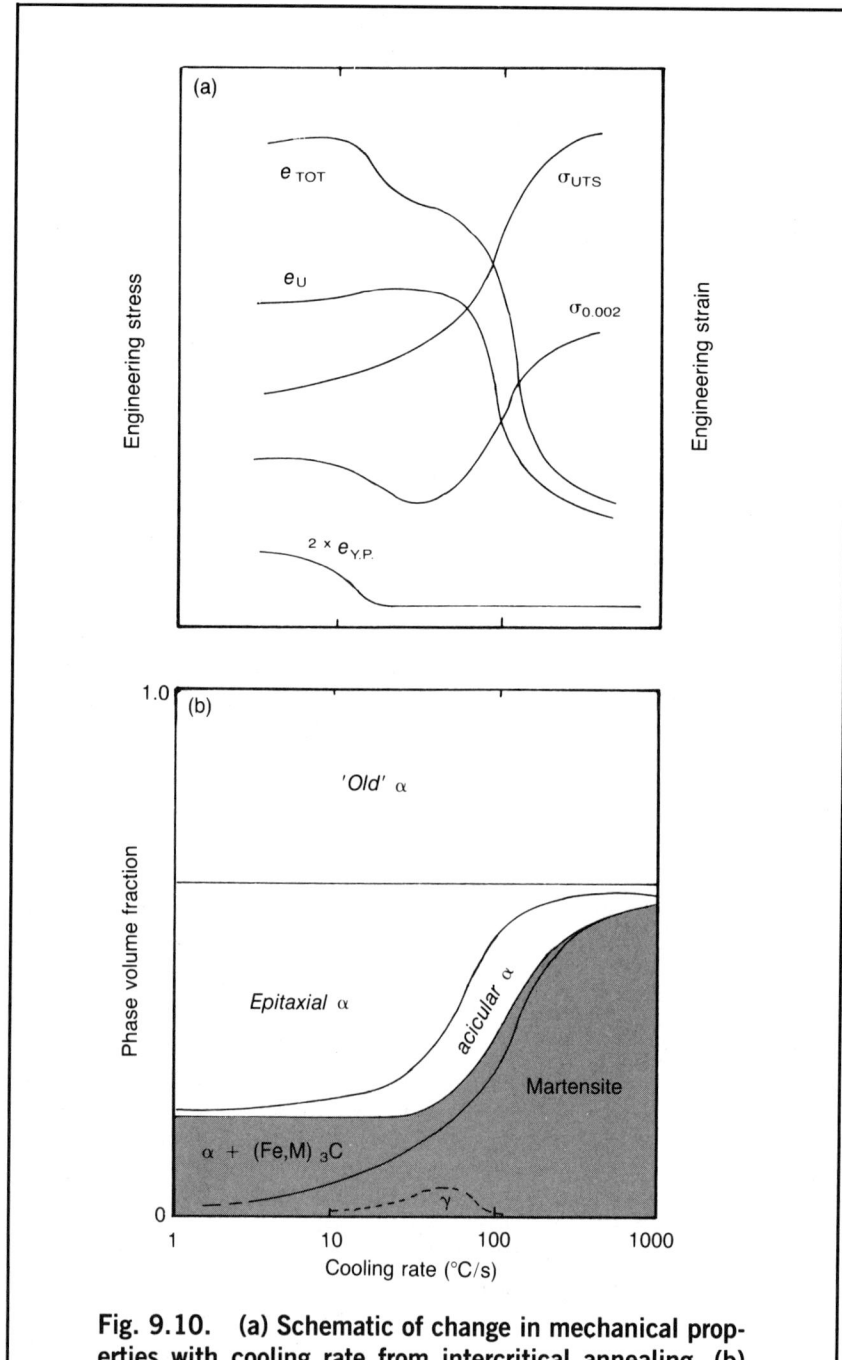

Fig. 9.10. (a) Schematic of change in mechanical properties with cooling rate from intercritical annealing. (b) Schematic map of microstructures formed by different cooling rates. (Ref 9.14)

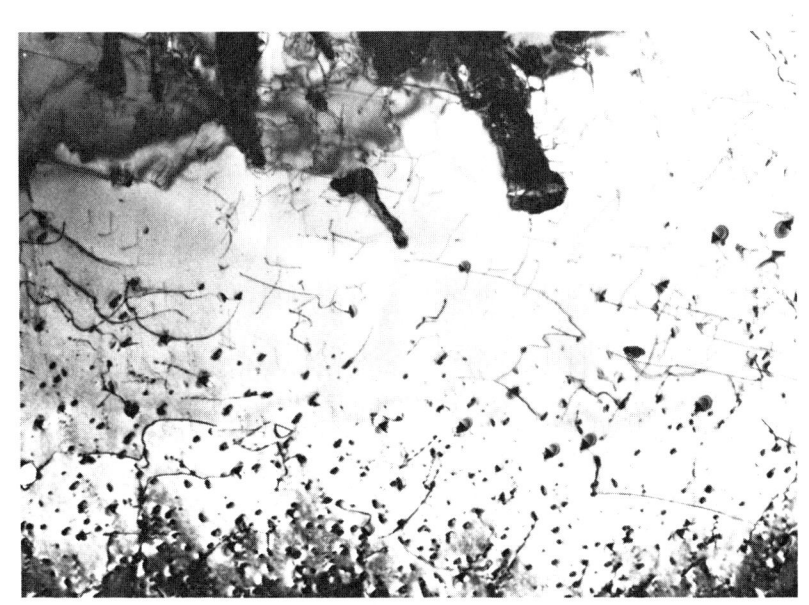

Fig. 9.11. Area adjacent to lamellar carbides formed during cooling a microalloyed steel from 810 °C (1490 °F). The ferrite immediately adjacent to the carbides is free of the precipitates that have developed in the retained ferrite matrix. Transmission electron micrograph. Magnification, 16 500×; shown here at 70%. Courtesy of M.D. Geib, Colorado School of Mines, Golden

References

9.1 A. Vinckier and A. Dhooge, Reheat Cracking in Welded Structures During Stress Relief Heat Treatments, *J Heat Treating*, Vol 1, 1979, p 72-80

9.2 R. Wilson, *Metallurgy and Heat Treatment of Tool Steels*, McGraw-Hill, London, 1975

9.3 M.A. Grossmann and E.C. Bain, *Principles of Heat Treatment*, 5th ed., American Society for Metals, Metals Park, OH, 1964, p 189-196

9.4 E.C. Rollason, *Fundamental Aspects of Molybdenum on Transformation of Steel*, Climax Molybdenum Co., London

9.5 E.S. Davenport, E.L. Roff, and E.C. Bain, *Trans ASM*, Vol 22, 1934, p 289-310

9.6 S.V. Radcliffe and E.B. Kula, in *Fundamentals of Deformation Processing*, W.A. Backofen *et al.* (Eds.), Syracuse University Press, New York, 1964, p 321-363

9.7 E.B. Kula, in *Strengthening Mechanisms—Metals and Ceramics*, J.J. Burke, N.L. Reed, and V. Weiss (Eds.), Syracuse University Press, New York, 1966, p 83-121

9.8 M. Azrin, G.B. Olson, E.B. Kula, and W.F. Markey, Jr., Soviet Progress in Thermomechanical Treatment of Metals, *J Appl Metalworking*, Vol 1, No. 2, 1989, p 5

9.9 V.F. Zackay, E.R. Parker, D. Fahr, and R. Busch, The Enhancement of Ductility in High-Strength Steels, *Trans ASM*, Vol 60, 1967, p 252-259

9.10 A. Magnee, J.M. Drapier, J. Dumont, D. Coutsouradis, and L. Habraken, *Cobalt-containing High-Strength Steels*. Centre d'information du Cobalt, Brussels, 1974

9.11 H.J. Henning, *Applications and Potential of Thermomechanical Treatment*, DMIC Memorandum 251, Defense Metals Information Center, Battelle Memorial Institute, Columbus, OH, Nov 1970

9.12 M.S. Rashid, "GM980x—Potential Applications and Review," Research Publication GMR-232, General Motors Corp, Warren, MI, 1977

9.13 S. Hayami and T. Furukawa, A Family of High Strength Cold Rolled Steels, *Microalloying 75*, Union Carbide Corp., New York, p 311-321

9.14 D.K. Matlock, G. Krauss, L.F. Ramos, and G.S. Huppi, A Correlation of Processing Variables with Deformation Behavior of Dual-Phase Steels, in Proceedings of Symposium *Structure and Properties of Highly Formable Dual-Phase HSLA Steels*, AIME, Warrendale, PA, 1980, p 62-90

9.15 R.D. Lawson, D.K. Matlock, and G. Krauss, An Etching Technique for Microalloy Dual-Phase Steels, *Metallography*, Vol 13, 1980, p 71-87

CHAPTER 10

Surface Hardening

Surface hardening is used to extend the versatility of certain steels by producing combinations of properties not readily attainable in other ways. For many applications, wear and the most severe stresses act only on the surface of a part. Therefore, the part may be fabricated from a formable low- or medium-carbon steel, and is surface hardened by a final heat treatment after all other processing has been accomplished. Surface hardening also reduces distortion and eliminates cracking that might accompany through hardening, especially in large sections. Localized hardening of selected areas is also possible by means of certain surface hardening techniques. This chapter describes two major approaches to surface hardening. One approach does not change composition and consists of hardening the surface by flame or induction heating. The other approach changes surface composition and includes the applications of such techniques as carburizing, nitriding, and carbonitriding.

Flame Hardening

Flame hardening consists of austenitizing the surface of a steel by heating with an oxyacetylene or oxyhydrogen torch and immediately quenching with water. A hard surface layer of martensite over a softer interior core with a ferrite-pearlite structure results. There is no change in composition, and therefore the flame-hardened steel must have adequate carbon content for the desired surface hardness. The rate of heating and the conduction of heat into the interior appear to be more important in establishing case depth than having a steel of high hardenability (Ref 10.1). Figure 10.1 shows hardness gradients produced by various rates of flame travel across a 1050 steel forging. The

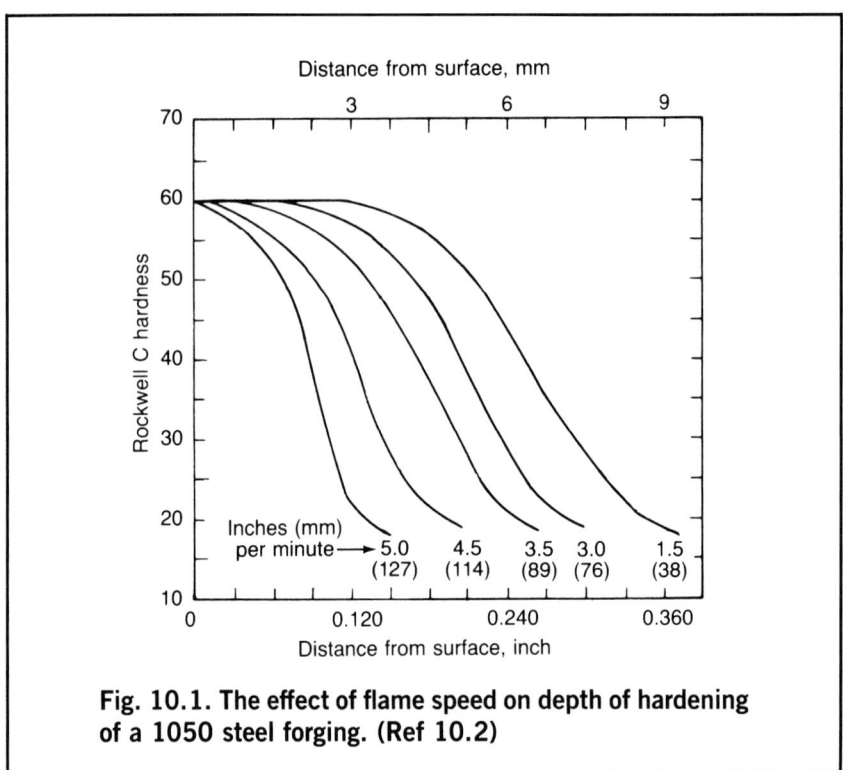

Fig. 10.1. The effect of flame speed on depth of hardening of a 1050 steel forging. (Ref 10.2)

slower the rate of travel, the greater the heat penetration and the depth of hardening.

A number of different methods of flame hardening have been developed (Ref 10.2). Localized or spot hardening may be performed by directing a stationary flame head to an area of a stationary work piece. Progressive methods where the torch travels over the work piece or the work piece travels under a stationary torch and quenching fixture are used for long bars. Spinning methods in which the work piece is rotated within an array of torches are often used for small rounds. In this method heating is performed first, then the flames are extinguished, and quenching is finally accomplished by water sprays or dropping the part into a quench tank. In all cases, the quenched parts are tempered to improve toughness and relieve stresses induced by the surface hardening.

Induction Heating

Induction heating is an extremely versatile method for hardening steel. Uniform surface hardening, localized surface hardening,

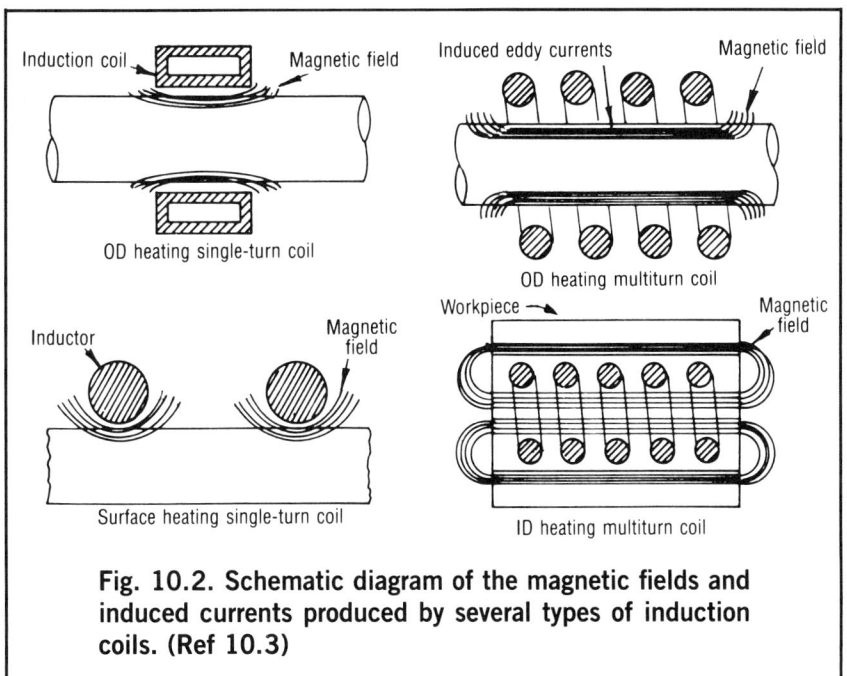

Fig. 10.2. Schematic diagram of the magnetic fields and induced currents produced by several types of induction coils. (Ref 10.3)

through hardening, and tempering of hardened pieces, may all be performed by induction heating. Heating is accomplished by placing a steel part in the magnetic field generated by high-frequency alternating current passing through an inductor, usually a water-cooled copper coil. The rapidly alternating magnetic field established within the coil induces current (I) within the steel. The induced currents then generate heat (H) according to the relationship $H = I^2R$, where R is the electrical resistance. Steel, consisting primarily of ferrite or bcc iron, is ferromagnetic up to its Curie temperature (768 °C or 1400 °F), and the rapid change in direction of the internal magnetization of domains in a steel within the field of the coil also generates considerable heat. When a steel transforms to austenite, which is nonmagnetic, this contribution to induction heating becomes negligible. A wide variety of heating patterns may be established by induction heating depending on the shape of the coil, the number of turns of the coil, the operating frequency, and the alternating current power input (Ref 10.3). Figure 10.2 shows examples of the heating patterns produced by various types of coils.

The depth of heating produced by induction is related to the frequency of the alternating current. The higher the frequency, the thinner or more shallow the heating. Therefore, deeper case depths and even through hardening are produced by using lower frequencies.

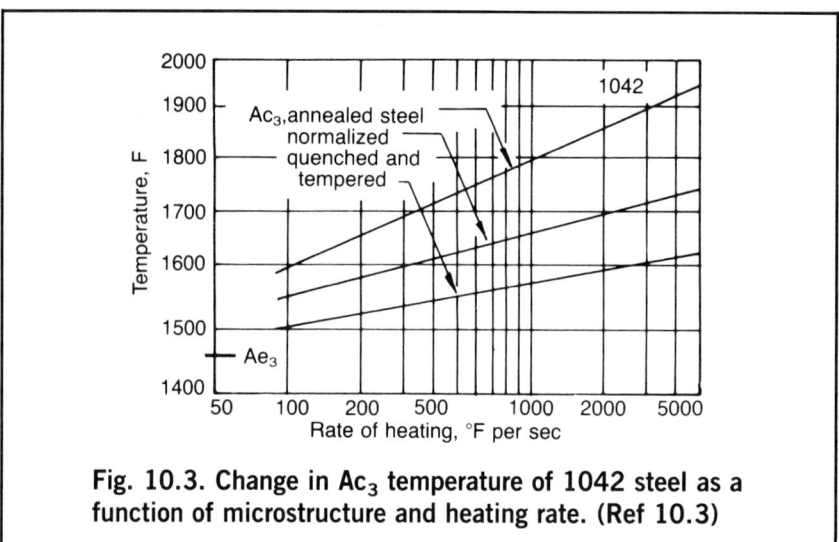

Fig. 10.3. Change in Ac_3 temperature of 1042 steel as a function of microstructure and heating rate. (Ref 10.3)

The various types of commercial equipment and the selection of operating conditions for a given application are fully described in Ref 10.3 to 10.5.

As in flame hardening, induction heating does not change the composition of a steel, and therefore a steel selected for induction hardening must have sufficient carbon content and alloying for the desired surface hardness distribution. Generally, medium- and high-carbon steels are selected because the high surface strengths and hardness attainable in these steels significantly improve fatigue and wear resistance. Induction hardening introduces residual compressive stresses into the surface of hardened parts. Therefore, the fatigue strengths of induction surface hardened parts may be higher than those of through hardened parts in which quenching develops residual surface tensile stresses that may only be partly relieved during tempering. The greater the depth of hardening by induction heating, however, the more the surface stress state approaches that of through hardening. Too deep a hardened case may in fact cause surface tensile stresses and even cracking of susceptible steels (Ref 10.3).

The duration of high-frequency induction heating cycles for surface hardening is extremely short, often only a few seconds. As a result the time for formation of austenite is limited, and compensation is made by increasing the temperature of austenitizing. Figure 10.3 shows how the Ac_3 temperature in a 1042 steel is affected by heating rate and microstructure. The high heating rates of induction heating substantially raise Ac_3. Microstructures with coarse car-

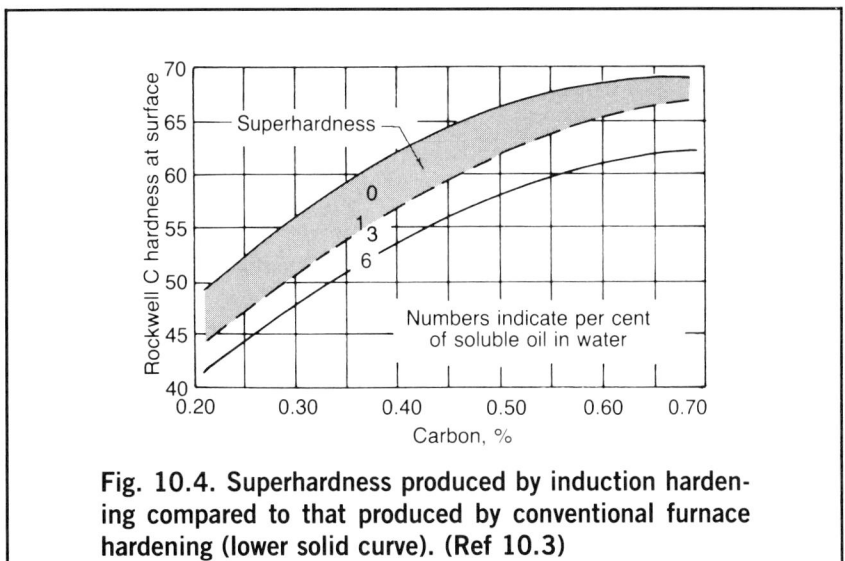

Fig. 10.4. Superhardness produced by induction hardening compared to that produced by conventional furnace hardening (lower solid curve). (Ref 10.3)

bides, such as the 1042 steel in the annealed condition as shown, or steels with coarse spheroidized microstructures or alloy carbides, require higher austenitizing temperatures for carbide solution than do steels with finer microstructures. Too high an austenitizing temperature, however, may result in austenite grain coarsening (see Chapter 7). An interesting consequence of the very short austenitizing time for induction surface hardening is the development of a hardness above that normally expected for through-hardened martensite. This higher hardness is sometimes referred to as superhardness (see Fig. 10.4) and, as discussed in earlier chapters, may be a result of martensite formed in very fine-grained, imperfect austenite produced by the short-time austenitizing treatments used in induction surface hardening.

Carburizing: Processing Principles

Carburizing is a heat treatment in which the carbon content of the surface of a low-carbon steel is increased by exposure to an appropriate atmosphere at a temperature in the austenite phase field. Hardening is accomplished when the high-carbon surface layer is quenched to form martensite. The Fe-C diagram (see Chapter 1) shows that the maximum solubility of carbon in austenite ranges from 0.8% at the eutectoid temperature to about 2% at the eutectic temperature. Although alloying elements reduce carbon solubility, more than enough carbon can be introduced into the austenite of plain carbon or alloy steels by carburizing to produce the maximum martensitic hardness after quenching. Complications of carbide formation, brittle martensite and retained austenite develop if carbon content is too high, and for these reasons the maximum carbon content in a carburized steel is generally controlled to between 0.8 and 1%. Carburizing is most frequently performed between 850 to 950 °C (1550 and 1750 °F), but sometimes higher temperatures are used to reduce cycle times and/or produce deeper depths of the high-carbon surface layer.

Two important processes influence the introduction of carbon into austenite during carburizing. One is the environmental reaction that causes carbon to be absorbed at the surface of the steel. Another is the rate at which carbon can diffuse from the surface to the interior of the steel. Carbon is introduced by the use of gaseous atmospheres (gas carburizing), salt baths (liquid carburizing), and solid compounds (pack carburizing) (Ref 10.6). All of these methods have limitations and advantages, but gas carburizing is used most often for large-scale production because it can be accurately controlled and involves a minimum of special handling.

Carbon is introduced into the surface of steel by gas-metal reactions between the various components of an atmosphere gas mixture and the solid solution austenite. Following Harvey (Ref 10.7), one of the most important carburizing reactions is

$$CO_2(g) + C \rightleftharpoons 2CO(g) \qquad (Eq\ 10.1)$$

where C is carbon introduced into the austenite. At equilibrium, a carbon ratio of CO_2 and CO has a certain carbon potential or maintains a certain level of carbon in the austenite. At any temperature, the relationship between the gaseous components and the carbon in solution of the austenite is given by the equilibrium constant K which for reaction 10.1 is written as:

$$K = \frac{P_{CO}^2}{a_c P_{CO_2}} \qquad (Eq\ 10.2)$$

where P_{CO} and P_{CO_2} are the partial pressures of CO and CO_2, respectively, and a_c is activity of carbon. The activity of carbon is related to the weight percent carbon in the austenite by the activity coefficient of carbon (f_c) by the following equation:

$$a_c = f_c\ wt.\%\ C \qquad (Eq\ 10.3)$$

K is a function of temperature, and for the reaction represented in Eq 10.1 is

$$\log K = \frac{-8918}{T} + 9.1148 \qquad \text{(Eq 10.4)}$$

where T is the absolute temperature in degrees Kelvin. The partial pressures of CO and CO_2 required to maintain a given surface austenite carbon content are given by combining Eq 10.2 and Eq 10.3 as follows:

$$\text{wt.\% C} = \frac{1}{Kf_c} P_{CO}^2 / P_{CO_2} \qquad \text{(Eq 10.5)}$$

If the CO content of an atmosphere exceeds the partial pressure required to maintain a given carbon content, the reaction represented in Eq 10.1, as written, will go to the left and carburizing will occur until a new equilibrium is reached. This is the case in commercial carburizing where the carbon content of a low-carbon steel is raised to some desirable higher level. On the other hand, if the CO_2 partial pressure is too high relative to the CO content, the reaction in 10.1 will go to the right and decarburization will occur. The latter condition is sometimes purposely introduced in commercial practice if initial carburizing produces too high a carbon content, say 1.2%, and it is desired to reduce surface carbon to a lower level, say 0.9%. This step in a carburizing cycle is referred to as a "diffusion step", since much of initially high carbon in the austenite immediately adjacent to the surface diffuses into interior of the part and produces a deeper case. Equation 10.5 requires a knowledge of the activity coefficient, which varies as a function of temperature and the composition of austenite. Harvey (Ref 10.7) tabulates relationships for the activity coefficients in ternary Fe-X-C systems where X may be the elements nickel, silicon, manganese, chromium, molybdenum, or vanadium. Harvey also presents a system for evaluating the activity coefficient and carburizing potentials for steels with more than three components.

The preceding discussion described the basic concept of gas equilibrium and carburizing. In conventional gas carburizing practice, carburizing atmospheres are produced by combustion of natural gas or other hydrocarbon gas in exothermic or endothermic gas generators and contain CO, CO_2, CH_4, H_2, H_2O, and N_2 (Ref 10.8). Therefore, there are many other reactions that may occur in addition to that represented in 10.1, including the following:

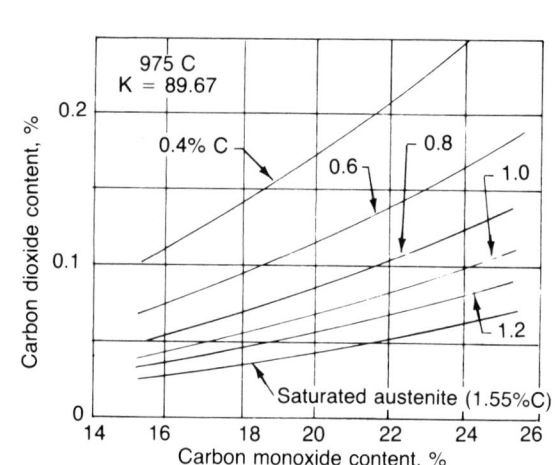

Fig. 10.5. Equilibrium percentages of carbon monoxide and carbon dioxide required to maintain various carbon concentrations at 975 °C (1790 °F) in plain carbon and certain low-alloy steels. (Ref 10.10)

$$CH_4 \rightleftharpoons C + 2H_2 \qquad (Eq\ 10.6)$$

$$CO + H_2O \rightleftharpoons CO_2 + H_2 \qquad (Eq\ 10.7)$$

Much of the current technology (Ref 10.6) of gas carburizing is based on relationships determined by Harris (Ref 10.9) for Eq 10.1, 10.6, and 10.7 according to the approach described above and with the assumption that the activity of carbon in saturated austenite (austenite with the carbon content given by the A_{cm} at a given temperature) is unity. For carbon concentration in the austenite less than saturation, the activity is assumed to be proportional to the degree of saturation. For example, if the carbon content of austenite at saturation is 1.33%, and one wants to know what partial pressures of CO and CO_2 will maintain 1.0% carbon in austenite, the activity is equal to 1.00/1.33 = 0.75. These assumptions appear to be valid for plain carbon and low-nickel steels, and within 10% for more highly alloyed steels (Ref 10.6). A number of curves, based on the Harris approach, relating the CO and CO_2 contents required to maintain various surface carbon contents are given in Ref 10.10 for carburizing temperatures between 825 and 1025 °C (1515 to 1875 °F) in 25 °C (76 °F) increments. Figure 10.5 shows such a curve for 975 °C (1790 °F) increments. As shown, a much higher

CO than CO_2 content is required for carburizing, especially for higher surface carbon contents.

Equation 10.7 gives the equilibrium between CO, CO_2, H_2O, and H_2 and is used as the basis of control and determination of the carbon potential of carburizing atmospheres. For the latter purposes, measurements of the contents of H_2O or CO_2 are often used. Carbon dioxide is measured by infrared analyzers. The dew point is defined as the temperature, at a given pressure, at which a gas mixture will precipitate its moisture content and is used to determine the H_2O content of an atmosphere. More recently, carbon potential has been determined by measurement of the partial pressure of oxygen with instruments referred to as oxygen probes. Once the oxygen content is known, the CO and CO_2 contents can be determined by means of the following equilibrium reaction:

$$CO + \tfrac{1}{2}O_2 \rightleftharpoons CO_2 \qquad \text{(Eq 10.8)}$$

The various approaches and instruments for control of carburizing potentials are described in detail in Ref 10.6.

After a given carbon potential is established, the depth of case produced in a given carburizing treatment is determined by the time-dependent diffusion of carbon from the surface to the interior of the steel part. Figure 10.6 shows carbon profiles calculated for an Fe-C alloy carburized at 925 °C (1700 °F) for times between 2 and 16 h, and Fig. 10.7 shows the effect of temperature on carburizing an Fe-C alloy when time is held constant at 8 h. These figures demonstrate the important effect of time and temperature on the depth of case produced by carburizing.

Following Goldstein and Moren (Ref 10.11), the curves of Fig. 10.6 and 10.7 were calculated by means of the Van-Ostrand-Dewey solution to the diffusion equation as follows:

$$\frac{C_c - C_s}{C_o - C_s} = \operatorname{erf}(X/2\sqrt{Dt}) \qquad \text{(Eq 10.9)}$$

where C_s is the surface concentration of carbon as maintained by the carbon potential of the atmosphere; C_o is the initial carbon level in the Fe-C alloy prior to carburizing; D is the diffusion coefficient for carbon in austenite; C_c is the carbon concentration as a function of distance from the surface; and t is the time after the start of carburizing. The diffusion coefficient was assumed to be independent of composition and to have an average value $D_c^\gamma = 0.12 \times \exp(-32\,000/RT)$ cm^2/s (Ref

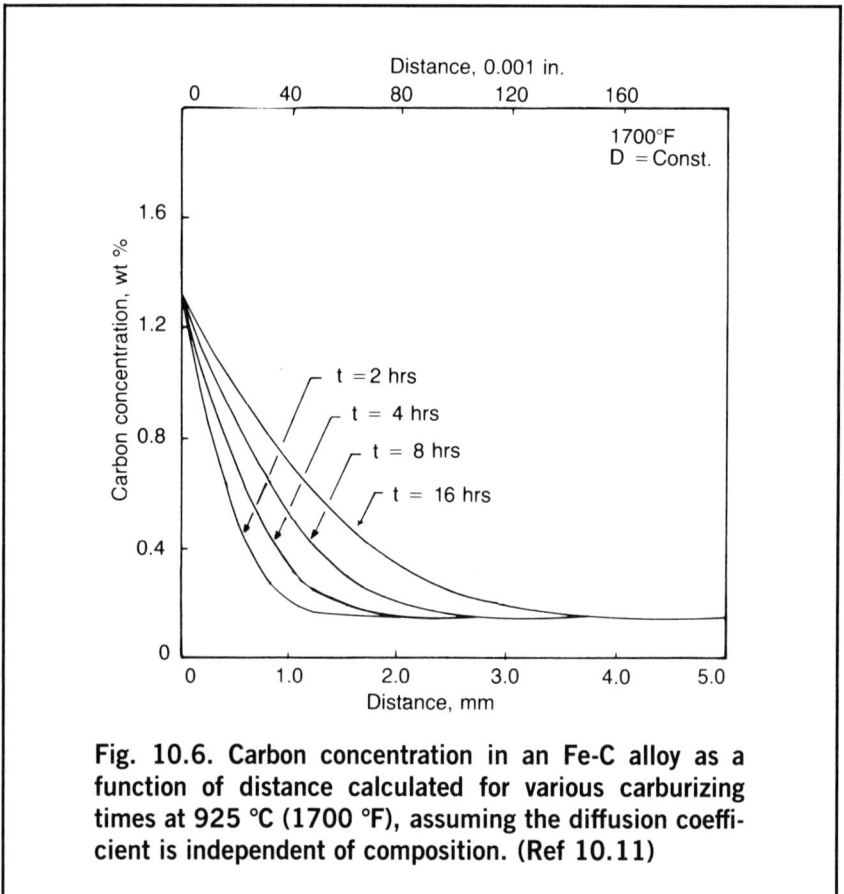

Fig. 10.6. Carbon concentration in an Fe-C alloy as a function of distance calculated for various carburizing times at 925 °C (1700 °F), assuming the diffusion coefficient is independent of composition. (Ref 10.11)

10.12). The diffusion coefficient in fact varies with carbon concentration of the austenite.

Equation 10.9, while it demonstrates the basic diffusion principles involved in carburizing, is highly idealized with respect to commercial practice. In particular, the diffusion coefficient of carbon in steels varies not only with carbon but also with alloy content, a situation for which Goldstein and Moren (Ref 10.11) have developed mathematical models which incorporate the effect of other alloying elements on the diffusion process. Apart from this recent approach, the equations based on empirical analysis of Eq 10.6 by Harris have proven adequate for plain carbon and alloy steels. At any temperature, Eq 10.9 reduces to:

$$X \text{ (case depth)} = K\sqrt{t} \qquad \text{(Eq 10.10)}$$

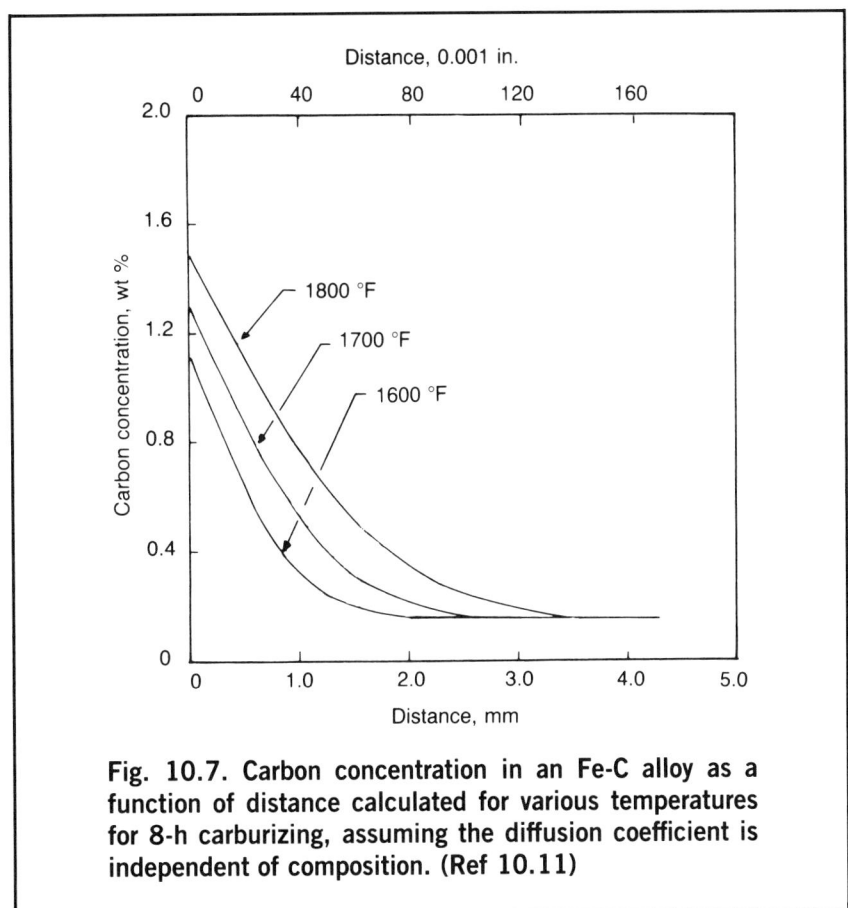

Fig. 10.7. Carbon concentration in an Fe-C alloy as a function of distance calculated for various temperatures for 8-h carburizing, assuming the diffusion coefficient is independent of composition. (Ref 10.11)

where K is a function of temperature and includes the temperature dependence of the diffusion coefficient. Table 10.1 (Ref 10.2) lists values of case depths at various times for three commonly used carburizing temperatures.

Carburizing: Properties and Structure

The objective of carburizing is to obtain a high-carbon martensitic case with good wear and fatigue resistance superimposed on a tough, low-carbon steel core. Carburizing steels usually have base-carbon contents around 0.2%. Therefore, if hardenability is low as is the case for plain carbon steels, the core microstructure will consist of ferrite and pearlite of relatively low strength. Many applications, however, require high core strength to support the case in heavy-duty applications. In addition, core strength is required where the stress gradients

Table 10.1. Case Depth Calculated by the Harris Equation (Ref 10.2)

Time, t, h	Case depth (in.) after carburizing at(a):		
	1600 °F	1650 °F	1700 °F
2	0.025	0.030	0.035
4	0.035	0.042	0.050
8	0.050	0.060	0.071
12	0.061	0.073	0.087
16	0.071	0.084	0.100
20	0.079	0.094	0.112
24	0.086	0.103	0.122
30	0.097	0.116	0.137
36	0.108	0.126	0.150

(a) Case depth = $0.025\sqrt{t}$ for 927 °C (1700 °F); $0.021\sqrt{t}$ for 899 °C (1650 °F); $0.018\sqrt{t}$ for 871 °C (1600 °F). For normal carburizing (saturated austenite at the steel surface while at temperature).

between the surface and interior of a part in service are high enough to cause subsurface crack initiation in an unhardened core. For these reasons, alloy steels with good core hardenability that form martensite throughout a carburized part are in wide use. Table 10.2, taken from a study of fracture resistance of carburized steels (Ref 10.13), shows the compositions and the hardenabilities in D_I of some commonly used

Table 10.2. Compositions, Grain Sizes, and Hardenabilities of Some Carburizing Steels (Ref 10.13)

	Group I		Group II		Group III			Group IV	
	EX24	SAE 8620	EX29	SAE 4320	20NiMoCr6	SAE 4817	SAE 4820	EX32	EX55
Composition, wt %:									
Carbon	0.20	0.20	0.20	0.21	0.22	0.17	0.19	0.19	0.17
Manganese	0.88	0.89	0.87	0.58	0.58	0.54	0.60	0.82	0.87
Silicon	0.34	0.34	0.34	0.33	0.54	0.33	0.28	0.27	0.28
Phosphorus	0.015(a)	0.015(a)	0.015(a)	0.015(a)	0.021	0.015(a)	0.016	0.017	0.015(a)
Silicon	0.02(a)	0.02(a)	0.02(a)	0.02(a)	0.027	0.02(a)	0.02	0.02	0.02(a)
Chromium	0.51	0.47	0.48	0.52	0.64	NA(b)	NA(b)	0.53	0.49
Molybdenum	0.26	0.21	0.34	0.26	0.31	0.27	0.27	0.52	0.74
Nickel	NA(b)	0.53	0.54	1.76	1.56	3.56	3.48	0.80	1.84
Aluminum	0.08(a)	0.08(a)	0.08(a)	0.08(a)	0.043	0.08(a)	0.075	0.082	0.08(a)
ASTM grain size	9½	9½	9½	9½	7½	9½	9½	9	9½
D_I hardenability:									
Inches	1.6	1.7	2.0	1.9	3.0	2.5	2.7	3.3	4.7
Millimeters	41	43	51	48	76	63	69	84	120

(a) Amount added. (b) NA = none added.

alloy carburizing steels. The EX grades of steels are exchange grade steels not yet assigned standard SAE designations (Ref 10.14). In the case of the carburizing grades, these steels have been developed to match different hardenability ranges of standard grades (as shown in Table 10.2) by means of adjusting alloy content.

Depending on hardenability, i.e., the composition of a carburizing steel, specimen size, and quench severity, quite different hardness gradients may be associated with a given carbon gradient. Maximum hardness at any given carbon level will be associated with a fully martensitic microstructure, but bainite and other low-hardness microstructures might also form if cooling is not sufficient to form martensite at that location. Wyss (Ref 10.15) has developed a scheme for calculating hardness gradients from carbon gradients. The first step is to calculate Jominy curves for a given alloy composition at several carbon contents. Then an equivalent distance from the quenched end of a Jominy specimen is obtained for a given bar diameter and quench severity according to Grossmann's method. That equivalent distance, i.e., an effective cooling rate, is then used to obtain hardness values as a function of carbon content from the various Jominy curves. Then if carbon content as a function of distance from the carburized surface is known or calculated, hardness as a function of distance can be plotted.

Once the proper surface carbon concentration and depth of case are established by control of the processing parameters, the carburized steel is quenched to form martensite and tempered. The part may be hardened directly after carburizing or may be cooled and reheated to refine the microstructure. Figure 10.8 shows the microstructure of an EX 24 steel carburized and diffused at 1050 °C (1920 °F) and cooled to 845 °C (1550 °F) prior to quenching in oil (Ref 10.16). This microstructure is typical of that produced in fine-grained steels by direct quenching of carburized cases containing about 1% carbon at the surface. The martensite has a plate morphology, and as much as 20 or 30% retained austenite might be present. Lower carbon contents in a case would produce a finer martensite tending to a lath morphology and would reduce retained austenite content. Figure 10.9 shows the case microstructure produced in an 8620 steel that has been reheated to a temperature below the A_{cm}. The microstructure is highly refined by this treatment, so much so that the matrix martensite and retained austenite are not resolvable in the light microscope. The fine white particles that are visible are carbides that have been retained because the case was not reheated into the single-phase austenite field.

Figures 10.8 and 10.9 show microstructures that are typical of the results of good commercial practice. Problems due to processing, however, may adversely affect microstructures and properties. One

294 / STEELS: HEAT TREATMENT AND PROCESSING PRINCIPLES

Fig. 10.8. Microstructure adjacent to surface of an EX 24 steel carburized and diffused at 1050 °C (1920 °F), cooled to 845 °C (1550 °F), and oil quenched. Magnification, 1000×. (Ref 10.16)

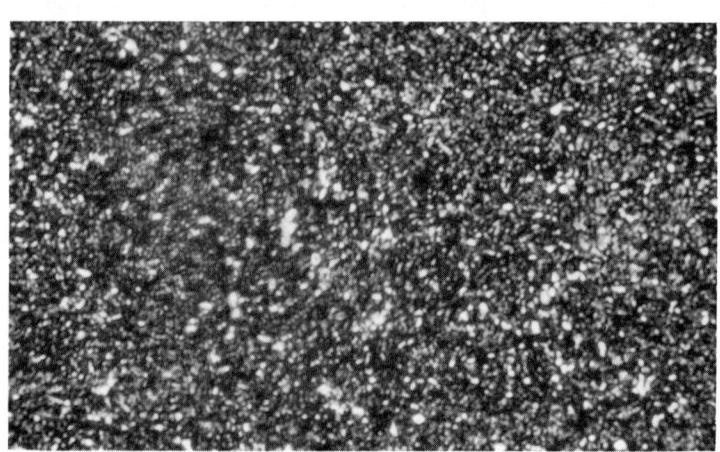

Fig. 10.9. Microstructure adjacent to surface of 8620 steel carburized at 1050 °C (1920 °F), oil quenched from 845 °C (1550 °F), and reheated to 845 °C (1550 °F). Magnification, 1000×. (Ref 10.16)

common feature of conventional gas carburizing is the internal oxidation that develops to a depth of about 0.001 in. (0.025 mm) from the carburized surface. Figure 10.10 shows an example of intergranular oxidation in a gas carburized specimen. Although many applications tolerate the presence of such intergranular oxidation, the oxides do adversely affect fatigue resistance, especially when the surface depletion of alloying elements by the formation of alloy oxides reduces the surface hardenability to the point where microstructures such as bainite or pearlite form instead of martensite (Ref 10.17, 10.18). The formation of nonmartensitic microstructures lowers surface compressive stresses and increases susceptibility to fatigue crack initiation (Ref 10.19). For critical applications, the oxide-containing layer is sometimes removed by grinding or machining. Since the oxidation is a direct result of the oxygen content of the carburizing gases, partial pressure or vacuum carburizing that significantly reduces oxygen content of a carburizing atmosphere is also used to eliminate the surface oxidation (Ref 10.20). A systematic study (Ref 10.21) of internal oxidation during carburizing showed that certain elements such as chromium, manganese, silicon, and titanium enhanced oxidation, and that removal of silicon from carburizing steels tended to eliminate the surface oxidation.

Figures 10.11 and 10.12 show the deleterious effects of too high a surface carbon concentration on case microstructure. The high surface carbon concentration may be a result of too short a diffusion cycle after carburizing has first saturated the austenite, or a result of geometry, where carbon has good surface access to a portion of a steel (especially specimen corners) but diffusion to the interior is restricted (Ref 10.22). Figure 10.11 shows a very high retained austenite content in an 8620 steel and Fig. 10.12 shows grain boundary carbides that have formed in an EX 24 steel that has been reheated for hardening after carburizing. The reheating has caused some agglomeration of the carbides and therefore reduced their detrimental effect on fatigue crack initiation (Ref 10.22).

Another microstructural feature that may adversely affect properties of carburized steels is the microcracking that develops in high-carbon martensitic microstructures because of plate impingement (Ref 10.23). Figure 10.13 shows examples of microcracks that have formed in the case of a coarse-grained 8620 steel (Ref 10.24). The micrograph shows an extreme example of microcracking. Microcracking is much reduced when martensite forms in fine-grain austenite (Ref 10.25); when the martensite morphology approaches that of lath martensite as carbon is reduced below 1% (Ref 10.26); and as a result of tempering (Ref 10.27). Many of the microstructural features de-

296 / STEELS: HEAT TREATMENT AND PROCESSING PRINCIPLES

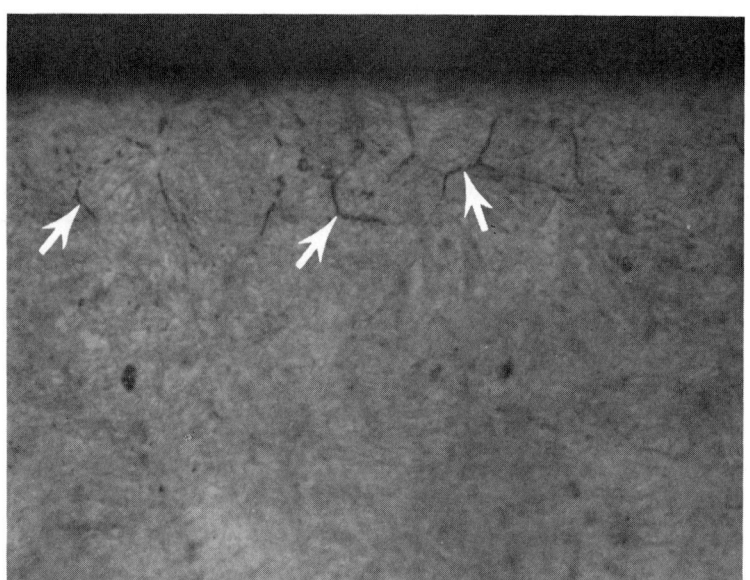

Fig. 10.10. Example of surface intergranular oxidation along austenite grain boundaries in a carburized steel. Magnification, 1000×. (G. Krauss, unpublished research, Colorado School of Mines, Golden)

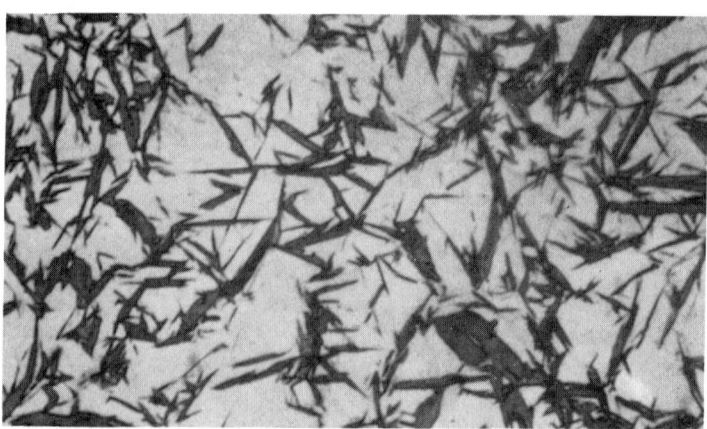

Fig. 10.11. High austenite content in corner of an 8620 steel carburized and diffused at 1050 °C (1920 °F) and cooled to 845 °C (1550 °F) before oil quenching. (Ref 10.22)

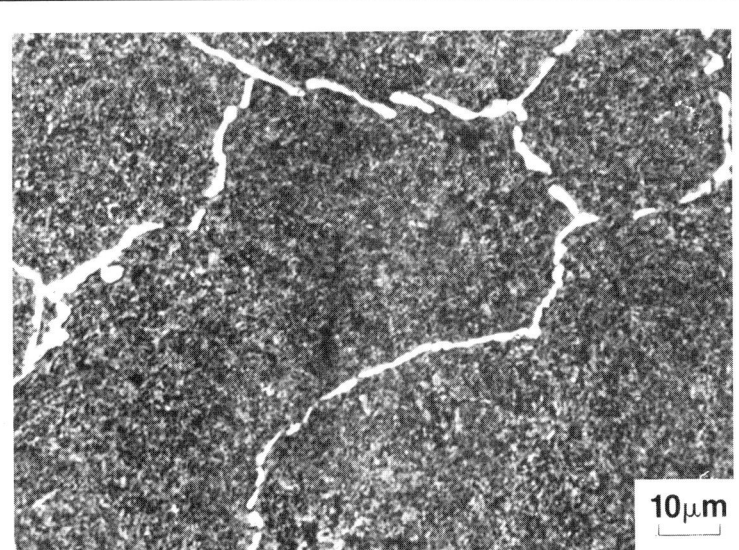

Fig. 10.12. Cementite network in corner of an EX 24 specimen carburized at 1050 °C (1920 °F), oil quenched, and reheated to 845 °C (1550 °F). (Ref 10.16)

scribed above and their influence on the properties of carburized parts have been extensively reviewed by Parrish (Ref 10.28).

Figure 10.14 shows the results of a study of the effect of martensite morphology, including the effects of microcracking, on fatigue resistance of a carburized coarse-grained 8620 steel (Ref 10.24). All specimens were chemically polished and, therefore, the influence of intergranular oxidation on fatigue cracking was removed from the study. The specimens directly quenched from the carburizing temperature had the coarsest structure and the highest density of microcracks, some of which were directly exposed on the specimen surfaces by the chemical polishing. The single reheat specimens had a finer austenite grain structure and therefore finer martensite plates and a lower density of microcracks. Since the retained austenite content and hardness profiles of the direct and single reheat specimens were identical, the improved fatigue resistance of the single reheat specimens is attributed to the smaller size of the microcracks and their lower density in the finer structure. The best fatigue resistance was shown by the double reheat specimens with case microstructure similar to that shown in Fig. 10.9. The virtual elimination of the surface microcracks, the much finer structure of the martensite, and

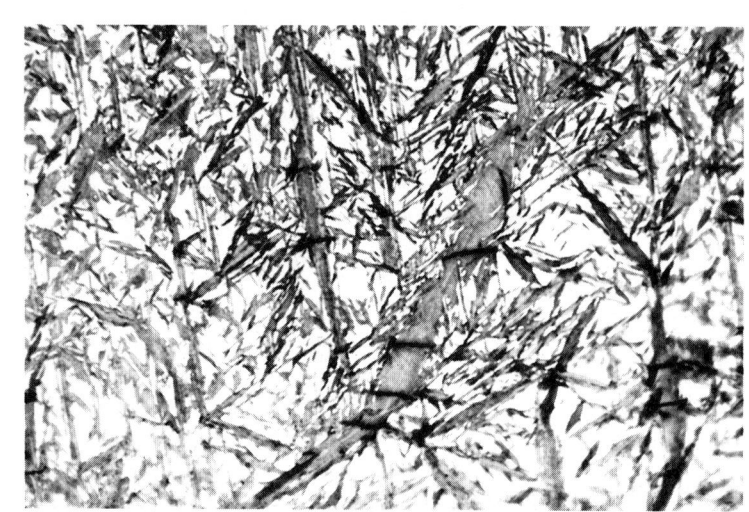

Fig. 10.13. Microcracks in the martensite of a carburized coarse-grained 8620 steel. (Ref 10.24)

the higher hardness and reduced retained austenite were probably all factors that contributed to the very high fatigue resistance of the double reheat specimens.

The study (Ref 10.24) of chemically polished carburized specimens showed that microcracks, or possibly an associated microstructural feature such as an embrittled austenite grain boundary, initiated fatigue fracture in specimens quenched from the austenite field. In specimens reheated to the austenite-cementite field, fatigue cracks were initiated at pits produced by the chemical polishing. A study of fatigue crack origins in a steel carburized to 0.7% carbon showed that cracks in electropolished specimens initiated mainly at prior austenite grain boundaries and sometimes at inclusion particles (Ref 10.29). In commercially carburized specimens not subjected to any surface polishing or removal treatment, fatigue cracks are most probably initiated at intergranular surface oxides or surface roughness in one form or another, such as grooves produced by machining. In fine-grained steels or steels carburized to a carbon level that yields lath martensite, microcracking would not be expected to be a major cause of fatigue crack initiation.

The study of the carburized coarse-grained 8620 steel also revealed several characteristics of the fracture of carburized steels. Figure 10.15 shows fatigue fracture origins in direct quenched and double reheat carburized specimens. The fatigue crack in the direct quenched specimen

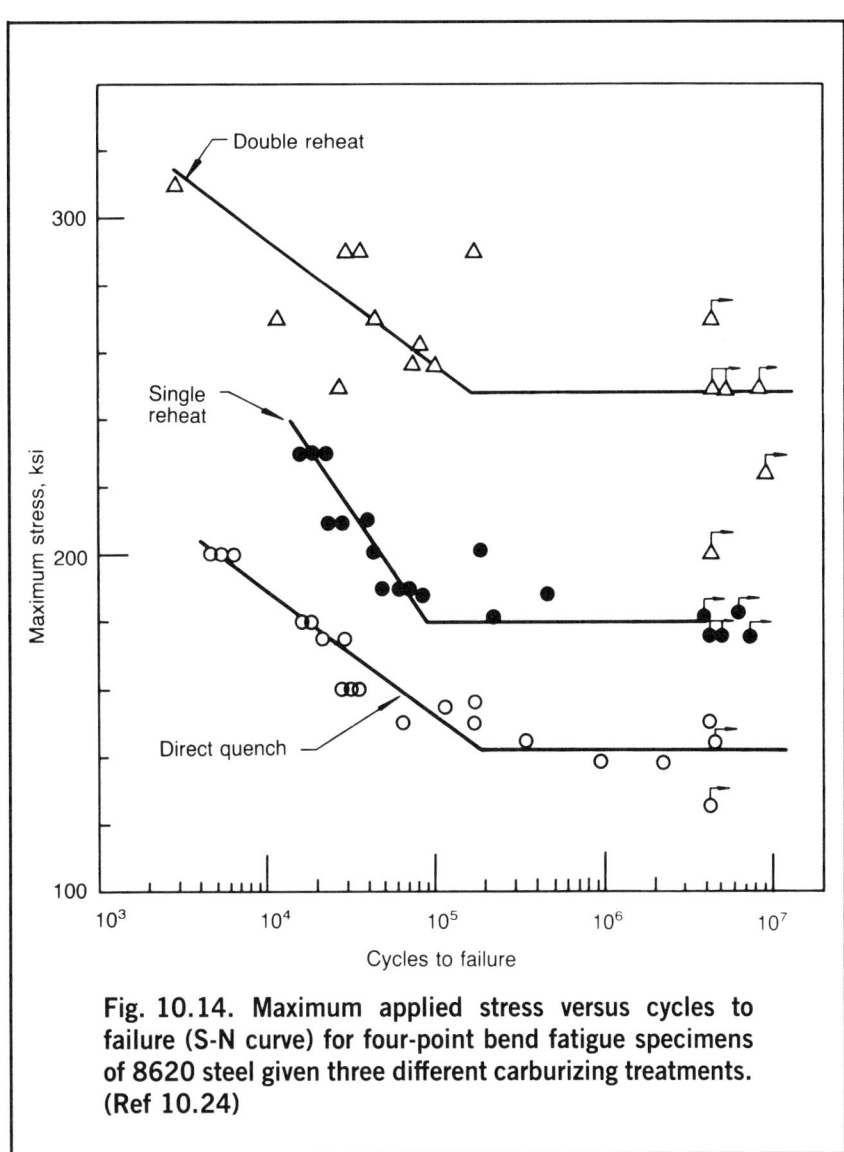

Fig. 10.14. Maximum applied stress versus cycles to failure (S-N curve) for four-point bend fatigue specimens of 8620 steel given three different carburizing treatments. (Ref 10.24)

(Fig. 10.15a) was initiated either at microcracks or an embrittled prior austenite grain boundary. The fatigue crack was quite smooth (area within dashed semicircle) but on overload and rapid propagation became intergranular. In contrast, both the fatigue and overload portions of the fracture surface of the double reheat specimen are transgranular and quite smooth. Figure 10.16 shows higher magnification scanning electron micrographs of overload fracture in direct quench and double reheated specimens. The intergranular fracture (see Fig. 10.16a) is frequently

300 / STEELS: HEAT TREATMENT AND PROCESSING PRINCIPLES

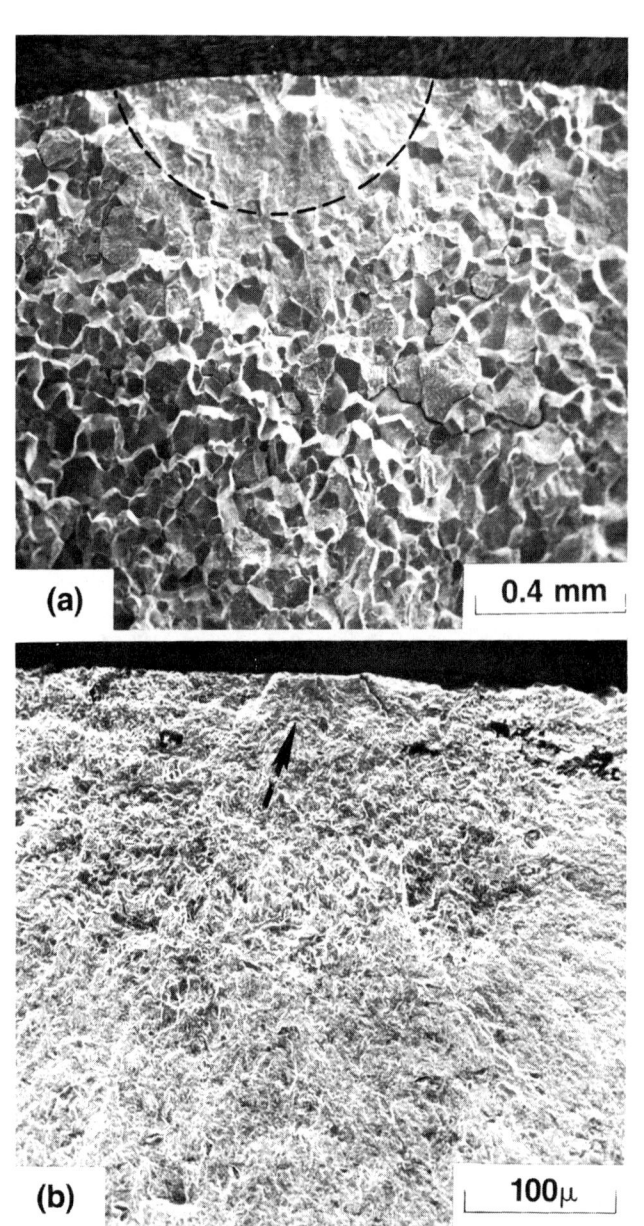

Fig. 10.15. Fatigue crack initiation in carburized coarse-grained 8620 steel (a) quenched directly from carburizing at 927 °C (1700 °F) and (b) reheated after carburizing to 788 °C (1450 °F). Both specimens tempered at 145 °C (300 °F). Scanning electron micrographs. (Ref 10.24)

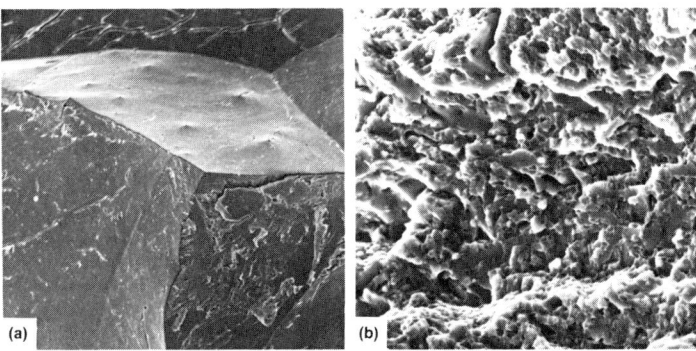

Fig. 10.16. Overload case fracture surfaces in carburized 8620 steel (a) quenched directly after carburizing at 927 °C (1700 °F) and (b) reheated to 788 °C (1450 °F). Both specimens tempered at 145 °C (300 °F). Scanning electron micrographs. (Ref 10.30)

observed in carburized steels and is determined by Auger analysis (Ref 10.30-10.32) to be associated with higher phosphorus and carbon concentrations than those found in the bulk of the steel. The shape of the Auger peak indicates that the carbon at the grain boundary is present in the form of cementite, and thus the intergranular fracture appears to be very similar to that associated with TME of medium-carbon steels. In carburized steels, however, the conditions for intergranular TME-type embrittlement appear to be present in as-quenched and tempered conditions that would normally be considered to be immune to TME in medium-carbon steels (Ref 10.30).

In addition to microstructure, the residual stresses developed during quenching of a carburized steel also favorably affect fatigue resistance. Koistinen (Ref 10.33) first explained that the compressive surface stresses formed in carburized steels were due to M_s gradients associated with the carbon gradients in carburized specimens. At the surface, M_s is a minimum, and as carbon decreases with distance into the specimen, M_s increases. On quenching, therefore, temperature is first lowered below the M_s at some point removed from the surface, and martensite forms at that location. Further on in the quenching cycle, the low M_s surface transforms to martensite, and because its expansion is restrained by the already transformed subsurface layer, it is put into compression. Ebert (Ref 10.34) has extensively discussed the development of residual stresses in carburized steels, and Fig. 10.17, taken from his work, compares the surface compressive stresses of a carbur-

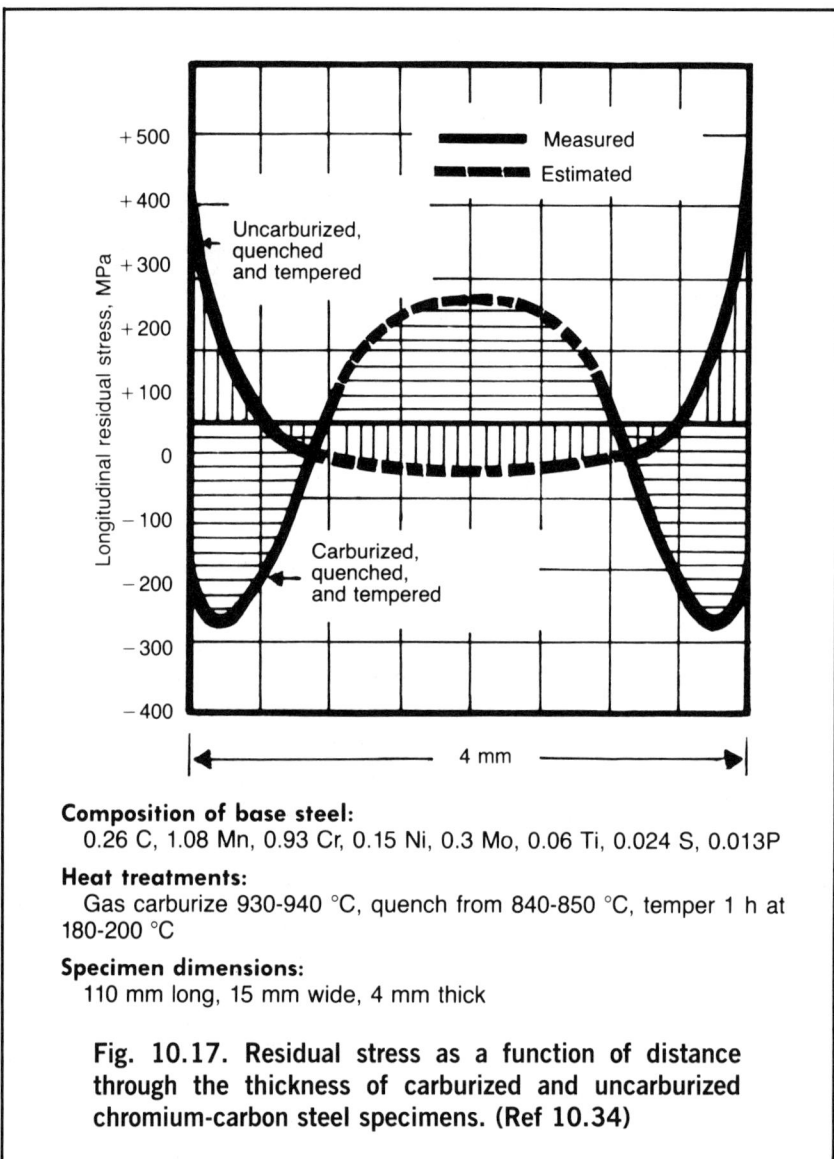

Fig. 10.17. Residual stress as a function of distance through the thickness of carburized and uncarburized chromium-carbon steel specimens. (Ref 10.34)

ized specimen to the surface tensile stresses developed in a through hardened specimen.

Carburizing: Fatigue and Fracture

This section describes additional observations concerning bending fatigue and fracture of carburized steels, and applies to carburized

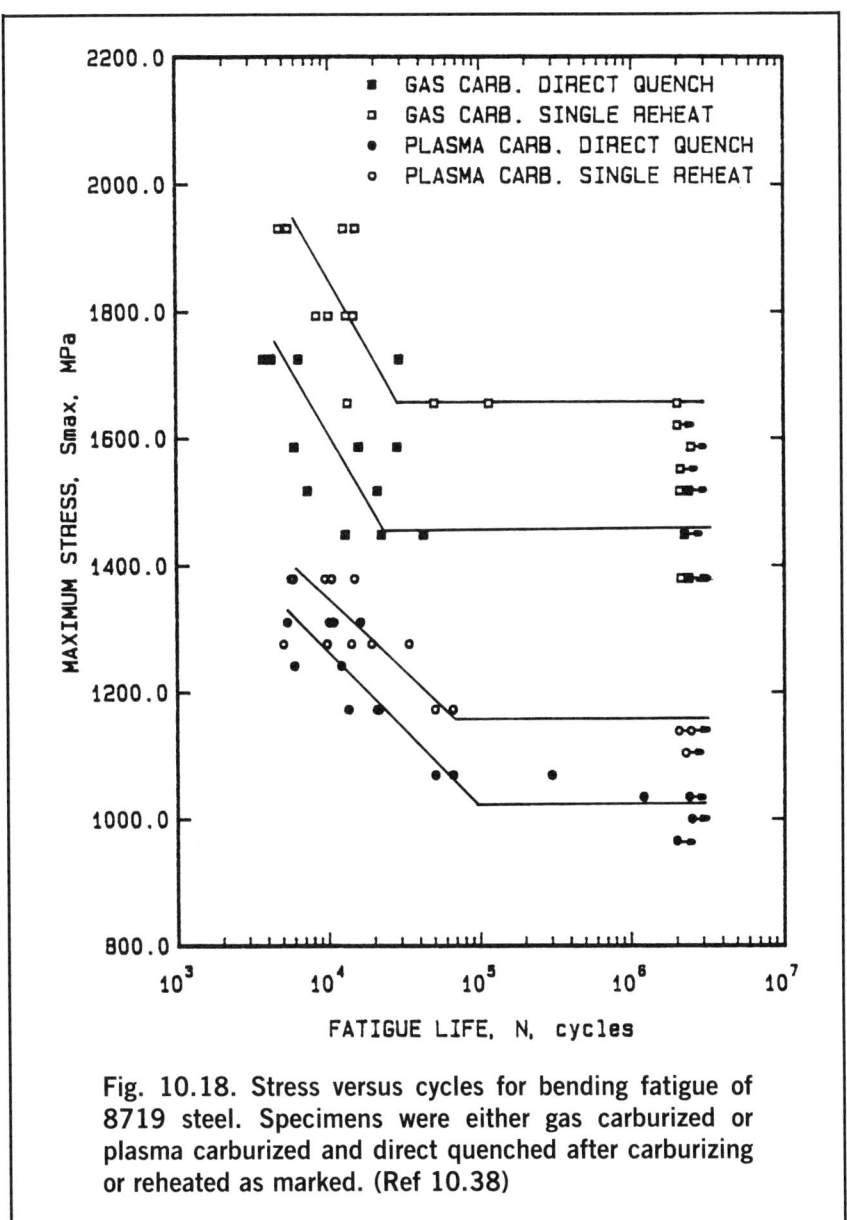

Fig. 10.18. Stress versus cycles for bending fatigue of 8719 steel. Specimens were either gas carburized or plasma carburized and direct quenched after carburizing or reheated as marked. (Ref 10.38)

microstructures of good commercial quality, as shown in Fig. 10.8 and 10.9. The two major types of fatigue fracture described in the previous section appear to be quite reproducible (Ref 10.35-10.39). The one (Fig. 10.15a) is associated with low to moderate fatigue strengths, while the other (Fig. 10.15b) is associated with very high fatigue strengths. The hardened case microstructures of carburized steels are in fact compos-

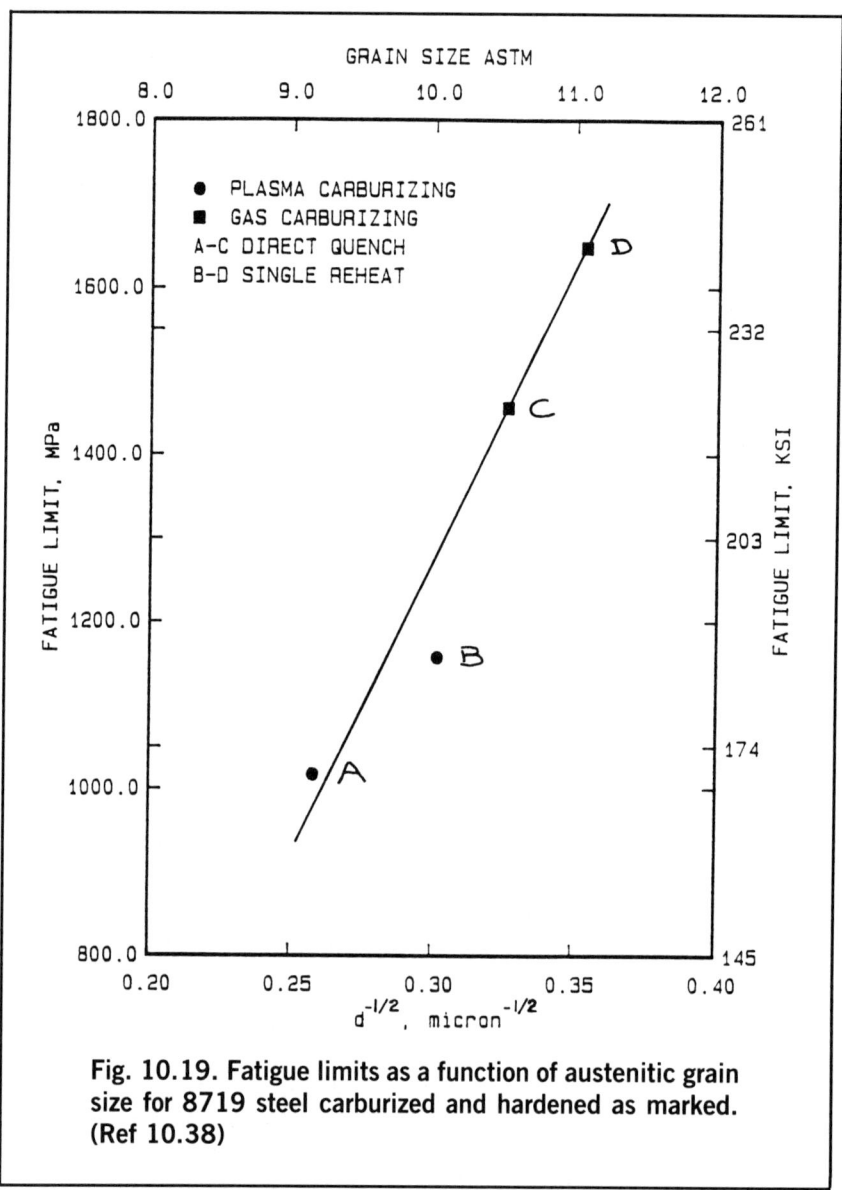

Fig. 10.19. Fatigue limits as a function of austenitic grain size for 8719 steel carburized and hardened as marked. (Ref 10.38)

ite microstructures consisting of retained austenite and tempered martensite, and both the amount of retained austenite and refinement of the martensite significantly influence the nucleation and growth of the two types of fatigue fracture.

The low-stress type of fracture characteristically forms under low-cycle, high-strain fatigue conditions and is related to high retained austenite contents and coarse austenite grain sizes, both microstructural features which favor plastic deformation. The fatigue crack invariably has its initiation at austenite grain boundaries (Ref 10.24, 10.29, 10.38, 10.39). Thus a sharp crack, one or two grain facets in size, initiates the fatigue crack. This grain boundary crack is related to boundary embrittlement caused by phosphorus segregation during austenitizing and cementite formation during cooling (Ref 10.30-10.32), and Zaccone (Ref 10.35) has shown that it forms in the first cycle of fatigue but is arrested by the transformation of retained austenite. The strain-induced formation of martensite induces favorable compressive stresses, and not only arrests the grain boundary crack but also slows the propagation of low-cycle fatigue cracks (Ref 10.39, 10.40). Nevertheless, the grain boundary crack initiates a flat, transgranular fatigue crack (Fig. 10.15a), which grows to a size commensurate with the relatively low fracture toughness of high-carbon hardened steel, 18 to 25 ksi$\sqrt{\text{in.}}$ (20 to 27 MPa$\sqrt{\text{m}}$) (Ref 10.36). When the critical flaw size is reached, overload fracture develops, again typically with a large fraction of intergranular fracture, the fracture mode characteristic of the microstructural state conducive to this mode of fatigue crack nucleation and growth.

The high-stress type of fatigue fracture is developed over many stress cycles, typically at surface defects, inclusions, or oxides (Ref 10.24, 10.29, 10.35, 10.36), in contrast to immediate grain boundary crack initiation in the first few cycles of loading. Thus microstructural conditions which resist plastic deformation and prevent grain boundary cracking—namely, low retained austenite content and fine austenite grain sizes, which translate into fine mixtures of retained austenite and tempered martensite—prevent the nucleation of fatigue cracks (i.e., contribute to high fatigue limits) or defer fatigue crack initiation to very high stress levels. The roles which austenitic grain size and retained austenite play in high-cycle fatigue were clearly demonstrated in experiments performed by Pacheco (Ref 10.37, 10.38). Figure 10.18 shows results of bending fatigue tests performed on gas- and plasma-carburized specimens of SAE 8719 steel. High fatigue limits correlated with fine austenitic grain sizes (Fig. 10.19) and low retained austenite contents (Fig. 10.20). The effects of the two microstructural parameters on fatigue limits are shown together in Fig. 10.21.

Nitriding

Nitriding is a surface hardening heat treatment that introduces nitrogen into the surface of steel while it is in the ferritic condition. Nitriding, therefore, is similar to carburizing in that surface composi-

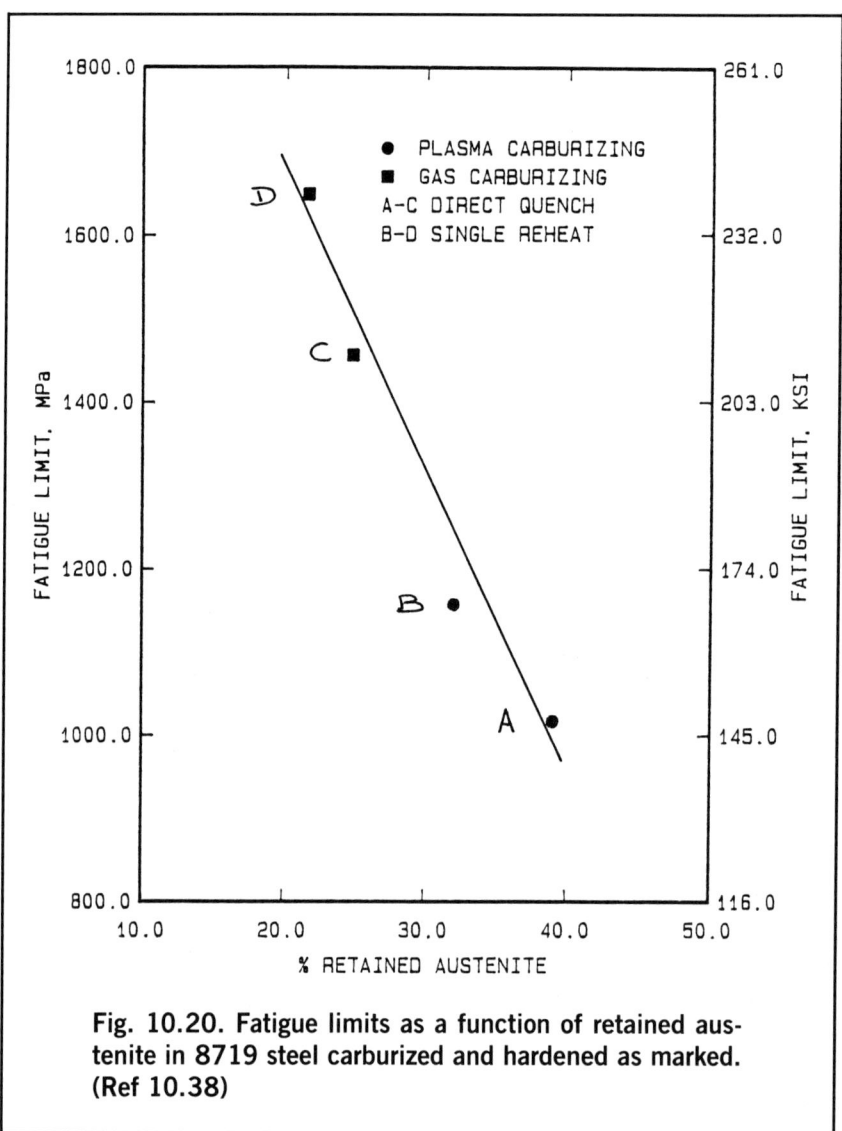

Fig. 10.20. Fatigue limits as a function of retained austenite in 8719 steel carburized and hardened as marked. (Ref 10.38)

tion is altered, but different in that the nitrogen is added into ferrite instead of austenite. The fact that nitriding does not involve heating into the austenite phase field and a subsequent quench to form martensite means that nitriding can be accomplished with a minimum of distortion and excellent dimensional control.

Steels that are nitrided are generally medium-carbon steels that contain strong nitride-forming elements such as chromium, aluminum, vanadium, and molybdenum. Aluminum especially, as already discussed

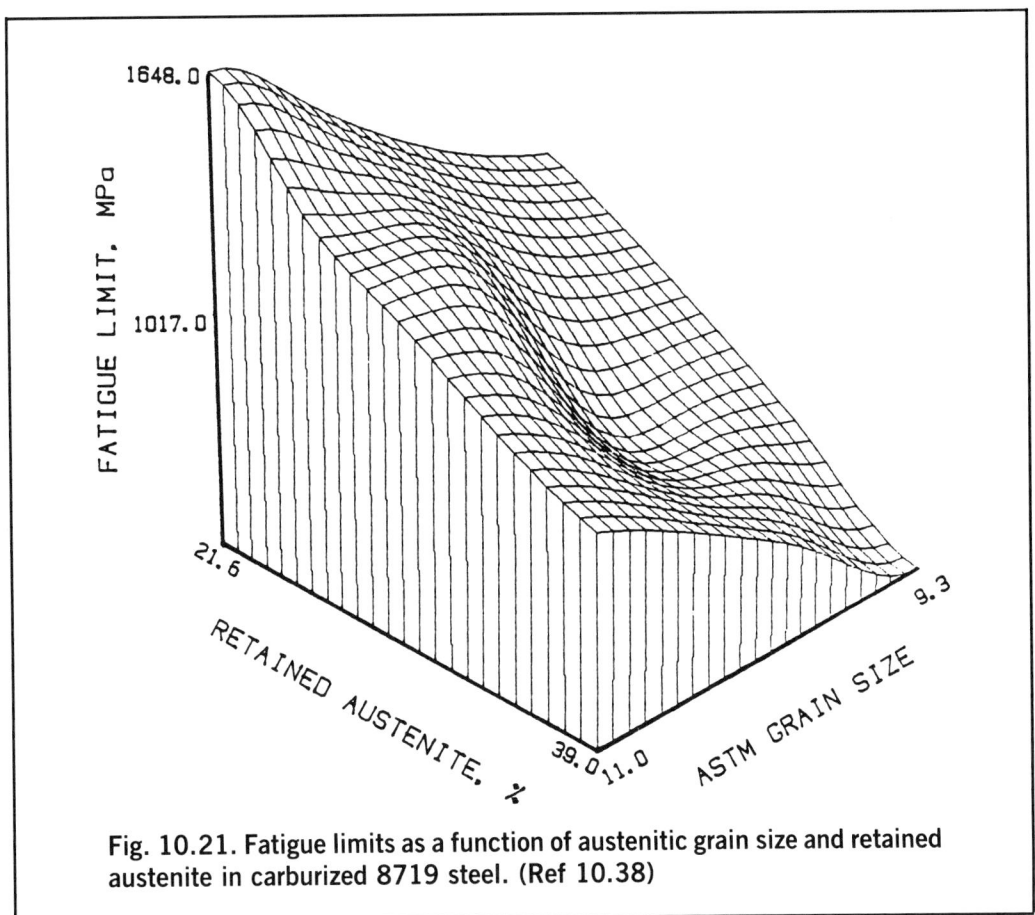

Fig. 10.21. Fatigue limits as a function of austenitic grain size and retained austenite in carburized 8719 steel. (Ref 10.38)

with respect to austenite grain size controls (see Chapter 7), is a very powerful nitride former and is used in amounts between 0.85 and 1.5% in nitriding steels (Ref 10.41). Prior to nitriding, the steels are austenitized, quenched, and tempered. Tempering is performed at temperatures between 540 and 750 °C (1000 and 1300 °F), a range above that at which the nitriding is performed. Tempering above the nitriding temperature provides a core structure that will be stable during nitriding.

Gas nitriding is accomplished with ammonia gas which dissociates on the surface of the steel according to the following reaction:

$$NH_3 \rightleftharpoons N + 3H \qquad \text{(Eq 10.11)}$$

The resulting atomic nitrogen is absorbed at the surface of the steel.

Depending on temperature and the concentration of nitrogen which diffuses into the ferrite of iron or plain carbon steels, a number

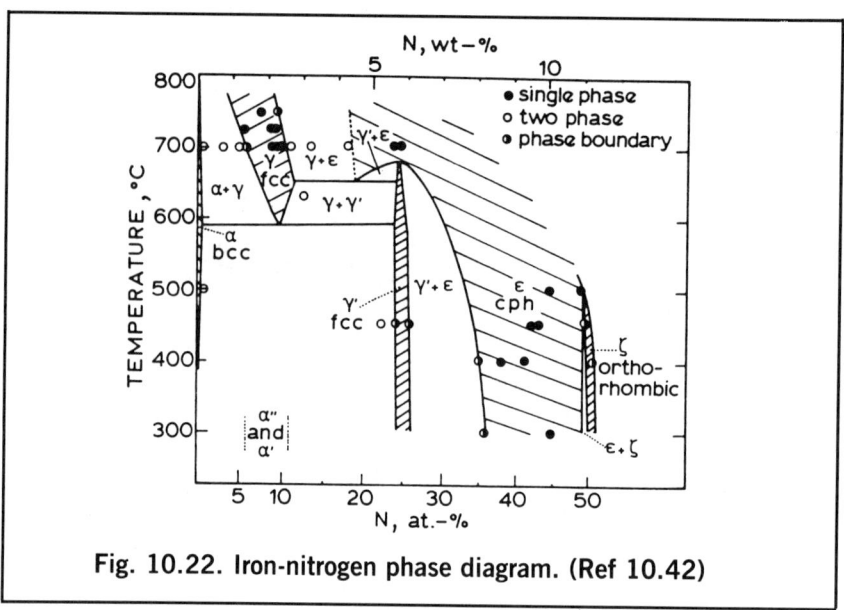

Fig. 10.22. Iron-nitrogen phase diagram. (Ref 10.42)

of phases may form. Low concentrations of nitrogen cause α'', $Fe_{16}N_2$, to precipitate from the ferrite in the form of fine, coherent precipitates. Higher concentrations of nitrogen produce γ', or Fe_4N, the phase which constitutes the brittle white layer of nitrided steels. Even higher concentrations of nitrogen produce ϵ nitride, which when combined with carbon, is considered to be a tribologically desirable phase.

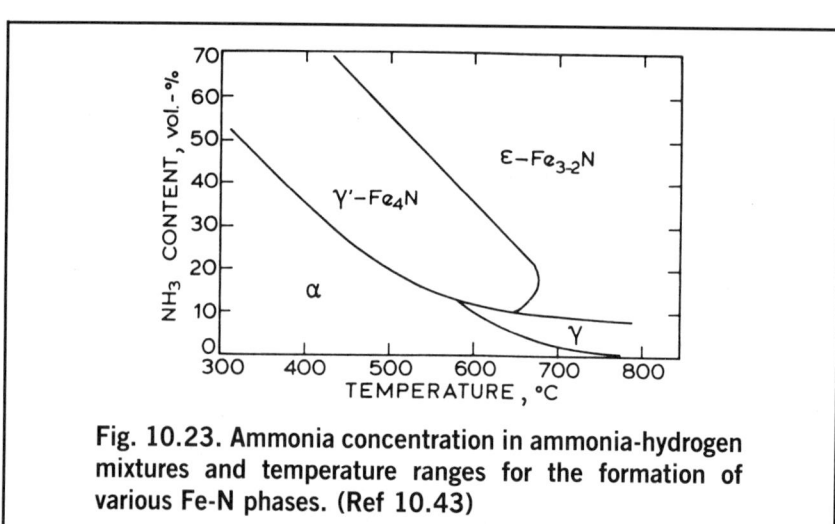

Fig. 10.23. Ammonia concentration in ammonia-hydrogen mixtures and temperature ranges for the formation of various Fe-N phases. (Ref 10.43)

Fig. 10.24. White layer and diffusion zone in nitrided steel. Steel is Nitralloy 135 Modified containing 0.4% carbon, 1.6% chromium, 0.35% molybdenum, and 1.13% aluminum. Base microstructure is tempered martensite of hardness R_c 30. Etched in 1.5% nital. Magnification, 500×. Courtesy of D. Stratford, Sundstrand Corp., Denver

Figures 10.22 and 10.23 show respectively the Fe-N diagram (Ref 10.42) and the NH_3 concentrations in NH_3-H_2 gas mixtures which produce the various nitride phases (Ref 10.43). Nitriding of alloy steels produces diffusion zones of fine precipitates as discussed below.

The nitriding may be accomplished by either a single-stage or double-stage process. The single-stage process is performed at 500 to 525 °C (930 to 975 °F) with 15 to 30% dissociation of the ammonia, i.e., with an atmosphere that contains 70 to 85% NH_3, the source of the nitrogen. This process produces brittle γ' iron nitride. A patented process developed by Floe (Ref 10.44) uses a two-stage process to minimize the thickness of the white layer. The first stage is similar to that described above, but in the second stage the dissociation is increased to 65 to 85%, thereby reducing the NH_3 content of the atmosphere that supplies nitrogen to the surface according to the reaction represented in Eq 10.11. As a result, the iron nitride does not grow as rapidly, and in fact dissolves as it supplies nitrogen into the interior of the steel. Nitriding times are quite long, anywhere from 10 to 130 h depending on the application (Ref 10.41), and the case depths are relatively shallow, usually less than 0.020 in. (0.5 mm).

The microstructure of a nitrided steel is shown in Fig. 10.24. The white layer is clearly visible. Below the white layer, fine alloy nitrides and/or zone structures have formed but are much smaller than can be resolved in the light microscope. Jack and his colleagues have studied many alloys of iron, nitrogen, and other substitutional alloying elements (Ref 10.45). At the temperatures at which commercial nitriding is performed, very fine clusters or precipitate zones of substitutional alloying elements and the interstitial nitrogen form. The substitutional-interstitial zones remain quite fine and do not coarsen readily on heating because of the sluggish diffusion of the substitutional atoms. At lower temperatures only iron nitrides form, and at higher temperatures alloy nitrides form. Nitrided cases are harder than those produced by carburizing and are quite stable in service up to the temperature of the nitriding process. Nitriding, therefore, produces excellent wear, seizing, and galling resistance under conditions where heat is generated by friction between moving parts in contact. Improved fatigue life is also an important benefit of nitriding.

Carbonitriding

Carbonitriding is a surface hardening heat treatment that introduces carbon and nitrogen into the austenite of steel. This treatment is therefore similar to carburizing in that the austenite composition is changed and high surface hardness is produced by quenching to form martensite. Carbonitriding surface hardening, however, is dependent to some extent on nitride as well as martensite formation.

The process of carbonitriding utilizes an atmosphere containing ammonia plus a carbon-rich gas or vaporized liquid hydrocarbon that is a source of carbon as in carburizing. The various gas interchange and gas-metal reactions involved in carbonitriding have been reviewed by Slycke and Ericsson (Ref 10.46). The ammonia dissociates on the surface of the steel and introduces atomic nitrogen. The nitrogen inhibits the diffusion of carbon and this factor, plus the fact that carbonitriding is performed at lower temperatures (705 to 900 °C, or 1300 to 1650 °F) and shorter times than carburizing, results in relatively shallow case depths, from 0.003 to 0.030 in. (0.075 to 0.75 mm). At higher temperatures, the thermal decomposition of ammonia is too rapid, limiting the supply of nitrogen. The lower carbonitriding temperatures are not often used because of the hazard of explosion and

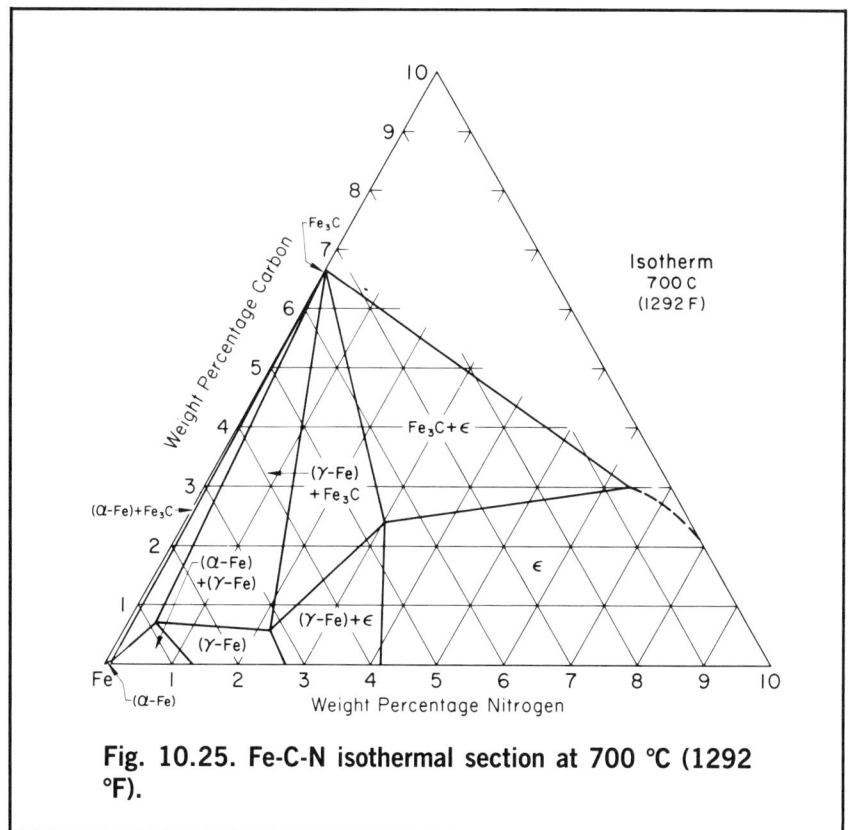

Fig. 10.25. Fe-C-N isothermal section at 700 °C (1292 °F).

the brittle structures formed at the lower temperatures (Ref 10.6). However, a lower temperature variant of carbonitriding, referred to as austenitic nitrocarburizing, now appears to be well developed (Ref 10.47). The latter process is optimally applied in the temperature range 675 to 775 °C (1247 to 1427 °F), and should be controlled to produce a surface compound layer of epsilon carbonitride. Figure 10.25 shows the carbon and nitrogen contents which produce the tribologically desirable (ϵ) epsilon phase at 700 °C (1292 °F).

The nitrogen in carbonitrided steels also enhances hardenability, and makes it possible to form martensite in plain carbon and low-alloy steels that initially have low hardenability. The nitrides formed due to the presence of nitrogen also contribute to the high hardness of the case. Nitrogen, similar to carbon, is an austenite stabilizer. Therefore, considerable austenite may be retained after quenching a carbonitrided part. If the retained austenite content is so high that it reduces hardness and wear resistance, it may be controlled by reducing the ammonia content of the carbonitriding gas either throughout the cycle or during the latter

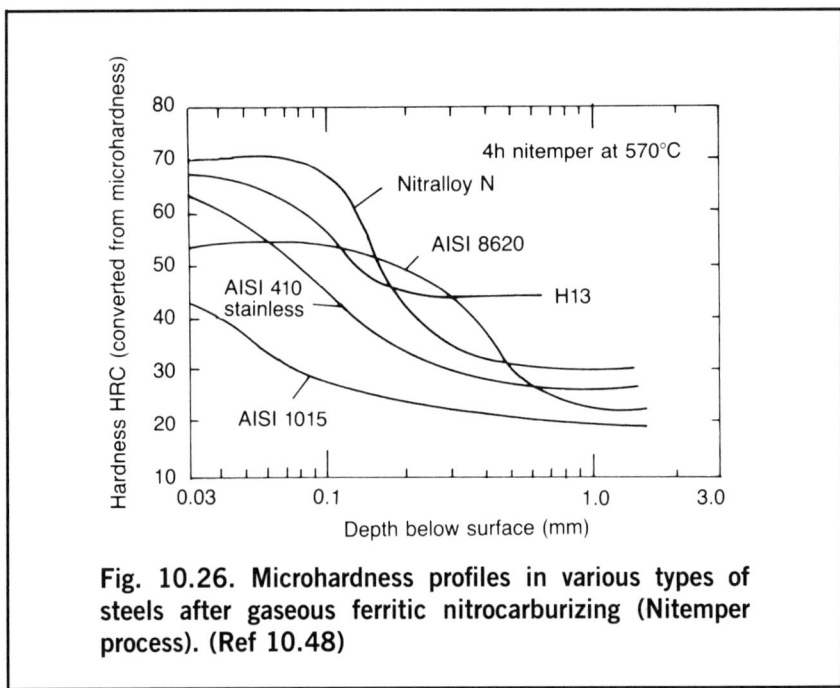

Fig. 10.26. Microhardness profiles in various types of steels after gaseous ferritic nitrocarburizing (Nitemper process). (Ref 10.48)

portion of the cycle (Ref 10.6). Another consequence of excessive nitrogen content in the carbonitrided case is porosity (Ref 10.46).

Ferritic Nitrocarburizing

Another type of surface hardening that involves the introduction of carbon and nitrogen into a steel is ferritic nitrocarburizing. In contrast to carbonitriding but similar to the nitriding process, the carbon and nitrogen are added to ferrite below the Ac_1 temperature. Bell (Ref 10.48) describes a number of ferritic nitrocarburizing processes, both liquid and gaseous. The common beneficial result in all the processes again is a very thin single-phase layer of epsilon carbonitride—a hexagonal ternary compound of iron, nitrogen, and carbon—formed between 450 and 590 °C (840 and 1095 °F) (Ref 10.48).

The epsilon carbonitride compound layer has excellent wear and antiscuffing properties and is produced with minimum distortion. The layer can be formed on inexpensive mild steels with ferrite-pearlite microstructures, thereby greatly improving their wear and fatigue resistance. Figure 10.26 shows the hardness profiles of various steels subjected to a gaseous ferritic nitrocarburizing treatment. The very

SURFACE HARDENING / 313

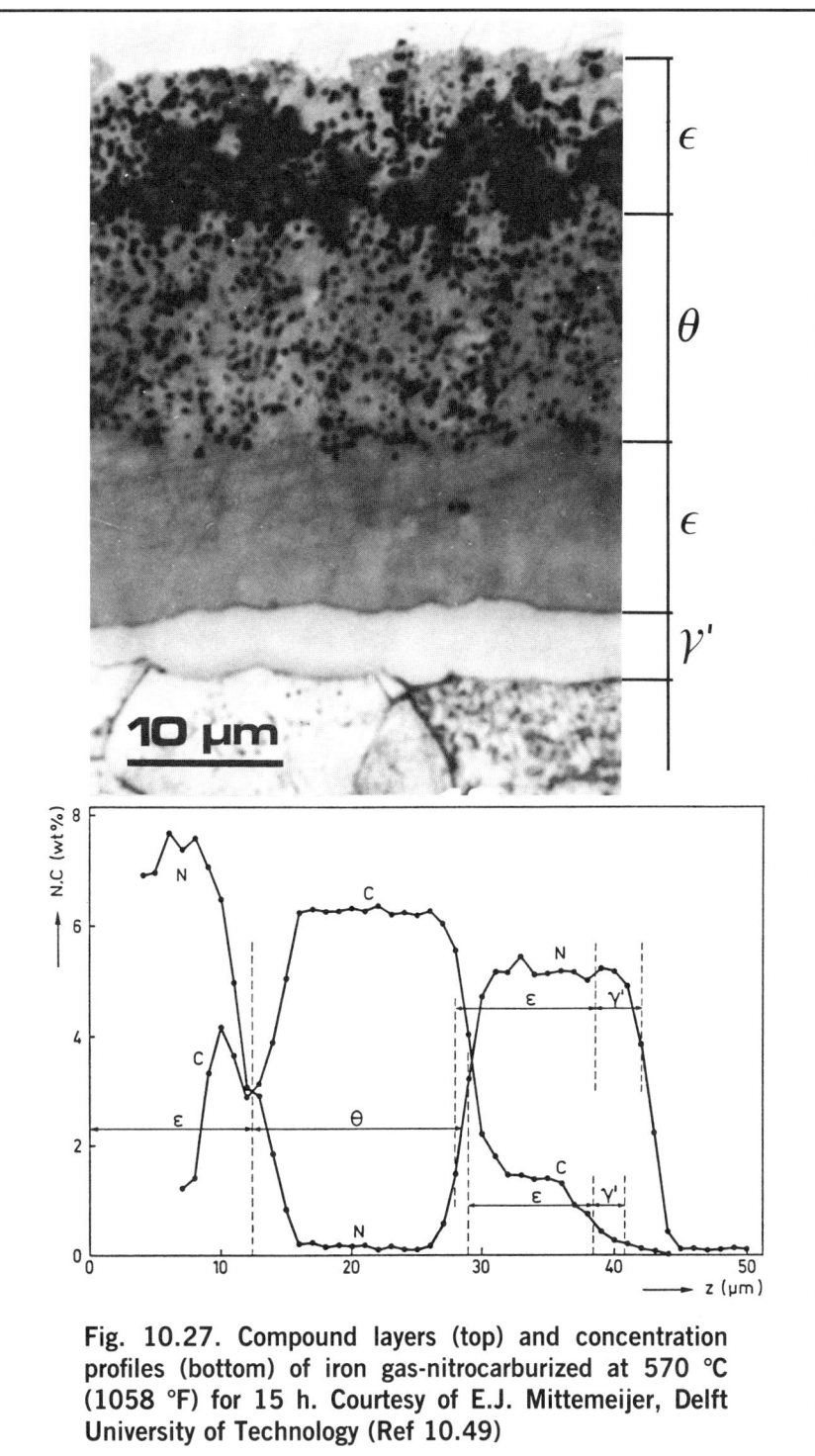

Fig. 10.27. Compound layers (top) and concentration profiles (bottom) of iron gas-nitrocarburized at 570 °C (1058 °F) for 15 h. Courtesy of E.J. Mittemeijer, Delft University of Technology (Ref 10.49)

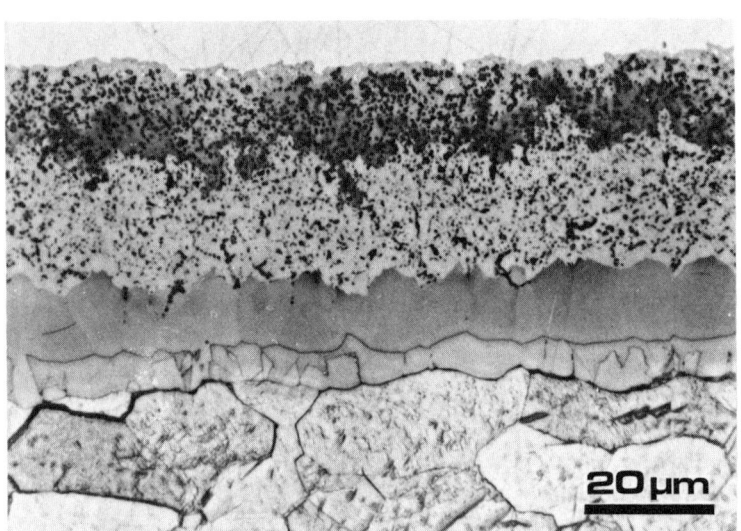

Fig. 10.28. Compound layers in sequence as in Fig. 10.27 in iron gas-nitrocarburized at 570 °C (1058 °F) for 24 h. Courtesy of E.J. Mittemeijer, Delft University of Technology (Ref 10.49)

high hardness and shallow case depths produced by the process in alloy steels are apparent, and even the plain steel benefits significantly. Some of the increased hardness of the case is due to a diffusion zone beneath the compound layer, especially in the more highly alloyed steels with strong nitride formers such as Nitralloy N (1.00 to 1.30% chromium, 0.85 to 1.20% aluminum, and 0.20 to 0.30% molybdenum). In this diffusion zone, nitrides or precipitate zones similar to those that form as a result of nitriding are developed as nitrogen diffuses into the interior of the steel from the compound layer. As a general rule, the compound layer gives good tribological properties, but a substantial diffusion zone is required for good fatigue resistance.

Somers and Mittemeijer (Ref 10.49) have documented the following complex sequence of layers formed by nitrocarburizing iron at 570 °C (1058 °F) in an atmosphere of 53.1 vol.% NH_3, 43.9 vol.% H_2, and 3 vol.% CO. Starting at the surface, the following layers form: ϵ carbonitride ($Fe_2(N,C)_{1-x}$), cementite (θ, or Fe_3C) with a high degree of porosity, ϵ carbonitride with a lower carbon content, and finally carbon-poor γ' carbonitride ($Fe_4(N,C)_{1-x}$) adjacent to the substrate iron. These layers and the nitrogen and carbon concentration profiles produced by nitrocarburizing 15 h are shown in Fig. 10.27, and a similar layered structure produced by nitrocarburizing 24 h is shown in Fig. 10.28. The work by Somers and Mittemeijer shows that compound formation starts with γ' nucleation and subsequent ϵ formation on the γ'. The formation of the cementite is caused by pores produced by the recombination of dissolved nitrogen atoms and preferential carbon uptake along pore channels. Further mechanisms and the kinetics of the compound layer formation are also described (Ref 10.49).

References

10.1 H.W. Grönegress, *Stahl und Eisen*, Vol 70, 1950, p 192
10.2 *Surface Hardening*, P.D. Harvey (Ed.), Metals Engineering Institute, American Society for Metals, Metals Park, OH, 1979
10.3 Induction Hardening and Tempering, in *Metals Handbook*, Vol 2, 8th ed., American Society for Metals, Metals Park, OH, 1964, p 167-188
10.4 S.L. Semiatin and D.E. Stutz, *Induction Heat Treatment of Steel*, American Society for Metals, Metals Park, OH, 1986
10.5 S. Zinn and S.L. Semiatin, *Elements of Induction Heating, Design, Control and Applications*, ASM INTERNATIONAL, Metals Park, OH, 1988
10.6 *Carburizing and Carbonitriding*, ASM Committee on Gas Carburizing, American Society for Metals, Metals Park, OH, 1977
10.7 F.J. Harvey, Thermodynamic Aspects of Gas-Metal Heat Treating Reactions, *Met Trans A*, Vol 9A, 1978, p 1507-1513
10.8 Furnace Atmospheres, in *Metals Handbook*, Vol 2, 8th ed., American Society for Metals, Metals Park, OH, 1964, p 67-84
10.9 F.E. Harris, Reactions Between Hot Steel and Furnace Atmospheres, *Metal Progress*, Vol 47 (No. 1), Jan 1945, p 84-89
10.10 Application of Equilibrium Data, in *Carburizing and Carbonitriding*, American Society for Metals, Metals Park, OH, 1977, p 14-15
10.11 J.I. Goldstein and A.E. Moren, Diffusion Modeling of the Carburization Process, *Met Trans A*, Vol 9A, 1978, p 1515-1525
10.12 C. Wells and R.F. Mehl, Rate of Diffusion of Carbon in Austenite in Plain Carbon, in Nickel, and in Manganese Steels, *Trans AIME*, Vol 140, 1940, p 279-306
10.13 D.E. Diesburg and G.T. Eldis, Fracture Resistance of Various Carburized Steels, *Met Trans A*, Vol 9A, 1978, p 1561-1570
10.14 W.T. Groves, How to Select the Right EX Steel, *Metal Progress*, Vol 102, 1972, p 89-99
10.15 U. Wyss, Kohlenstoff- und Härteverlauf in der Einsatzhärtungsschicht verschieden legierter Einsatzstähle, *Härterei-Technische Mitteilungen*, Vol 43, 1988, p 27-35

10.16 K.D. Jones and G. Krauss, Microstructure and Fatigue of Partial Pressure Carburized SAE 8620 and EX 24 Steels, *J Heat Treating*, Vol 1, 1979, p 64-71

10.17 S. Gunnarson, Structure Anomalies in the Surface Zone of Gas-carburized, Case-Hardened Steel, *Metal Treatment and Drop Forging*, June 1963, p 219-229 [also published in *Jernkontoretz Annaler*, Vol 145 (No. 5), 1962]

10.18 T. Naito, H. Veda, and M. Kikuchi, Fatigue Behavior of Carburized Steel with Internal Oxides and Nonmartensitic Microstructure Near the Surface, *Met Trans A*, Vol 15A, 1984, p 1431-1436

10.19 B. Hildenwall and T. Ericsson, Residual Stresses in the Soft Pearlite Layer of Carburized Steel, *J Heat Treat*, Vol 1 (No. 3), 1980, p 3-13

10.20 H.C. Child, Vacuum Carburizing, *Heat Treat Met*, Vol 3, 1976, p 60-65

10.21 R. Chatterjee-Fischer, Internal Oxidation During Carburizing and Heat Treating, *Met Trans A*, Vol 9A, 1978, p 1553-1560

10.22 K.D. Jones and G. Krauss, Effects of High-carbon Specimen Corners on Microstructure and Fatigue of Partial Pressure Carburized Steels, *Proceedings of the Metals Society, Heat Treatment '79*, Birmingham, England, 22-24 May 1979

10.23 A.R. Marder and A.O. Benscoter, Microcracking in Fe-C Acicular Martensite, *Trans ASM*, Vol 61, 1968, p 293-299

10.24 C.A. Apple and G. Krauss, Microcracking and Fatigue in a Carburized Steel, *Met Trans*, Vol 4, 1973, p 1195-1200

10.25 R.P. Brobst and G. Krauss, The Effect of Austenite Grain Size on Microcracking in Martensite of an Fe-1.22 C Alloy, *Met Trans*, Vol 5, 1975, p 457-462

10.26 M.G. Mendiratta, J. Sasser, and G. Krauss, Effect of Dissolved Carbon on Microcracking in Martensite of an Fe-1.39 C Alloy, *Met Trans*, Vol 3, 1972, p 351-353

10.27 T.A. Balliett and G. Krauss, The Effect of the First and Second Stages of Tempering on Microcracking in Martensite of an Fe-1.22 C Alloy, *Met Trans A*, Vol 7A, 1976, p 81-86

10.28 G. Parrish, The Influence of Microstructure on the Properties of Case-Carburized Components, series of articles in *Heat Treatment of Metals*, Vol 3, 4, 1976-1977; published in book form by American Society for Metals, Metals Park, OH, 1980

10.29 L. Magnusson and T. Ericsson, Initiation and Propagation of Fatigue Cracks in Carburized Steel, *Proceedings of the Metals Society, Heat Treatment '79*, Birmingham, England, 22-24 May 1979

10.30 G. Krauss, The Microstructure and Fatigue of a Carburized Steel, *Met Trans A*, Vol 9A, 1978, p 1527-1535

10.31 H.K. Obermeyer, The Effects of Heat Treatment and Phosphorus Content on Fracture of a 0.85% Carbon Alloy Steel, M.S. thesis, Colorado School of Mines, Golden, 1978

10.32 T. Ando and G. Krauss, The Effect of Phosphorus Content on Grain Boundary Cementite Formation in AISI 52100 Steel, *Met Trans A*, Vol 12A, 1981, p 1283-1290

10.33 D.P. Koistinen, The Distribution of Residual Stresses in Carburized Cases and Their Origin, *Trans ASM*, Vol 50, 1958, p 227-241

10.34 L.J. Ebert, The Role of Residual Stresses in the Mechanical Performance of Case Carburized Steel, *Met Trans A*, Vol 9A, 1978, p 1537-1551
10.35 M.A. Zaccone, Flow Properties of High Carbon Tempered Martensite, M.S. thesis, Colorado School of Mines, Golden, 1987
10.36 B. Kelley, The Effect of Chromium on the Microstructure and Bending Fatigue Behavior of 0.82 pct C, 1.75 pct Ni, and 0.75 pct Mo Steels, M.S. thesis, Colorado School of Mines, Golden, 1984
10.37 J.L. Pacheco, Fatigue Resistance of Plasma and Gas Carburized SAE 8719 Steel, M.S. thesis, Colorado School of Mines, Golden, 1988
10.38 J.L. Pacheco and G. Krauss, Microstructure and High Bending Fatigue Strength in Carburized Steel, in *Carburizing: Processing and Performance*, G. Krauss (Ed.), ASM INTERNATIONAL, Metals Park, OH, 1989, p 227-238
10.39 M.A. Zaccone, B. Kelley, and G. Krauss, Strain Hardening and Fatigue in Simulated Case Microstructures, in *Carburizing: Processing and Performance*, G. Krauss (Ed.), ASM INTERNATIONAL, Metals Park, OH, 1989, p 239-248
10.40 M.M. Shea, Impact Properties of Selected Gear Steels, Reprint No. 780772, Society of Automotive Engineers, Warrendale, PA, 1978
10.41 Gas Nitriding, in *Metals Handbook*, Vol 2, 8th ed., American Society for Metals, Metals Park, OH, 1964, p 149-163
10.42 K.H. Jack, The Occurrence and the Crystal Structure of α''-iron Nitride; A New Type of Interstitial Alloy Formed During the Tempering of Nitrogen-Martensite, *Proc Royal Soc,* Vol A208, 1951, p 216-224
10.43 E. Lehrer, Uber das Eisen-Wasserstoff-Ammoniak-Gleichgewicht, *Zeitsch Elektrochem*, Vol 36, 1930, p 383-392
10.44 C.F. Floe, A Study of the Nitriding Process: Effect of Ammonia on Case Depth and Structure, *Trans ASM*, Vol 32, 1943, p 134
10.45 K.H. Jack, Nitriding, *Proceedings of the Metals Society, Heat Treatment '73*, London, 1973, p 39-50
10.46 J. Slycke and T. Ericsson, A Study of Reactions Occurring During the Carbonitriding Process, *J Heat Treat,* Vol 2 (No. 1), 1981, p 3-19
10.47 T. Bell, M. Kinali, and G. Munstermann, Physical Metallurgy Aspects of Austenitic Nitrocarburizing Process, Paper presented at 5th International Congress on Heat Treatment of Materials, Budapest, 1986
10.48 T. Bell, Ferritic Nitrocarburizing, *Heat Treatment of Metals*, Vol 2, 1975, p 39-49
10.49 M.A.J. Somers and E.J. Mittemeijer, Formation and Growth of Compound Layer on Nitrocarburizing Iron: Kinetics and Microstructural Evaluation, *Surface Eng*, Vol 3 (No. 2), 1987, p 123-137

CHAPTER 11

Surface Modification

The newer surface modification processes are a logical extension of heat treatment technology to steels. The objectives remain the same as in the established surface heat treatments: enhanced surface wear, fatigue, corrosion, and oxidation resistance. The use of high-energy beams, plasmas, and vapor deposition techniques in vacuum environments, however, offers the potential of much more controlled, higher quality surface modifications than heretofore possible. Also, the possibility of creating new engineered surface structures and surface-core systems are almost limitless because of the ability to deposit almost any material by vapor deposition and the ultrahigh heating and cooling rates of thin surface layers heated by electron and laser beams. This chapter describes some of the newer processes used to apply thermochemical modifications, coatings, and high-energy density surface modifications to steels. Included are descriptions of plasma nitriding, plasma carburizing, ion implantation and mixing, physical vapor deposition techniques, chemical vapor deposition, and transformation hardening and melting by laser and electron beams.

Introduction

Surface and near-surface microstructures in load-bearing machine components, tools, and other structural components manufactured from steels and cast irons are directly subjected to much higher static and cyclic stresses, friction, wear, and corrosive environments than interior microstructures. Thus there are compelling economic and engineering reasons to develop and apply surface modifications which prevent surface-related failures and which extend the range of operating conditions for bulk engineering materials. Long-established but still evolving surface modification techniques such as induction hardening, gas carburizing, and gas nitriding have already been discussed in Chapter 10. However, in the 1980s a large number of new surface modification processes have been developed. The various techniques in this new generation of processes often use high-energy beams (electron, laser, and ion), electric or magnetic fields, and vacuum environments, and thus constitute a significant increase in technology applied to surface modification. Many of the newer techniques were first developed and are still extensively used for thin-film electronic applications (Ref 11.1). However, the new techniques are increasingly being applied to steels and irons, as well as to aluminum, titanium, and high-temperature alloys.

The structural changes and property improvements due to the new surface technologies are process dependent and quite diverse. Types of surface modifications include surface melting and cladding, plating or the formation of hard layers on a substrate, and modification of work piece subsurface chemistry. The proper selection and application of the various surface processes have led to the development of the interdisciplinary activity of surface engineering, defined as follows (Ref 11.2):

"Surface engineering involves the application of traditional and innovative surface technologies to engineering components and materials in order to produce a composite material with properties unattainable in either the base or surface material. Frequently, the various technologies are applied to existing designs of engineering components but ideally, surface engineering involves the design of the component with a knowledge of the surface treatment to be employed."

At least two journals are currently exclusively devoted to surface engineering (Ref 11.3, 11.4), and a large number of books and conference proceedings describe the many new surface modification techniques (Ref 11.1-11.12).

The following sections of this chapter briefly describe processing principles, surface characteristics, and examples of applications of some of the newer surface modification techniques.

Plasma Nitriding

The objectives of plasma nitriding, also referred to as ion nitriding, are the same as that of gas nitriding described in Chapter 10; i.e., the process should produce a hardened surface zone, typically on the order of 0.1 mm in depth, in a variety of steels. Nitrogen is adsorbed on

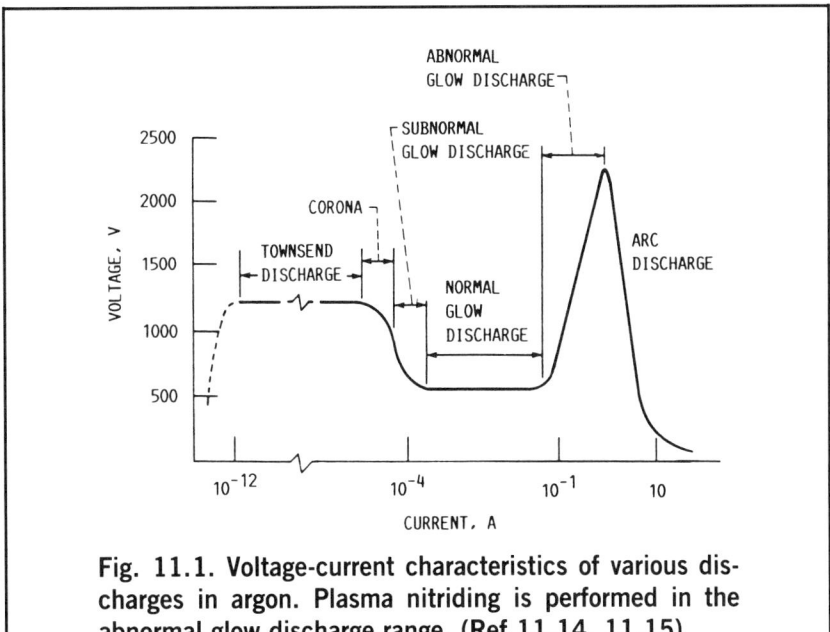

Fig. 11.1. Voltage-current characteristics of various discharges in argon. Plasma nitriding is performed in the abnormal glow discharge range. (Ref 11.14, 11.15)

the surface of the steel, diffuses inward at temperatures around 500 °C (930 °F), and hardening is accomplished by precipitation of very fine nitride particles in the diffusion zone. Depending on nitrogen concentration, surface layers of face-centered cubic γ' Fe_4N or close-packed hexagonal ϵ Fe_2N_{1-x} may form, as shown in the iron-nitrogen phase diagram (Fig. 10.22) in Chapter 10 (Ref 11.13). In view of the diffusion-controlled increase in nitrogen content of a steel surface produced by plasma nitriding, the process is categorized as thermochemical, just as is gas nitriding.

Plasma nitriding, however, uses much different equipment than does gas nitriding. Processing is accomplished in a vacuum-tight, cold-wall chamber with the work load made the cathode (negative) and the chamber walls the anode and grounded. First hydrogen and then a mixture of nitrogen and hydrogen are added to the chamber. An applied direct current potential across the cathode work piece and chamber wall creates a plasma, defined as a gaseous state of matter with good electrical conductivity and consisting of ions, electrons, and charged and neutral atoms and molecules. The initial hydrogen stage creates a glow discharge which heats and cleans the surface of the work pieces, and the addition of the nitrogen initiates and sustains the nitriding action.

Figure 11.1 shows voltage-current relationships for nitrogen-hydrogen mixtures (Ref 11.14, 11.15). Plasma nitriding is performed with current densities in the abnormal glow discharge range. Under these conditions current increases with voltage, and a uniform purple glow is established around the cathodic work pieces. The visible glow is caused by collisions of electrons with gas molecules in the electric field immediately adjacent to the cathode, i.e., in the cathode fall region (Ref 11.16). In this region, ions and neutral atoms acquire high kinetic energy and are accelerated to the cathode while electrons are accelerated to the anode. Generally the heating of the work piece is generated by the plasma-driven impact of the nitrogen ions, and no additional heat source is required. However, in newer system designs, convective heating is used to reduce cycle times (Ref 11.17).

A major advantage of plasma nitriding is the enhanced mass transfer of high-energy nitrogen molecules and ions to the surface of the steel under the action of the electric field (Ref 11.16, 11.17). The kinetics of nitrogen penetration into the steel of course remain controlled by solid-state diffusion and nitride precipitation. Safety, reduced gas consumption, reduced energy consumption in cold-wall chambers, clean environmental operation, and good control of γ' white layer structures are other advantages of plasma nitriding (Ref 11.16, 11.17). Temperature variations due to radiation losses in parts stacked closest to cold chamber walls, and the "hollow cathode" effect, a concentration of plasma which causes local overheating at holes and cavities, may be problems in commercial plasma nitriding operations (Ref 11.6).

Plasma nitriding is widely applied and is the oldest plasma surface technology used commercially. Improvements in friction, scuffing resistance, and fatigue resistance are produced on a wide variety of materials, especially alloy and stainless steels, in a wide variety of machine components which require high surface hardness and good dimensional control (Ref 11.3, 11.5, 11.6). Plasma or ion carbonitriding, in which methane as well as hydrogen and nitrogen are added to the plasma, is also used where ternary Fe-C-N compound surface layers with good scuffing resistance are desired (Ref 11.19).

Plasma Carburizing

Plasma carburizing, similar to plasma nitriding, is another thermochemical glow discharge surface treatment. Carbon is brought to the surface of low-carbon steel in the austenitic state and diffuses to the interior of the work piece to produce a high-carbon case as described in Chapter 10. The work pieces in plasma carburizing are made cathodes in a dc electric circuit. In the presence of a carburizing gas, the resulting glow discharge plasma increases the mass transfer of carbon to the steel surface. Thus some acceleration of the carburizing process is possible (Ref 11.16, 11.17, 11.20). Case depth is still largely controlled by solid-state diffusion of carbon in the steel, a time-temperature-dependent process which proceeds independently of the plasma. Unlike plasma nitriding, where the impact of ions in the glow discharge is sufficient to heat work pieces to nitriding temperatures (around 500 °C or 930 °F) without an external source of heat, plasma carburizing must be performed in internally heated chambers because of the higher temperatures (around 930 °C or 1700 °F) required to produce austenite with its high solubility for carbon. Relatively little heating of the specimen is generated by energy input of the plasma at normal current densities of about 1.3 mA/in.^{2} (0.2 mA/cm^2) (Ref 11.17).

Plasma carburizing is accomplished by establishing a vacuum, heating the load to carburizing temperatures while sputter cleaning in a hydrogen plasma, and carburizing in a hydrocarbon (propane or methane)-hydrogen-argon plasma (Ref 11.17) at relatively low gas partial pressures (0.1 to 10 torr). The carbon transfer is complex, but it appears that ionized and neutral hydrocarbon molecules strike the steel surface where they are first weakly attached (physisorbed) by Van der Waal's forces and then chemically bonded by the loss of a hydrogen atom from the gas molecule (chemisorbed) (Ref 11.21). Hydrogen atoms are then released from the molecules, leaving carbon atoms which diffuse into the steel. Following the carburizing, which may include several alternate carburize-diffuse cycles, the specimens are oil quenched in a tank incorporated into the furnace (Ref 11.17).

In addition to reduced carburizing times, plasma carburizing produces very uniform case depths even in parts with irregular surfaces (Ref 11.17, 11.20). This uniformity is caused by the glow discharge plasma which closely envelops the specimen surface, provided recesses or holes are not too small (Ref 11.20). Also, because plasma carburizing is performed in a vacuum, there is no surface oxidation. Figure 11.2 compares plasma-carburized and gas-carburized 8719 steel (1.0% Mn, 0.5% Cr, 0.5% Ni, and 0.17% Mo) specimens (Ref 11.22). Both specimens are as-polished and have been nickel plated to prevent edge rounding during mechanical polishing. Below the nickel plate, no oxidation is visible in the plasma-carburized specimen, while a well-developed surface oxide layer has formed in the gas-carburized specimen. Surface oxidation may be detrimental to fatigue, especially if surface compressive residual stresses are reduced, but other factors

324 / STEELS: HEAT TREATMENT AND PROCESSING PRINCIPLES

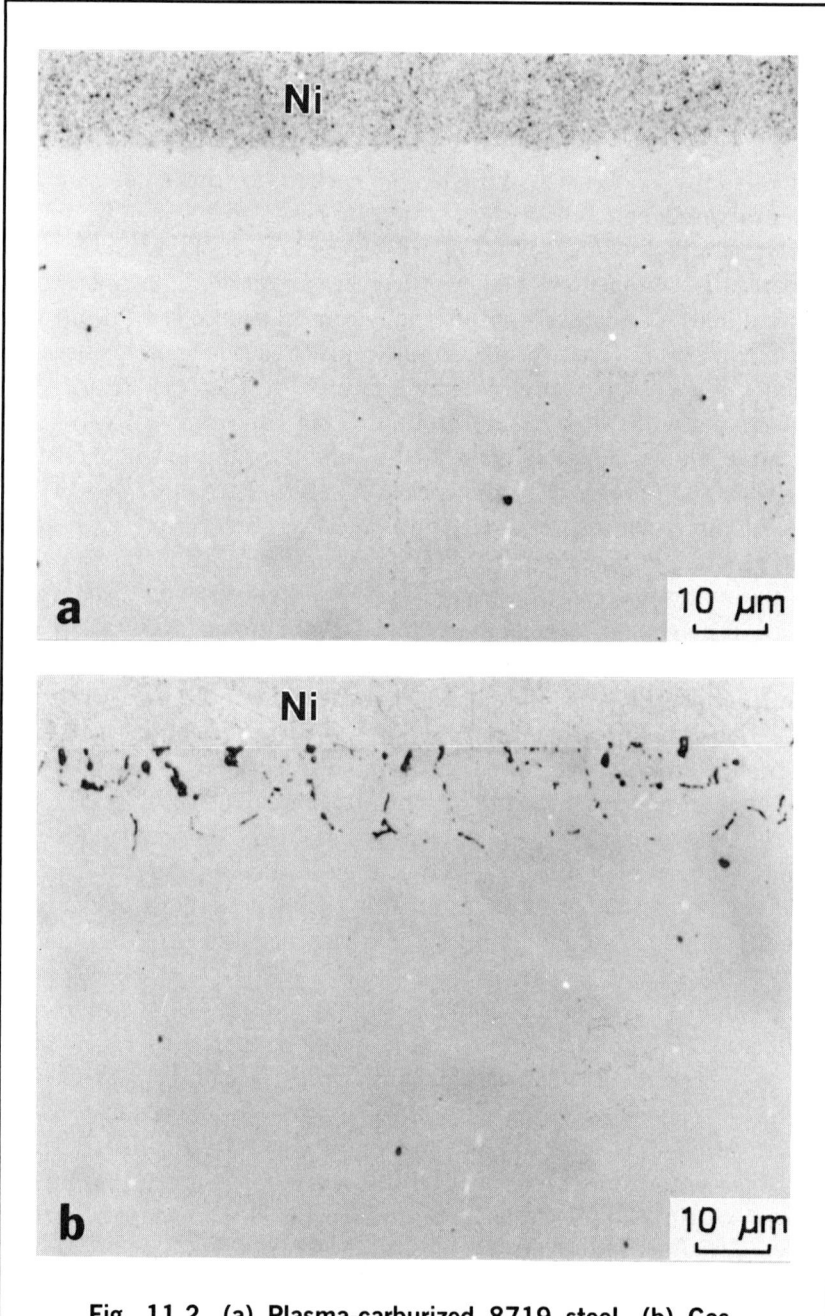

Fig. 11.2. (a) Plasma-carburized 8719 steel. (b) Gas-carburized 8719 steel. Both specimens Ni-plated and unetched. Light micrographs. (Ref 11.22)

such as a very fine austenite grain size can offset the detrimental effect of oxides, provided only martensite forms on quenching as discussed in Chapter 10.

Ion Implantation and Ion Mixing

Ion implantation is a surface modification process in which ions with very high energy are driven into a substrate (Ref 11.14, 11.23, 11.24). Ions of almost any atom species can be implanted, but nitrogen is widely used to improve corrosion resistance and tribological properties of steels and other alloys. Although the nitrogen content of alloy surfaces is increased by both nitrogen ion implantation and plasma nitriding, major differences exist between the two processes and the surface modification which they create.

Ion implantation machines accelerate ions, generated by specially designed sources (Ref 11.24), at very high energies, from 10 to 500 keV. In contrast, the energy of ions and atoms in plasma nitriding is much lower, less than 1 keV. Ion implantation is carried out with the substrate close to room temperature, thus minimizing diffusion-controlled formation of precipitates and coarsening of the subsurface microstructure. The low temperature of application and the fact that the process is carried out in accelerators with very good vacuums (10^{-5} torr or better) ensure clean surfaces and reduce undesirable surface chemical reactions such as oxidation. Ion implantation is a line-of-sight process, i.e., only relatively small areas directly exposed to the ion beam are implanted. For coverage of areas larger than the beam, either the specimen must be translated or the ion beam must be rastered over the specimen surface.

Figure 11.3 shows ion-implanted nitrogen distributions in iron as a function of ion beam energy (Ref 11.14). The nitrogen concentrations are quite high and have a slightly skewed gaussian distribution. However, the depth of ion penetration is relatively shallow, generally less than 0.25 μm, compared to gas- or plasma-nitrided case depths, which are 100 μm and deeper. This difference in case depth is due to diffusion-controlled case formation in the nitriding processes and the virtual absence of this mechanism in ion implantation.

Compensating for the shallow case depths of ion implantation are very high strengths or hardness of the nitrogen-implanted surface layers. Ion implantation is a complex, nonequilibrium process which creates significant lattice damage in the form of vacancies and interstitial point defects. Concentrations of implanted species much higher than equilibrium solubility limits may be introduced. In fact, the

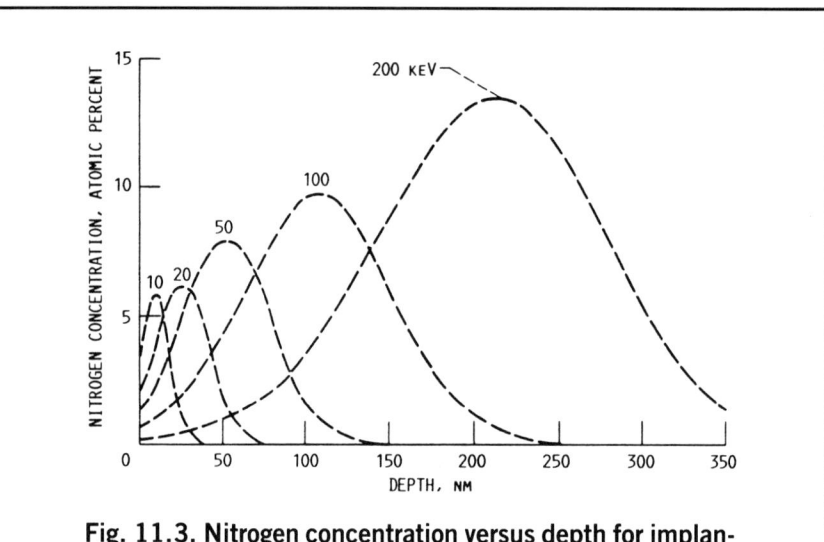

Fig. 11.3. Nitrogen concentration versus depth for implantation of iron performed at various beam energies. (Ref 11.14)

incorporation of high densities of atoms of significantly different size compared to the substrate lattice may produce amorphous structures or metastable phases (Ref 11.24). Figure 11.4 shows schematically some characteristics of ion implantation by 100-keV nitrogen ions in iron (Ref 11.25). Each ion creates a large number of point defects, and the lower right of Fig. 11.4 illustrates the formation of a cascade of vacancy-interstitial pairs or Frenkel defects by a nitrogen ion. The implanted ions, the lattice defects, and the resulting compressive stresses all act to produce very high strength and hardness of the implanted layer.

The properties of ion-implanted surfaces and shallow case depths make ion implantation suitable for very special applications. Since the surface of the part itself is modified, there are no adhesion problems as are sometimes encountered with coated layers of high hardness. Also, since ion implantation is usually accomplished with very little heating, dimensional stability is excellent. Examples of applications of ion implantation include surface hardening of razor blades (Ref 11.17) and knives (Ref 11.24), a variety of tool steel applications (Ref 11.26), and implantation of 52100 and 440C bearings with titanium and/or nitrogen to improve rolling contact fatigue resistance (Ref 11.27-11.29). In the latter applications titanium was found to reduce the coefficient of friction and nitrogen was found to raise hardness by intermetallic compound formation.

SURFACE MODIFICATION / 327

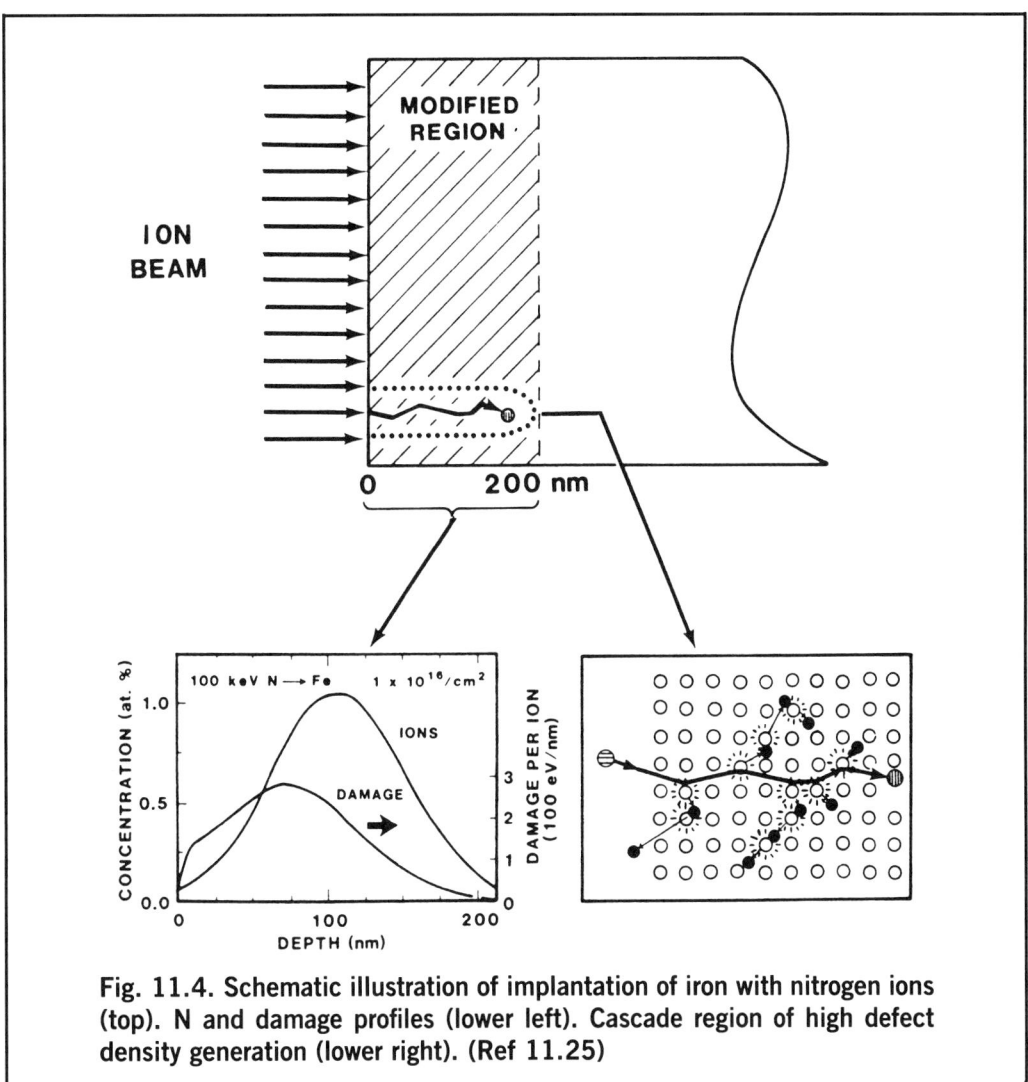

Fig. 11.4. Schematic illustration of implantation of iron with nitrogen ions (top). N and damage profiles (lower left). Cascade region of high defect density generation (lower right). (Ref 11.25)

Ion-beam mixing is an extension of the use of ion beams to produce coating-ion implanted systems with enhanced properties (Ref 11.14, 11.30). In this process, a surface layer or coating, deposited on a substrate by another technique, is subjected to a high-energy ion beam, as shown schematically in Fig. 11.5 (Ref 11.14). Ions are deposited and transmitted through the coating and into the substrate, producing the same changes produced by direct ion implantation. Moreover, the atoms of the coating are mixed with the substrate, leading to improved adhesion of the coating on the

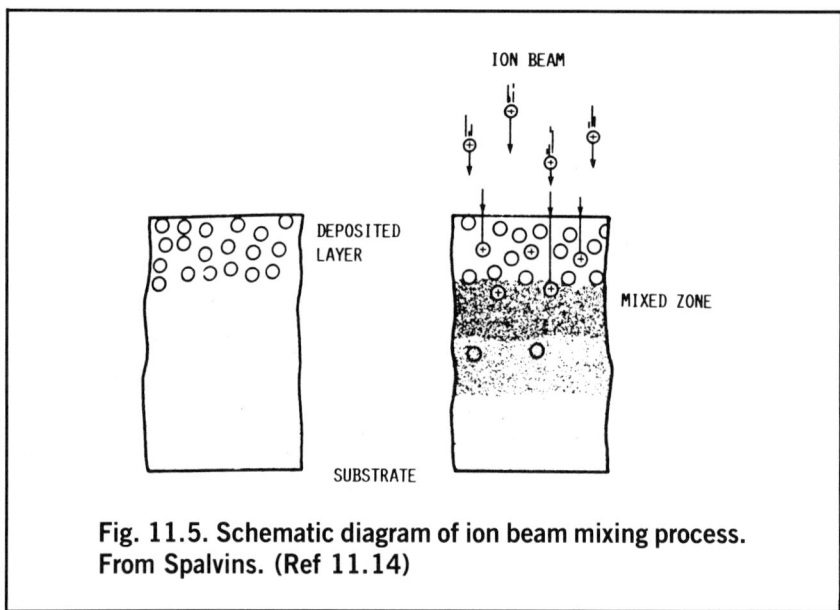

Fig. 11.5. Schematic diagram of ion beam mixing process. From Spalvins. (Ref 11.14)

substrate. Ion-beam mixing for improved wear and corrosion performance has been applied to a variety of metal substrates (Ref 11.30). Applications for steels include ion mixing of gold layers on 15.5 PH stainless steel for improved resistance to fretting corrosion and ion mixing of layers on high-carbon steels for improved oxidation resistance.

Physical Vapor Deposition: Processing

Physical vapor deposition (PVD) surface modification techniques produce layers or coatings on substrates, in contrast to the previously discussed thermochemical techniques which modify substrate surface composition and structure without layer formation. Physical vapor deposition techniques are used to enhance properties of a great variety of materials, and the technology has again been driven by thin film electronic applications (Ref 11.1). However, these techniques are being increasingly applied to steels, in particular to extend the life of tools and dies which are used to machine and form steels and other materials.

The term physical vapor deposition refers to any process which physically generates and deposits atoms or molecules on a substrate in a high-vacuum environment (Ref 11.31). The atom flux which impinges on a substrate may be generated by evaporation, sputtering, or

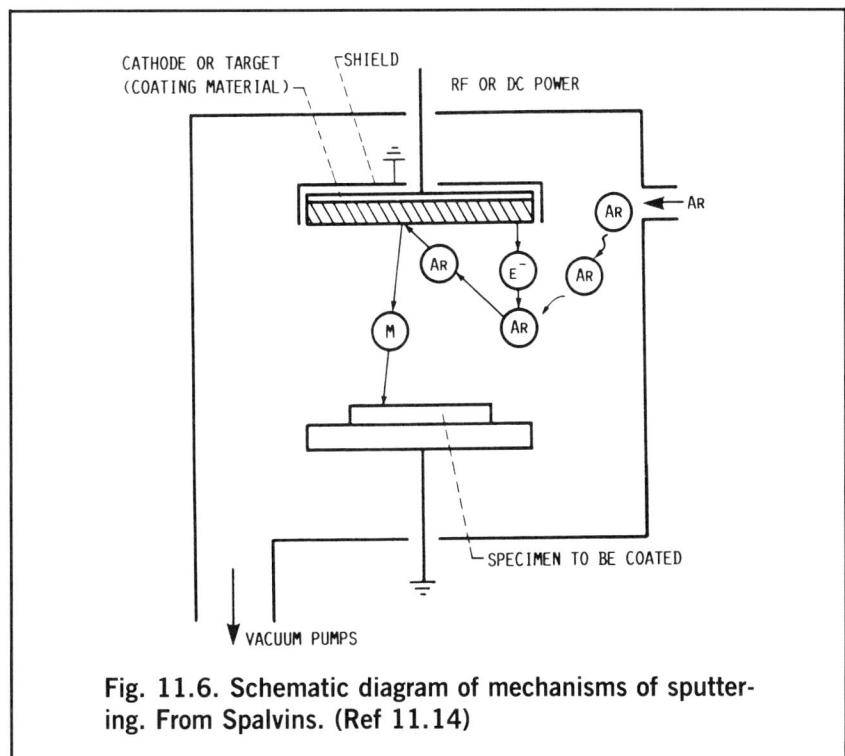

Fig. 11.6. Schematic diagram of mechanisms of sputtering. From Spalvins. (Ref 11.14)

ion plating. These mechanisms and their modifications lead to a multiplicity of PVD processes (Ref 11.14, 11.16, 11.31). Evaporation is accomplished by heating source materials in high vacuums (7.5×10^{-6} torr or better). At sufficiently high temperatures, atoms or molecules are thermally evaporated from the source, travel through the vacuum, and deposit on a substrate (Ref 11.31). For many applications, deposition processes which are based solely on thermal evaporation are being replaced by sputtering and ion plating, more efficient processes which use glow discharge plasmas.

Sputtering is a coating process in which atoms are ejected mechanically from a target by the impact of ions or energetic neutral atoms. Figure 11.6 shows schematically the mechanism of sputtering in a simple diode system (Ref 11.14). The chamber is initially evacuated, back filled with argon gas, and the target is made cathodic or negative by the application of a dc potential (between -500 and -5000 V). A low-pressure glow discharge plasma is created around the target cathode, creating positively charged argon ions which are accelerated to the target. The momentum transfer due to the impact of the argon ions is sufficient to eject target atoms which travel to the substrate and

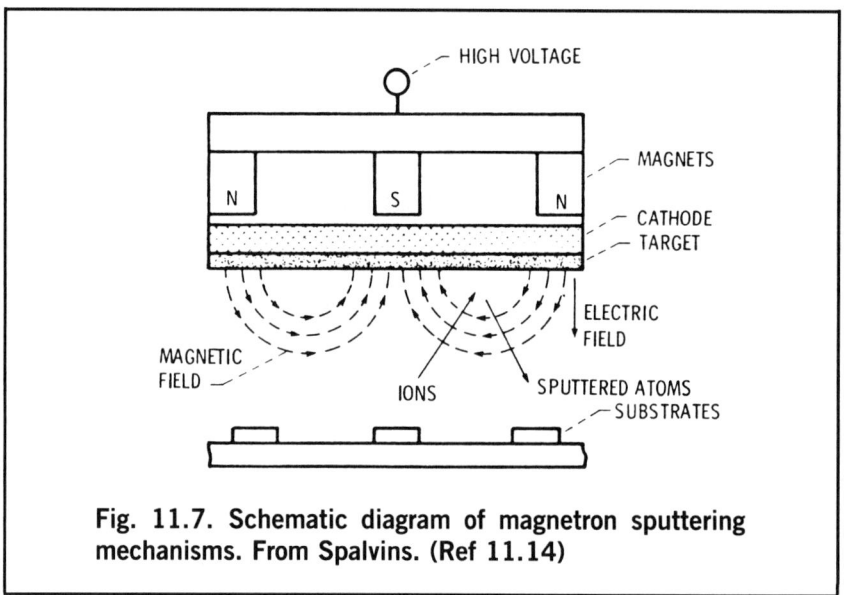

Fig. 11.7. Schematic diagram of magnetron sputtering mechanisms. From Spalvins. (Ref 11.14)

other parts of the chamber. The mechanical transfer of atoms by sputtering is more readily controlled than thermal transfer by evaporation, and sputtered atoms have higher energies than thermally evaporated atoms (Ref 11.31). Even insulators or semiconductors may be sputtered, but in order to eliminate charge buildup on a nonconducting target, radio frequency (RF) voltages are applied to the system rather than dc voltages.

Simple diode sputtering systems have relatively low rates of deposition. Thus improved sputtering systems, with magnetic fields applied at the targets, have been developed. The resulting sputtering process is referred to as magnetron sputtering, and is schematically illustrated in Fig. 11.7 (Ref 11.14). The magnetic fields, applied by permanent magnets, trap secondary electrons generated by the target and greatly increase ionization in the cathode plasma. Thus more argon ions strike the target and sputtering and deposition rates are significantly increased relative to diode sputtering.

Ion plating, also referred to as plasma-assisted PVD or evaporative-source PVD coating, generates coating atoms by thermal evaporation from an appropriate source (Ref 11.14, 11.16, 11.31). The source may be an electrically heated wire, an electron beam, or of hollow cathode design, and is made the anode in the system. The substrate is made the cathode by the application of a dc or RF voltage ranging from −500 to −5000 V. In the resulting substrate cathode glow discharge, atoms and ions are accelerated at high energies to the substrate coating. Dense coatings and excellent adhesion because of the bombardment of highly energetic particles, and good coverage because of the cathode glow discharge, are important characteristics of coatings produced by ion plating systems.

The diode ion plating systems have been further improved by designs which enhance ionization with ion currents that can be controlled independently of the bias voltage between the evaporative source and the substrate (Ref 11.14). These modified designs are referred to as triode ion plating systems.

In most PVD sputtering and ion plating systems, gases such as nitrogen, methane, or oxygen may be introduced to react with metal atoms generated by sputtering or evaporative source. Such reactive PVD deposition produces metal nitride, carbide, or oxide ceramic coatings.

Physical Vapor Deposition: Microstructures

Coating microstructures produced by PVD processes tend to be quite fine because of extremely rapid effective quench rates. Moreover, coating morphology and adhesion are quite variable depending on such factors as substrate temperature, pressure of the sputtering gas, and the energy of the incident atoms (Ref 11.31-11.33). The coatings are created by nucleation and growth processes which first involve the adsorption of incident atoms, referred to as adatoms, on a substrate surface. The adatoms then diffuse on the substrate surface to preferred bonding sites such as ledges or vacancies or to growing clusters or nuclei. Three different coating nucleation and growth processes, as reviewed by Rigsbee (Ref 11.31), have been identified: (1) three-dimensional island or Volmer-Weber growth, (2) two-dimensional layer-by-layer or Frank-van der Merwe growth, and (3) initial layer-by-layer growth followed by island growth. This mixed-mode growth is referred to as Stranski-Krastanov growth. The first mode consists of the formation of clusters, the growth of clusters which reach critical size, and the eventual impingement of the islands to produce a continuous film. Layer growth is typical of systems where adatoms have high surface mobility and bind more strongly to substrate atoms than to each other. The mixed-mode growth may be due to initial epitaxial layer growth terminated by the build-up of residual stresses which eventually produce defect sites for island nucleation and growth.

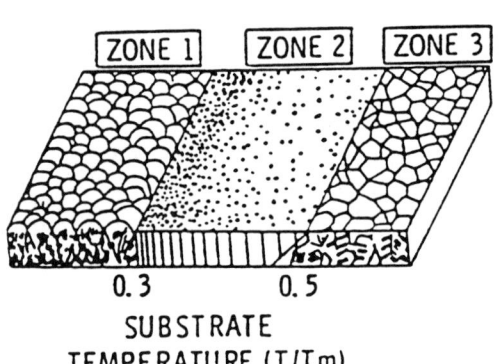

Fig. 11.8. Schematic diagram of structural zone model for coating growth as a function of deposition temperature as proposed by Movchan and Demchishin. (Ref 11.32, 11.35)

Two useful diagrams which classify the effects of processing conditions on PVD coating microstructure and morphology have been developed. The first diagram, Fig. 11.8, shows schematically the effect of substrate temperature, T, relative to the melting temperature, Tm,

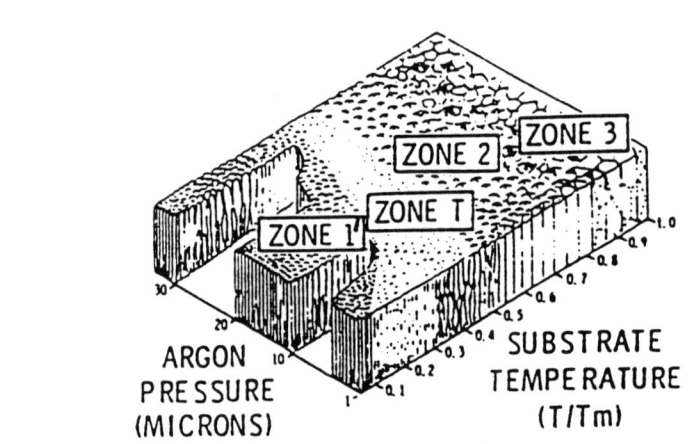

Fig. 11.9. Schematic diagram of structural zone model for coating growth as a function of deposition temperature and argon pressure as proposed by Thornton. (Ref 11.33, 11.35)

of the coating material. This diagram was developed by Movchan and Demchishin (Ref 11.32) for evaporated metal and oxide coatings. The second diagram, Fig. 11.9, incorporates the effect of sputtering system gas pressure and was developed for sputtered coatings by Thornton (Ref 11.33, 11.34). The schematic diagrams show that various zones or ranges of operating conditions produce quite different coating morphologies. Recognizing that thermal evaporation and sputtering may produce quite different fine structures within similar microstructures, zone 1 in the Movchan and Demchishin diagram is sometimes referred to as zone 1' in the Thornton diagram (Ref 11.35).

The coating morphologies which develop in zone 1, at low substrate temperatures, are porous and consist of conical arrays of crystallites which taper from narrow clusters of nucleating crystallites to broader, dome-shaped arrays with increasing film thickness. The rounded tips of these tapered arrays produce relatively rough coating surfaces. The unique tapered morphology is a result of low adatom surface diffusivity at low substrate temperatures (Ref 11.31, 11.35). As a result, relatively few nuclei develop. Those nuclei which do grow effectively shield or shadow intervening areas from incident atoms, and consequently a high degree of porosity is incorporated between the growing tapered crystallite arrays. Moreover, there is accumulating evidence that the tapered zone microstructures contain a very high density of atomic scale microvoids, and that sputtering produces a more imperfect fine structure than does thermal evaporation (Ref 11.35). Higher argon pressures shift the boundaries of zone 1' to higher temperatures (Fig. 11.9), because vapor scattering of incident sputtered atoms effectively reduces adatom energy and therefore adatom surface mobility (Ref 11.31, 11.34).

Zone T microstructures, formed at higher substrate temperatures, are denser and the coating surfaces are smoother than zone 1 or zone 1' microstructures, because of increased adatom surface diffusion. These transition microstructures are still very fine and characterized by columnar growth. Increased substrate surface diffusion and bulk diffusion in the coatings is reflected in zone 2, where coarser, columnar crystals nucleate and grow, and in zone 3, where coarse columnar and even equiaxed grains may grow. Increasing crystal perfection within the growing crystallites and grains accompanies increasing substrate temperatures.

Very high residual stresses develop in zone 1 type PVD coatings, and may be either tensile or compressive. Thermal evaporation, a process where the atoms possess thermal energy but relatively low kinetic energy, produces exclusively tensile stresses, while sputtering may produce either tensile or compressive stresses, depending on

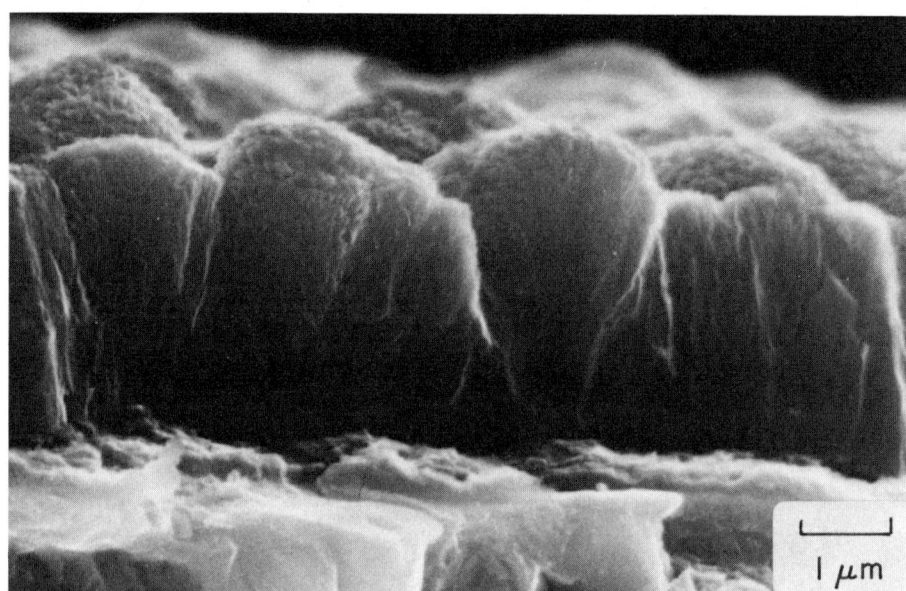

Fig. 11.10. $(Ti_{33}Al_{17})N$ coating deposited by triode ion plating at low substrate current density. Scanning electron micrograph. Courtesy of A.S. Korhonen, Helsinki University of Technology. (Ref 11.39, 11.40)

sputtering conditions (Ref 11.34-11.36). It now appears that the stress state of most sputtered coatings undergoes a transition from tension to compression with decreasing sputtering gas pressure or with increasing sputtering current density at constant pressure. The tensile and compressive stresses are both quite high, on the order of several GPa, and may lead to coating cracking or buckling, respectively (Ref 11.35). The processing and mechanistic reasons for this transition are complex (Ref 11.34-11.36), but the compressive stresses are associated with dense, less columnar coating structures formed by the deposition of highly energetic atoms. Thus, low sputtering gas pressures, which minimize incident atom energy loss by scattering and collisions within the gas phase, favor the formation of dense films and compressive stresses by maintaining high rates of adatom surface diffusion. Likewise, minimizing, by bias sputtering, the content of impurity atoms which limit adatom diffusivity is also beneficial. Thus processing conditions which produce dense coatings at low substrate temperatures appear to yield the most favorable residual stresses. Deposition at higher substrate temperatures, while producing denser and more

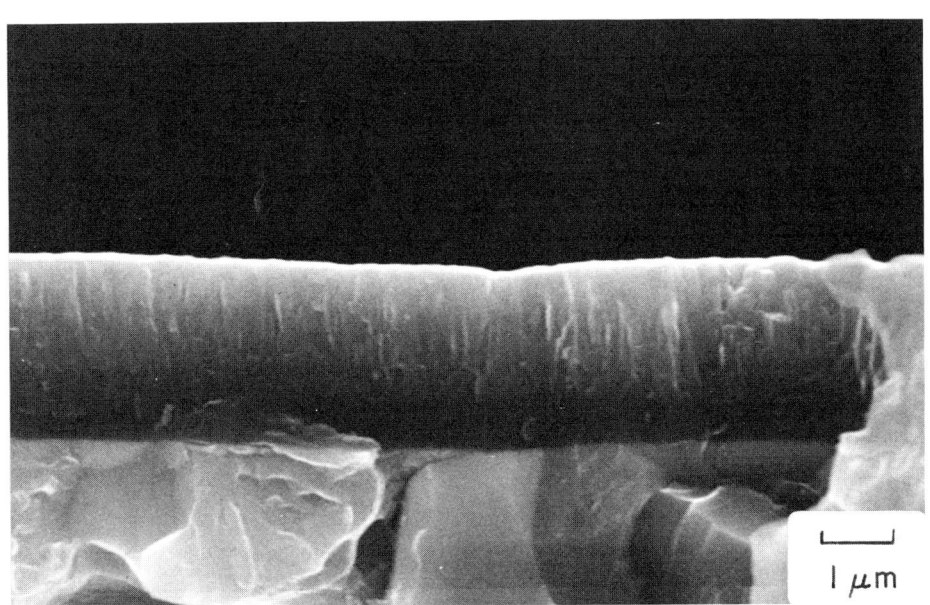

Fig. 11.11. (Ti,Al)N coating deposited by triode ion plating at high substrate current density. Courtesy of A.S. Korhonen, Helsinki University of Technology. (Ref 11.39, 11.40)

perfect coatings as discussed relative to Fig. 11.9 and 11.10, result in coarser microstructures and lower residual stresses.

The most striking application of PVD coatings in ferrous metallurgy is the titanium nitride (TiN) coating of high-speed steels for cutting and machining and tool steels for hot and cold working molds and dies. Many of the commercial PVD processes, dramatic improvements in TiN-coated tool performance, and causes of variability of TiN coatings have been reviewed by Matthews (Ref 11.37). The coatings produced by the various processes range from less than 1 μm to as thick as 6 μm and give the tools a uniform gold color. Deposition temperatures are 500 °C (930 °F) or lower. Hardness of the coatings increases with nitrogen content, with a peak hardness associated with off-stoichiometric compositions of about 40 at.% nitrogen (Ref 11.38). Typical hardness of commercial TiN coatings is about 2500 HV compared to the typical hardness of hardened tool steels of about 800 HV (Ref 11.38). Thus significant improvements in wear resistance are possible with properly applied coatings.

336 / STEELS: HEAT TREATMENT AND PROCESSING PRINCIPLES

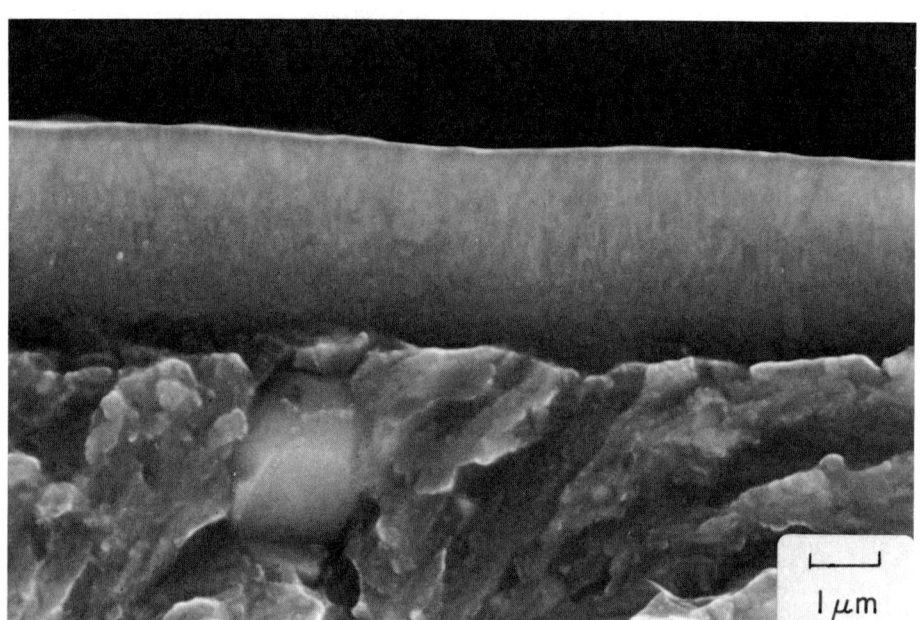

Fig. 11.12. ZrN coating deposited by triode ion plating. Scanning electron micrograph. Courtesy of A.S. Korhonen, Helsinki University of Technology. (Ref 11.39, 11.40)

Figures 11.10 to 11.12 show examples of PVD metal nitride coatings which have been deposited on high-speed steel and Type 304 stainless steel substrates by reactive triode ion plating (Ref 11.38-11.40). The metal atoms were evaporated into a chamber with an atmosphere of nitrogen and argon by heating suitable metal targets with an electron beam gun. The substrates were biased negatively and a glow discharge was created and controlled with heated tungsten filaments. The coatings are between 2 and 3 μm thick, and fracture cross sections of the coatings and adjacent underlying substrates are shown. Figure 11.10 shows a typical zone 1 microstructure of a $(Ti_{33}Al_{17})N$ coating. The crystallite arrays are tapered with rounded tops and the coating surface is relatively rough. As discussed above, increasing the substrate current density while holding all other processing parameters constant (Ref 11.40) produces a much denser, smoother $(Ti_{33}Al)N$ coating (Fig. 11.11). The latter coating consists of two layers. The surface layer structure is columnar while the layer adjacent to the substrate has a virtually featureless fracture surface. Figure 11.12 shows a zirconium nitride (ZrN) coating which is very smooth and almost featureless within the resolution of the scanning electron microscope. High-resolution transmission electron microscopy shows that the structure of the ZrN coating consists of very fine columnar grains, 30 to 60 nm in diameter and about 200 nm long (Ref 11.40). Tests of cutting performance ranked ZrN, (Ti,Al)N, and TiN coatings in order of decreasing performance (Ref 11.39).

Chemical Vapor Deposition

Chemical vapor deposition (CVD) is a coating process in which all reactants are gases. A chemical reaction takes place in the vapor phase adjacent to or on a substrate, depositing the reaction products on the substrate. An example of a CVD process is the deposition of tungsten according to the following reaction (Ref 11.41):

$$WF_6(\text{vapor}) + 3H_2(\text{gas}) \xrightarrow{\text{heated substrate}} W\ (\text{deposit}) + 6HF\ (\text{gas})$$

Usually the substrate must be heated to thermally activate the reaction.

Chemical vapor deposition is used to apply metal and ceramic compound coatings for a variety of electronic, corrosion protection, oxidation-resistant, heat-resistant, and machining applications (Ref 11.41, 11.42). The throwing power or ability to cover complex shapes by CVD is good compared to some of the line-of-sight PVD processes, but the chemicals used might be quite toxic.

The CVD technique has been used to coat cemented carbides and tool steels for machining applications. A pertinent study (Ref 11.43) has evaluated wear and failure of ceramic CVD coatings applied to cemented carbide tools used to machine stainless steels. The coatings consisted of various amounts of Al_2O_3, ZrO_2, TiC, and TiN, and ranged in hardness from 1540 to 1850 HV. Commonly used TiN CVD coatings are produced by the reaction of titanium tetrachloride and ammonia substrates heated to about 1000 °C (1830 °F) (Ref 11.44). As a result of the high substrate temperatures, CVD-coated tools must be hardened after coating, a sequence of operations which produces some distortion. In contrast, PVD coatings, as described in the previous section, are applied to hardened tools at 500 °C (930 °F) and the coated parts need not be heat treated subsequent to deposition.

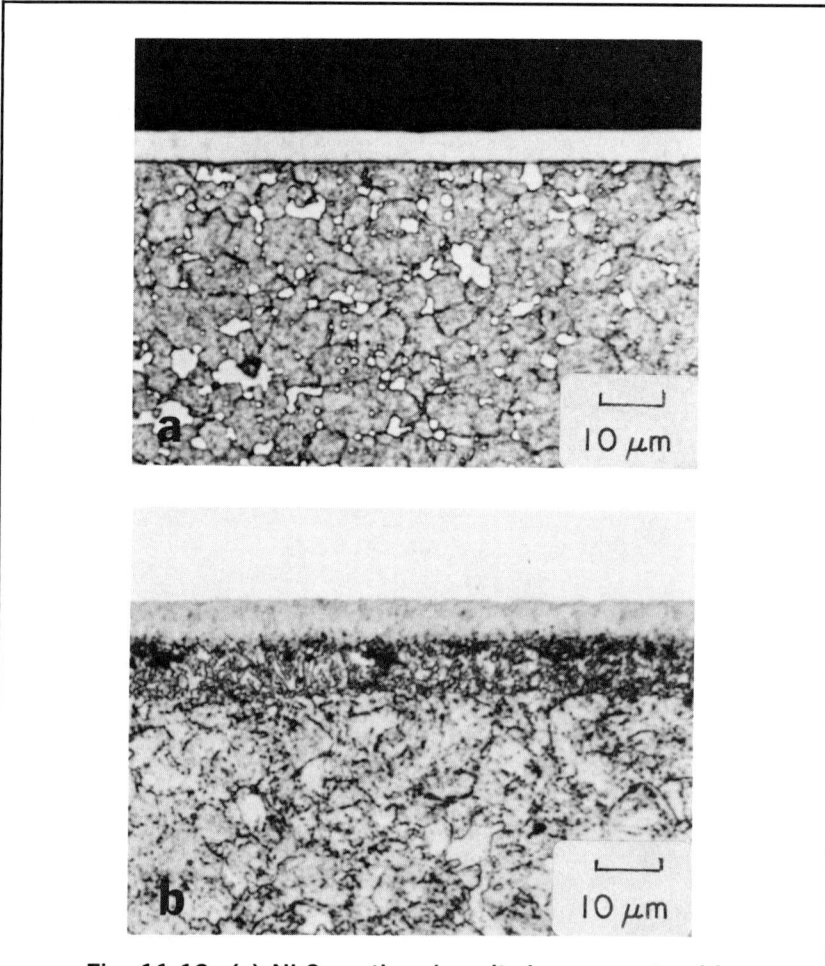

Fig. 11.13. (a) NbC coating deposited on a martensitic stainless steel by a salt bath process. (b) Chromium carbonitride coating deposited on nitrided AISI 1045 steel by a salt bath process. Light micrographs. Courtesy of T. Arai, Toyota Research Laboratories, Nagakute-Cho, Aichi

Salt Bath Coating Process

Hard alloy carbide, nitride, and carbonitride coatings can be applied to steels by means of salt bath processing. One such technique, the TD process (Toyota Diffusion coating process), uses molten borax with additions of carbide-forming elements, such as vanadium, niobium, titanium, or chromium, which combine with carbon from the substrate steel to produce alloy carbide layers (Ref 11.45-11.48). The growth of the layers is therefore dependent on carbon diffusion, and the process requires a relatively high temperature, from 800 to 1250 °C (1472 to 2282 °F), to maintain adequate coating rates. Carbide coating thicknesses of 4 to 7 μm are produced in 10 min to 8 h, depending on bath temperature and type of steel, and coating hardnesses over 3000 HV have been reported for VC and NbC layers (Ref 11.47). The coated steels may be cooled and reheated for hardening or the bath temperature may be selected to correspond to the steel austenitizing temperature, permitting the steel to be quenched directly after coating. In order to lower salt bath deposition temperatures, techniques to produce alloy carbonitride coatings have been developed (Ref 11.48). Such coatings are applied to hardened and nitrided steels in vanadium-containing chloride baths at temperatures of 550 to 600 °C (1022 to 1112 °F).

Figure 11.13 shows examples of salt-bath-applied coatings. The NbC coating, Fig. 11.13(a), was produced on a martensitic stainless steel (14% Cr, 1.5% C, 0.6% Mo, 0.4% Co) by immersion for 4 h in a borax bath containing 20% ferro-niobium. The NbC-coated steel was air cooled after coating and then reheated to 1060 °C (1940 °F) for hardening. The Cr(C,N) coating, Fig. 11.13(b), was produced by immersion of a previously hardened and nitrided (austenitized at 850 °C or 1562 °F, tempered at 600 °C or 1112 °F, and nitrided at 570 °C or 1058 °F) AISI 1045 steel in a chloride bath containing 15% chromium at 570 °C (1058 °F). At the latter temperatures, carbide growth is negligible, and the growth rate of coatings is accelerated by salt treatment of nitrided steels. This approach produces a coating layer and a diffusion layer as shown in Fig. 11.13(b).

The salt bath coating methods produce hard coatings for applications similar to those for which CVD and PVD coatings are considered. In particular, the TD coatings have significantly improved the life of mold and die steels for sheet steel stamping, aluminum die casting, and cold forging and extrusion.

Laser and Electron Beam Surface Modification

Laser and electron beams provide very high energy, directed sources of heat, and are used for many surface modification techniques. Depending on power input, high-energy beams may be used for cutting and welding, surface melting and alloying, and localized surface heat treatment (Ref 11.7, 11.8, 11.10). Welding and cutting require the highest power, and the ability to focus laser and electron beams makes possible very deep, narrow welds of high quality. This technology is highly developed and has followed the continuous development of high-energy density power sources (Ref 11.8, 11.10). Of the laser and electron beam surface modification techniques, localized surface heat treatment is the most highly developed and commercially applied. Surface melting and alloying, accompanied by very rapid solidification rates, offer unique opportunities for surface modification and now constitute a highly active field of research and development (Ref 11.7).

Heating by laser and electron beams is accomplished by photon interactions of the incident radiation with the electronic structures of the substrate material. The incident energy is very rapidly converted into heat just below the surface, on the order of a few tens of nanometers for laser light and a few microns for electron beams depending on the accelerating voltage, which generally varies between 10 and 100 keV (Ref 11.25, 11.49). Electron beam treatments must be conducted in vacuum, but laser light is not subject to this constraint and therefore offers considerable flexibility in manufacturing operations.

The term laser stands for "light amplification by stimulated emission of radiation," and three different types of lasers have been developed: Nd:YAG (neodymium dissolved in yttrium aluminum garnet), CO_2, and excimer. An excellent review of the operation and characteristics of the various types of lasers is presented by Bass (Ref 11.49). The Nd:YAG lasers operate at wavelengths of 1.06 μm and are widely used for welding and drilling applications. The CO_2 lasers have the highest power commercially available and operate in the infrared range, frequently at a wavelength of 10.6 μm, while the more recently developed excimer lasers operate at wavelengths in the near-ultraviolet range, between 0.193 and 0.351 μm. Laser light may be reflected, depending on material and wavelength. Therefore, for effective laser heating, wavelengths which are absorbed by the work piece must be selected or the irradiated work piece must be coated with a light-absorbing material.

Laser surface heat treatment is widely used to harden localized areas of steel and cast iron machine components (Ref 11.50). The heat generated by the absorption of the laser light is controlled to prevent melting, and therefore is used to selectively austenitize local surface regions which transform to martensite as a result of rapid cooling by the conduction of heat into the bulk of the work piece. This process is sometimes referred to as laser transformation hardening to differentiate it from laser surface melting phenomena. There is no chemistry change produced by laser surface heat treatment, and

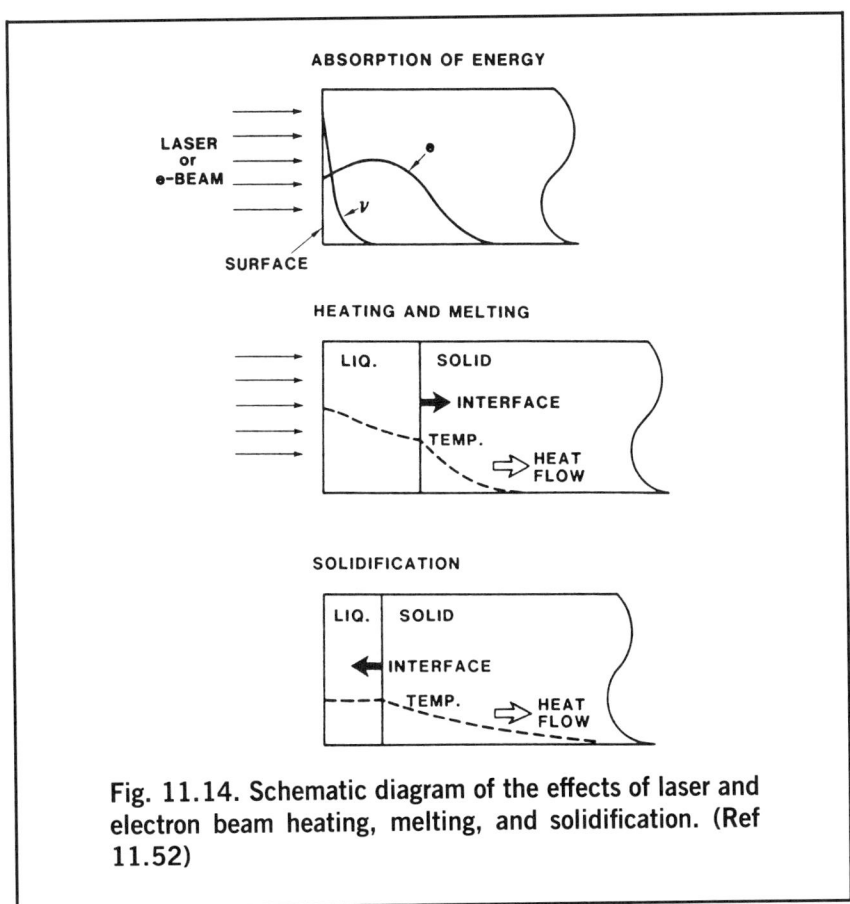

Fig. 11.14. Schematic diagram of the effects of laser and electron beam heating, melting, and solidification. (Ref 11.52)

laser heating presents, in addition to induction and flame hardening, an effective processing technique to selectively harden ferrous materials.

Laser heat treatment produces thin surface zones which are heated and cooled very rapidly, resulting in very fine martensitic microstructures, even in steels with relatively low hardenability. High hardness and good wear resistance with less distortion result from this process. The laser can be located at some distance from the work pieces, unlike induction and flame heating, and the laser light is reflected by mirrors to the focusing lens where the width of the heated spot or track is controlled (Ref 11.49, 11.50).

Molian (Ref 11.50) has tabulated the characteristics of 50 applications of laser transformation hardening. The materials hardened include plain carbon steels (1040, 1050, 1070), alloy steels (4340, 52100), tool steels, and cast irons (gray, malleable, ductile). The

Table 11.1. Hierarchy of Equilibrium (Ref 11.54)

I. **Full diffusional (global) equilibrium**
   A. No chemical potential gradients (compositions of phases are uniform)
   B. No temperature gradients
   C. Lever rule applicable

II. **Local interfacial equilibrium**
   A. Phase diagram gives compositions and temperature only at liquid-solid interface
   B. Includes corrections made for interface curvature (Gibbs-Thomson Effect)

III. **Metastable local interfacial equilibrium**
   A. Relevant if stable phase cannot nucleate or grow sufficiently fast
   B. Interface conditions given by a metastable phase diagram that is a true thermodynamic phase diagram constrained to be missing the stable phase or phases
   C. Also relevant if phases are constrained by elastic stresses

IV. **Interfacial nonequilibrium**
   A. Phase diagram fails to give temperature and compositions at interface
   B. Chemical potentials are not equal at interface
   C. Free-energy functions of phases useful to yield criteria for the impossible

energy-absorbing coatings are listed, and typical case depths for steels are 250 to 750 μm and for cast irons about 1000 μm. The flexibility of laser delivery systems, low distortion, and high surface hardness have made lasers very effective in selective hardening of wear and fatigue-prone areas on irregularly shaped machine components such as camshafts and crankshafts (Ref 11.49). Electron beams, similar to laser heat treatment, are also used to harden the surfaces of steels. The processing considerations, microstructures, and property changes produced by electron beam hardening of steel have been reviewed by Zenker et al. (Ref 11.51).

A completely different spectrum of surface modifications results when lasers and electron beams are used to melt the surface of a material. Figure 11.14 shows this process schematically (Ref 11.25, 11.52). Differences in energy absorption of electron (e) and laser (V) beams are shown qualitatively in the top illustration, and the melting and resolidification of the surface layers are shown in the bottom illustrations. Heating may be extremely rapid, on the order of nanoseconds, and cooling, accomplished by thermal conduction into the unheated mass of the substrate, is similarly very rapid. Exact rates of

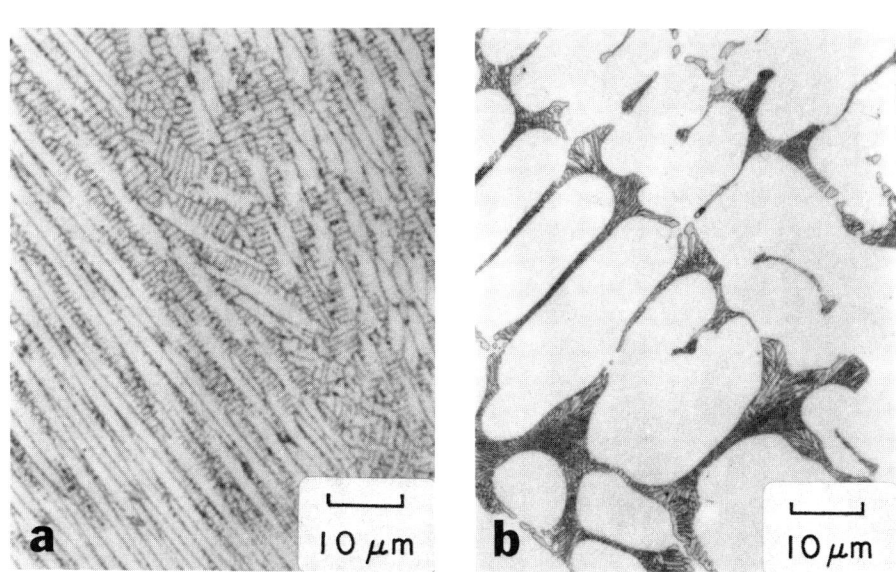

Fig. 11.15. (a) Laser-melted dendritic structure of M42 tool steel. (b) Chill-cast dendritic structure of M42 steel. Light micrographs. Courtesy of T. Bell, University of Birmingham. (Ref 11.56)

heating and cooling are dependent on many factors, such as power input, time of irradiation, laser pulsing, and surface and bulk characteristics of the heated substrate (Ref 11.52, 11.53). The very high heating and cooling rates attainable, 10^8 to 10^{10} °C/s, produce extremely rapid solidification and therefore make possible very fine nonequilibrium microstructures. In the extreme, new metastable crystalline phases, glassy or amorphous structures, or highly supersaturated phases may be developed in rapidly cooled surface layers. The various degrees of equilibrium possible in materials systems have been characterized by Perepezko and Boettinger (Ref 11.54) (see Table 11.1). True equilibrium, where coexisting phases have uniform compositions according to the equilibrium phase diagrams, is achieved only by high-temperature, long-time annealing or by very slow cooling. During rapid cooling, equilibrium may be attained only at the interfaces between stable or metastable phases, and in the extreme even such local equilibrium breaks down, greatly modifying or suppressing phase transformations dependent on diffusion. Thus many new microstructures with unique properties may be produced by rapidly solidified surface-melted alloys.

344 / STEELS: HEAT TREATMENT AND PROCESSING PRINCIPLES

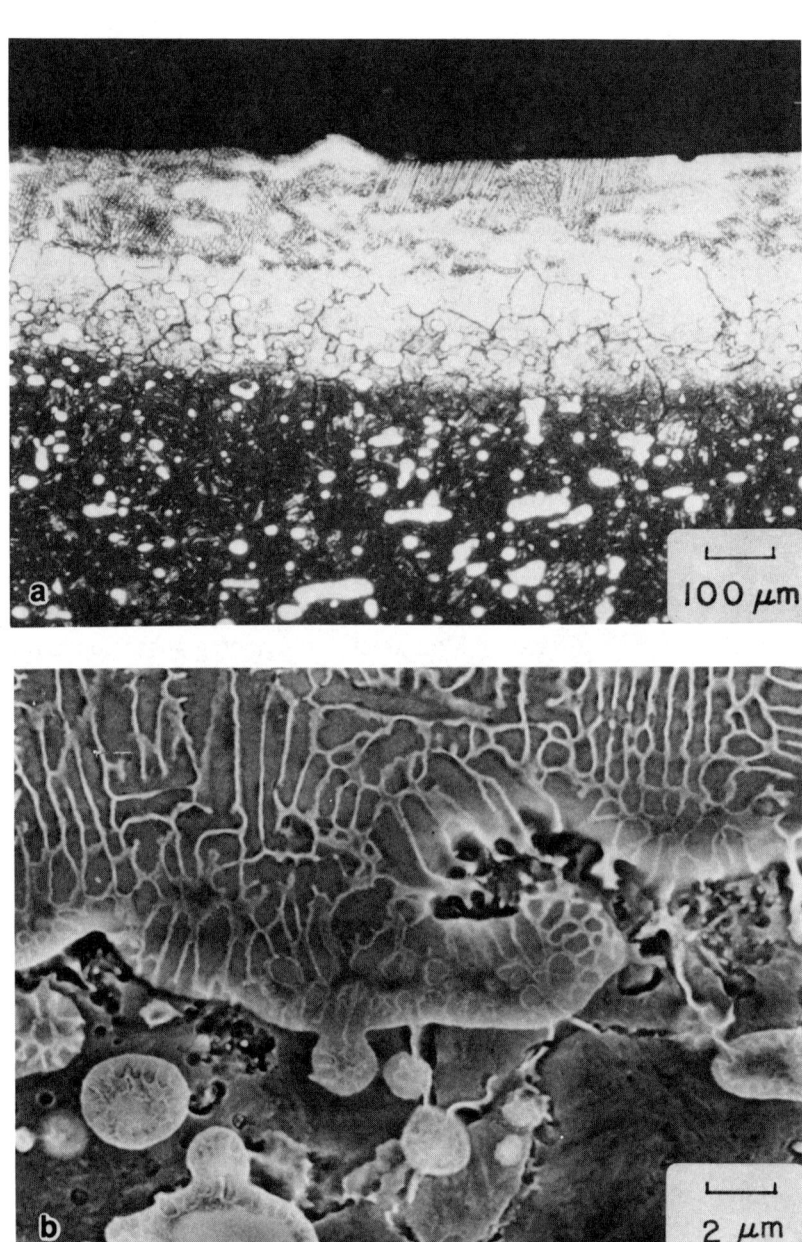

Fig. 11.16. (a) Laser-melted surface layer on M42 tool steel. (b) Higher magnification view of (a) showing partial melting of carbides at melt interface. Light micrographs. Courtesy of T. Bell, University of Birmingham. (Ref 11.56)

Laser surface alloying incorporates laser surface melting and cooling as described above, but in addition changes composition to effect changes in surface structure and properties. The alloying may be accomplished by producing a surface layer of chemistry different from the substrate by another surface modification technique prior to laser melting or by injecting powders into the laser-melted zone (Ref 11.54).

When the object of laser treatment, either by direct laser melting or by laser surface alloying, is to produce an amorphous or glassy layer, that process is referred to as laser glazing. Glass formation in Si, Pd-Cu-Si, and Fe-Ni-P-B alloys is readily accomplished, but is much more difficult in metals and alloys. For example, a study (Ref 11.55) of laser glazing of iron and tool steels, which had been pack boronized prior to laser glazing to promote amorphization, did not produce evidence of amorphous structures. Apparently nucleation of crystalline phases occurs too readily at the interface of the molten layer and the unmolten substrate crystal structure. Nevertheless, the laser surface alloying produced layers of very high hardness (2100 HV), containing very fine boride particles, on the tool steels. Cracking and porosity were problems sometimes encountered in the laser glazing study.

The dramatic changes in surface microstructure produced by laser surface melting of M42 high-speed tool steel are shown in Fig. 11.15 and 11.16 (Ref 11.56). The M42 steel contains nominally 1% carbon, 8% cobalt, 1.5% tungsten, 1.1% vanadium, 3.75% chromium, and 9.5% molybdenum, and because of the high content of carbide-forming elements the wrought microstructure contains a high volume fraction of coarse primary carbides. Figure 11.15 shows the considerable refinement of the microstructure of laser-melted M42 relative to chill-cast M42, and shows the absence of primary carbides in the laser melt zone. Dissolution of the carbides was a function of traverse speed, and at higher speeds carbides were not dissolved. Figure 11.16 shows the laser-melted surface and melting around primary carbides in the matrix below the fine solidification structure of the melt zone. Melting of the carbides is due to a low melting eutectic reaction. The laser-melted surface layer, produced by slow traverse speeds, apparently because of greater solution of alloying elements for subsequent carbide precipitation, showed much higher peak hardness after triple tempering than conventionally treated steel.

Summary

This chapter has briefly reviewed a number of the newer, high-technology surface modification treatments which are beginning to be applied to ferrous alloys. Process principles, terminology, coating characteristics, and examples of application have been described in order to provide a comparison of the various techniques. For more information regarding either the equipment and processing details, applications, or theoretical aspects of the deposition process, the reader is referred to the references cited throughout the text. That literature is only a sampling and is certain to grow at an increasing rate in view of the very active interdisciplinary research and development now in progress regarding surface modification techniques of all types.

A number of surface modification processes, in various stages of maturity and growth, have not been discussed. These processes have evolved somewhat removed from traditional heat treatment, and include such processes as electrodeposition, hot dipping, cementation, cladding, thermal spraying, and hardfacing. Each of the processes has a highly developed technology and may be useful in solving a given wear, corrosion, or high-temperature oxidation problem.

Many of the newer surface modification processes are only now being applied to ferrous alloys. Plasma nitriding, laser hardening of machine components, and PVD coating of cutting tools and dies appear to be the most widely applied at this time. Paralleling the electronics industry, the first applications are to high-value or critical components where the increased costs of higher technology are justified by higher quality and improved performance. Nevertheless, applications of the newer techniques to low-carbon sheet steels are in progress. For example, large-scale, high-deposition-rate electron beam evaporation of aluminum on strip steel has been used to produce aluminum-coated steel with corrosion characteristics comparable to tin plate (Ref 11.57).

The future should see much wider application to a great variety of structural applications. For optimum application and economy of any surface modification technique, it should be incorporated at the start into the mechanical and materials design and manufacturing sequence of a given component.

References

11.1 L.I. Maissel and R. Glang (Eds.), *Handbook of Thin Film Technology*, McGraw-Hill, New York, 1970

11.2 T. Bell, A. Bloyce, and J. Lanagan, Surface Engineering of Light Metals, in *Heat Treatment and Surface Engineering*, G. Krauss (Ed.), ASM INTERNATIONAL, Metals Park, OH, 1988, p 1-7

11.3 T. Bell (Ed.), *Surface Engineering*, Institute of Metals/Wolfson Institute for Surface Engineering, in association with the Surface Engineering Society, 1 Carlton House Terrace, London SW1Y 5DB, England

11.4 B.D. Sartwell and A. Matthews (Eds.), *Surface and Coatings Technology*, Elsevier Science, New York, NY 10017

11.5 T. Spalvins (Ed.), *Ion Nitriding*, ASM INTERNATIONAL, Metals Park, OH, 1987

11.6 *Plasma Heat Treatment, Science and Technology*, PYC Edition, Paris, 1987

11.7 L.E. Rehn, S.T. Picraux, and H. Wiedersich (Eds.), *Surface Alloying by Ion, Electron and Laser Beams*, ASM INTERNATIONAL, Metals Park, OH, 1987

11.8 Y. Arata, *Plasma, Electron and Laser Beam Technology*, American Society for Metals, Metals Park, OH, 1986

11.9 R.F. Hochman (Ed.), *Ion Plating and Implantation*, American Society for Metals, Metals Park, OH, 1986

11.10 E.A. Metzbower and D. Hauser, *Power Beam Processing, Electron, Laser, Plasma-Arc*, ASM INTERNATIONAL, Metals Park, OH, 1988

11.11 K.N. Strafford, P.K. Datta, and C.G. Googan (Eds.), *Coating and Surface Treatment for Corrosion and Wear Resistance*, Ellis Harwood Ltd, Chichester, U.K., 1984

11.12 R.D. Sisson, Jr. (Ed.), *Surface Modifications and Coatings*, American Society for Metals, Metals Park, OH, 1986

11.13 K.H. Jack, The Occurrence and the Crystal Structure of Iron Nitride; A New Type of Interstitial Alloy Formed During Tempering of Nitrogen-Martensite, *Proc Royal Soc*, Vol A208, 1951, p 216-224

11.14 T. Spalvins, Plasma Assisted Surface Coating/Modification Processes: An Emerging Technology, in *Ion Nitriding*, T. Spalvins (Ed.), ASM INTERNATIONAL, Metals Park, OH, 1987, p 1-8

11.15 B. Edenhofer, Physical and Metallurgical Aspects of Ionitriding, *Heat Treat Metals*, Vol 1 (No. 1), 1974, p 23-28

11.16 T. Bell and P.A. Dearnley, Plasma Surface Engineering, in *Plasma Heat Treatment, Science and Technology*, PYC Edition, 1987, p 13-53

11.17 B. Edenhofer, M.H. Jacobs, and J.N. George, Industrial Processes, Applications and Benefits of Plasma Heat Treatment, *Plasma Heat Treatment, Science and Technology*, PYC Edition, 1987, p 399-415

11.18 J.P. Lebrun, Technical Developments and Industrial Applications of Ion Nitriding, in *Plasma Heat Treatment, Science and Technology*, PYC Edition, 1987, p 425-444

11.19 H. Michael, M. Foos, and M. Gantois, Metallurgical Characterization of Plasma Induced Epsilon-iron Carbonitride Layers, in *Ion Nitriding*, T. Spalvins (Ed.), ASM INTERNATIONAL, Metals Park, OH, 1987, p 117-125

11.20 W.L. Grube and J.G. Gay, High-Rate Carburizing in a Glow-Discharge Methane Plasma, *Met Trans A*, Vol 91, 1978, p 1421-1429

11.21 A.C. Dexter, T. Farrell, M.I. Lees, and B.J. Taylor, The Physical and Chemical Processes of Vacuum and Glow Discharge Carburizing, in *Plasma Heat Treatment, Science and Technology*, PYC Edition, 1987, p 58-71

11.22 J. Pacheco, Fatigue Resistance of Plasma and Gas Carburized SAE 8719 Steel, M.S. Thesis T-3750, Colorado School of Mines, Golden, 1988; and J. Pacheco and G. Krauss, in *Carburizing, Processing and*

Performance, ASM INTERNATIONAL, Metals Park, OH, 1989, p 227-238

11.23 R.F. Hochman, Surface Modification by Ion Processes—An Engineering Technology, in *Ion Plating and Implantation*, R.F. Hochman (Ed.), American Society for Metals, Metals Park, OH, 1986, p 1-6

11.24 G. Dearnaley, Ion Implantation and Ion Assisted Coatings for Wear Resistance in Metals, *Surface Eng*, Vol 2, 1986, p 213-221

11.25 L.E. Rehn, S.T. Picraux, and H. Wiedersich, Overview of Surface Alloying by Ion, Electron and Laser Beams, in *Surface Alloying by Ion, Electron and Laser Beams,* ASM INTERNATIONAL, Metals Park, OH, 1987, p 1-17

11.26 J.K. Hirvonen, The Industrial Applications of Ion Beam Processes, in *Surface Alloying by Ion, Electron and Laser Beams*, L.E. Rehn, S.T. Picraux, and H. Wiedersich (Eds.), ASM INTERNATIONAL, Metals Park, OH, 1987, p 373-388

11.27 D.L. Williamson, F.M. Kustas, and D.F. Fobare, Mossbauer Study of Ti-Implanted 52100 Steel, *J Appl Phys*, Vol 60, 1986, p 1493-1500

11.28 F.M. Kustas, M.S. Misra, and D.L. Williamson, Microstructural Characterization of Nitrogen Implanted 440C Steel, Nuclear Instruments and Methods in Physics Research B31, North-Holland, 1988, p 393-401

11.29 F.M. Kustas, M.S. Misra, and P. Sioshansi, Effects of Ion Implantation on the Rolling Contact Fatigue of 440C Stainless Steel, in *Ion Implantation and Ion Beam Processing of Materials*, G.K. Hubler, O.W. Holland, and C.R. Clayton (Eds.), MRS Symposia Proceedings, Vol 27, 1984, p 675-690

11.30 J.K. Hirvonen, Applications of Ion Beam Mixing, in *Ion Plating and Implantation*, R.F. Hochman (Ed.), American Society for Metals, Metals Park, OH, 1986, p 49-53

11.31 J.M. Rigsbee, Physical Vapor Deposition, in *Surface Modification Engineering*, R.P. Kossowsky (Ed.), CRC Press, Boca Raton, FL, 1989, p 231-255

11.32 B.A. Movchan and A.V. Demchishin, Study of the Structure and Properties of Thick Vacuum Condensates of Nickel, Titanium, Tungsten, Aluminum Oxide and Zirconium Oxide, *Fiz Metal Metalloved,* Vol 28, 1969, p 653

11.33 J.A. Thornton, High Rate Thick Film Growth, *Ann Rev Mater Sci*, 1977, p 239-260

11.34 J.A. Thornton, The Microstructure of Sputter-Deposited Coatings, *J Vac Sci Technol*, Vol A4, 1986, p 3059-3065

11.35 D.W. Hoffman and R.C. McCune, Microstructural Control of Plasma-Sputtered Refractory Coatings, in *Plasma-Based Processing*, J.J. Cuomo, S.M. Rossnagel, and W.D. Westwood (Eds.), Noyes, Park Ride, NJ, 1989

11.36 D.W. Hoffman, Fine Tuning the Structures and Properties of Thin Films, in *Proceedings of the Joint Symposium of the Chinese and American Vacuum Societies*, Beijing, 8-10 Sept 1987

11.37 A. Matthews, Titanium Nitride PVD Coating Technology, *Surface Eng*, Vol 1, 1987, p 93-103

11.38 A.S. Korhonen, J.M. Molarius, S. Osenius, and M.S. Sulonen, Ion Plasting of Tools and Dies, in *Tool Materials For Molds and Dies*, G.

Krauss and H. Nordberg (Eds.), Colorado School of Mines Press, Golden, p 217-230

11.39 J.M. Molarius, A.S. Korhonen, E. Harju, and R. Lappalainen, Comparison of Cutting Performance of Ion-Plated NbN, ZrN, TiN, and (Ti,Al)N Coatings, *Surface Coatings Technol*, Vol 33, 1987, p 117-131

11.40 I. Penttinen, J.M. Molarius, A.S. Korhonen, and R. Lappalainen, Structure and Composition of ZrN and (Ti,Al)N Coatings, *J Vac Sci Technol*, Vol A6, 1988, p 2158-2161

11.41 R.F. Bunshah, Deposition Technologies: An Overview, in *Deposition Technologies for Films and Coatings*, R.F. Bunshah *et al*. (Eds.), Noyes, Park Ride, NJ, 1982, p 1-16

11.42 A. Kolb-Telieps, Introduction to Surface Engineering for Corrosion Protection, *Surface Eng*, Vol 2, 1986, p 203-212

11.43 P.A. Dearnley and V. Thompson, Evaluation of Failure Mechanisms of Ceramics and Coated Carbides Used for Machining Stainless Steels, *Surface Eng*, Vol 2, 1986, p 191-202

11.44 M.H. Jacobs, Process and Engineering Benefits of Sputter Ion Plated Titanium Nitride Coatings, Paper 8512-010 in *1985 International Conference on Surface Modifications and Coatings*, Toronto, American Society for Metals, Metals Park, OH, 1985

11.45 T. Arai, Carbide Coating Process by Use of Molten Borax Bath in Japan, *J Heat Treat*, Vol 1 (No. 2), 1979, p 15-22

11.46 T. Arai, H. Fujita, Y. Sugimoto, and Y. Ohta, Diffusion Carbide Coatings Formed in Molten Borax Systems, *J Mater Eng*, Vol 9, 1987, p 183-189

11.47 T. Arai, Carbide Coating Process by Use of Molten Borax Bath, *Wire*, 31 July 1981, p 102-104, 208-210

11.48 T. Arai, H. Fujita, Y. Sugimoto, Y. Ohta, Vanadium Carbonitride Coating by Immersing into Low Temperature Salt Bath, in *Heat Treatment and Surface Engineering*, G. Krauss (Ed.), ASM INTERNATIONAL, Metals Park, OH, 1988, p 49-53

11.49 M. Bass, Lasers and Electron Beams, in *Surface Alloying by Ion, Electron and Laser Beams*, L.E. Rehn, S.T. Picraux, and H. Wiedersich (Eds.), ASM INTERNATIONAL, Metals Park, OH, 1987, p 357-372

11.50 P.A. Molian, Engineering Applications and Analysis of Hardening Data for Laser Heat Treated Ferrous Alloys, *Surface Eng*, Vol 2, 1986, p 19-28

11.51 R. Zenker and M. Mueller, Electron Beam Hardening. Part I, Principles, Process Technology and Properties, *Heat Treat Metals*, Vol 15 (No. 4), 1988, p 79-88; and R. Zenker, W. John, D. Rathjen, and G. Fritsche, "Electron Beam Hardening. Part 2, Influence on Microstructure and Properties," *Heat Treat Metals*, Vol 16 (No. 2), 1989, p 43-51

11.52 D.M. Follstaedt and S.T. Picraux, Microstructures of Surface-Melted Alloys, in *Surface Alloying by Ion, Electron and Laser Beams*, L.E. Rehn, S.T. Picraux, and H. Wiedersich (Eds.), ASM INTERNATIONAL, Metals Park, OH, 1987, p 175-221

11.53 C.W. White and M.J. Aziz, Energy Deposition, Heat Flow and Rapid Solidification During Pulsed-Laser and Electron Beam Irradiation of Materials, in *Surface Alloying by Ion, Electron and Laser Beams*, L.E.

Rehn, S.T. Picraux, and H. Wiedersich (Eds.), ASM INTERNATIONAL, Metals Park, OH, 1987, p 19-50

11.54 J.H. Perepezko and W.J. Boettinger, Kinetics of Resolidification, in *Surface Alloying by Ion, Electron, and Laser Beams*, L.E. Rehn, S.T. Picraux, and H. Wiedersich (Eds.), ASM INTERNATIONAL, Metals Park, OH, 1987, p 51-90

11.55 P.A. Molian and H.S. Rajasekhara, Laser Glazing of Boronized Iron and Tool Steels, *Surface Eng*, Vol 2, 1986, p 269-276

11.56 T. Bell, I.M. Hancock, and A. Bloyce, Laser Surface Treatment of Tool Steels, in *Tool Materials for Molds and Dies*, G. Krauss and H. Nordberg (Eds.), Colorado School of Mines Press, Golden, 1987, p 197-216

11.57 S. Schiller, G. Beister, M. Neumann, and G. Jaesch, Vacuum Coating of Large Areas, *Thin Sold Films*, Vol 96, 1982, p 199-216

CHAPTER 12

Stainless Steels

Stainless steels are a large group of special alloys developed primarily to withstand corrosion. Other desirable features may include excellent formability, high room-temperature and cryogenic toughness, and good resistance to scaling, oxidation, and creep at elevated temperatures. Chromium is the alloying element which imparts corrosion resistance to stainless steels, but many other elements may be added to stabilize other phases, provide added corrosion resistance, or produce enhanced mechanical properties. Austenitic, ferritic, and duplex stainless steels cannot be hardened by heat treatment, and therefore alloying and thermomechanical processing are designed to minimize the formation of phases detrimental to corrosion resistance or toughness. In austenitic stainless steels, strength is also developed by cold work and strain-induced martensite formation. Martensitic stainless steels can be heat treated by quench and tempering to high hardness and strength. Precipitation-hardening grades of stainless steel have also been developed. This chapter describes alloy design, microstructure, and thermomechanical processing used for optimum performance of the various classes of stainless steels.

Alloy Design and Phase Equilibria

Chromium in excess of 12% by weight is required to impart "stainless" characteristics to iron alloys. Enhanced corrosion resistance relative to other steels is attributed to the ability of chromium to produce tightly adherent oxide layers on stainless steel surfaces. The layer is very thin, on the order of only a few atom layers in thickness, and effectively protects or passivates stainless steels in many corrosive environments (Ref 12.1, 12.2). Thus all stainless steels contain large amounts of chromium, and an important starting place to understand

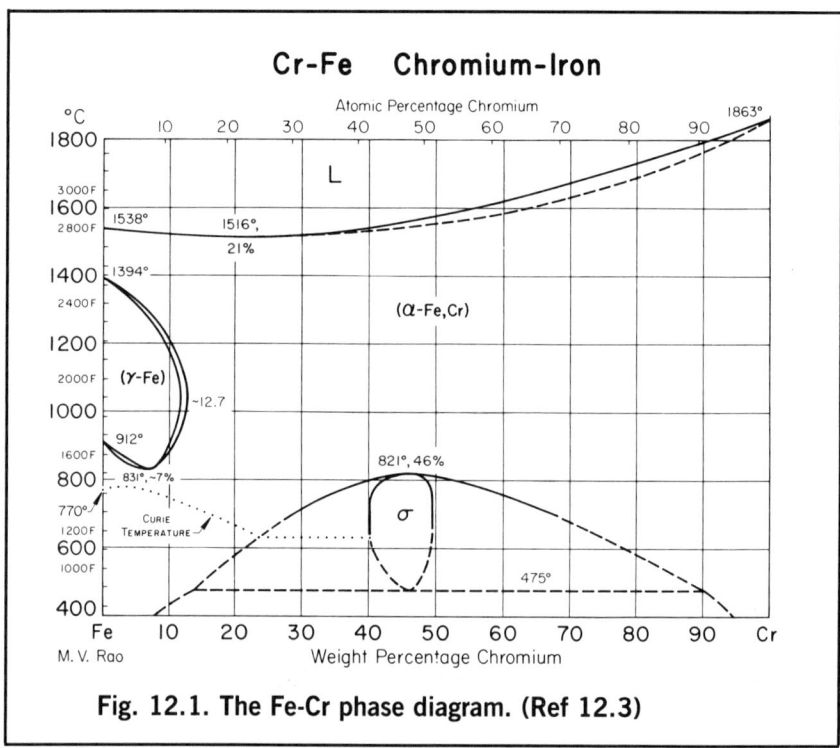

Fig. 12.1. The Fe-Cr phase diagram. (Ref 12.3)

the phase relationships and microstructures in stainless steels is the iron-chromium (Fe-Cr) equilibrium phase diagram.

Figure 12.1 shows the Fe-Cr equilibrium phase diagram. As in the Fe-C system, the allotropic forms of iron constitute the iron end of the diagram. Chromium is an element which stabilizes the body-centered cubic ferrite structure of iron; therefore, with increasing chromium content the high-temperature and low-temperature delta and alpha ferrite fields expand. At around 12% chromium, body-centered cubic ferrite is completely stable from room temperature up to the melting point. As the ferrite field expands, the austenite field contracts, producing what is often referred to as the gamma (γ) loop. Figure 12.2 shows that other ferrite stabilizing elements such as vanadium and molybdenum act similarly to chromium when alloyed with iron and also form gamma loops.

The Fe-Cr diagram directly produces the basis for martensitic and ferritic stainless steels. The martensitic stainless steels must be able to form austenite, which will transform to martensite on cooling. Therefore, the compositions of martensitic stainless steels must lie within the gamma loop (as expanded by other alloying elements) and contain sufficient chromium to impart stainless corrosion behavior. Ferritic

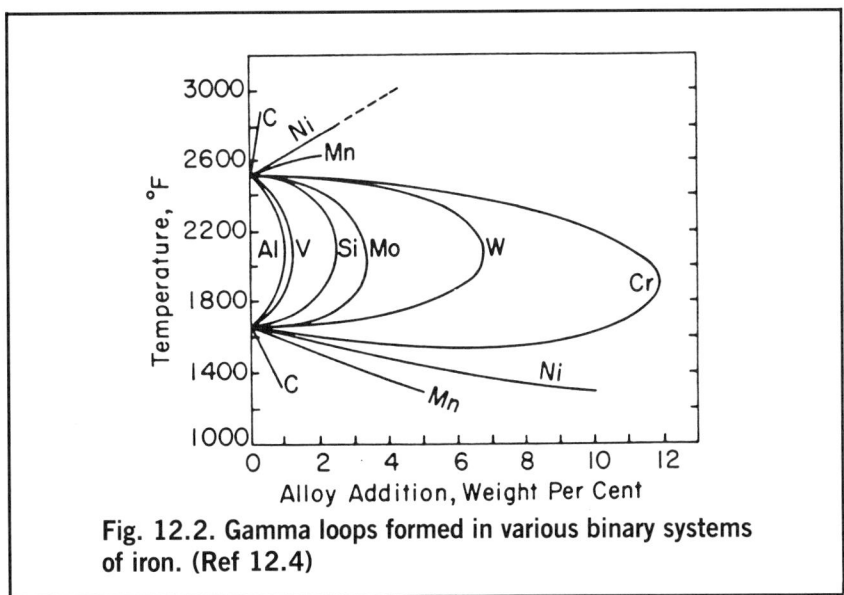

Fig. 12.2. Gamma loops formed in various binary systems of iron. (Ref 12.4)

stainless steels are alloyed with much higher amounts of chromium than martensitic stainless steels; therefore, ferrite is stable at all temperatures, as shown in Fig. 12.1. Martensitic and ferritic stainless steels contain alloying elements other than chromium; the effects of these elements will be discussed in following sections.

Next to chromium, nickel is the alloying element which most strongly influences alloy design of certain classes of stainless steels. Nickel stabilizes the face-centered cubic structure of iron and therefore expands the austenite or gamma phase field when alloyed with iron. The iron-nickel (Fe-Ni) equilibrium phase diagram (Fig. 12.3) shows that with sufficient nickel, austenite is stable at all temperatures above room temperature. In binary Fe-Ni alloys, about 30 wt.% nickel is required to completely stabilize austenite, partly because close to room temperature the diffusion of iron and nickel is too sluggish to form a mixture of ferrite and austenite. However, if chromium is also present, in amounts sufficient for stainless corrosion behavior, much less nickel is required to stabilize austenite. Thus alloys containing typically 18 wt.% chromium and 8 wt.% nickel are fully austenitic from well below room temperature to melting temperatures. The latter types of steels constitute the very important group of alloys designated as austenitic stainless steels.

Almost all stainless steels have three or more components, and therefore their phase relationships as a function of temperature and composition are represented by ternary phase diagrams. In the case of

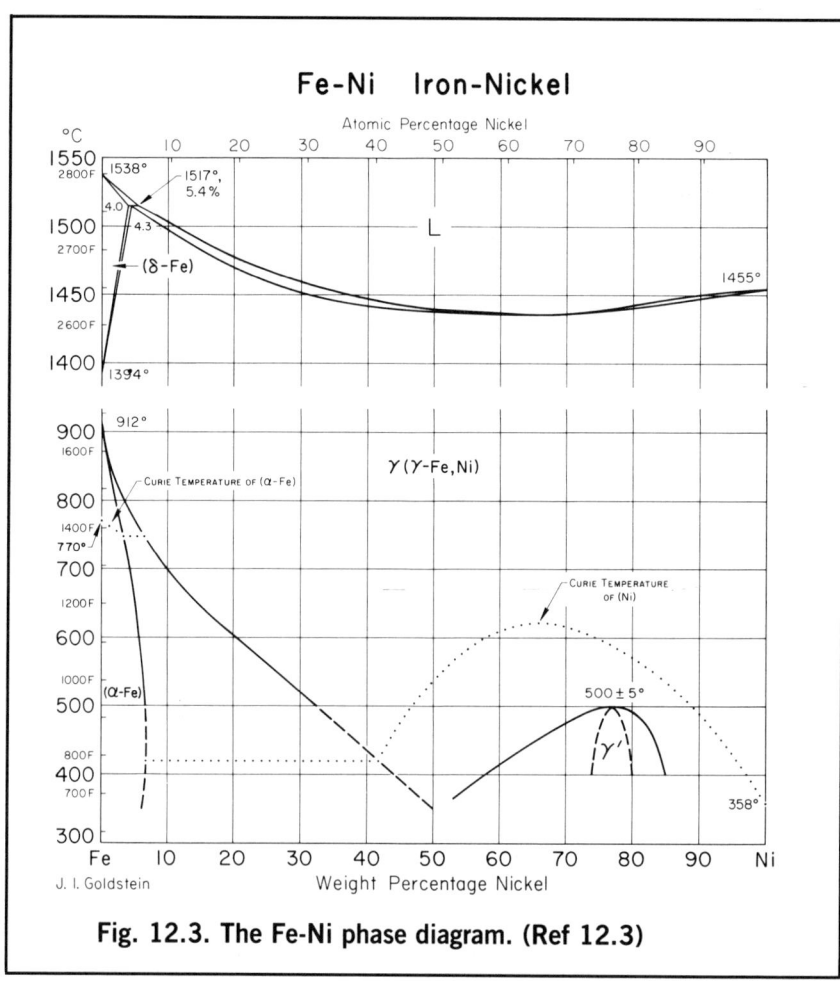

Fig. 12.3. The Fe-Ni phase diagram. (Ref 12.3)

systems with more than three components, diagrams which combine the various austenite and ferrite stabilizing elements are established and the phases present at room temperature are related to the two groups of alloying elements. Frequently vertical sections through ternary systems, in which the amount of a given component is held constant, are used to establish hot working and heat treatment schedules. For example, Fig. 13.4(b) in Chapter 13 shows a vertical section through the Fe-C-Cr system at 13% chromium. This vertical section is therefore useful in rationalizing the microstructures of martensitic stainless steels as a function of carbon content and temperature.

Figure 12.4 shows projections of liquidus and solidus surfaces of the Fe-Ni-Cr system for the composition ranges of interest to stainless steels, and Fig. 12.5 shows vertical sections at various constant iron

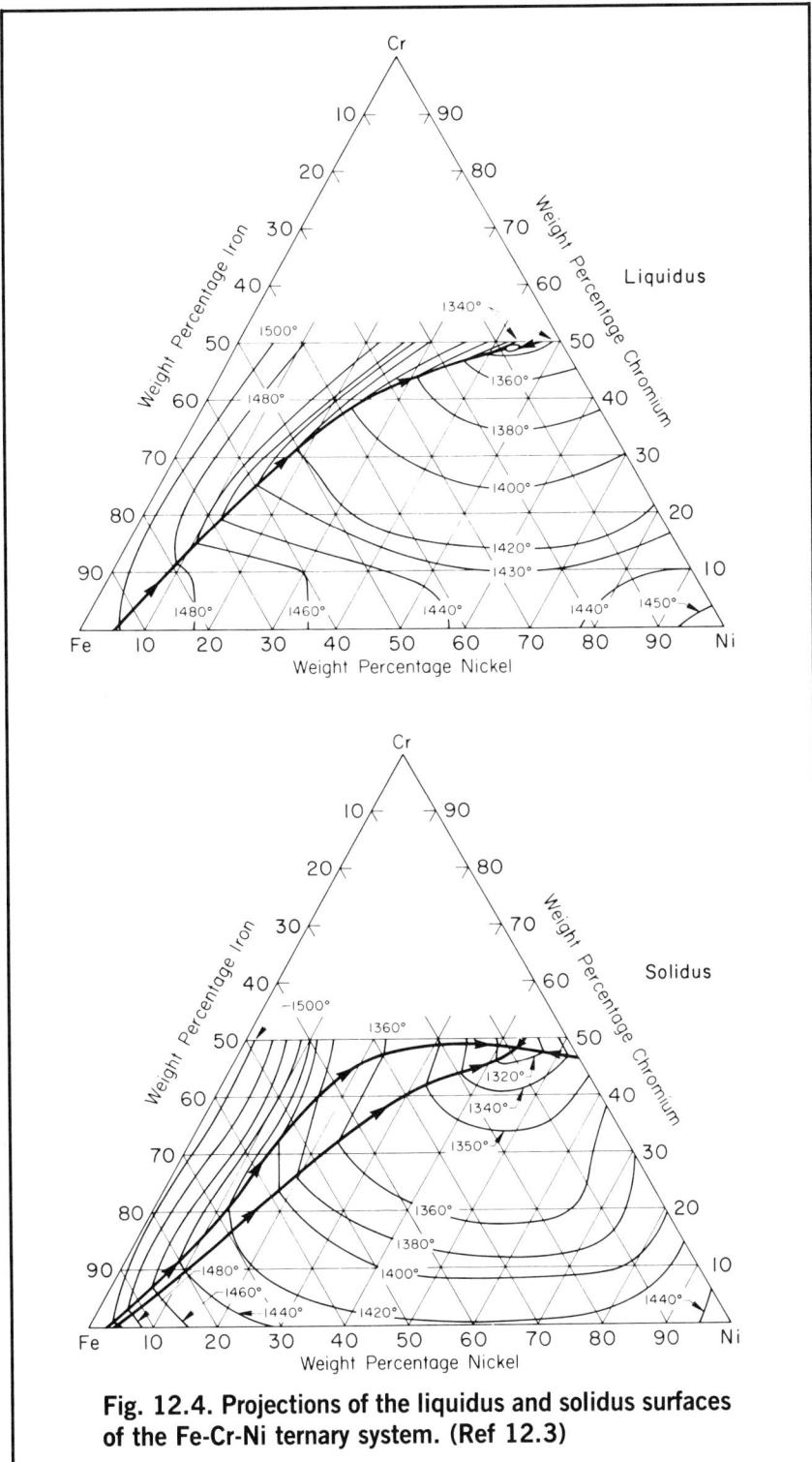

Fig. 12.4. Projections of the liquidus and solidus surfaces of the Fe-Cr-Ni ternary system. (Ref 12.3)

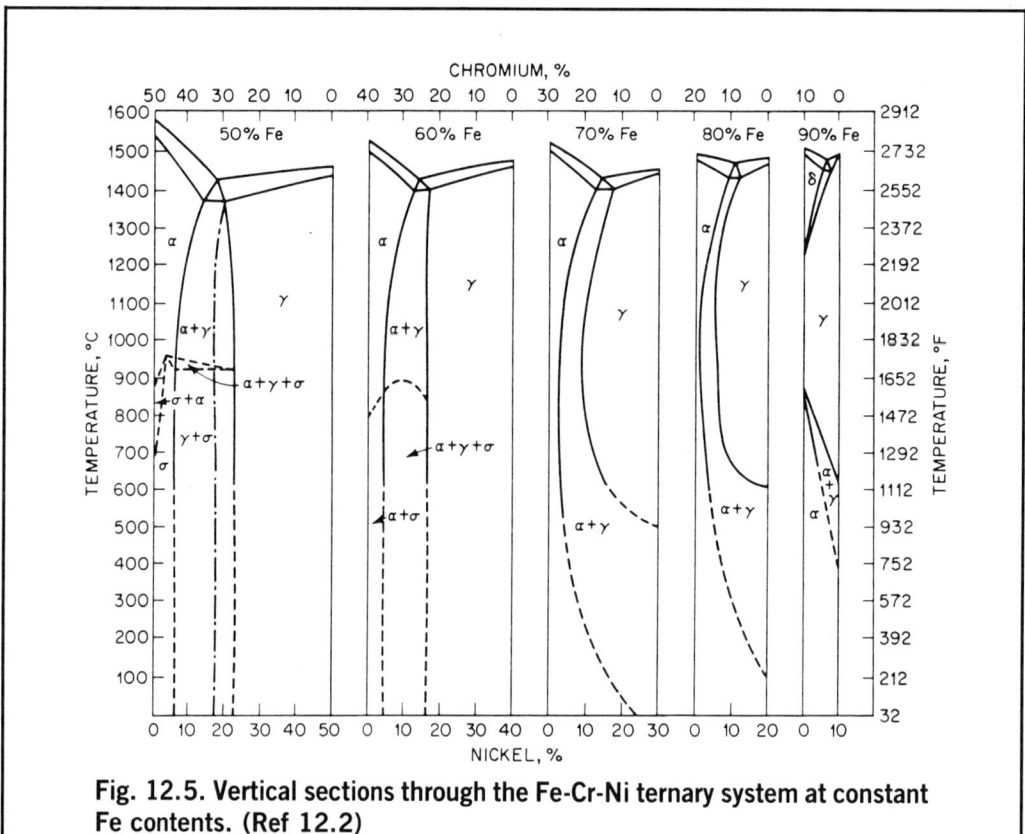

Fig. 12.5. Vertical sections through the Fe-Cr-Ni ternary system at constant Fe contents. (Ref 12.2)

contents through the same systems. Alloys rich in chromium solidify as ferrite and alloys rich in nickel solidify as austenite. However, many stainless Fe-Ni-Cr alloys solidify as two-phase, austenite-ferrite mixtures, and liquid coexists with austenite and ferrite during solidification. These three-phase fields are shown as the three-sided phase fields which contact the liquid phase field at liquidus surface minima on the vertical sections of Fig. 12.5. These three-phase fields are also defined by the intersections of constant iron planes with the heavy dark lines in Fig. 12.4. Thus ferrite-austenite microstructures frequently develop in stainless steels. In fact, the duplex stainless steels described later are designed to have microstructures which are about 50% ferrite and 50% austenite.

Ferrite-austenite microstructures are also frequently encountered in austenitic stainless steel weld metal and cast austenitic stainless steels. In the latter materials, nonequilibrium solidification and alloying effects combine to produce ferrite-austenite microstructures which

STAINLESS STEELS / 357

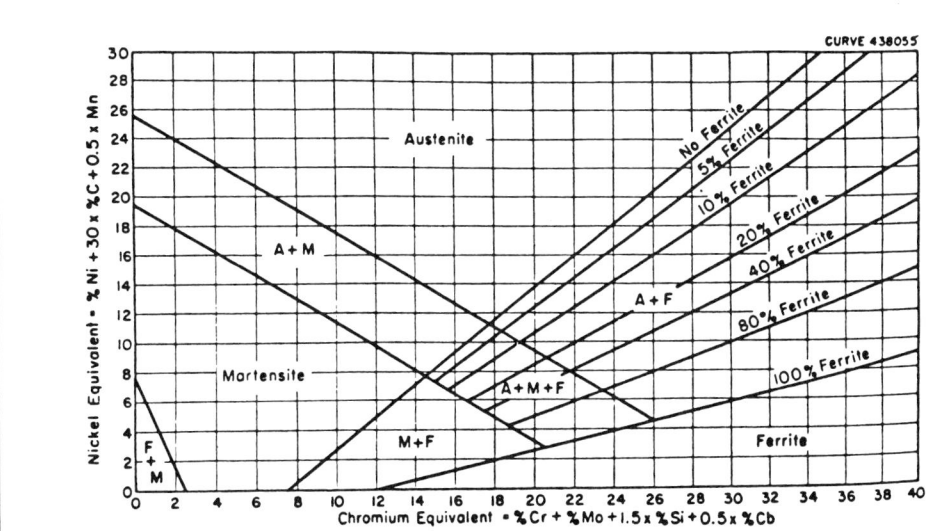

Fig. 12.6. The Schaeffler constitution diagram (1949) for stainless steel weld metal. (Ref 12.6, 12.7)

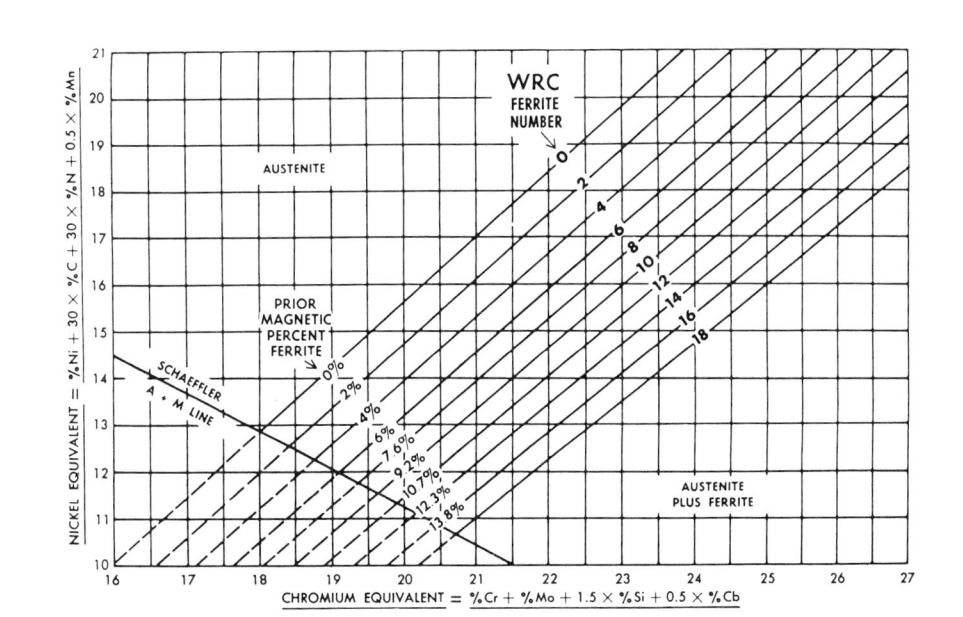

Fig. 12.7. The Delong constitution diagram (1974) with Welding Research Council ferrite number system for weld metal. (Ref 12.6, 12.8)

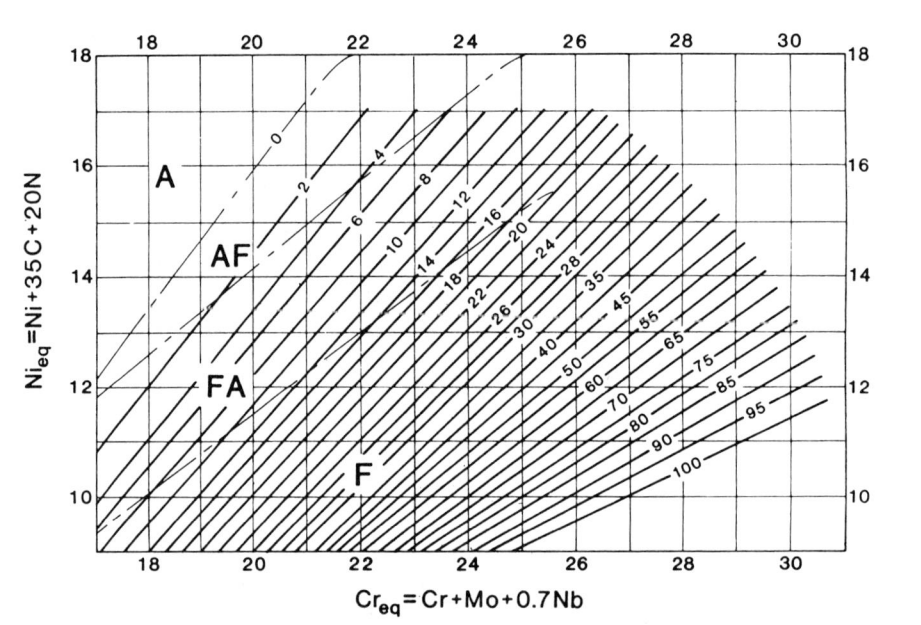

Fig. 12.8. The Siewert, McCowan, Olson constitution diagram (1988) for stainless steels. Ferrite numbers are plotted and A, AF, FA, and F indicate compositions which solidify by austenite, austenite followed by ferrite, ferrite followed by austenite, or ferrite formation, respectively. (Ref 12.6)

might not be present in wrought stainless steels of the same composition (Ref 12.5). Small amounts of ferrite are desirable in austenitic stainless steel weld metal because ferrite has a higher solubility for phosphorus and sulfur, elements which cause fissuring in fully austenitic microstructures (Ref 12.6). Therefore, ferrite reduces susceptibility to hot cracking or hot tearing. On the other hand, too high a ferrite content in austenitic stainless steel welds or castings may lower corrosion resistance and toughness.

In view of the importance of phase stability and composition, especially with respect to welding of austenitic stainless steels, several diagrams have been developed to show the effects of various combinations of austenite- and ferrite-stabilizing elements on ferrite content in stainless steels. The ferrite-stabilizing elements similar to chromium are molybdenum, silicon, and niobium, while the austenite-stabilizing elements similar to nickel are manganese, carbon, and nitrogen. Thus nickel and chromium equivalents are calculated according to the various strengths of the elements stabilizing austenite or ferrite.

Table 12.1. Compositions of Selected AISI Type 300 Austenitic Stainless Steels. (Ref 12.9)

AISI type no.	Nominal composition, %				
	C	Mn	Cr	Ni	Others
301	0.15 max	2.0	16–18	6.0–8.0	
302	0.15 max	2.0	17–19	8.0–10	
304	0.08 max	2.0	18–20	8.0–12	
304L	0.03 max	2.0	18–20	8.0–12	
309	0.20 max	2.0	22–24	12–15	
310	0.25 max	2.0	24–26	19–22	
316	0.08 max	2.0	16–18	10–14	2–3Mo
316L	0.03 max	2.0	16–18	10–14	2–3Mo
321	0.08 max	2.0	17–19	9–12	(5 × %C) Ti min
347	0.08 max	2.0	17–19	9–13	(10 × %C) Nb-Ta min

Face-centered cubic, nonmagnetic, not heat treatable

Figure 12.6 shows the Schaeffler diagram (Ref 12.7), Fig. 12.7 shows the DeLong diagram (Ref 12.8), and Fig. 12.8 shows the Siewert, McCowan, and Olson diagram (Ref 12.6). The equations used to calculate the chromium and nickel equivalents are shown on the axes of the various diagrams, and the compositions which produce austenite, martensite, ferrite, or mixtures of various phases are indicated. Ferrite number (FN), which can be calibrated with magnetic attraction since austenite is nonmagnetic and ferrite is magnetic, was selected by the Welding Research Council to correlate with ferrite content, and is plotted in Fig. 12.7 and 12.8. The diagrams have evolved with the accumulation of more data regarding compositions and microstructure, and the most recent diagram is based on a large number of alloys covering a wide range of compositions, including duplex stainless steels. The Siewert *et al.* diagram also indicates several solidification ranges which relate to the phase diagrams shown in Fig. 12.4 and 12.5. Complete austenite solidification is indicated by A, primary austenite followed by austenite plus ferrite solidification by AF, primary ferrite followed by austenite plus ferrite solidification by FA, and complete ferrite solidification by F.

Austenitic Stainless Steels

Table 12.1 shows the nominal compositions of AISI type 300 austenitic stainless steels. This table and the others which follow for other groups of stainless steels are taken from an article by Fischer and Maciag (Ref 12.9). The important grades of steels and the nominal

360 / STEELS: HEAT TREATMENT AND PROCESSING PRINCIPLES

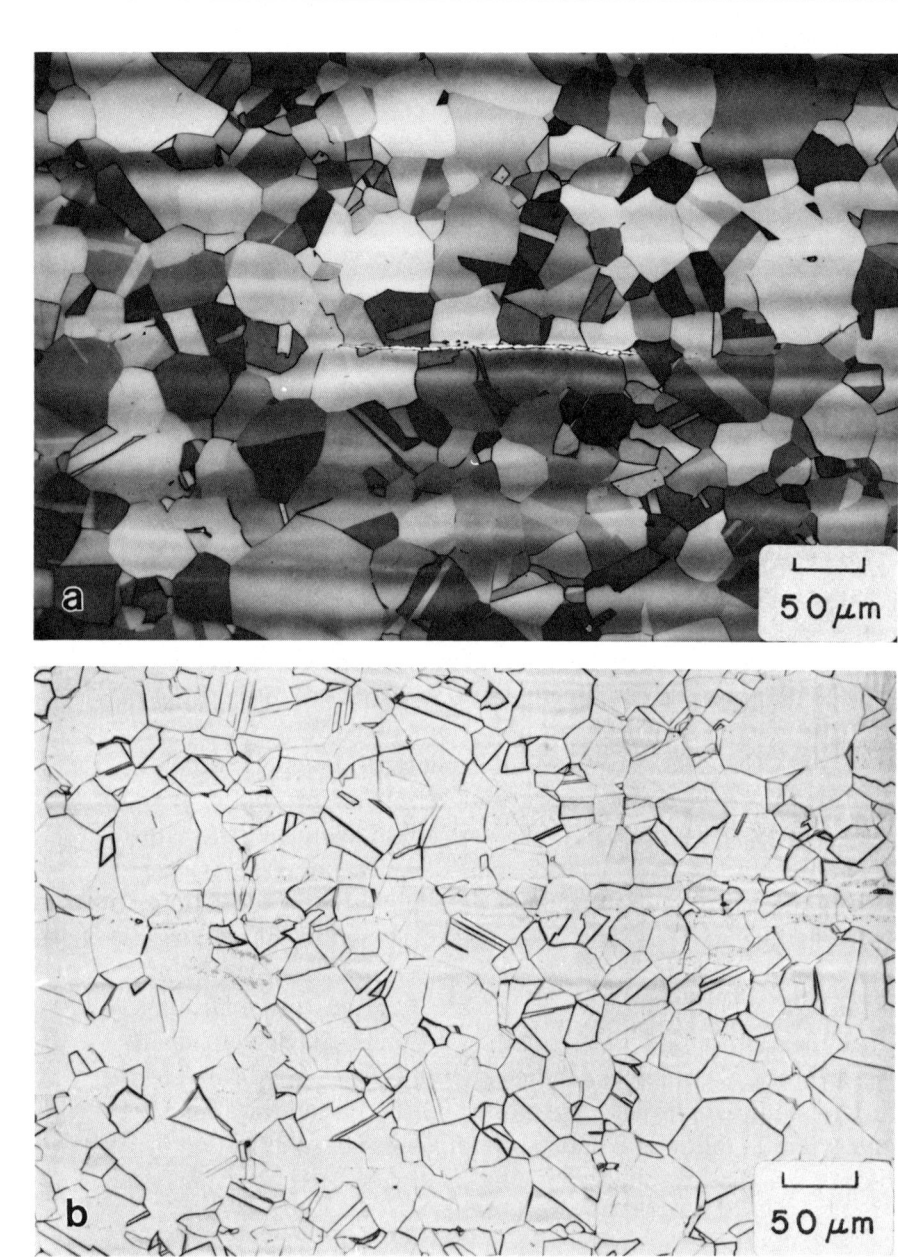

Fig. 12.9. Microstructure of annealed type 316L austenitic stainless steel. (a) Etched in 20% HCl, 2% NH_4FHF, 0.8% PMP (Ref 12.12, 12.13). (b) Etched in waterless Kalling's reagent (Ref 12.12, 12.13). Light micrographs. Courtesy of G. Vander Voort, Carpenter Technology Corp., Reading, PA

amounts of the most important alloying elements are clearly indicated. More extensive tables listing other stainless steels, Unified Numbering System (UNS) alloy numbers, and other information, are given elsewhere (Ref 12.10, 12.11).

Stainless steels are selected for corrosion resistance to the atmosphere, seawater, and a vast variety of chemical environments, and the reader is referred to other sources (Ref 12.1, 12.2, 12.10, 12.11) and manufacturers' literature to help select the proper steel for a given environment. Apart from resistance to specific corrosive environments, the austenitic stainless steels have evolved around the following physical metallurgical principles: varying austenite stability relative to martensite formation during cold work (AISI types 301, 302, and 304), reduction of carbon and alloying to eliminate chromium carbide formation and intergranular corrosion (AISI types 304L, 316L, 321, and 347), alloying with molybdenum to increase pitting resistance (AISI type 316), and heavily alloying with chromium and nickel to produce high-temperature strength and scaling resistance (AISI types 309 and 310).

Figure 12.9 shows the microstructure of annealed type 316L austenitic stainless steel. The microstructure in Fig. 12.9(a) is etched to produce contrast between grains of different orientation, and the microstructure in Fig. 12.9(b) is etched only to show grain boundaries (Ref 12.12, 12.13). The single-phase austenite is present as many equiaxed grains, many of which contain annealing twins. The twins are identified as bands with parallel sides and are formed when changes in the stacking of atoms on close-packed (111) planes occur during recrystallization and grain growth. Austenitic stainless steels are usually annealed at high temperatures to accomplish recrystallization and carbide solution. Water quenching follows annealing to prevent carbide formation during cooling as discussed in the next section. Typical minimum mechanical properties of annealed austenitic stainless steels are yield strengths of 30 ksi (205 MPa), ultimate tensile strengths of 75 ksi (515 MPa), and elongations of 40% (Ref 12.10). Properly processed wrought austenitic stainless steels are truly single phase, without carbides, ferrite, or other phases, and with all alloying elements in solid solution. For the latter condition, corrosion resistance for a given grade is at its best.

Intergranular Carbides in Austenitic Stainless Steels

Figure 12.10 shows the microstructure of type 304 austenitic stainless steel in which chromium carbides have precipitated on grain

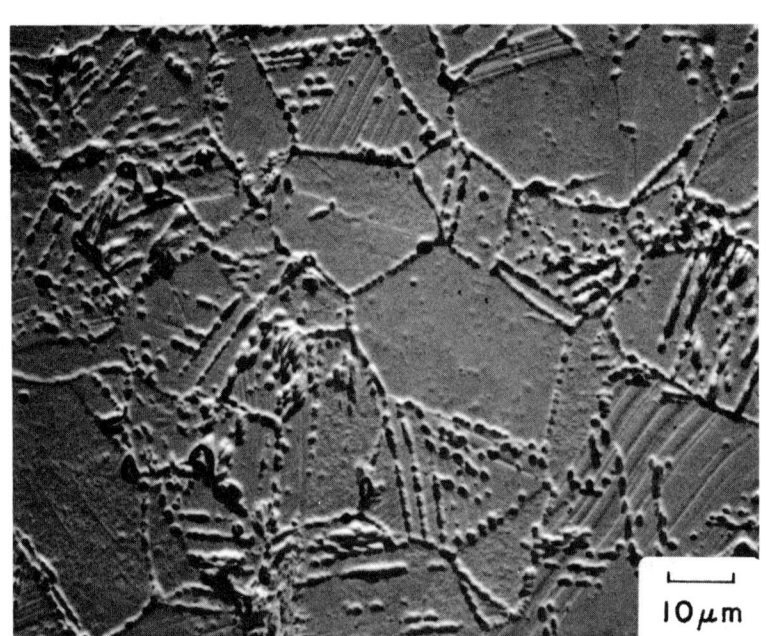

Fig. 12.10. Microstructure of type 304 stainless steel with chromium carbide precipitation on grain boundaries. ASTM A262 Practice A oxalic acid etch. Scanning electron micrograph. Courtesy of G. Vander Voort, Carpenter Technology Corp., Reading, PA

boundaries. In this condition the steel is said to be sensitized and is highly susceptible to catastrophic intergranular corrosion (Ref 12.1, 12.14). In metallographic sections, the chromium carbide precipitation is revealed by deep grain boundary attack by certain etching procedures such as the use of electrolyte oxalic acid etching (Ref 12.12). Generally the carbides are too fine to be resolved by the light microscope, but are indirectly revealed by deep etching of affected grain boundaries. In contrast, as shown in Fig. 12.9, grain boundaries in metallographic specimens of austenitic stainless without chromium carbide precipitation are well defined and not deeply etched.

The susceptibility to severe intergranular corrosive attack is caused by the depletion of chromium due to chromium carbide formation on austenite grain boundaries. The carbides have been identified as $M_{23}C_6$, where M denotes the metal atom content of the carbide, which may include iron and molybdenum as well as chromium. However, the high concentration of chromium in the $M_{23}C_6$ particles

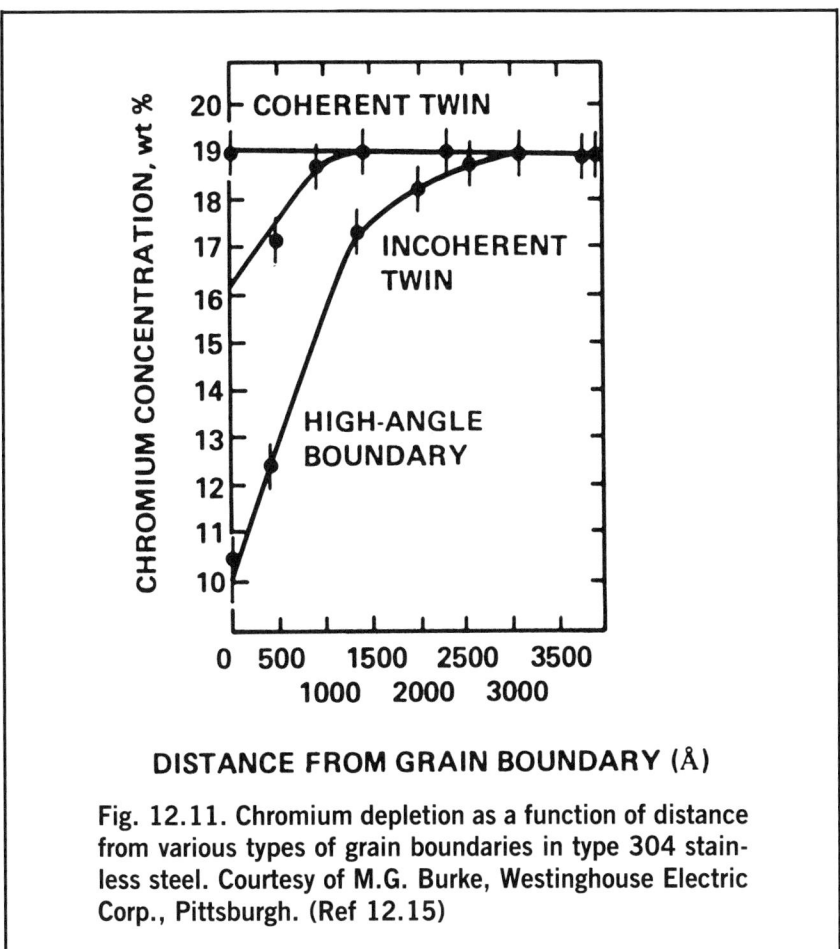

Fig. 12.11. Chromium depletion as a function of distance from various types of grain boundaries in type 304 stainless steel. Courtesy of M.G. Burke, Westinghouse Electric Corp., Pittsburgh. (Ref 12.15)

locally lowers the chromium content of the austenite to below the 12% required for stainless corrosion behavior. Figure 12.11 shows chromium concentration gradients, obtained by analytical transmission electron microscopy, in austenite adjacent to $M_{23}C_6$ particles at twin and high-angle grain boundaries in a 304 stainless steel (Ref 12.15). High-angle grain boundaries are preferred sites for precipitation and diffusion because of the relatively high atomic disorder where grains of different orientations meet. Thus $M_{23}C_6$ particles readily nucleate and grow, severely depleting the adjacent austenite of chromium, from 19 wt.% to about 10 wt.% in the case of the data shown in Fig. 12.11. Twin boundaries have much better atomic matching than most high-angle boundaries and therefore are not as favorable for nucleation and growth of $M_{23}C_6$ particles.

364 / STEELS: HEAT TREATMENT AND PROCESSING PRINCIPLES

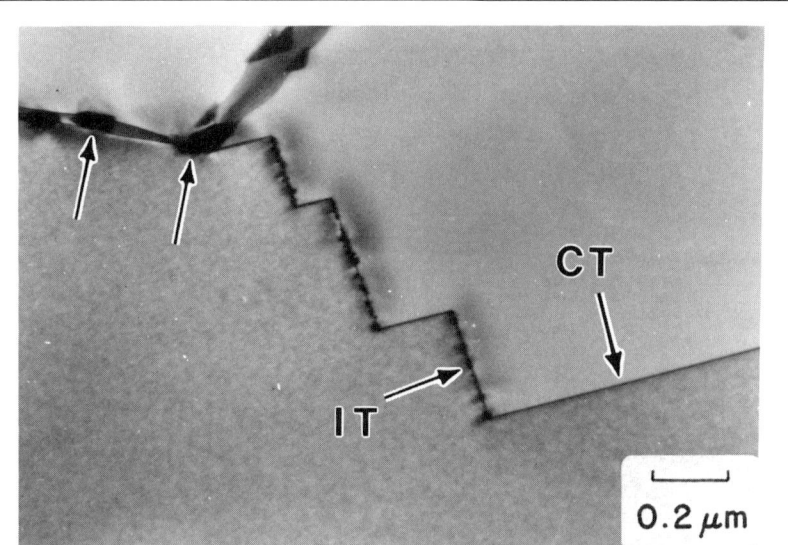

Fig. 12.12. Chromium carbide precipitation on various types of boundaries in type 304 stainless steel. Arrows in upper left point to large carbides on a high-angle grain boundary, and IT and CT refer to incoherent and coherent twin boundaries, respectively. Transmission electron micrograph. Courtesy of M.G. Burke, Westinghouse Electric Corp., Pittsburgh

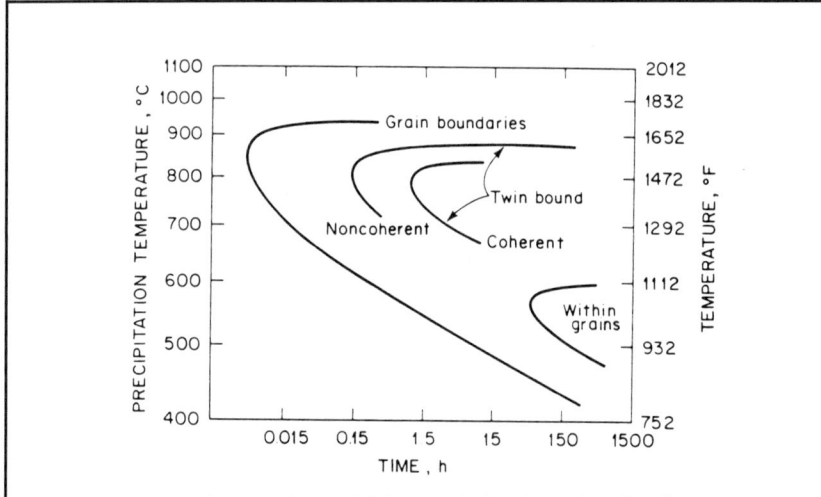

Fig. 12.13. $M_{23}C_6$ carbide precipitation kinetics in type 304 stainless steel containing 0.05% carbon and originally quenched from 1250 °C (2282 °F). (Ref 12.14)

Figure 12.12 shows a TEM micrograph with examples of $M_{23}C_6$ carbides and three types of boundaries in a sensitized 304 stainless steel. In agreement with Fig. 12.11, the carbides are largest on the high-angle grain boundaries (arrows, upper left), quite small on the incoherent twin boundaries (center), and absent on the coherent twin boundaries (normal to the incoherent twin boundaries). In addition to reduced corrosion resistance adjacent to $M_{23}C_6$ precipitation, the lowered chromium content raises the M_s temperatures and may result in martensite formation (Ref 12.15, 12.16).

The catastrophic consequences of intergranular corrosion due to chromium carbide precipitation has led to a number of heat treatment and alloying approaches to minimize or eliminate this problem. One approach is simply to select an extra-low carbon modification of 304 or 316 austenitic stainless steels. These modifications are designated as types 304L and 316L and have an upper limit of 0.03% carbon. Although chromium carbide formation may not be completely suppressed, it is greatly reduced, and the low-carbon grades are adequate for many applications.

Figure 12.13 shows $M_{23}C_6$ precipitation curves for type 304 stainless steel (Ref 12.14). The kinetics show "C" curve behavior with most rapid precipitation occurring between 800 and 900 °C (1472 and 1652 °F). Above 950 °C (1742 °F), chromium and carbon are dissolved as atoms in the crystal structure of austenite and there is no thermodynamic driving force for chromium carbide formation. Below 500 °C (932 °F), the diffusion of chromium atoms required for $M_{23}C_6$ formation is too sluggish and carbide formation essentially stops. Based on the $M_{23}C_6$ precipitation kinetics, wrought austenitic stainless steel products are annealed or solution treated at temperatures between 1040 and 1150 °C (1904 and 2102 °F) and quenched to eliminate sensitization. The solution treatment dissolves the $M_{23}C_6$ carbides, and the rapid cooling prevents the repre- cipitation of $M_{23}C_6$ in the critical temperature range around the nose of the C-curve. This approach is also effective in welded austenitic stainless assemblies where $M_{23}C_6$ carbides may have precipitated in heat-affected zones adjacent to weld metal.

Another approach used to eliminate chromium carbide precipita- tion is to alloy austenitic stainless steels with very strong carbide- forming elements such as titanium, niobium, or tantalum. Such austenitic stainless steels (types 321 and 347, Table 12.1) are referred to as stabilized grades. The alloying additions form carbides such as TiC and NbC and reduce the carbon available for $M_{23}C_6$ precipitation. Stabilizing heat treatments, performed at temperatures between 840 and 900 °C (1544 and 1652 °F), are designed to produce the most effective intragranular dispersion of the alloy carbides (Ref 12.14).

Under most conditions stabilized austenitic stainless steels are effective in reducing chromium carbide formation and intergranular attack. However, the very high temperatures adjacent to welds may cause even TiC and NbC carbides to redissolve and make possible the precipitation of $M_{23}C_6$ if the weldments are held in or slowly cooled through the $M_{23}C_6$ precipitation temperature range. This may lead to the localized corrosion referred to as "knife-line attack," and may be remedied by subjecting weldments to a final stabilizing heat treatment (Ref 12.1, 12.10, 12.14).

Martensite Formation in Austenitic Stainless Steels

Martensite may form in austenitic stainless steels during cooling below room temperature, i.e., thermally, or in response to cold work, i.e., mechanically. Eichelman and Hull (Ref 12.17) have developed the following equation for M_s, the temperature at which martensite first forms on cooling, of austenitic stainless steels:

$$M_s \text{ (°F)} = 75(14.6 - Cr) + 110(8.9 - Ni) + 60(1.33 - Mn) \\ + 50(0.47 - Si) + 3000[0.068 - (C + N)] \quad \text{(Eq 12.1)}$$

This equation shows that the substitutional alloying elements chromium and nickel have a moderate effect on the M_s compared to the very strong effect of carbon and nitrogen. Residual nitrogen contents of austenitic stainless steel are usually in the range of 300 to 700 ppm (0.03 to 0.07 wt.%) (Ref 12.14), and thus when combined with carbon may have a strong effect on stabilizing austenite with respect to martensite formation. When $M_{23}C_6$ carbides form at austenite grain boundaries, both carbon and chromium are removed from the adjacent austenite, M_s is locally raised, and martensite may form at grain boundaries, as mentioned earlier (Ref 12.15, 12.16). This phenomenon is in fact used as one approach to develop martensitic structures in semiaustenitic precipitation-hardening stainless steels, as discussed later.

Two types of martensite form spontaneously on cooling austenitic stainless steels below room temperature: hexagonal close-packed epsilon (ϵ) martensite and body-centered cubic alpha prime (α') martensite. The epsilon martensite forms on close-packed (111) planes in the austenite and, except for size, is morphologically very similar to deformation twins or stacking fault clusters, which also form on (111) planes (Ref 12.18, 12.19). The α' martensite forms as plates with (225)

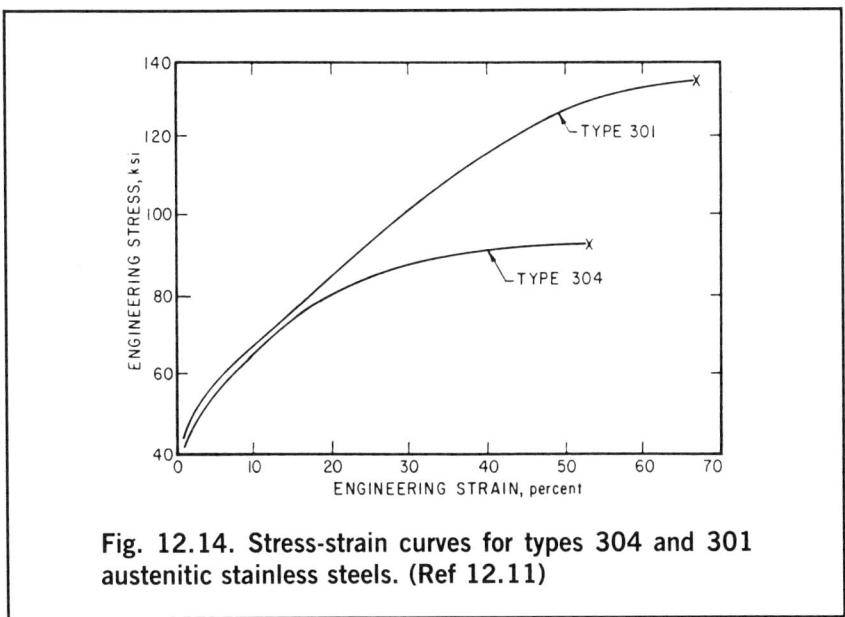

Fig. 12.14. Stress-strain curves for types 304 and 301 austenitic stainless steels. (Ref 12.11)

habit planes in groups bounded by faulted sheets of austenite on (111) planes (Ref 12.18). The nucleation of α' martensite and its relationship to ε martensite has been difficult to resolve; evidence for α' formation

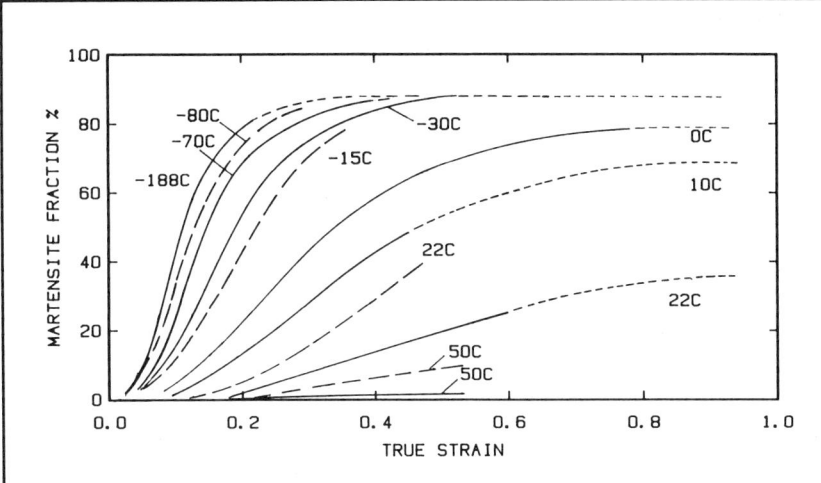

Fig. 12.15. Strain-induced martensite formation as a function of strain at various temperatures (Ref 12.24). Solid lines are original data of Angel (Ref 12.20), dashed lines are data of Hecker *et al.* (Ref 12.25), and dotted extrapolations are from Olson's analysis (Ref 12.26).

directly from austenite and with ϵ as an intermediate phase is reviewed in Ref 12.14.

Deformation-induced or strain-induced martensite formation is another unique feature of austenitic stainless steels. Strain-induced martensite forms at higher temperatures than does martensite, which forms on cooling, and the parameter M_D, the highest temperature at which a designated amount of martensite forms under defined deformation conditions, is used to characterize austenite stability relative to deformation. Angel (Ref 12.20) has published the following correlation of M_D to the composition of austenitic stainless steels:

$$M_{D30}(°C) = 413 - 462(C + N) - 9.2(Si) - 8.1(Mn) - 13.7(Cr) - 9.5(Ni) - 18.5(Mo) \qquad \text{(Eq 12.2)}$$

where M_{D30} is defined as the temperature at which 50% martensite is formed by 30% true strain in tension. Again, carbon and nitrogen have a very strong effect on austenite stability and the extra-low carbon grades such as 304L are quite sensitive to strain-induced martensite formation, a characteristic which may render them susceptible to reduced performance in high-pressure hydrogen (Ref 12.21). Deformation-induced martensite, however, significantly enhances strength generated by cold work, and types 301 and 302 stainless steels are designed to have lower chromium and nickel contents in order to exploit this strengthening mechanism. The effectiveness of this approach is demonstrated in the comparison of the stress-strain curves of types 301 and 304 stainless steels shown in Fig. 12.14 (Ref 12.22). The much more stable type 304 does not strain harden nearly as much as the type 301 stainless steel.

The extent of strain-induced transformation of austenite to martensite is dependent on temperature strain rate, and strain, in addition to composition (Ref 12.23). Figure 12.15 shows the effect of temperature and strain on strain-induced martensite formation in type 304 stainless steels. Large amounts of martensite form at low strains during low-temperature deformation, and the amount of strain-induced transformation becomes negligible above room temperature. Figure 12.16 shows stress-strain curves obtained in constant temperature baths for a type 304 stainless steel (Ref 12.24). The strong effect of strain-induced martensite formation at lower temperatures is marked by noticeable inflections in the stress-strain curves. The strain hardening associated with these inflections produces very high ultimate tensile strengths, and as the strain-induced transformation decreases, the ultimate tensile strength also decreases.

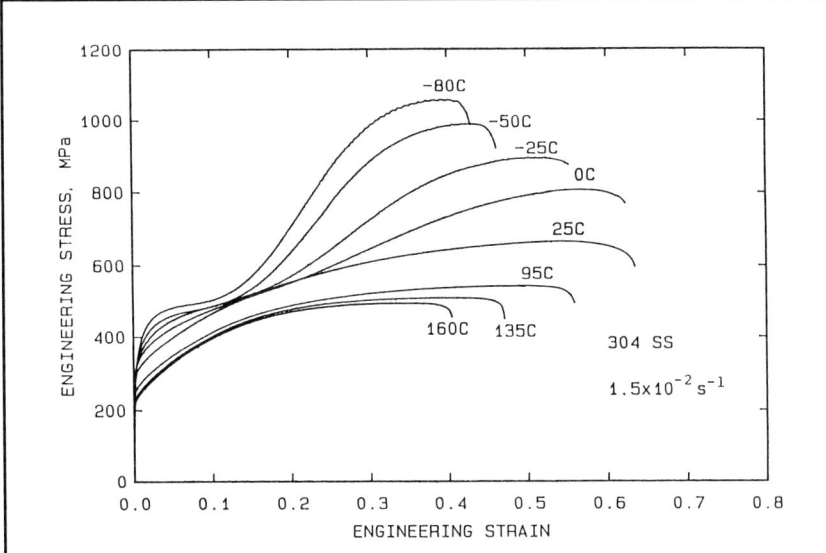

Fig. 12.16. Engineering stress-strain curves for type 304 stainless steels at various temperatures. (Ref 12.24)

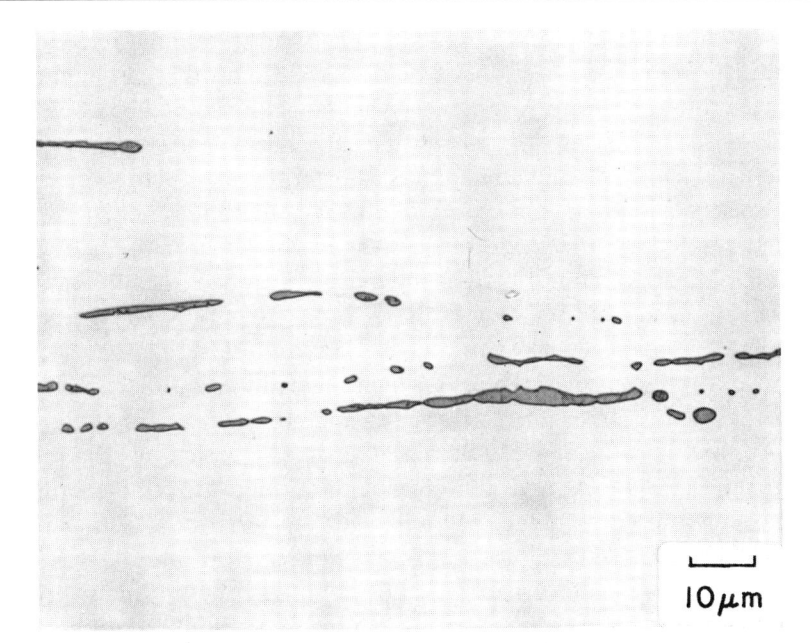

Fig. 12.17. Ferrite in a plate of type 304L stainless steel. Light micrograph. Courtesy of S. Yun, Colorado School of Mines, Golden

Figure 12.16 shows that the total ductility of the type 304 stainless steels goes through a maximum between 0 and 25 °C (32 and 77 °F). During deformation in this temperature range, the strain-induced martensite transformation is delayed to high strains, where the associated strain hardening is useful in delaying necking and increasing post-uniform elongation. Effectively, as soon as necking starts, the additional strain causes that area to locally transform and harden, retarding necking and causing deformation to be displaced to lower strength regions of the specimen. Above room temperature or as a result of specimen temperature increases associated with deformation heating and necking at room temperature, strain-induced martensite transformation and the associated strain hardening become negligible, even at high strains, and ductility decreases.

Other Phases in Austenitic Stainless Steels

Ideally, austenitic stainless steels have microstructures consisting only of polycrystalline austenite. However, because of segregation during solidification, ferrite tends to form and is commonly found in austenitic stainless steel castings and welds, as discussed earlier. Wrought austenitic stainless steels are homogenized by hot work, and smaller section sizes and sheet usually show uniform austenitic structures. However, heavier plates and forgings, which receive less hot work, frequently show some ferrite within the austenitic microstructure (Ref 12.27). Figure 12.17 shows delta ferrite in a 304L stainless steel. The ferritic areas have been flattened and elongated as a result of plate rolling.

The presence of ferrite in austenitic stainless steels may lead to the formation of sigma (σ) phase, which may adversely affect ductility, toughness, and corrosion resistance of austenitic stainless steels (Ref 12.14). Sigma phase dominates the central portion of the Fe-Cr diagram (Fig. 12.1), and therefore is an important factor in the processing and performance of highly alloyed ferritic stainless steels. As shown by the Fe-Cr phase diagram, sigma phase is an intermetallic phase with composition centered about equal amounts of iron and chromium. The crystal structure is complex, body-centered tetragonal, with 30 atoms per unit cell, and may contain other elements such as molybdenum.

In austenitic stainless steels containing ferrite, sigma phase forms in ferritic areas because chromium, which is a major component of sigma, is already concentrated in the ferrite as a result of partitioning during solidification. The transformation of ferrite to sigma is sluggish and depends on alloy composition. Therefore, sigma phase is often found in austenitic stainless steel components which have been subjected to long-time service at temperatures in the range of 500 to 700 °C (932 to 1292 °F) (Ref 12.14).

Sigma phase tends to nucleate and grow preferentially at ferrite-austenite interfaces, but intragranular sigma formation has also been observed in Type 321 stainless steel held 17 years at around 600 °C (1112 °F) (Ref 12.28). In austenitic stainless steels subjected to 10,000 h aging at temperatures around 600 °C (1112 °F), Gray et al. (Ref 12.29) have found that austenite formation accompanies the formation of sigma from delta ferrite. Thus the partitioning of alloying elements for sigma formation appears to be accomplished by the solid-state reaction $\delta \rightarrow \sigma + \gamma$. Sigma phase formation from delta ferrite can be accelerated by strain, as documented by a forging study of austenitic 21-6-9 stainless steel (Ref 12.27).

A number of phases other than ferrite, sigma, and $M_{23}C_6$ may form in austenitic stainless steels. These phases include various alloy carbides and nitrides and intermetallic phases such as Laves and chi. The formation of these phases is alloy specific and dependent on processing and service conditions. The reader is referred to the article by Novak (Ref 12.14) for a comprehensive review of the literature regarding the formation of such phases in austenitic stainless steels.

Other Austenitic Stainless Steels

In addition to the wrought austenitic stainless steel containing roughly 18% chromium and 8% nickel, several other groups of austenitic stainless steels are available for specific applications or processing requirements. Each of the wrought austenitic stainless steels has a counterpart cast alloy with a specific cast alloy designation (Ref 12.30). For example, CF-3, CF-8, CF-3M, and CF-8M correspond to the wrought types 304L, 304, 316L, and 316, respectively. The cast austenitic stainless steels are designed for good castability, and therefore the composition ranges may vary from those of their counterpart wrought steels. In particular, the chromium and silicon contents are higher and the nickel contents lower in cast alloys compared to wrought alloys. As a result, delta ferrite, which reduces hot cracking as discussed earlier, is usually found in cast austenitic stainless steels.

Many heat-resisting grades of stainless steel have austenitic structures. The heat-resisting grades have much higher chromium

Table 12.2. Compositions of AISI Type 200 Austenitic Stainless Steels. (Ref 12.9)

AISI type no.	Nominal composition, %				
	C	Mn	Cr	Ni	Others
201	0.15 max	7.5	16–18	3.5–5.5	0.25N max
202	0.15 max	10.0	17–19	4.0–6.0	0.25N max
Face-centered cubic, nonmagnetic, not heat treatable					

and nickel contents for scaling resistance and high-temperature strength compared to the 18Cr-8Ni types of stainless steel. Again there are counterpart wrought and cast grades of heat-resisting stainless steels (for example, types 309 and 310, and HH and HK, respectively). There are, however, many other cast grades of heat-resistant alloys, and these alloys have much higher carbon contents (0.20 to 0.75%) than do the wrought grades (Ref 12.30). Thus alloy carbides, which contribute substantially to creep resistance, are an important component of the microstructure of the cast austenitic high-temperature alloys. The heat-resistant austenitic stainless steels are used at temperatures as high as 1100 °C (2012 °F), sometimes in very aggressive gaseous environments, and are expected to provide many years of service. Thus temperature-induced microstructural changes, creep-rupture mechanisms, scaling and oxidation, carburization, decarburization, and sulfidation are critical phenomena which affect selection and performance of heat-resistant austenitic stainless steels (Ref 12.31, 12.32).

Other groups of austenitic stainless steels include those in which substitutions for nickel are made. Type 200 austenitic stainless steels (Table 12.2) are alloyed with manganese and nitrogen, both austenite-stabilizing elements, to replace nickel. The 200 series austenitic stainless steels have properties and work hardening characteristics similar to types 301 and 302 steels. Higher strength austenitic stainless steels with high manganese and nitrogen and reduced nickel contents (Table 12.3) have also been developed (Ref 12.33). Several of these steels have the trademark Nitronic, and are sometimes referred to by their composition in nominal amounts of chromium, nickel, and manganese. The yield strengths of these steels range from 50 to 70 ksi (345 to 480 MPa), significantly above those attainable in annealed type 300 stainless steels. An even newer type of austenitic stainless steel, "super nitrogen" stainless steel, contains up to 1 wt.% nitrogen (Ref 12.34). This level of nitrogen exceeds the atmospheric solubility of nitrogen in austenite, and is made possible by pressurized-electroslag-

Table 12.3. Compositions of High-Strength Manganese Austenitic Stainless Steels. (Ref 12.33)

Trade designation	Typical composition, %	Tensile strength, MPa (ksi)	Yield strength, MPa (ksi)(a)	Elongation in 50 mm (2 in.), %
Nitronic 32(b)	0.10 C, 12.0 Mn, 18.0 Cr, 1.6 Ni, 0.32 N	690 (100)	380 (55)	30
Nitronic 40	0.03 C, 9.0 Mn, 21.0 Cr, 7 Ni, 0.03 N	550 (80)	345 (50)	45
Nitronic 50	0.04 C, 5.0 Mn, 21.2 Cr, 12.5 Ni, 0.30 N, 2.50 Mo, 0.20 Cb, 0.20 V	690 (100)	380 (55)	35
Nitronic 60	0.07 C, 8.0 Mn, 17.0 Cr, 8.5 Ni, 0.14 N, 4.0 Si	655 (95)	345 (50)	35
Tenelon	0.08 C, 15.0 Mn, 18.0 Cr, 0.75 Ni, 0.35 N	860 (125)	480 (70)	40
Type 216	0.04 C, 8.0 Mn, 21.0 Cr, 6.0 Ni, 0.27 N, 2.3 Mo	690 (100)	415 (60)	40

(a) At 0.2% offset. (b) Nitronic is a trademark of Armco, Inc., Middletown, OH.

remelting. Even higher strengths at good levels of toughness appear to be attainable with these ultrahigh-nitrogen steels.

Heat Treatment of Austenitic Stainless Steels

The above review of the alloying and physical metallurgy of austenitic stainless steels shows that they cannot be strengthened by heat treatment such as quenching to form martensite or by precipitation hardening. Strengthening must be accomplished by alloying, in particular by solid solution strengthening, and by cold working. Strengthening by cold working may involve strain-induced martensite formation.

The heat treatments applied to austenite stainless steels, therefore, include annealing, treatments to prevent chromium carbide precipitation, and stress relief (Ref 12.10, 12.11, 12.14). Since austenite stainless steels are very ductile, they are readily wrought to thin sheet or fine-diameter tubing and wire by sequential cold working and annealing cycles. The annealing causes recrystallization of the strain-hardened microstructure and restoration of ductility for subsequent working operations. Heat treatments to prevent sensitization may include solution treatments to dissolve chromium carbides or stabilization treatments to cause the precipitation of alloy carbides and

Table 12.4. Compositions of AISI Type 400 Ferritic Stainless Steels. (Ref 12.9)

AISI type no.	Nominal composition, %			
	C	Mn	Cr	Others
430	0.08 max	1.0	16.0–18.0	
430F	0.12 max	1.25	16.0–18.0	0.6Mo max
430F Se	0.12 max	1.25	16.0–18.0	0.15Se min
446	0.20 max	1.5	23.0–27.0	0.25N max

Body-centered cubic, magnetic, not heat treatable.

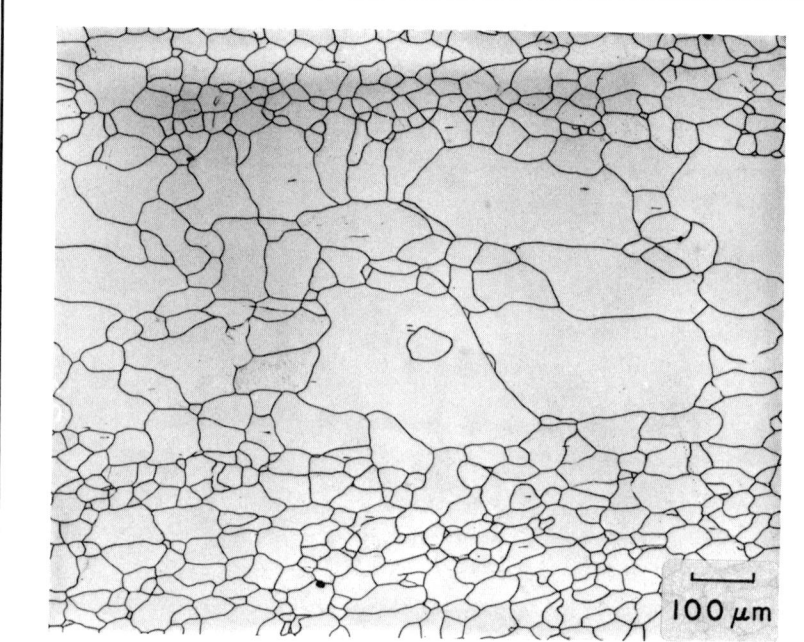

Fig. 12.18. Microstructure of annealed ferritic stainless steel (E-Brite 26-1 containing 26% chromium and 1% molybdenum). Etched electrolytically in 60% HNO_3-H_2O. Light micrograph. Courtesy of G. Vander Voort, Carpenter Technology Corp., Reading, PA

thereby a reduction of the carbon available for chromium carbide precipitation. Finally, stress relief treatments are applied to weldments, but care must be taken not to stress relieve sensitive alloys in the chromium carbide precipitation temperature range (Ref 12.14).

Ferritic Stainless Steels

Table 12.4 lists the compositions of some common type 400 ferritic stainless steels. Chromium, in amounts sufficient to completely stabilize ferrite (Fig. 12.1), is the major alloying element. Carbon is restricted both to maintain high toughness and ductility and to prevent austenite formation related to the expansion of the gamma loop by carbon. Figure 12.18 shows the polycrystalline, single-phase microstructure of an annealed ferritic stainless steel. Annealing for recrystallization of cold worked structures is performed in the temperature range 760 to 966 °C (1400 to 1750 °F). Rapid cooling after annealing is required for the more highly alloyed ferritic grades in order to prevent the formation of phases detrimental to ductility and toughness.

The ferritic stainless steels, similar to the austenitic stainless steels, cannot be strengthened by heat treatment. Also, since the strain hardening rates of ferrite are relatively low and since cold work significantly lowers ductility, the ferritic stainless steels are not often strengthened by cold work (Ref 12.33). Typical annealed yield and tensile strengths for the steels listed in Table 12.4 are 35 to 55 ksi (240 to 380 MPa) and 60 to 85 ksi (415 to 585 MPa), respectively. Ductilities tend to range between 20 and 35%. Higher strengths, up to 75 ksi (515 MPa) yield and 95 ksi (655 MPa) tensile, are obtained in the more highly alloyed ferritic steels such as 29Cr-4Mo-2Ni and 27Cr-3.5Mo-1.2Ni.

The ductility and toughness of ferritic stainless steels are affected by many factors (Ref 12.35). Fundamentally, the strength and ability of the body-centered cubic ferrite structure to sustain plastic deformation are very temperature dependent, especially below room temperature. Strength increases rapidly and ductility drops sharply with decreasing temperature, apparently because screw dislocations lose their ability to cross slip in the bcc structure (Ref 12.36). As a result, ferritic steels undergo a transition from ductile fracture, characterized by microvoid coalescence, to brittle fracture, characterized by cleavage. The temperature at which this fracture transition occurs is referred to as the ductile-to-brittle transition temperature (DBTT), and the cleavage fracture may be initiated by intergranular cracking or strain-induced cracking of second-phase particles (Ref 12.37, 12.38). In contrast, austenitic stainless steels do not undergo a ductile to brittle transition, and maintain good ductility and toughness to temperatures well below room temperature. In ferritic stainless steels the DBTT may be well above room temperature. Figure 12.19 shows the DBTT as a function of section thickness for several ferritic

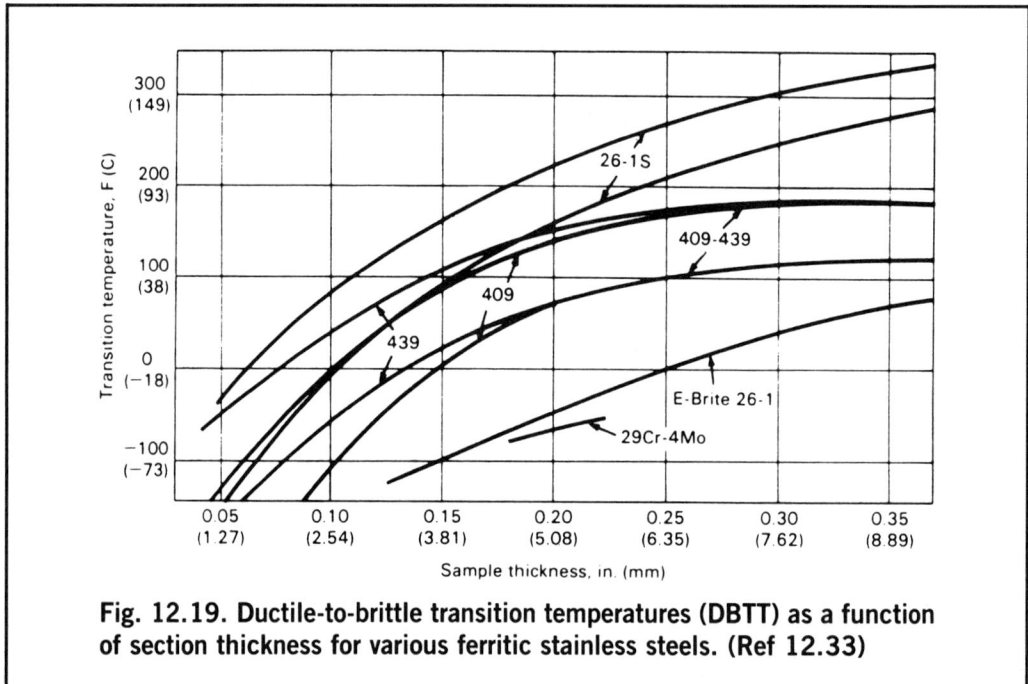

Fig. 12.19. Ductile-to-brittle transition temperatures (DBTT) as a function of section thickness for various ferritic stainless steels. (Ref 12.33)

stainless steels (Ref 12.33). The thicker sections offer more constraint to plastic flow, and consequently brittle fracture occurs without exception above room temperature as section size increases. In contrast, thin sheets in which yielding can take place through the thickness remain ductile and highly formable well below room temperature.

Other factors which influence the DBTT of ferritic stainless steels are grain size, interstitial carbon and nitrogen content, and the presence of various types of second phases. Thus, fine grain size, low interstitial element contents, and the elimination of second phases by

Table 12.5. Characteristics of Intermetallic Phases in Ferritic Stainless Steels. (Ref 12.41)

Type of phase	Structure	Lattice parameters	Comments
Sigma (σ)	Body-centered tetragonal $D8_b$, 30 atoms/cell	$a = 0.88–0.91$ nm $c = 0.45–0.46$ nm	Parameters of phases collected from superalloys
Chi (χ)	Cubic A12 58 atoms/cell	$a = 0.884–0.893$ nm	From various steels
Laves	Hexagonal C14 or C36	$a = 0.475–0.495$ nm $c = 0.770–0.815$ nm	Parameters of phases collected from superalloys

proper heat treatment all enhance ductility and toughness (Ref 12.33, 12.35). Improved melting practices, including argon-oxygen-decarburization (AOD) and vacuum melting, and stabilization by additions of titanium or niobium have been extremely important approaches to lowering carbon and nitrogen contents and associated carbide and nitride precipitates detrimental to toughness of ferritic stainless steels (Ref 12.39).

Intermetallic Phases in Ferritic Stainless Steels

Ferritic stainless steels are highly alloyed and may form a number of brittle intermetallic phases when exposed to operating temperatures or processing conditions, such as slow cooling of heavy sections, in the temperature range between 500 and 1000 °C (932 and 1832 °F). Prominent among these phases is sigma phase, which dominates the central portion of the Fe-Cr phase diagram (Fig. 12.1). The various intermetallic phases form by the arrangement of iron, chromium, molybdenum, and other transition metal atoms into crystal structures which accommodate atomic size and electronic differences that limit the low-temperature solid solubility of alloying elements in the bcc ferritic structure (Ref 12.40). The crystal structures tend to be complex, and the phases are characterized by large unit cells containing many atoms. Table 12.5 lists characteristics of the sigma, chi, and Laves phases which may form in ferritic stainless steels.

Examples of intermetallic phases which have formed at 850 °C (1562 °F) in a 25Cr-3Mo-4Ni ferritic stainless steel are shown in Fig. 12.20. The various phases have formed at grain boundaries and within grains, and sigma phase dominates the microstructure after holding 300 min at 850 °C (1562 °F). A special consequence of the nickel content of this ferritic stainless steel is austenite formation adjacent to sigma phase. Apparently, the depletion of the chromium and molybdenum and concentration of nickel in areas adjacent to the sigma phase decrease the stability of the ferrite phase to where it is replaced by austenite.

The formation of the intermetallic phases follows "C" curve kinetics which are influenced by alloy composition. Figure 12.21 shows such curves for the 25Cr-3Mo-4Ni ferritic stainless steel (Ref 12.41). Leaner alloys would tend to have longer incubation times for the formation of the intermetallic phases. The "C" curves are useful in that they define temperature ranges which can be used to dissolve the

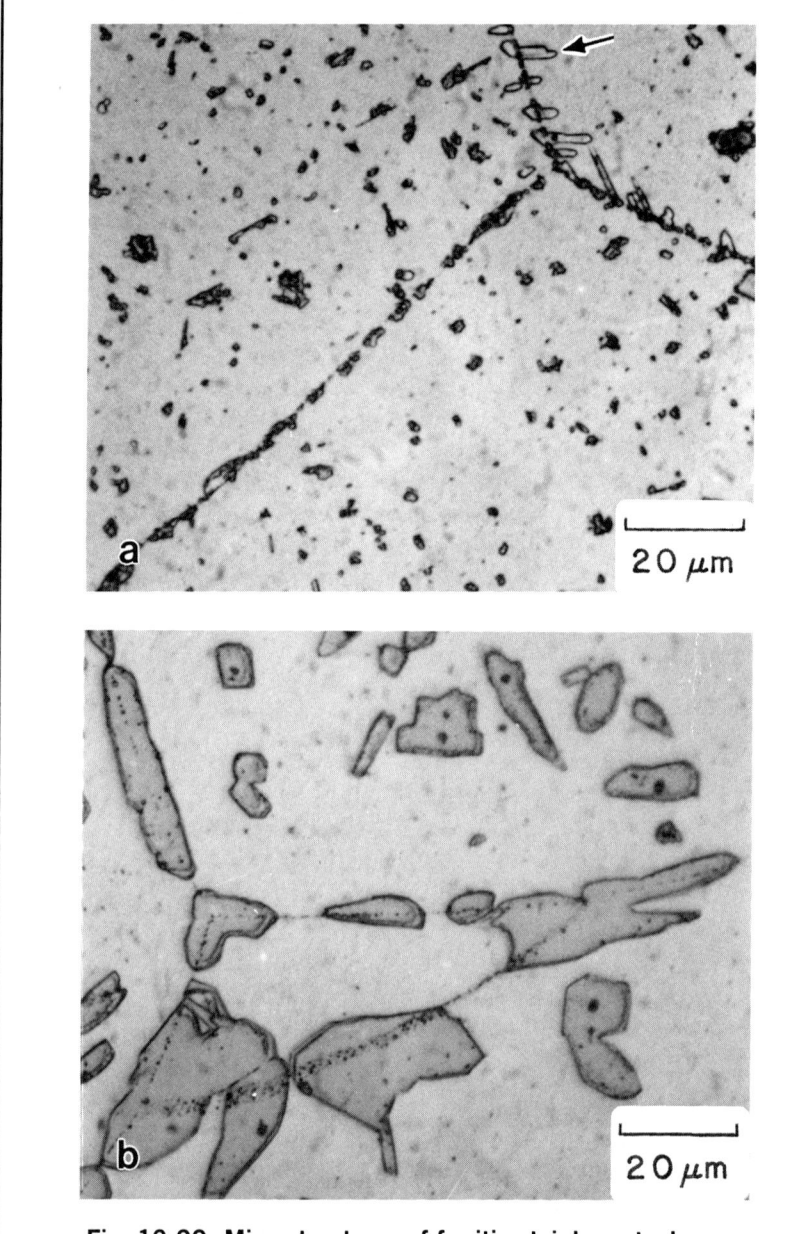

Fig. 12.20. Microstructures of ferritic stainless steel containing 24.5% chromium, 3.54% molybdenum, 3.90% nickel, 0.17% niobium, and 0.32% aluminum annealed at 850 °C (1562 °F). (a) Annealed 100 min. Arrow points to chi phase. (b) Annealed 300 min. Dominant second phase (etched gray) is sigma. Light micrographs. (Ref 12.41)

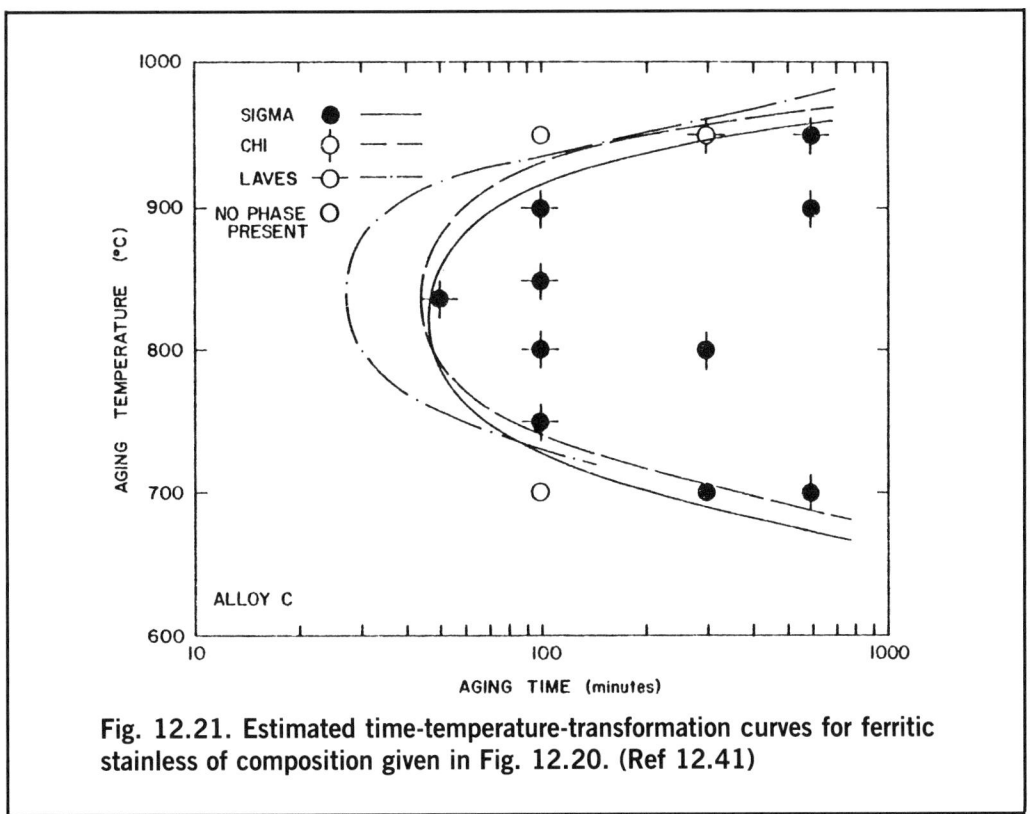

Fig. 12.21. Estimated time-temperature-transformation curves for ferritic stainless of composition given in Fig. 12.20. (Ref 12.41)

intermetallic phases and through which specimens must be rapidly cooled to avoid reprecipitation of the phases. The "C" curves also identify operating temperatures which should be avoided for application of ferritic stainless steels.

475 °C (885 °F) Embrittlement in Ferritic Stainless Steels

In addition to the high-temperature embrittlement phenomena related to the intermetallic phases described above, high-chromium ferritic stainless steels also may undergo a lower temperature embrittlement. This phenomenon, termed 475 °C or 885 °F embrittlement, develops in the temperature range 400 to 550 °C (752 to 1022 °F), and its causes were identified only with difficulty because light microscopy and x-ray diffraction showed no evidence of any structural changes. The embrittlement is associated with very fine-scale precipitation, resolvable only by transmission electron microscopy, and the precipi-

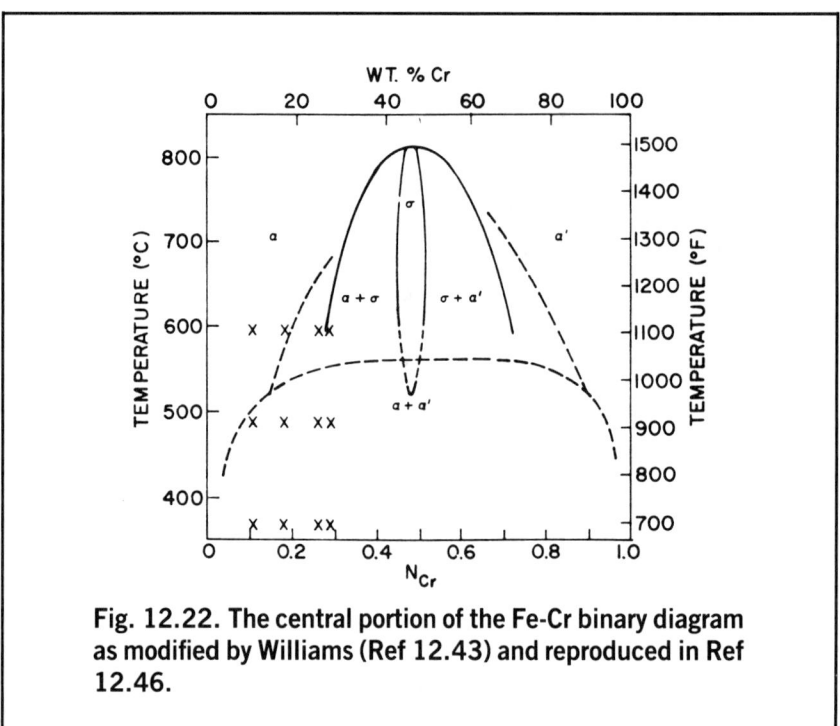

Fig. 12.22. The central portion of the Fe-Cr binary diagram as modified by Williams (Ref 12.43) and reproduced in Ref 12.46.

tating phase, designated α, has a bcc structure with almost the same lattice parameter as the parent bcc ferrite (Ref 12.42).

The discovery of the α' phase necessitated a modification of the central portion of the Fe-Cr phase diagram. Williams (Ref 12.43) proposed the modification shown in Fig. 12.22. The dashed lines identify a miscibility gap in the bcc ferritic solid solution. At high temperatures chromium and iron atoms are randomly distributed on the bcc crystal lattice, but at compositions and temperatures below the dashed lines the iron and chromium atoms tend to separate and cluster. Therefore, subject to the constraint of diffusion, closely spaced areas become either rich in iron atoms (these areas remain the matrix α phase) or rich in chromium (these areas become the α' phase). The α' may form by nucleation and growth of discrete particles, or it may develop by a mechanism referred to as spinodal decomposition, which produces fine clusters of iron and chromium without well-defined interfaces. Regardless of the mechanism of formation, the α' phase is quite fine, on the order of a few tens of nanometers in size (Ref 12.42-12.45).

The formation of α' phase at temperatures around 475 °C (885 °F) causes significant changes in mechanical properties of ferritic stainless

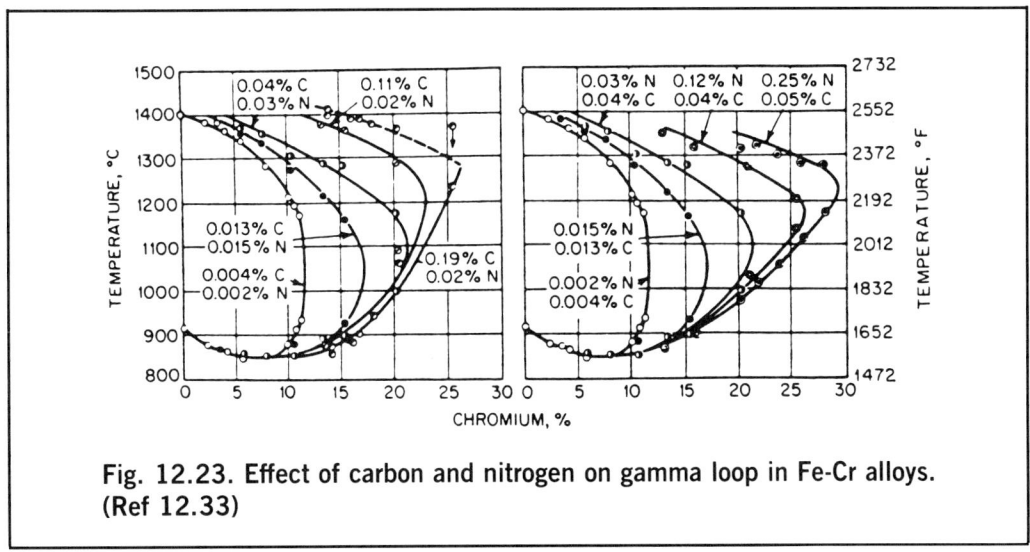

Fig. 12.23. Effect of carbon and nitrogen on gamma loop in Fe-Cr alloys. (Ref 12.33)

steels. Most striking is the dramatic increase in the ductile-to-brittle transition temperature as identified by the change in fracture appearance (FATT) (Ref 12.46). Hardness and yield strength also show significant increases and tensile elongation shows significant decreases as the α' precipitation develops. These property changes are accompanied by changes in deformation from that characterized by uniform dislocation cell formation and cross slip to that characterized by planar dislocation arrays. The study by Nichol et al. (Ref 12.46) showed that high-purity, high-chromium ferritic stainless steel (29Cr-4Mo-2Ni) was the most sensitive, stabilized grades less sensitive, and a titanium-stabilized 11% chromium almost immune to 475 °C (885 °F) embrittlement.

Table 12.6. Compositions of AISI Type 400 Martensitic Stainless Steels. (Ref 12.9)

AISI type no.	C	Mn	Cr	Ni	Others
403	0.15 max	1.0	11.5–13		
410	0.15 max	1.0	11.5–13		
416	0.15 max	1.2	12–14	...	0.15S min
420	0.15 min	1.0	12–14		
431	0.20 max	1.0	15–17	1.2–2.5	
440A	0.60–0.75	1.0	16–18	...	0.75Mo max
440B	0.75–0.95	1.0	16–18	...	0.75Mo max
440C	0.95–1.20	1.0	16–18	...	0.75Mo max

Body-centered cubic, magnetic, heat treatable

Martensitic Stainless Steels

Martensitic stainless steels can be forged and then heat treated by austenitizing, martensite formation, and tempering for many applications which require not only corrosion resistance but also good edge retention, high strength, high hardness, and wear resistance. This processing approach is made possible by balancing chromium content between that required for stainless corrosion properties and that required to ensure full transformation to austenite within the gamma loop on heating. The gamma loop of the Fe-Cr system is expanded by carbon and nitrogen, both austenite-stabilizing elements, as shown in Fig. 12.23. Thus higher carbon and nitrogen contents make possible higher chromium contents in martensitic stainless steels.

Table 12.6 (Ref 12.9) lists the compositions of a number of commonly used AISI 400 grades of martensitic stainless steels. As discussed above, carbon and chromium are balanced to ensure that full austenitization can be achieved. Several of the grades have low carbon content and therefore are limited to a maximum hardness of about 45 HRC. Types 403 and 410 are comparable, except that type 403 is a special-quality grade for applications such as turbine blades. Higher hardness, to 60 HRC, is attainable in the type 440 grades, which have significantly higher specified carbon ranges. More chromium is required in the higher carbon martensitic stainless steels to offset the chromium tied up in chromium carbide particles. Because of the high chromium content, all of the martensitic stainless steels have good hardenability and may be oil quenched or air cooled for hardening.

The martensitic stainless steels are process annealed to microstructures of ferrite and spheroidized carbides for maximum ductility and machinability. Annealing is accomplished by subcritical heating at temperatures of 650 to 760 °C (1202 to 1400 °F) or by heating to higher temperatures and slow cooling (Ref 12.47). Figure 12.24 shows spherical carbides dispersed in ferrite in annealed type 403 and 416 martensitic stainless steels. The microstructure of the annealed type 416 steel, which contains a deliberate addition of sulfur, also contains a high density of coarse MnS particles, which are introduced to improve machinability.

Martensitic stainless steels are hardened by austenitizing to between 925 and 1065 °C (1697 and 1949 °F) and oil quenching or air cooling. The austenitizing temperature selected depends on the degree of carbide solution desired and the necessity to avoid delta ferrite formation by overheating. Higher austenitizing temperatures result in greater carbide dissolution and better corrosion resistance and

STAINLESS STEELS / 383

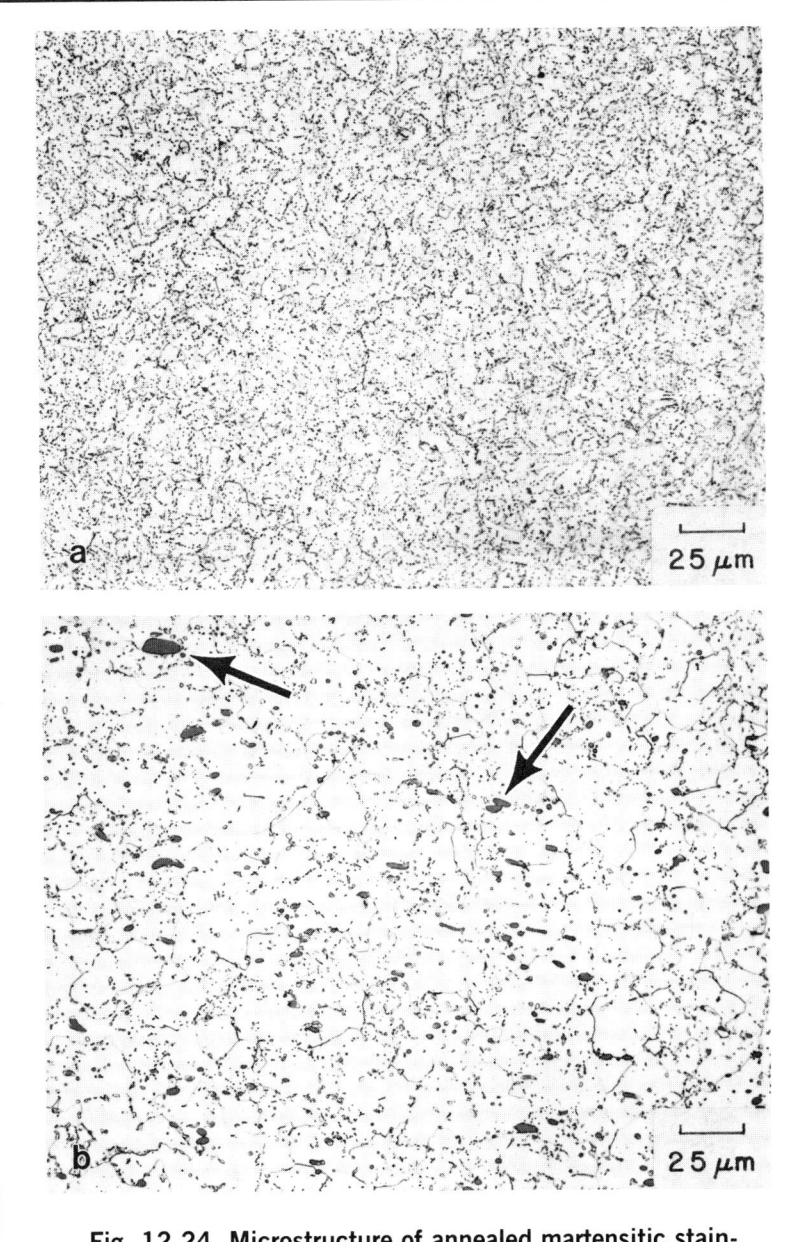

Fig. 12.24. Microstructure of annealed martensitic stainless steels. Fine particles are spheroidized carbides. (a) Type 403 stainless steel etched in 4% picral-HCl. (b) Type 416 stainless steel etched with Vilella's reagent. Arrows point to sulfide particles for machinability. Light micrographs. Courtesy of G. Vander Voort, Carpenter Technology Corp., Reading, PA

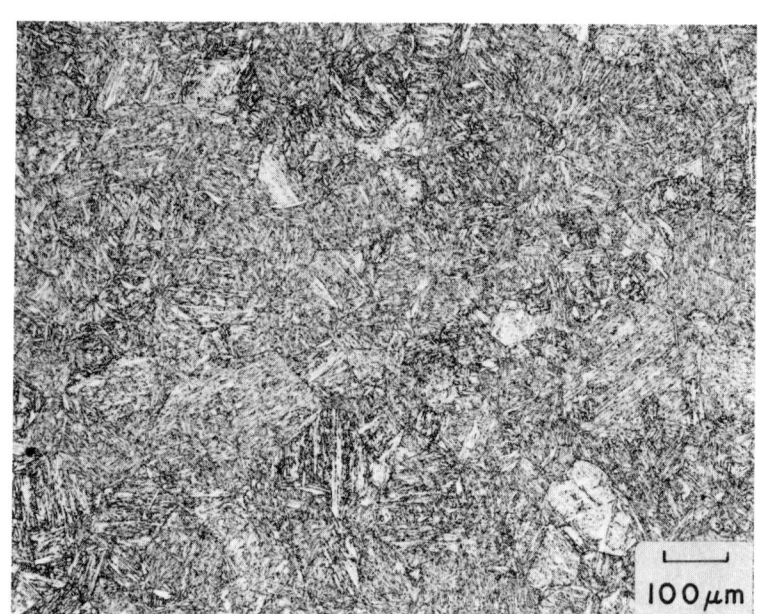

Fig. 12.25. Lath martensite microstructure of hardened type 403 stainless steel. 4% picral-HCl etch. Light micrograph. Courtesy of G. Vander Voort, Carpenter Technology Corp., Reading, PA

strength (Ref 12.47). The thermal conductivity of the chromium-containing martensitic stainless steels is considerably lower than that of carbon steels, and therefore preheating of parts with complex shapes or section changes between 760 and 790 °C (1400 and 1454 °F) may be desirable to equalize temperature and thereby minimize distortion or cracking on heating to the final austenitizing temperature.

Figure 12.25 shows the martensitic structure of hardened type 403 martensitic stainless steel. The microstructure consists entirely of lath martensite. Alloys with higher chromium and carbon contents, such as the type 440 stainless steels, may contain significant volume fractions of alloy carbides and retained austenite after hardening. The mechanical property changes produced by tempering an oil-quenched type 410 martensitic stainless steel are shown in Fig. 12.26 (Ref 12.33). As-quenched hardness and strength are maintained well after tempering up to 450 °C (842 °F) and then drop rapidly. Higher austenitizing temperatures dissolve more chromium, and therefore the secondary hardening peak is sharpened in the specimens austenitized at 1010 °C

STAINLESS STEELS / 385

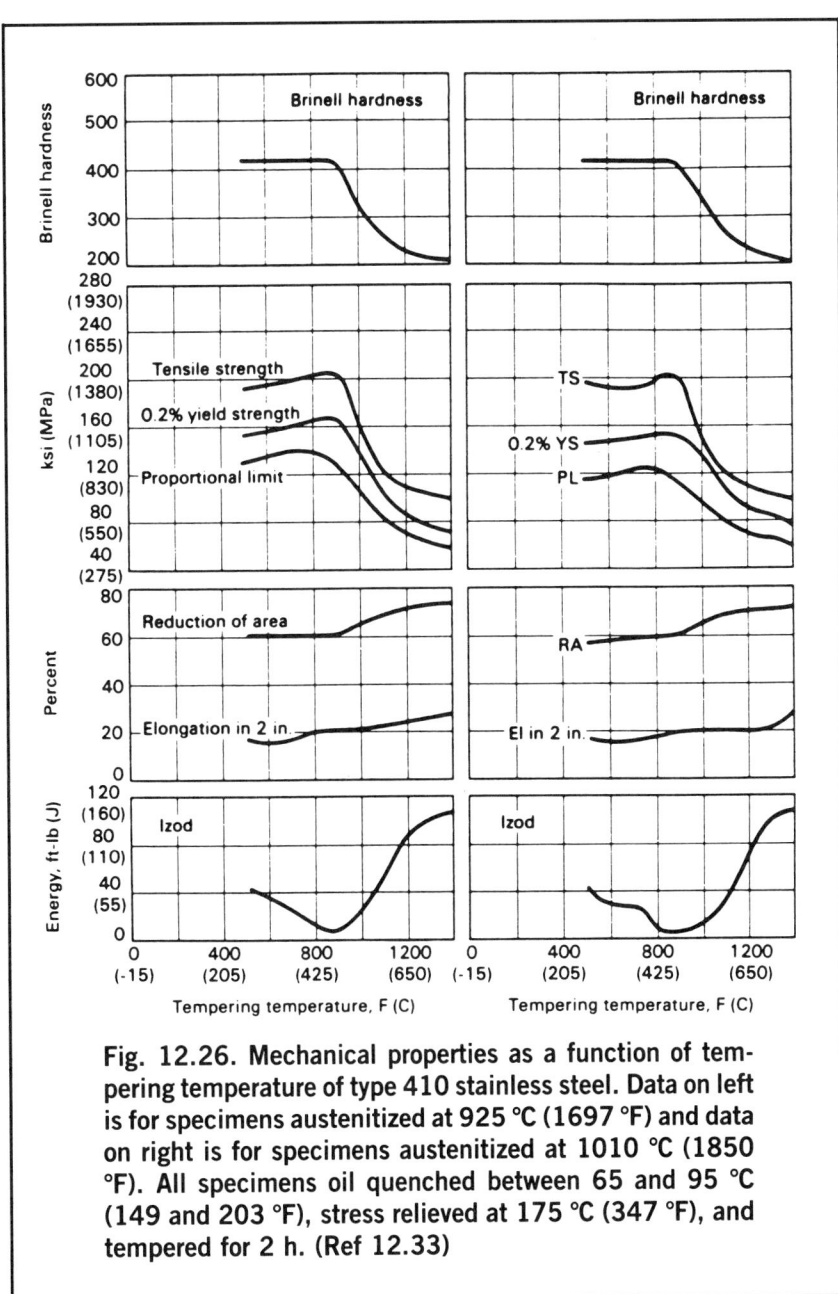

Fig. 12.26. Mechanical properties as a function of tempering temperature of type 410 stainless steel. Data on left is for specimens austenitized at 925 °C (1697 °F) and data on right is for specimens austenitized at 1010 °C (1850 °F). All specimens oil quenched between 65 and 95 °C (149 and 203 °F), stress relieved at 175 °C (347 °F), and tempered for 2 h. (Ref 12.33)

(1850 °F). Temper embrittlement develops during tempering between 425 and 565 °C (797 and 1049 °F) (Ref 12.33), as shown by the minimum in Izod impact toughness, and tempering in this temperature range should be avoided for impact-sensitive applications.

Table 12.7. Compositions of Selected Precipitation-Hardening Stainless Steels. (Ref 12.33, 12.48)

AISI type no.	Trade name/producer	C	Cr	Ni	Mo	Cu	Al	Ti	Other
Martensitic grades									
635	Stainless W (U.S. Steel)	0.06	16.75	6.25			0.2	0.8	
630	17.4 PH (Armco Steel)	0.04	16.0	4.2		3.4			0.25Cb
...	PH 13–8 Mo (Armco Steel)	0.04	12.7	8.2	2.2		1.1		
Semiaustenitic grades									
631	17.7 PH (Armco Steel)	0.07	17.0	7.1			1.2		
633	AM 350 (Allegheny Ludlum)	0.05	16.5	4.25	2.75				0.10N
Austenitic grades									
600	A-286 (Allegheny Ludlum)	0.05	14.75	25.25	1.30		0.15	2.15	0.30V, 0.005B

Precipitation-Hardening Stainless Steels

Precipitation-hardening stainless steels have been developed to provide high strength and toughness while maintaining the corrosion resistance of stainless steels. This alloy category was necessitated by limits to strengthening austenitic and ferritic stainless steels by solid-solution and strain hardening, and the limited ductility and toughness of the carbon-strengthened martensite of high-hardness martensitic stainless steels. Strengthening in precipitation-hardening stainless steels is accomplished by the precipitation of intermetallic compounds such as Ni_3Al in austenitic or ductile low-carbon martensitic matrices.

The alloying elements in precipitation-hardening stainless steels may be balanced to produce martensitic structures at room temperature, metastable austenite which can be converted readily to martensite, or completely stable austenite. Thus alloy design makes possible three classes of precipitation-hardening stainless steels: martensitic, semiaustenitic, and austenitic. The aging or precipitation-hardening treatments applied to all classes are similar and are accomplished at relatively low temperatures, around 500 °C (930 °F) for the martensitic and semiaustenitic grades and around 700 °C (1290 °F) for the

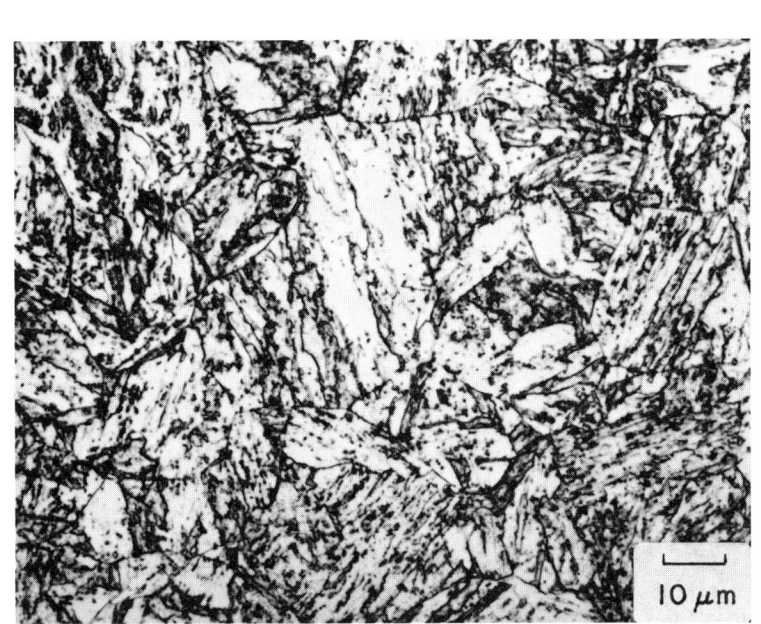

Fig. 12.27. Solution-treated and aged microstructure of martensitic precipitation-hardening stainless steel PH 13-8 Mo. Etched in Fry's reagent. Light micrograph. Courtesy of G. Vander Voort, Carpenter Technology Corp., Reading, PA

austenitic grades. Table 12.7 lists representative precipitation-hardening stainless steels in each of the three classes. The steels are given the AISI type 600 category, but are most commonly referred to by their trade names. Many more grades of precipitation-hardening stainless steels and specific processing information are presented in other references (Ref 12.33, 12.47, 12.48).

Martensitic precipitation-hardening stainless steels were first developed by Smith, Wyche, and Gore (Ref 12.49), and tend to be characterized by low carbon, low nickel, and stabilizing additions which further lower carbon in solution. Thus the austenite has relatively low stability and transforms to low-carbon martensite at room temperature. This martensite is then strengthened by lower temperature aging, which forms fine precipitates of nickel-containing intermetallic compounds such as Ni_3Al. Figure 12.27 shows the solution treated and aged microstructure of PH 13-8 Mo steel. The precipitation cannot be resolved, but the tempering of the lath mar-

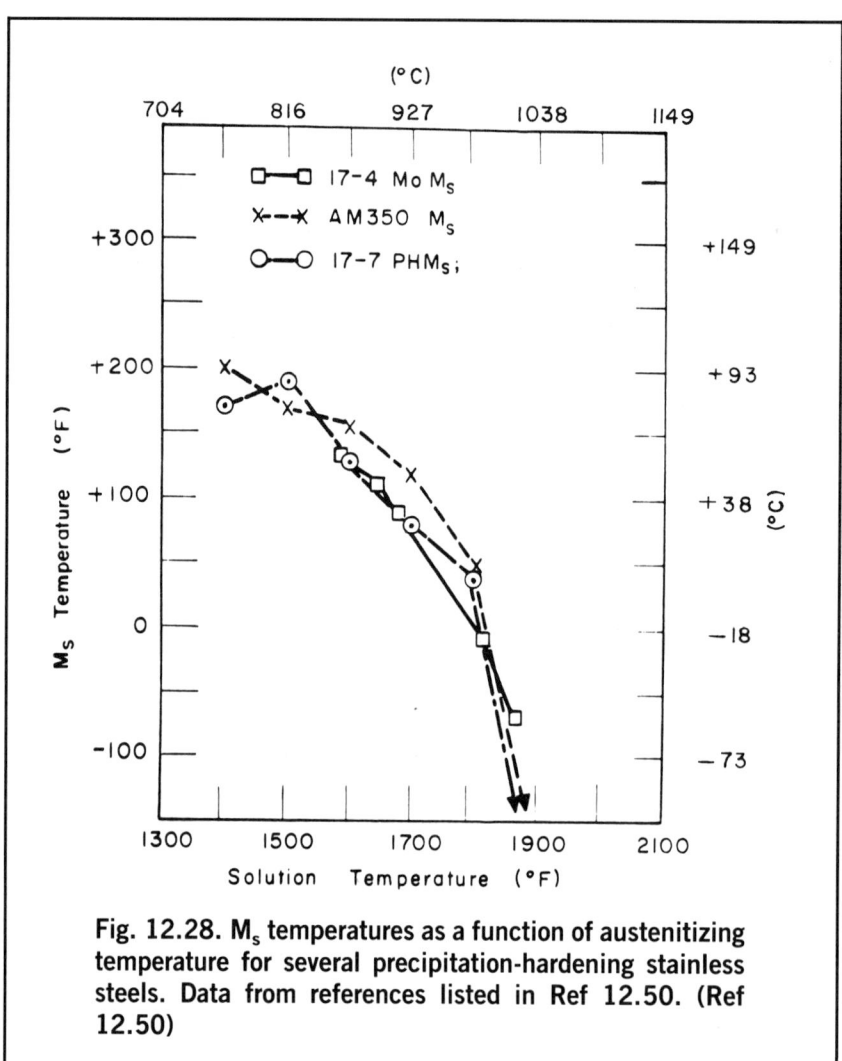

Fig. 12.28. M_s temperatures as a function of austenitizing temperature for several precipitation-hardening stainless steels. Data from references listed in Ref 12.50. (Ref 12.50)

tensite, which proceeds concurrently with the precipitation reaction, is apparent.

The semiaustenitic precipitation-hardening stainless steels are alloyed to have low austenite stabilities which can be further modified by conditioning or solution treatments. Often some delta ferrite is a component of the microstructure. Figure 12.28 shows M_s as a function of austenitizing temperature for several precipitation-hardening steels (Ref 12.50). At high austenitizing temperatures, all elements are in solution, M_s temperatures are well below room temperature, and the microstructure is largely austenitic at room temperature. In this condition, the

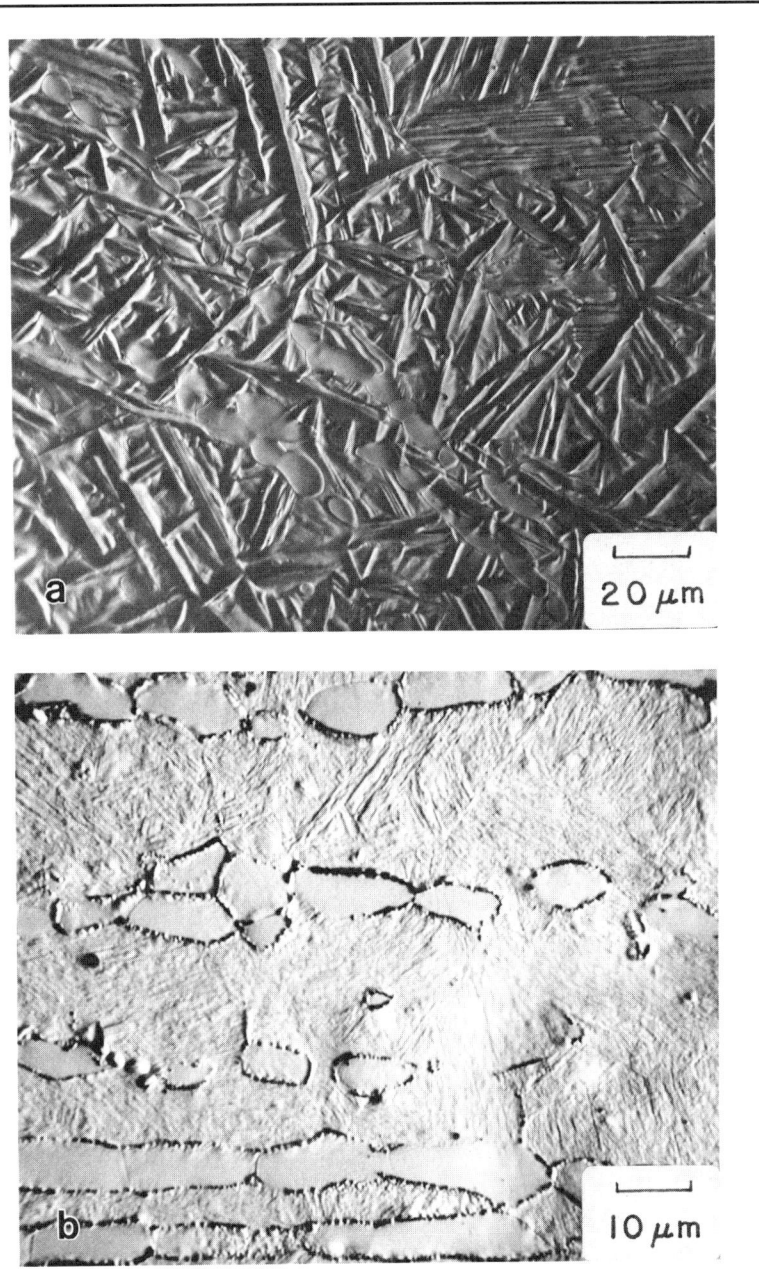

Fig. 12.29. Microstructure of 17-7 PH. (a) Surface tilting caused by martensite formation on refrigeration to −73 °C (−100 °F). (b) Refrigerated and aged at 480 °C (896 °F). Electropolished and etched in chrome-acetic acid electrolyte. Light micrographs. (Ref 12.50)

semiaustenitic precipitation-hardening stainless steels can be readily formed by cold work. At low austenitizing temperatures, alloy carbides precipitate, the austenite stability is lowered, and the M_s increases. As a result, the austenite transforms to martensite on cooling to room temperature. Another approach to producing a martensitic microstructure for aging is to refrigerate parts below room temperature.

Figure 12.29 shows the microstructure of a 17-7 PH steel in several heat treated conditions after solution treatment at 980 °C (1796 °F). The tilting surface relief formed by martensite plates formed by refrigeration at −73 °C (−100 °F) is shown in Fig. 12.29(a). The specimen surface was polished flat prior to refrigeration, and at that point the microstructure consisted of austenite and delta ferrite. Figure 12.29(b) shows the microstructure after aging at 480 °C (896 °F) for 5.75 h to 43 HRC. The interfaces of the delta ferrite and the matrix etch heavily because of carbide precipitation. Some of the martensite reverts to austenite during aging, but this phenomenon and the strengthening precipitates are not resolvable in the light microscope.

The austenitic grades of precipitation-hardening stainless steels are heavily alloyed to completely stabilize austenite. For example, A-286 contains 15% chromium and 26% nickel in addition to other elements which contribute to precipitation reactions. The strengthening precipitate is the ordered fcc γ' phase $Ni_3(Ti,Al)$, but secondary precipitates which may form during aging include eta (η) or Ni_3Ti, alloy carbides such as $(Ti,Mo)C$, and a boride, M_3B_2 (Ref 12.51, 12.52). Figure 12.30 shows very fine γ' precipitates in JBK-75, an austenitic precipitation-hardening stainless steel similar to A-286 but developed for better weldability (Ref 12.51).

The martensitic and semiaustenitic precipitation-hardening stainless steels can be readily heat treated to yield and ultimate strengths exceeding 200 ksi (1400 MPa) with good ductility. Solution and conditioning treatments, subzero cooling, and aging treatments are alloy specific and appropriate handbook and producer recommendations for the various alloys should be followed. The austenitic precipitation-hardening stainless steels do not reach the same high strength levels as the other two classes, but they have excellent elevated-temperature performance and are widely used in the aerospace industry.

Duplex Stainless Steels

Duplex stainless steels are alloyed and processed to develop microstructures consisting of roughly equal amounts of ferrite and austenite.

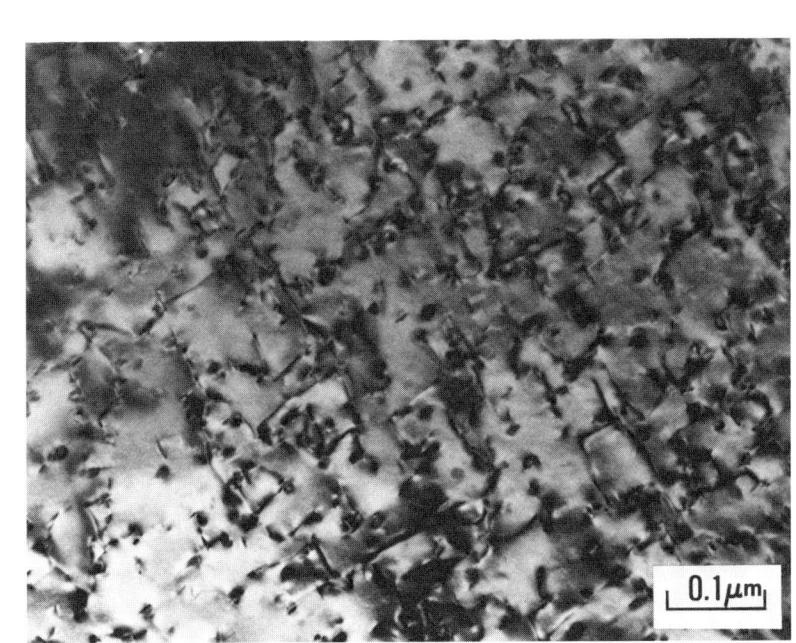

Fig. 12.30. Fine, disc-shaped γ' precipitates in an aged austenitic precipitation-hardening stainless steel, JBK-75. Transmission electron micrograph. (Ref 12.52)

Ferrite in austenitic stainless steels has already been discussed as a weld metal or casting alloy component which minimizes hot cracking, as a precursor for sigma phase formation in wrought alloys, and as a conse-

Table 12.8. Compositions of Selected Duplex Stainless Steels. (Ref 12.54, 12.55, 12.57)

Trade name/producer	Nominal composition, %				
	C	Cr	Ni	Mo	Other
Al 2205 (Allegheny Ludlum)	0.02	22.0	5.5	3.0	0.15N
7-Mo PLUS (Carpenter Technology Corp.)	0.02	26.0	4.7	1.4	0.20N
Ferralium 255 (Cabot Corp.)	<0.08	25.5	5.5	3.0	>0.1N, 1.3–1.4Cu
DPI (Sumitomo)	<0.03	18.5	4.75	2.75	
U45 (Creusot-Loire Steels)	<0.03	22.0	5.75	3.0	<0.2N

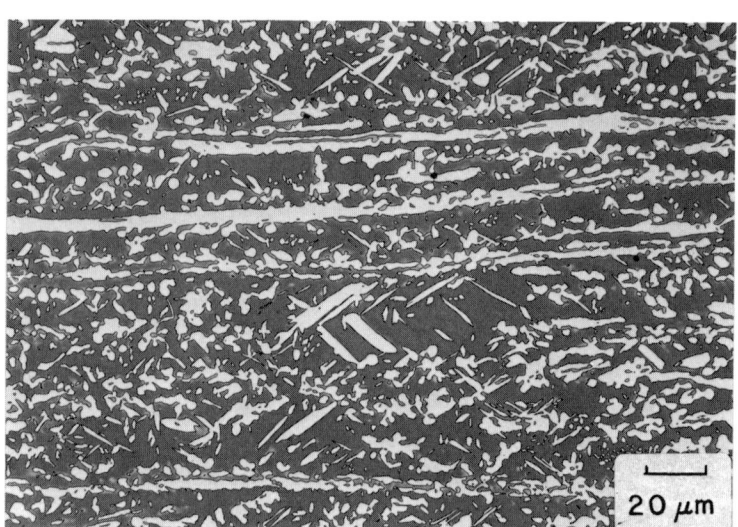

Fig. 12.31. Microstructure of duplex stainless steel 7Mo-Plus (UNS 532950). Gray phase is ferrite and white phase is austenite. Etched electrolytically in 20% NaOH. Light micrograph. Courtesy of G. Vander Voort, Carpenter Technology Corp., Reading, PA

quence of alloying for semiaustenitic precipitation-hardening stainless steels. Duplex stainless steels constitute a new class of materials because by design they have nearly balanced amounts of ferrite and austenite. This microstructure and a chemistry which is relatively high in chromium and molybdenum produce good corrosion resistance in pitting, crevice, sulfide stress, and chloride stress corrosion environments at strength levels about double that of annealed austenitic stainless steels (Ref 12.53). As for ferritic stainless steels, improved steelmaking processes such as argon-oxygen-decarburization (AOD) have made possible the production of low interstitial content duplex stainless steels with good ductility and toughness. Both wrought and cast grades of duplex stainless steels have been developed (Ref 12.53).

Table 12.8 lists some selected examples of duplex stainless steels. A much more extensive listing is given by Solomon and Devine (Ref 12.54). Compositions of duplex stainless steels range from 17 to 30% chromium and 3 to 13% nickel, with typically chromium on the high side and nickel on the low side of the ranges to ensure adequate ferrite stability. Molybdenum, a ferrite stabilizer, is also typically present.

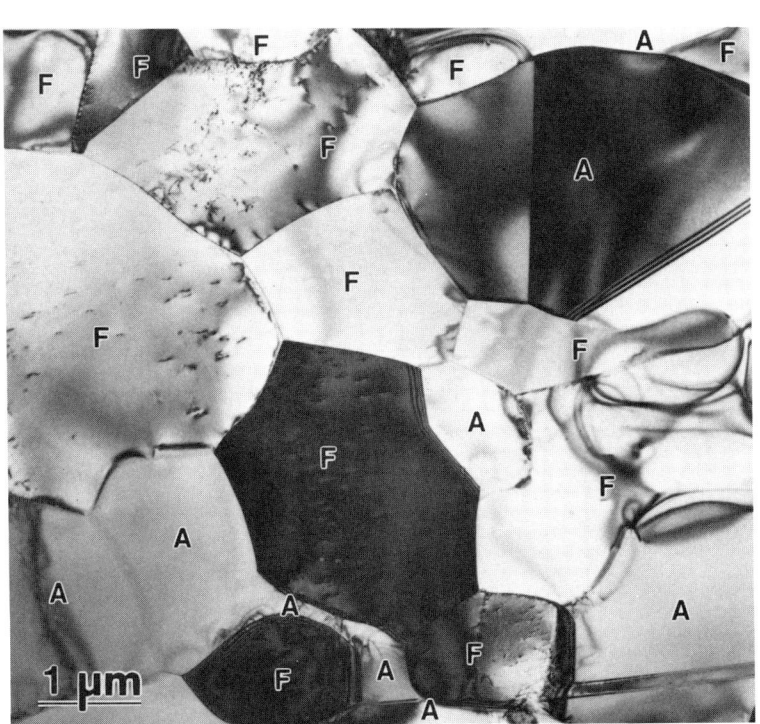

Fig. 12.32. Ferrite (F) and austenite (A) grains in duplex stainless steel Al 2205 (UNS 531803). Transmission electron micrograph. Courtesy of S.W. Thompson, Colorado School of Mines, Golden

Thermomechanical processing of wrought duplex stainless steels is accomplished in the two-phase ferrite austenite fields shown in the verticle sections of Fig. 12.5. Since the boundaries of the ferrite-austenite fields shift with temperature, the amounts of ferrite and austenite formed during hot working or annealing are a function of temperature. Higher temperatures produce larger amounts of ferrite. Therefore, hot working temperatures must be controlled, usually between 1000 and 1200 °C (1832 and 2192 °F), to maintain the desired balance of ferrite and austenite.

Figure 12.31 shows the microstructure of a 7Mo-Plus (UNS 532950) duplex stainless steel (Ref 12.55). In this alloy, the white phase is austenite and the continuous gray matrix phase is ferrite (Ref 12.13). In other stainless steels austenite might be the majority, or matrix, phase. Widmanstätten or acicular grains of austenite have

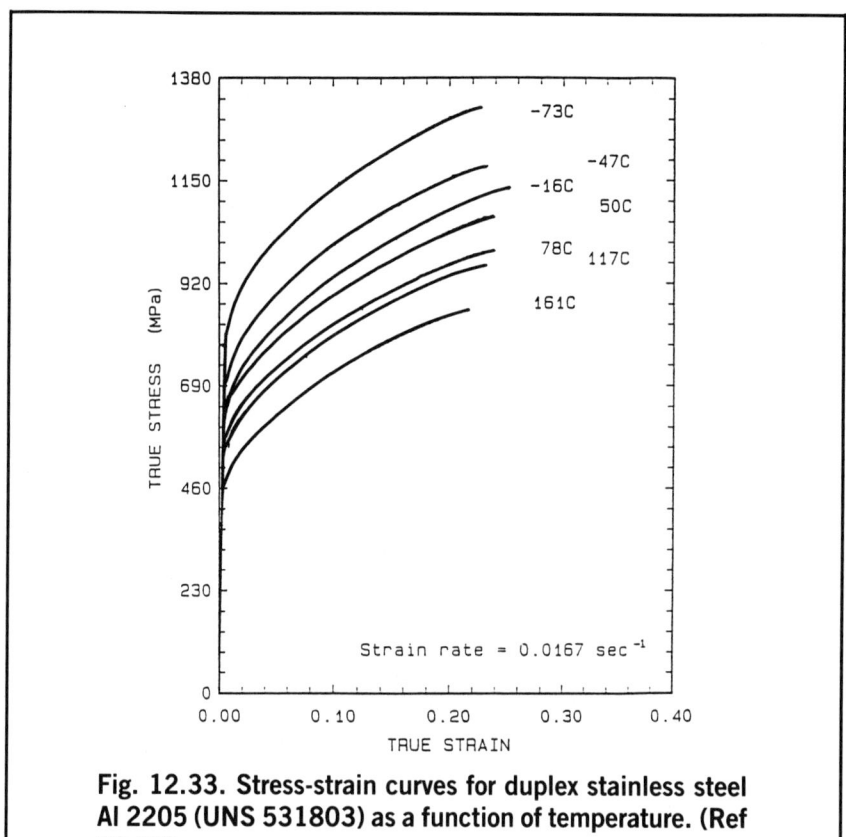

Fig. 12.33. Stress-strain curves for duplex stainless steel Al 2205 (UNS 531803) as a function of temperature. (Ref 12.57)

been observed to form on air cooling from temperatures above the austenite solvus (Ref 12.56), and this mechanism of transformation may account for the needle-like morphology of some of the austenite in Fig. 12.31.

As a result of hot working or annealing of cold worked shapes, both the ferrite and austenite areas of duplex stainless steels are polycrystalline. Etching to show the grain structure of the highly alloyed ferrite and austenite is extremely difficulty, and most published light micrographs show only the two phases without delineating grain structure. Figure 12.32 shows the grain structure in cold rolled and annealed Al 2205 (UNS 531803) duplex stainless steel. This micrograph was made by transmission electron microscopy and each grain was identified as bcc ferrite (F) or fcc austenite (A) by electron diffraction. Some of the austenite grains contain annealing twins, and the grain size of both phases is quite fine. Both phases have recrystallized with a very low density of dislocations, and many of the grain boundaries have established equilibrium triple points characterized by 120° dihedral angles. However, some boundaries remain highly curved, especially those between ferrite and austenite grains, reflecting the more sluggish kinetics of establishing phase equilibrium by partitioning of chromium, nickel, and molybdenum between the two phases before interfacial energies can be minimized.

The mechanical properties of duplex stainless steels are a function of the deformation behavior of ferrite and austenite. Figure 12.33 shows stress-strain curves obtained by testing sheet tensile specimens of an Al 2205 duplex stainless steel (Ref 12.57). As test temperature decreases, yield and ultimate strengths increase significantly but ductility remains almost the same. This behavior reflects the strength and strong temperature dependence of the flow strength of bcc ferrite, i.e., very sharp increases in flow stress with decreasing temperature below room temperature (Ref 12.36), which would normally be accompanied by sharp decreases in ductility. However, the fcc austenite compensates for the lower ductility of the ferrite, in part by providing increased rates of strain hardening by strain-induced transformation of austenite to martensite (Ref 12.57).

The highly alloyed duplex stainless steels, similar to austenitic and ferritic stainless steels, are subject to the deleterious effects of phases other than ferrite and austenite on corrosion resistance, ductility, and toughness (Ref 12.53-12.55). The ferrite phase is susceptible to 475 °C (885 °F) embrittlement and sigma phase formation at temperatures between 600 and 950 °C (1112 and 1742 °F), and chromium carbides may form between 560 and 1050 °C (1040 and 1922 °F) (Ref 12.55). As a result, rapid cooling through critical temperature ranges may be required during processing and fabrication, and upper temperature limits for service must be recognized.

Summary

This chapter has described the alloying, microstructure, and thermomechanical processing of the major types of stainless steels. These alloys meet many critical needs for materials which require excellent corrosion resistance and structural integrity. A dynamic search continues for improved processing and grades with better corrosion resistance, strength, high-temperature performance, formability, and toughness. The success of stainless steel application and development depends in large part on understanding the complex phase relationships in these highly alloyed materials. This understanding should result in the application of process controls and the selection of steels for service conditions which exploit beneficial microstructures and limit the formation of damaging phases.

References

12.1 M.G. Fontana and N.D. Greene, *Corrosion Engineering*, McGraw-Hill, New York, 1967
12.2 D. Peckner and I.M. Bernstein, *Handbook of Stainless Steels*, McGraw-Hill, New York, 1977
12.3 *Metals Handbook*, Vol 8, 8th ed., *Metallography, Structures and Phase Diagrams*, American Society for Metals, Metals Park, OH, 1973
12.4 P.G. Nelson, *Constitution and Heat Treatment of Stainless Steels*, Metals Engineering Institute, American Society for Metals, Metals Park, OH, 1969
12.5 J.C. Lippold and W.F. Savage, Solidification of Austenitic Stainless Steel Weldments: Part I-A Proposed Mechanism, *Weld Res Suppl*, 1979, p 362s-374s
12.6 T.A. Siewert, C.N. McCowan, and D.L. Olson, Ferrite Number Prediction to 100FN in Stainless Steel Weld Metal, *Weld Res Suppl*, 1988, p 289s-298s
12.7 A. Schaeffler, Constitution Diagram for Stainless Steel Weld Metal, *Metal Prog*, Vol 56 (No. 11), 1949, p 680-680B
12.8 W.T. DeLong, Ferrite in Austenitic Stainless Steel Weld Metal, *Weld J*, Vol 53 (No. 7), 1974, p 273s-286s
12.9 G.J. Fischer and R.J. Maciag, The Wrought Stainless Steels, in *Handbook of Stainless Steels*, McGraw-Hill, New York, p 1-1 to 1-10
12.10 Stainless Steels, J.D. Redmond (Ed.), in *Metals Handbook, Desk Edition*, American Society for Metals, Metals Park, OH, 1985, p 15-1 to 15-21
12.11 *Metals Handbook*, Vol 3, 9th ed., *Properties and Selection: Stainless Steels, Tool Materials, and Special-Purpose Metals*, American Society for Metals, Metals Park, OH, 1980, p 1-185
12.12 G.F. Vander Voort, *Metallography: Principles and Practice*, 1984, McGraw-Hill, New York
12.13 G.F. Vander Voort, The Metallography of Stainless Steels, *J Metals*, Vol 41 (No. 3), 1989, p 6-11
12.14 C.J. Novak, Structure and Constitution of Wrought Austenitic Stainless Steels, in *Handbook of Stainless Steels*, McGraw-Hill, New York, 1977, p 4-1 to 4-78
12.15 E.P. Butler and M.G. Burke, Preferential Formation of Martensite in Type 304 Stainless Steel: A Microstructural and Compositional Investigation, in *Solid-Solid Phase Transformations*, H.J. Aaronson et al. (Eds.), TMS-AIME, Warrendale, PA, 1982, p 1403-1407
12.16 S.R. Thomas and G. Krauss, Cyclic Martensitic Transformation and the Structure of a Commercial 18 Cr-8 Ni Stainless Steel, *Trans TMS-AIME*, Vol 239, 1967, p 1136-1142

12.17 A.H. Eichelman, Jr. and F.C. Hull, The Effect of Composition on the Temperature of Spontaneous Transformation of Austenite to Martensite in 18-8 Type Stainless Steel, *Trans ASM*, Vol 45, 1953, p 77-104

12.18 R.P. Reed, The Spontaneous Martensitic Transformations in 18 pct Cr, 8 pct Ni Steels, *Acta Met*, Vol 10, 1962, p 865-877

12.19 M.C. Mataya, M.J. Carr, and G. Krauss, The Bauschinger Effect in a Nitrogen-strengthened Austenitic Stainless Steel, *Mater Sci Eng*, Vol 57 (No. 2), 1983, p 205-222

12.20 T. Angel, Formation of Martensite in Austenitic Stainless Steels, *J Iron Steel Inst*, Vol 177, 1954, p 165-174

12.21 R.M. Vennett and G.S. Ansell, The Effect of High-Pressure Hydrogen upon the Tensile Properties and Fracture Behavior of 304L Stainless Steel, *Trans ASM*, Vol 60, 1967, p 242-251

12.22 K.G. Brickner, Stainless Steels for Room and Cryogenic Temperatures, in *Selection of Stainless Steels*, American Society for Metals, Metals Park, OH, 1968, p 1-29

12.23 J.P. Bressanelli and A. Moskowitz, Effects of Strain Rate, Temperature and Composition on Tensile Properties of Metastable Austenitic Stainless Steels, *Trans ASM*, Vol 59, 1966, p 223-239

12.24 G.L. Huang, D.K. Matlock, and G. Krauss, Martensite Formation, Strain Rate Sensitivity, and Deformation Behavior of Type 304 Stainless Steels Sheet, *Met Trans A*, Vol 20A, 1989, p 1239-1246

12.25 S.S. Hecker, M.G. Stout, K.P. Staudhammer, and J.L. Smith, Effects of Strain State and Strain Rate on Deformation-Induced Transformation in 304 Stainless Steel: Part I and Part II, *Met Trans A*, Vol 13A, 1982, p 619-626, 627-635

12.26 G.B. Olson, Transformation Plasticity and the Stability of Plastic Flow, in *Deformation, Processing and Structure*, G. Krauss (Ed.), American Society for Metals, Metals Park, OH, 1984, p 391-424

12.27 M.C. Mataya and M.J. Carr, Characterization of Inhomogeneities in Complex Austenitic Stainless Steel Forgings, in *Deformation, Processing and Structure*, G. Krauss (Ed.), American Society for Metals, Metals Park, OH, 1984, p 445-501

12.28 J. Bentley and J.M. Leitnaker, Stable Phases in Aged Type 321 Stainless Steel, in *The Metal Science of Stainless Steels*, E.W. Collings and H.W. King (Eds.), TMS-AIME, Warrendale, PA, 1979, p 70-71

12.29 R.J. Gray, V.K. Sikka, and R.T. King, Detecting Transformation of Delta-Ferrite to Sigma-Phase in Stainless Steels by Advanced Metallographic Techniques, *J Metals*, 1978, p 18-26

12.30 E.A. Schoeffer, The Cast Stainless Steels, in *Handbook of Stainless Steels*, McGraw-Hill, New York, 1977, p 2-1 to 2-18

12.31 *Materials Technology in Steam Reforming Processes*, C. Edeleanau (Ed.), Pergamon Press, London, 1966

12.32 L. Dillinger, R.D. Buchheit, J.A. VanEcho, D.B. Roach, and A.M. Hall, *Microstructures of Heat-Resistant Alloys*, Alloy Casting Institute Division, Steel Founders' Society of America, Cleveland, 1970

12.33 R.A. Lula, *Stainless Steel*, American Society for Metals, Metals Park, OH, 1986

12.34 R.P. Reed, Nitrogen in Austenitic Stainless Steels, *J Metals*, Vol 41 (No. 3), 1989, p 16-21

12.35 R.A. Lula (Ed.), *Toughness of Ferritic Stainless Steels*, STP 706, American Society for Testing and Materials, Philadelphia, 1980

12.36 W.C. Leslie, *The Physical Metallurgy of Steels*, McGraw-Hill, New York, 1981

12.37 J.F. Grubb, R.N. Wright, and P. Farrar, Jr., Micromechanisms of Brittle Fracture in Titanium-Stabilized and α'-embrittled Ferritic Stainless Steels, in *Toughness of Ferritic Stainless Steels*, R.A. Lula (Ed.), STP 706, American Society for Testing and Materials, Philadelphia, 1980, p 56-76

12.38 M.K. Veistinen and V.K. Lindroos, Cleavage Fracture Strength of a 26 Cr-1 Mo Ferritic Stainless Steel, in *New Developments in Stainless Steel Technology*, R.A. Lula (Ed.), American Society for Metals, Metals Park, OH, 1985, p 29-43

12.39 R.Q. Barr (Ed.), *Stainless Steel '77*, Climax Molybdenum Company, Greenwich, CT, 1977

12.40 C.S. Barrett and T.B. Masselski, *Structure of Metals*, 3rd ed., McGraw-Hill, New York, 1966

12.41 E.L. Brown, M.E. Burnett, P.T. Purtscher, and G. Krauss, Intermetallic Phase Formation in 25 Cr-3 Mo-4 Ni Ferritic Stainless Steel, *Met Trans A*, Vol 14A, 1983, p 791-800

12.42 R.M. Fisher, E.J. Dolis, and K.G. Carroll, Identification of the Precipitate Accompanying 885 °F Embrittlement in Chromium Steels, *Trans AIME*, Vol 197, 1953, p 690-695

12.43 R.O. Williams, Further Studies of the Iron-Chromium System, *Trans AIME*, Vol 212, 1958, p 497-502

12.44 R. Lagneborg, Metallography of the 475 C Embrittlement in an Iron-30 pct Chromium Alloy, *Trans ASM*, Vol 60, 1967, p 67-68

12.45 P.J. Grobner, The 885 °F (475 °C) Embrittlement of Ferritic Stainless Steels, *Met Trans*, Vol 4, 1973, p 251-260

12.46 T.J. Nichol, A. Datta, and G. Aggen, Embrittlement of Ferritic Stainless Steels, *Met Trans A*, Vol 11A, 1980, p 573-585

12.47 P.M. Unterweiser, H.E. Boyer, and J.J. Kubbs (Eds.), *The Heat Treater's Guide*, American Society for Metals, Metals Park, OH, 1982

12.48 D.C. Perry and J.C. Jasper, Structure and Constitution of Wrought Precipitation-Hardenable Stainless Steels, in *Handbook of Stainless Steels*, McGraw-Hill, New York, 1977, p 7-1 to 7-18

12.49 R. Smith, F.H. Wyche, and W. Gore, A Precipitation Hardening Stainless Steel of the 18 percent Chromium, 8 percent Nickel Type, *Trans AIME*, Vol 167, 1946, p 313

12.50 G. Krauss, Jr. and B.L. Averbach, Retained Austenite in Precipitation Hardening Stainless Steels, *Trans ASM*, Vol 52, 1960, p 434-450

12.51 T.J. Headley, M.M. Karnowsky, and W.R. Sorenson, Effect of Composition and High Energy Rate Forging on the Onset of Precipitation in an Iron-Base Superalloy, *Met Trans A*, Vol 13A, 1982, p 345-353

12.52 M.C. Mataya, M.J. Carr, and G. Krauss, Flow Localization and Shear Band Formation in a Precipitation Strengthened Austenitic Stainless Steel, *Met Trans A*, Vol 13A, 1982, p 1263-1274

12.53 R.A. Lula (Ed.), *Duplex Stainless Steels*, American Society for Metals, Metals Park, OH, 1983

12.54 H.D. Solomon and T.M. Devine, Jr., Duplex Stainless Steels—A Tale of Two Phases, in *Duplex Stainless Steels*, R.A. Lula (Ed.), American Society for Metals, Metals Park, OH, 1983, p 693-756

12.55 T.A. DeBold, Duplex Stainless Steel—Microstructure and Properties, *J Metals*, Vol 41 (No. 3), 1989, p 12-15

12.56 H.D. Solomon, Age Hardening in a Duplex Stainless Steel, in *Duplex Stainless Steels*, R.A. Lula (Ed.), American Society for Metals, Metals Park, OH, 1983, p 41-69

12.57 C.L. Beech, Effect of Temperature and Strain Rate on the Mechanical Properties and Deformation Behavior of a Duplex Stainless Steel, M.S. thesis, Colorado School of Mines, Golden, 1989

CHAPTER 13

Tool Steels

Tool steels are the steels used to form and machine other materials, and therefore are designed to have high hardness and durability under severe service conditions. Tool steel heat treatment is similar to that of the hardenable low-alloy steels discussed earlier; i.e., final properties are produced by austenitizing, martensite formation, and tempering. However, most tool steels are highly alloyed, and special precautions must be taken throughout processing to achieve the proper balance of alloy carbides in a matrix of tempered martensite for a given tool application. This chapter describes the alloy and process design of the various classes of tool steel, and the microstructures produced during tool steel heat treatment.

Introduction

Tool steels are a very large group of complex alloys which have evolved for many diverse hot and cold forming applications. Their industrial importance and complexity has led to a considerable text and handbook literature about their development, processing, and application (Ref 13.1-13.9). Details of processing, such as recommended cooling rates and heat treatment times and temperatures for specific steels, are found not only in the literature but also in detailed information distributed by major manufacturers of tool steels. The purpose of this chapter is to develop the principles of tool steel alloy design and to describe the role that heat treatment plays in the evolution of tool steel microstructure and properties.

Classification of Tool Steels

The various types of tool steels are categorized into a number of classes each of which according to essentially identical classification systems adopted by the American Iron and Steel Institute (AISI) and the Society of Automotive Engineers (SAE) is identified by a letter representing a unique characteristic, chemistry, or use of that class of steels. Table 13.1 lists nominal chemistries and the various classes of tool steels (Ref 13.10). The listed tool steel designations will be used throughout the balance of this chapter, and a brief summary of the major features of each class follows.

The water-hardening tool steels, AISI type W, have the lowest alloy content and therefore the lowest hardenability of any of the tool steels. As a result, the W tool steels frequently require water quenching, and heavy sections harden only to shallow depths. Thin sections can be hardened by oil quenching to minimize quenching cracking and distortion.

The shock-resistant tool steels, AISI type S, have lower carbon content and somewhat higher alloy content than the W steels. The medium carbon content improves toughness and makes the type S steels good for applications with shock and impact loading.

Tool steels for cold work include three classes of steels: AISI types O, A, and D. All classes have high carbon content for high hardness and high wear resistance in cold work applications, but differ in alloy content, which affects hardenability and the carbide distributions incorporated into the hardened microstructures. The relatively low-alloyed oil hardening grades, O, are oil quenched, but the high-alloyed A and D grades are hardenable by air cooling and therefore are less susceptible to distortion and cracking during hardening. The high chromium and molybdenum contents of the A and D tool steels also contribute to high carbide particle contents and excellent wear resistance. The low-alloy special-purpose tool steels, type L, by virtue of their somewhat lower carbon content, have higher toughness than do the O grades.

Tool steels used for dies to mold plastics, AISI type P, are exposed to less severe wear than metal-working steel, and therefore have low carbon content. A key requirement is good polishability and excellent surface finish. Type 420 martensite stainless steel is also used for plastic molds when corrosion might be a factor limiting performance of lower alloyed P steels.

Hot work tool steels, AISI type H, fall into groups which have either chromium, tungsten, or molybdenum as the major alloying element. The H steels are used for hot forging, extrusion, and metal die-casting dies. The medium carbon content and relatively high alloy content make the H steels air hardenable and resistant to impact and softening during repeated exposure to hot working operations.

The high-speed tool steels are very highly alloyed, with tungsten, and molybdenum as the major alloying elements in the T and M

Table 13.1. Classification and Approximate Compositions of Principal Types of Tool Steels (Ref 13.10)

AISI	UNS	\- Identifying elements, % \-								
		C	Mn	Si	Cr	V	W	Mo	Co	Ni
Water-Hardening Tool Steels										
W1	T72301	0.60-1.40(a)	...	...	...	...	...	...	...	...
W2	T72302	0.60-1.40(a)	...	...	...	0.25	...	...	...	...
W5	T72305	1.10	...	...	0.50	...	...	...	...	...
Shock-Resisting Tool Steels										
S1	T41901	0.50	...	...	1.50	...	2.50	...	...	...
S2	T41902	0.50	...	1.00	...	...	...	0.50	...	...
S5	T41905	0.55	0.80	2.00	...	...	...	0.40	...	...
S6	T41906	0.45	1.40	2.25	1.50	...	...	0.40	...	...
S7	T41907	0.50	...	...	3.25	...	...	1.40	...	...
Oil-Hardening Cold Work Tool Steels										
O1	T31501	0.90	1.00	...	0.50	...	0.50	...	...	...
O2	T31502	0.90	1.60	...	...	...	...	...	...	...
O6(b)	T31506	1.45	0.80	1.00	...	...	...	0.25	...	...
O7	T31507	1.20	...	...	0.75	...	1.75	...	...	...
Air-Hardening Medium-Alloy Cold Work Tool Steels										
A2	T30102	1.00	...	...	5.00	...	...	1.00	...	...
A3	T30103	1.25	...	...	5.00	1.00	...	1.00	...	...
A4	T30104	1.00	2.00	...	1.00	...	...	1.00	...	...
A6	T30106	0.70	2.00	...	1.00	...	...	1.25	...	...
A7	T30107	2.25	...	...	5.25	4.75	1.00(c)	1.00	...	...
A8	T30108	0.55	...	...	5.00	...	1.25	1.25	...	...
A9	T30109	0.50	...	...	5.00	1.00	...	1.40	...	1.50
A10(b)	T30110	1.35	1.80	1.25	...	...	...	1.50	...	1.80
High-Carbon High-Chromium Cold Work Steels										
D2	T30402	1.50	...	...	12.00	1.00	...	1.00	...	...
D3	T30403	2.25	...	...	12.00	...	...	...	...	...
D4	T30404	2.25	...	...	12.00	...	...	1.00	...	...
D5	T30405	1.50	...	...	12.00	...	...	1.00	3.00	...
D7	T30407	2.35	...	...	12.00	4.00	...	1.00	...	...
Low-Alloy Special-Purpose Tool Steels										
L2	T61202	0.50-1.10(a)	...	...	1.00	0.20	...	...	...	...
L6	T61206	0.70	...	...	0.75	...	...	0.25(c)	...	1.50
Mold Steels										
P2	T51602	0.07	...	...	2.00	...	...	0.20	...	0.50
P3	T51603	0.10	...	...	0.60	...	...	...	...	1.25
P4	T51604	0.07	...	...	5.00	...	...	0.75	...	...
P5	T51605	0.10	...	...	2.25	...	...	...	...	...
P6	T51606	0.10	...	...	1.50	...	...	...	...	3.50
P20	T51620	0.35	...	...	1.70	...	...	0.40	...	...
P21	T51621	0.20	1.20(Al)	...	...	...	...	...	...	4.00

(continued)

Table 13.1. (continued)

| AISI | UNS | \multicolumn{8}{c}{Identifying elements, %} |
		C	Mn	Si	Cr	V	W	Mo	Co	Ni
Chromium Hot Work Tool Steels										
H10	T20810	0.40	...	...	3.25	0.40	...	2.50	...	...
H11	T20811	0.35	...	...	5.00	0.40	...	1.50	...	...
H12	T20812	0.35	...	...	5.00	0.40	1.50	1.50	...	...
H13	T20813	0.35	...	...	5.00	1.00	...	1.50	...	...
H14	T20814	0.40	...	...	5.00	...	5.00	...	...	...
H19	T20819	0.40	...	...	4.25	2.00	4.25	...	4.25	...
Tungsten Hot Work Tool Steels										
H21	T20821	0.35	...	...	3.50	...	9.00	...	...	...
H22	T20822	0.35	...	...	2.00	...	11.00	...	...	...
H23	T20823	0.30	...	...	12.00	...	12.00	...	...	...
H24	T20824	0.45	...	...	3.00	...	15.00	...	...	...
H25	T20825	0.25	...	...	4.00	...	15.00	...	...	...
H26	T20826	0.50	...	...	4.00	1.00	18.00	...	...	...
Molybdenum Hot Work Tool Steels										
H42	T20842	0.60	...	...	4.00	2.00	6.00	5.00	...	...
Proprietary Hot Work Tool Steels										
6G	...	0.55	0.80	0.25	1.00	0.10	...	0.45	...	...
6F2	...	0.55	0.75	0.25	1.00	...	...	0.30	...	1.00
6F3	...	0.55	0.60	0.85	1.00	0.10	...	0.75	...	1.80
Tungsten High-Speed Tool Steels										
T1	T12001	0.75(a)	...	...	4.00	1.00	18.00	...	...	...
T2	T12002	0.80	...	...	4.00	2.00	18.00	...	...	...
T4	T12004	0.75	...	...	4.00	1.00	18.00	...	5.00	...
T5	T12005	0.80	...	...	4.00	2.00	18.00	...	8.00	...
T6	T12006	0.80	...	...	4.50	1.50	20.00	...	12.00	...
T8	T12008	0.75	...	...	4.00	2.00	14.00	...	5.00	...
T15	T12015	1.50	...	...	4.00	5.00	12.00	...	5.00	...
Molybdenum High-Speed Tool Steels										
M1	T11301	0.80(a)	...	...	4.00	1.00	1.50	8.00	...	...
M2	T11302	0.85-1.00(a)	...	...	4.00	2.00	6.00	5.00	...	...
M3 Class 1	T11313	1.05	...	...	4.00	2.40	6.00	5.00	...	...
M3 Class 2	T11323	1.20	...	...	4.00	3.00	6.00	5.00	...	...
M4	T11304	1.30	...	...	4.00	4.00	5.50	4.50	...	...
M6	T11306	0.80	...	...	4.00	2.00	4.00	5.00	12.00	...
M7	T11307	1.00	...	...	4.00	2.00	1.75	8.75	...	...
M10	T11310	0.85-1.00(a)	...	...	4.00	2.00	...	8.00	...	...
M30	T11330	0.80	...	...	4.00	1.25	2.00	8.00	5.00	...
M33	T11333	0.90	...	...	4.00	1.15	1.50	9.50	8.00	...
M34	T11334	0.90	...	...	4.00	2.00	2.00	8.00	8.00	...
M36	T11336	0.80	...	...	4.00	2.00	6.00	5.00	8.00	...
Ultrahard High-Speed Tool Steels										
M41	T11341	1.10	...	...	4.25	2.00	6.75	3.75	5.00	...
M42	T11342	1.10	...	...	3.75	1.15	1.50	9.50	8.00	...

(continued)

Table 13.1. (continued)

AISI	UNS	C	Mn	Si	Cr	V	W	Mo	Co	Ni
M43	T11343	1.20	...	...	3.75	1.60	2.75	8.00	8.25	...
M44	T11344	1.15	...	...	4.25	2.00	5.25	6.25	12.00	...
M46	T11346	1.25	...	...	4.00	3.20	2.00	8.25	8.25	...
M47	T11347	1.10	...	...	3.75	1.25	1.50	9.50	5.00	...

Maraging Steels

Type	C	Mn	Si	Al	Ti	Mo	Co	Ni
Grade 90	0.03 max	0.10 max	0.12 max	0.10	0.30	3.25	8.50	18.00
Grade 110	0.03 max	0.10 max	0.12 max	0.10	0.50	4.85	7.75	18.00
Grade 125	0.03 max	0.10 max	0.12 max	0.10	0.70	5.00	9.00	18.00

(a) Other carbon contents may be available. (b) Contains free graphite in the microstructure to improve machinability. (c) Optional

grades, respectively. The tungsten, molybdenum, chromium, and vanadium in these steels produce very high densities of stable carbides. As a result, the high-speed tool steels are capable of retaining hardness at temperatures as high as 600 °C (1112 °F) and are widely used for high-speed cutting and machining applications.

Maraging steels are also listed in Table 13.1, and are sometimes selected for tool and die applications. The maraging steels develop high strength and hardness by quite different mechanisms (Ref 13.11, 13.12) than steels dependent on carbon content for strength. Despite low carbon content, the high cobalt and nickel content of the maraging steels ensures that martensite forms on air cooling. The low-carbon, low-strength martensite is then hardened by fine-scale precipitation of intermetallic compounds, such as Ni_3Mo, by aging around 480 °C (896 °F). Excellent combinations of high strength and toughness are associated with the aged low-carbon martensitic microstructures and the maraging steels are used for many structural applications, including tools and dies, which require ultrahigh strength and toughness. Brandis and Haberling (Ref 13.8) describe maraging steels for hot work and plastic mold applications, and new grades of maraging steels, free of cobalt, are currently under active development (Ref 13.12).

Tool Steel Alloy Design

Tool steel alloy design is in large part based on alloying steel with strong carbide-forming transition elements such as chromium, molybdenum, tungsten, and vanadium. These elements partition between

Table 13.2. Transition Metal Carbides (Ref 13.13)

CARBIDE FORMATION IS FAIRLY COMMON AMONG THE TRANSITION ELEMENTS, EXCEPT FOR THE SECOND AND THIRD ROWS OF GROUP VIII(a).

III	IV	V	VI	VII	VIII		
$Sc_{2-3}C$, ScC_2, Sc_2C_3	TiC	V_2C, VC	$Cr_{23}C_6$, Cr_7C_3, Cr_3C_2	$Mn_{23}C_6$, Mn_3C, Mn_5C_2, Mn_7C_3	Fe_3C	Co_3C, Co_2C	Ni_3C
Y_2C, Y_2C_3, YC_2	ZrC	Nb_2C, NbC	Mo_2C, Mo_3C_2, MoC_{1-x}	TcC	Ru ✕	Rh ✕	Pd ✕
LaC_2	HfC	Ta_2C, TaC	W_2C, W_3C_2, WC	ReC	OsC	Ir ✕	Pt ✕

(a) ✕ indicates no carbide formation for this element.

carbides and the austenitic matrix during solidification, hot work, annealing, and austenitizing for hardening. During hardening, the alloy carbides formed in austenite are retained and the austenite matrix transforms to martensite. Further alloy-element partitioning occurs during tempering as retained austenite transforms and fine alloy carbides precipitate in tempered martensite. Strengthening and wear resistance are provided by all elements of the microstructure: the retained carbides, the tempered martensite, and the carbides formed on tempering.

Table 13.2 shows the carbides formed by the transition elements in the various groups of the periodic table (Ref 13.13). The transition metal carbides have very high hardness, high melting points, and unique electrical properties, and are often used in pure form (Ref 13.13). Tungsten carbide (WC) is the major component of cemented carbide cutting tools, and transition metal carbide and nitride coatings (as discussed in Chapter 11) are increasingly being used to improve wear resistance of tool steels.

In steels, the transition metal carbide crystal structures incorporate iron atoms as well as several major carbide-forming elements, and therefore the letter M is used to designate the total metal atom component of a carbide. Table 13.3 lists the types of carbides, crystal lattice, and some characteristics of each of the various carbides found in tool steels (Ref 13.2). The crystal structures of the carbides are described in detail by Jack and Jack (Ref 13.14).

The wear resistance of tool steels increases with increasing carbide volume fraction and carbide hardness. Figures 13.1 and 13.2 are two graphical comparisons of the hardness of various alloy carbides

Table 13.3. Characteristics of Alloy Carbides Found in Tool Steels (Ref 13.2)

Type of carbide	Lattice type	Remarks
M_3C	Orthorhombic	This is a carbide of the cementite (Fe_3C) type, M, maybe Fe, Mn, Cr with a little W, Mo, V.
M_7C_3	Hexagonal	Mostly found in Cr alloy steels. Resistant to dissolution at higher temperatures. Hard and abrasion resistant. Found as a product of tempering high-speed steels.
$M_{23}C_6$	Face-centered cubic	Present in high-Cr steels and all high-speed steels. The Cr can be replaced with Fe to yield carbides with W, and Mo.
M_6C	Face-centered cubic	Is a W- or Mo-rich carbide. May contain moderate amounts of Cr, V, Co. Present in all high-speed steels. Extremely abrasion resistant.
M_2C	Hexagonal	W- or Mo-rich carbide of the W_2C type. Appears after temper. Can dissolve a considerable amount of Cr.
MC	Face-centered cubic	V-rich carbide. Resists dissolution. Small amount which does dissolve reprecipitates on secondary hardening.

relative to the hardness of martensite and cementite, Fe_3C, the carbide typically found in plain carbon and low-alloy carbon steels. As shown, the transition metal carbides attain very high hardness, and thus contribute significant wear resistance to tool steels which are alloyed to contain large volume fractions of carbides. For example, high-speed tool steels may contain as much as 30 vol.% of carbides consisting of a mixture of MC, $M_{23}C_6$, and M_6C (Ref 13.1).

The amount and type of carbides in a tool steel depend on carbon content, alloy content, and temperature. Isothermal and vertical sections through ternary Fe-X-C systems (where X represents a transition metal such as chromium) are available to predict the carbide phases which will form in a given ternary alloy (Ref 13.1). However, tool steels are more complex than ternary alloys, often containing three or four major alloying elements, as shown in Table 13.1. In the more complex steels, the carbides are identified and their amounts quantified by a variety of experimental techniques, including metallography, selective etching, x-ray diffraction and chemical analysis of extracted carbide residues, electron microprobe analysis, and scanning transmission electron microscopy. Another approach to char-

408 / STEELS: HEAT TREATMENT AND PROCESSING PRINCIPLES

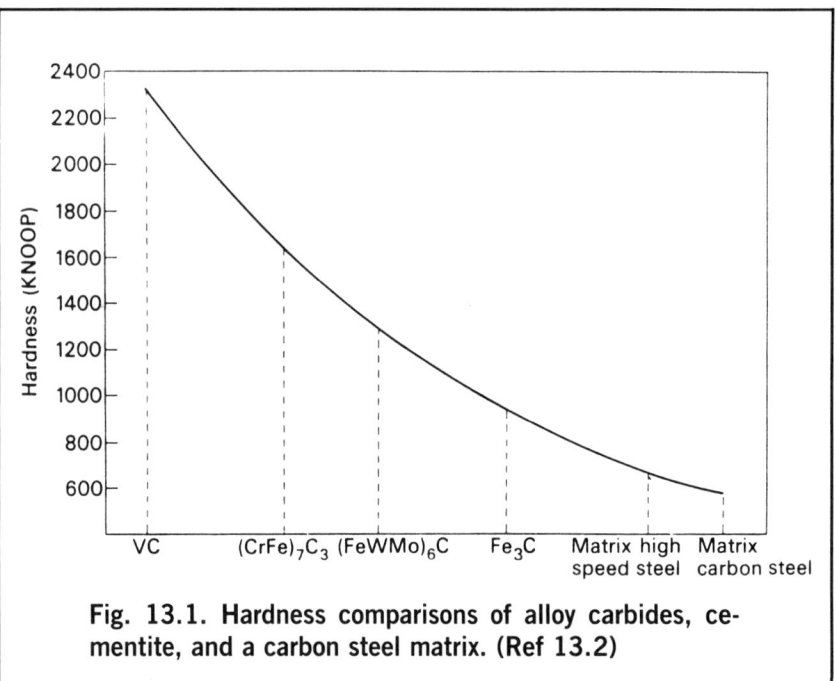

Fig. 13.1. Hardness comparisons of alloy carbides, cementite, and a carbon steel matrix. (Ref 13.2)

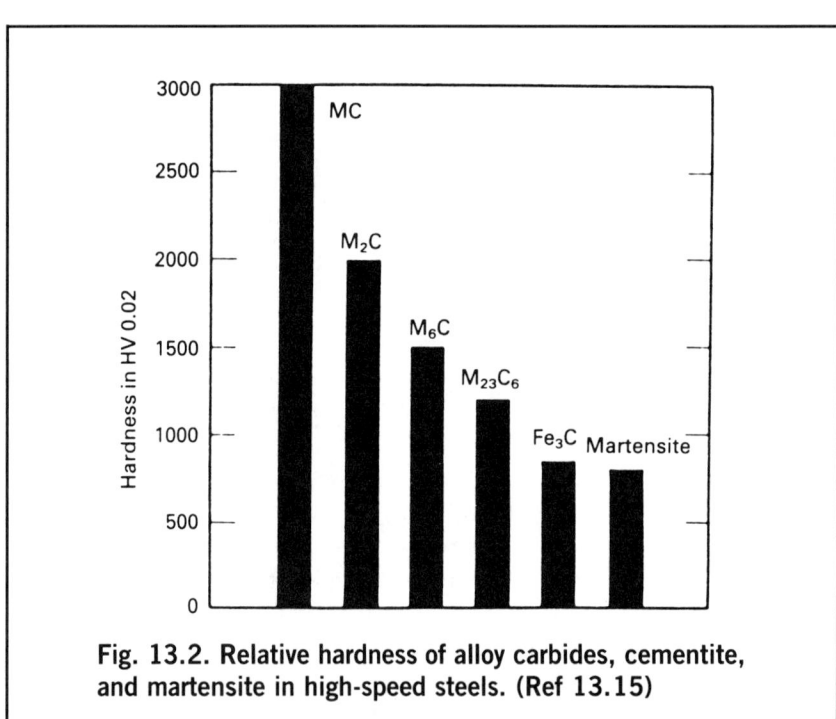

Fig. 13.2. Relative hardness of alloy carbides, cementite, and martensite in high-speed steels. (Ref 13.15)

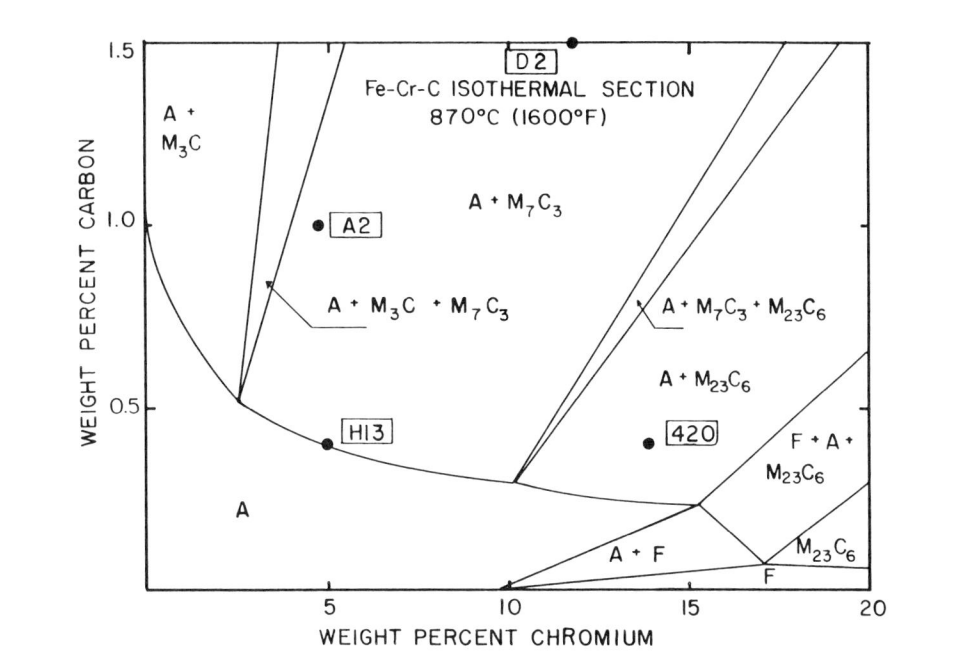

Fig. 13.3. Isothermal section of iron-rich corner of the Fe-Cr-C system at 870 °C (1598 °F). Compositions of alloys indicated are based only on chromium and carbon contents, but the alloys contain other elements which may introduce other phases. (Ref 13.23, 13.25)

acterizing the phases in tool steels is based on computing techniques which use thermodynamic functions to predict both carbide and matrix chemistry in complex systems (Ref 13.16-13.22). A discussion of the thermodynamic calculations is outside the scope of this chapter, but the selected references show the growing use of these methods. The calculations are coupled to experimental determination of thermodynamic parameters and experimental verification of the phases and their compositions.

Examples of isothermal and vertical sections of portions of the Fe-Cr-C system are shown in Fig. 13.3 and 13.4 (Ref 13.23, 13.24). Compositions of some typical tool steels are plotted on the diagrams, and although the compositions are understated with respect to vanadium and molybdenum contents, it is instructive to relate the alloy compositions to the phase relationships demonstrated in the diagrams. The isothermal section shows the various carbides which coexist with austenite at 870 °C (1598 °F). As chromium content increases, the carbide chemistry and crystal structure changes from M_3C to M_7C_3 to

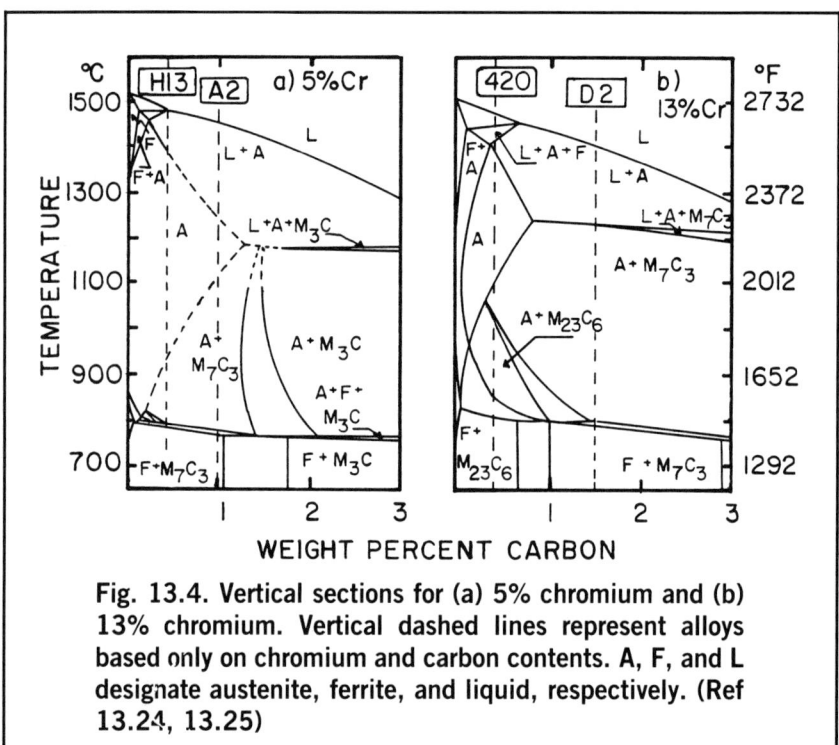

Fig. 13.4. Vertical sections for (a) 5% chromium and (b) 13% chromium. Vertical dashed lines represent alloys based only on chromium and carbon contents. A, F, and L designate austenite, ferrite, and liquid, respectively. (Ref 13.24, 13.25)

$M_{23}C_6$ to accommodate increasing amounts of chromium atoms. The H-13, A-2, and D-2 tool steels all contain M_7C_3 carbides in equilibrium with austenite at 870 °C (1598 °F), and as the carbon content of the steels increases from that of H-13 (0.4% carbon) to D-2 (1.50% carbon), the amount of M_7C_3 carbide increases.

The vertical sections (Fig. 13.4) are for Fe-Cr-C alloys containing 5 and 13 wt.% chromium. Compositions of the coexisting phases usually lie outside the vertical section, but the sections show temperature ranges over which the various carbides coexist with austenite and ferrite. The latter information is useful in designing hot work schedules and heat treatments for annealing and hardening.

Primary Processing of Tool Steels

Figures 13.5 and 13.6 show schematically the thermomechanical processing and final heat treatment schedules applied to tool steels (Ref 13.25). Processing starts with melting and solidification. Tool steels are melted in electric arc or induction furnaces from carefully

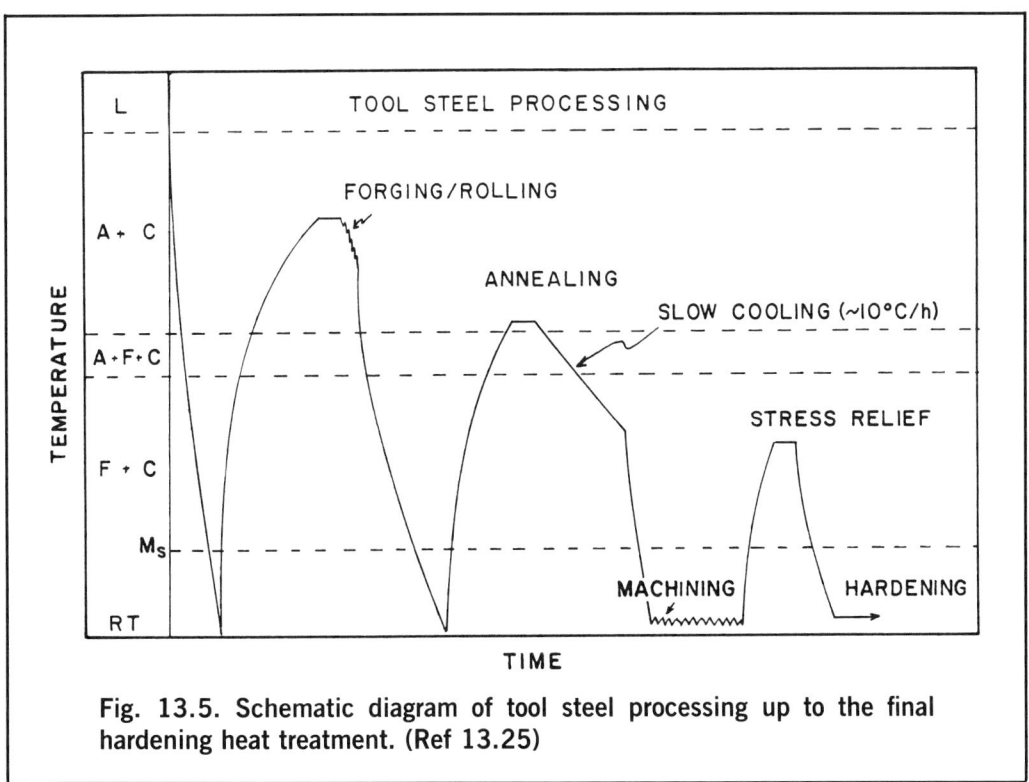

Fig. 13.5. Schematic diagram of tool steel processing up to the final hardening heat treatment. (Ref 13.25)

selected scrap and alloying additions, and for special high-quality grades, previously cast ingots may be subjected to consumable electrode vacuum or electroslag remelting and solidification (Ref 13.1, 13.2, 13.5). Ingot size is generally kept small to reduce dendrite arm spacing and segregation during cooling. New processing techniques such as atomization of tool speed melts into fine powders and subsequent compaction are also being applied to provide highly homogeneous microstructures (Ref 13.2, 13.5).

The high alloy content of multicomponent tool steels results in significant segregation and primary alloy carbide formation during solidification. The high-speed tool steels, by virtue of their very high alloy content, have the most complex solidification sequence (Ref 13.5, 13.15, 13.26). Figures 13.7 and 13.8 show some of the microstructural features developed during solidification of an M2 high-speed steel. Solidification begins with the formation of primary dendrites of delta ferrite. Such a dendrite is shown in the center of Fig. 13.7, an x-ray microradiograph taken from a thin specimen of the M2 steel. The micrograph is produced by exposure of an x-ray film to the transmitted x-ray radiation, and the light areas represent structure where alloying

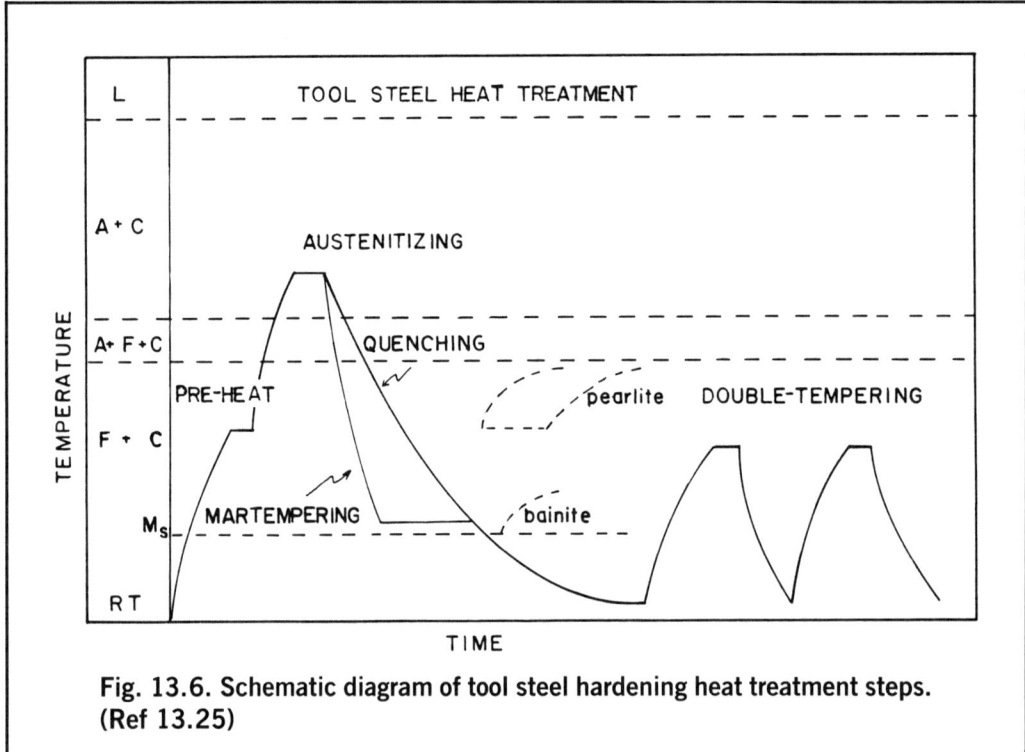

Fig. 13.6. Schematic diagram of tool steel hardening heat treatment steps. (Ref 13.25)

elements with high absorption coefficients are concentrated. The delta ferrite later transforms by a eutectoid reaction to an austenite-carbide aggregate. Following the formation of the primary ferrite crystals, a peritectic-type reaction causes some liquid to combine with the ferrite to form a rim of austenite around the dendrite arms. The remaining liquid is enriched in carbide-forming elements and final solidification occurs by complex eutectic solidification, which produces several morphologies of carbides and austenite. Figure 13.8 shows examples of feathery (austenite plus M_6C and MC), herringbone (austenite plus M_6), and blocky (austenite plus MC) eutectic morphologies in M2 high-speed steel (Ref 13.26).

Following ingot solidification, tool steels are soaked at high temperatures and hot worked by forging, extrusion, or rolling in the temperature range of austenite or austenite-carbide stability. Roberts (Ref 13.27) has reviewed the dynamic recovery and recrystallization phenomena which affect the flow behavior of austenite during hot work of stainless and tool steels. Hot working not only reduces section size but also reduces segregation produced during solidification. Also, in high-speed steels hot work breaks up the interconnected carbide

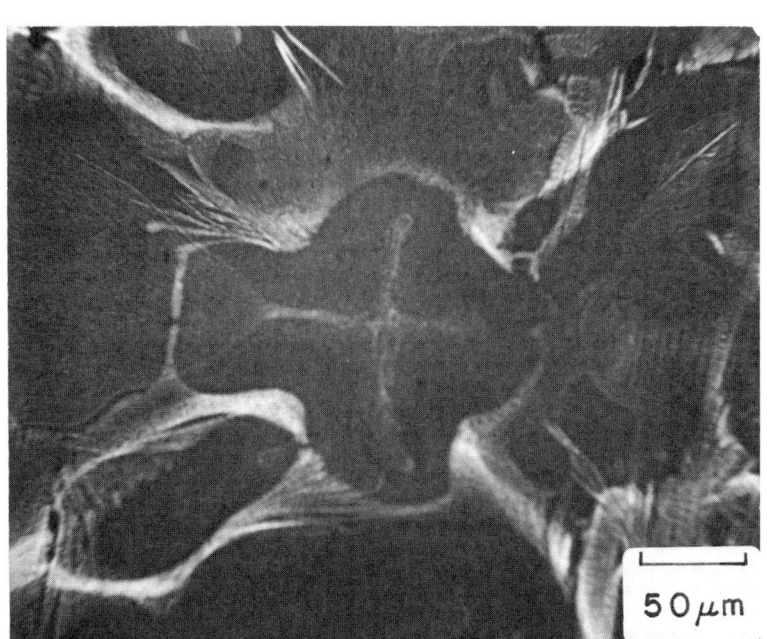

Fig. 13.7. Primary dendrite with axis normal to the plane of polish in an M2 high-speed steel. Microradiograph. Dark areas are due to exposure of x-ray film by transmitted radiation and white areas are structures of higher absorption. Courtesy of R.H. Barkalow and R.W. Kraft, Lehigh University, Bethlehem, PA. (Ref 13.26)

structures formed by eutectic solidification. However, until sufficient homogenization has occurred, the carbides spheroidize and align in bands, causing anisotropy in hot ductility. The aligned alloy carbides may be sites of void formation and cracking, and consequently highly alloyed tool steels require careful hot work to prevent cracking.

Annealing of Tool Steels

Following hot work to bar or plate, tool steels are machined into tools and dies of required shape. Annealing (Fig. 13.5) is required to put the hot work microstructures into a condition suitable for machining and subsequent hardening.

The objective of tool steel annealing treatments is to produce a microstructure consisting of uniformly dispersed spheroidized car-

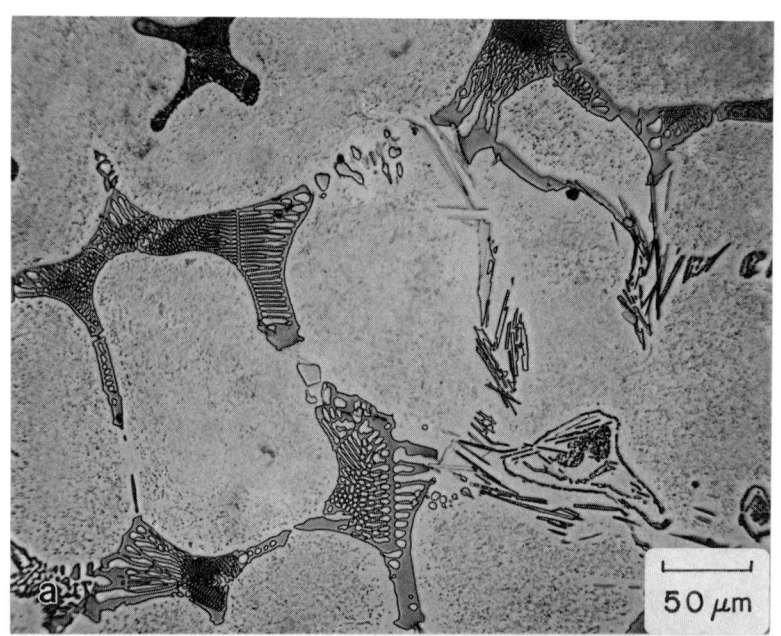
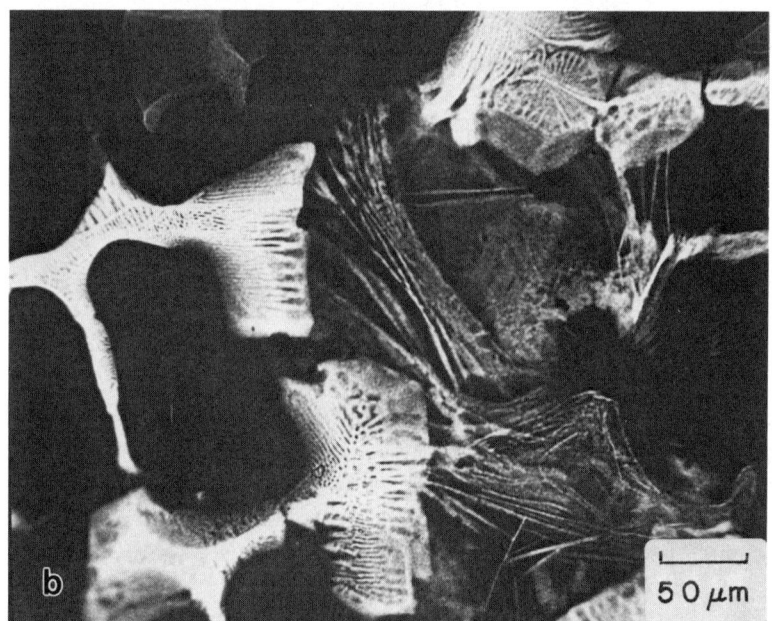

Fig. 13.8. Feathery, herringbone, and MC eutectics in M2 high-speed steel. (a) Light micrograph. $KMnO_4$ etch. (b) Microradiograph of same area. Chromium radiation. Courtesy of R.H. Barkalow and R.W. Kraft, Lehigh University, Bethlehem, PA. (Ref 13.26)

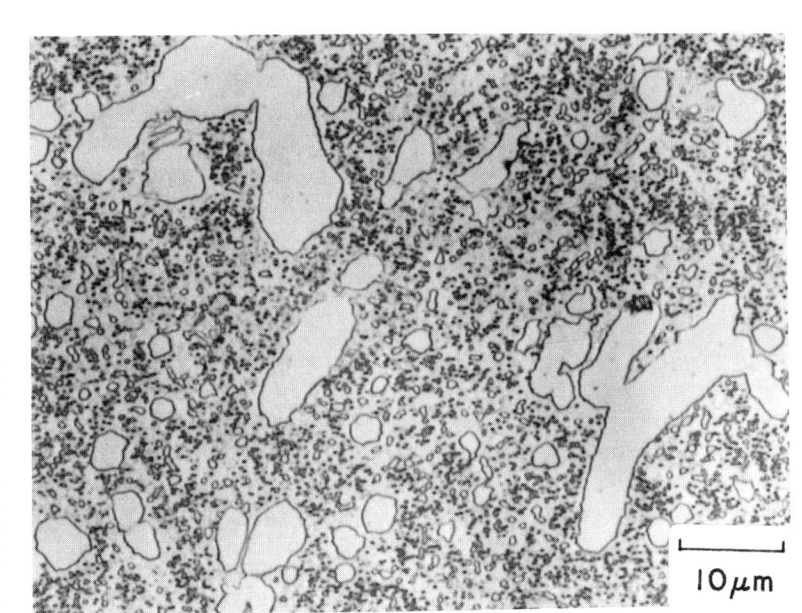

Fig. 13.9. Annealed microstructure of D2 tool steel. Light micrograph. Courtesy of J.R.T. Branco, Colorado School of Mines, Golden

bides in a matrix of ferrite. Such a microstructure has low hardness which renders it machinable and reduces wear on cutting tools. Annealing also refines coarse-grained structures which may have formed during high-temperature hot work, eliminates hard martensite or pearlite microstructure which may have formed during cooling after hot work, and homogenizes the effects of nonuniform deformation which may have developed during hot work of complex or heavy sections.

Figure 13.9 shows the annealed microstructure of type D2 tool steel. The high alloy content of D2 causes two distributions of carbide particles to develop. The coarse particles are primary M_7C_3 carbides which form during melting and are dispersed during hot work. The finer spheroidized particles are a result of secondary low-temperature precipitation or phase transformations. Tool steels with lower alloy content than D2 would have only the finer spheroidized carbides and lower carbide densities.

Annealing is accomplished by heating just to the temperature where all ferrite transforms to austenite. Carbide particles are retained and spheroidized, and the austenite transforms to ferrite and additional spheroidized carbides on cooling. If tool steels are annealed at too high a temperature, the alloy carbides dissolve and the enriched austenite may form carbides on austenitic grain boundaries or transform to pearlite or martensite on cooling, producing too hard a microstructure for good machinability. Similarly, the high hardenability of tool steels makes slow cooling from annealing temperatures essential to ensure that the austenite transforms to ductile ferrite-spheroidized carbide microstructures instead of pearlite or martensite.

Stress Relief of Tool Steels

Residual stresses may be introduced into tool steels by plastic deformation, which accompanies metal removal during machining operations. The residual stresses may cause distortion during heating and hardening, and therefore must be removed by a low-temperature, subcritical heat treatment (Fig. 13.5). Stress relief is typically performed at 650 °C (1202 °F), where ferrite and carbides are stable. The carbides are largely unaffected by stress relief, but high dislocation densities in ferrite strained by machining are reduced by recovery or eliminated by recrystallization of the ferrite.

Heavy sections should be very slowly cooled from 650 °C (1202 °F) to at least 300 °C (572 °F) after stress relief according to Thelning (Ref 13.6). This precaution reduces temperature gradients between the surface and centers of heavy sections and thereby avoids the development of new residual stresses.

Hardening of Tool Steels

The final processing of tool steels consists of heat treating to produce the required hardness and other properties for a given steel and applications. Figure 13.6 shows that final hardening consists of a number of steps, including preheating, austenitizing, cooling or quenching, and tempering. The goal of this processing sequence most often is to produce a microstructure of tempered martensite. Sometimes martempering to equalize temperature prior to martensite formation, or austempering to a microstructure of lower bainite, is performed by holding at temperatures above M_S (Chapter 9).

Preheating and Austenitizing

Highly alloyed tool steels, because of their high hardness and complex microstructures even in the annealed and stress relieved state, are susceptible to distortion and cracking on heating if temperature gradients develop through a cross section. Such gradients cause expansion because of heating and contraction because of austenite formation to occur in different locations of a part. The resulting gradients in volume create stresses which sometimes are high enough to cause cracking, especially in tool steels with low ductility and low resistance to fracture. For these reasons, preheating is applied to alloy tool steels to establish thermal equilibrium prior to heating to the final austenitizing temperature (Ref 13.2, 13.6).

Austenitizing is a very critical step in the hardening of a tool steel. It is in this step where the final alloy element partitioning between the austenitic matrix (which will transform to martensite) and the retained carbides occurs. This partitioning fixes the chemistry, volume fraction, and dispersion of the retained carbides. The retained alloy carbides not only contribute to wear resistance, but also control austenitic grain size. The finer the carbides and the larger the volume fraction of carbides, the more effectively austenitic grain growth is controlled. Thus if austenitizing is performed at too high a temperature, undesirable grain growth may occur as the alloy carbides increasingly coarsen or dissolve into the austenite. A special case of grain coarsening is associated with rehardening high-speed tool steel. Kula and Cohen (Ref 13.28) showed that the discontinuous growth that leads to coarse-grained intergranular fracture or "fish-scale" fracture was due to this dissolution of fine carbides formed from martensitic or bainitic microstructures during a second austenitizing. The discontinuous coarsening did not occur in high-speed tool steels with spheroidized carbide-ferrite microstructures.

The alloying elements not tied up in retained carbides are in solution in the austenite, and thus the carbides provide an important mechanism by which austenite composition is fixed. The austenite composition then sets the hardenability, M_s temperature, retained austenite content, and secondary hardening potential of a tool steel.

Figure 13.10 shows the effect of increasing austenitizing temperature on the as-quenched, as-quenched and subzero cooled, and tempered hardness of an A2 tool steel (Ref 13.6). The highest as-quenched hardness is produced by austenitizing at 950 °C (1742 °F), the recommended austenitizing temperature for A2. In this condition after quenching the retained austenite content is finely dispersed and at a minimum, and therefore subzero cooling has little effect on hardness. With increasing austenitizing temperature, more alloying elements go into solution, the M_s temperature drops, and more austenite is retained at room temperature. As a result, the as-quenched room

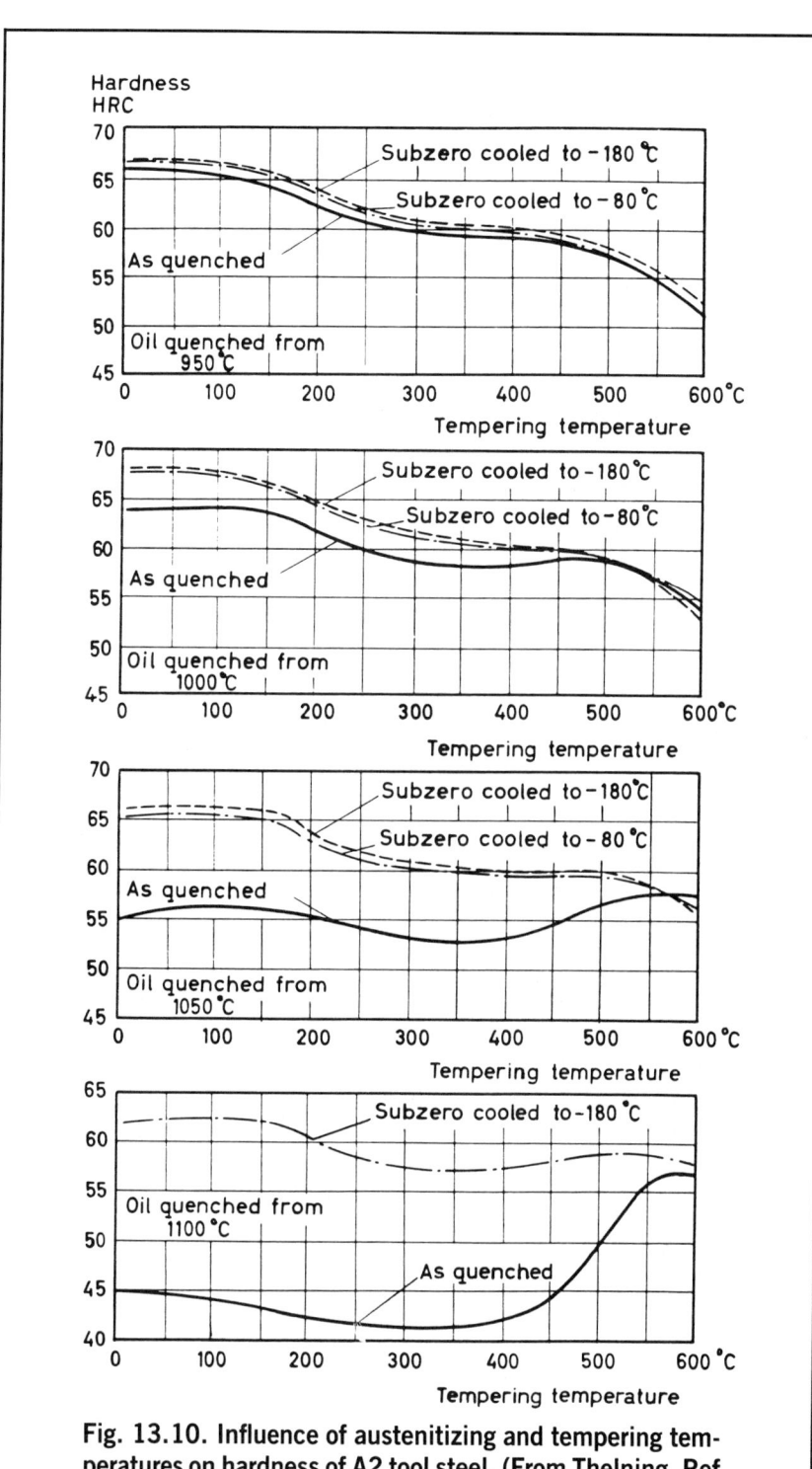

Fig. 13.10. Influence of austenitizing and tempering temperatures on hardness of A2 tool steel. (From Thelning, Ref 13.6)

temperature hardness decreases and subzero cooling has a greater effect as more of the large volume fraction of retained austenite transforms to martensite on subzero cooling. Figure 13.10 shows that eventually tempering, by a combination of retained austenite transformation and secondary hardening, will also raise the hardness of as-quenched structures with large amounts of retained austenite. Not shown is the deleterious increase in austenite grain size which develops as more and more carbides dissolve at the higher austenitizing temperatures (Ref 13.6).

Hardenability and Martensite Formation

The hardenability of most tool steels is high; therefore, oil quenching or air cooling, depending on alloy composition, austenitizing conditions, and section size, is sufficient to produce required microstructures and properties with a minimum of distortion and quench cracking. Medium-carbon, low-alloy steels have been extensively evaluated for hardenability as discussed in Chapter 6, and Jatczak (Ref 13.29) has evaluated the effects of various alloying elements on the hardenability of high-carbon steels. An important difference between medium-carbon hardenable steels and high-carbon tool steels is the strong effect of austenitizing and retained carbide content on hardenability. For example, high austenitizing temperatures decrease alloy carbide content, increase the alloy element content of the matrix austenite, and consequently increase hardenability relative to low austenitizing temperatures applied to the same steel. Therefore, the effects of alloying elements on hardenability in high-carbon steels may be quite different depending on austenitizing conditions (Ref 13.29).

Martensite forms in tool steels when cooling conditions and hardenability are sufficient to prevent diffusion-controlled transformation to proeutectoid carbides, pearlite, and bainite. The matrix austenite composition determines the morphology of the martensite microstructure (Chapter 3). Figure 13.11 shows lath martensite formed in H-13 steels, and Fig. 13.12 shows plate martensite formed in A-2 steel. In both microstructures, retained austenite and retained carbides are present, but to a much lesser extent in the H-13 steel than in the A-2 steel. The retained austenite is present in thin sheets between the parallel martensite laths of the H-13 steel, and as triangular regions between the nonparallel plates of the A-2 steel. Although there are many crystallographic variants of the martensite laths or plates formed in a given austenite grain, all of the austenite retained in a

Fig. 13.11. Lath martensite formed in H-13 tool steel. (a) Bright-field image. (b) Dark-field image of same area illuminating interlath retained austenite. Transmission electron micrographs. Courtesy of J.R.T. Branco, Colorado School of Mines, Golden

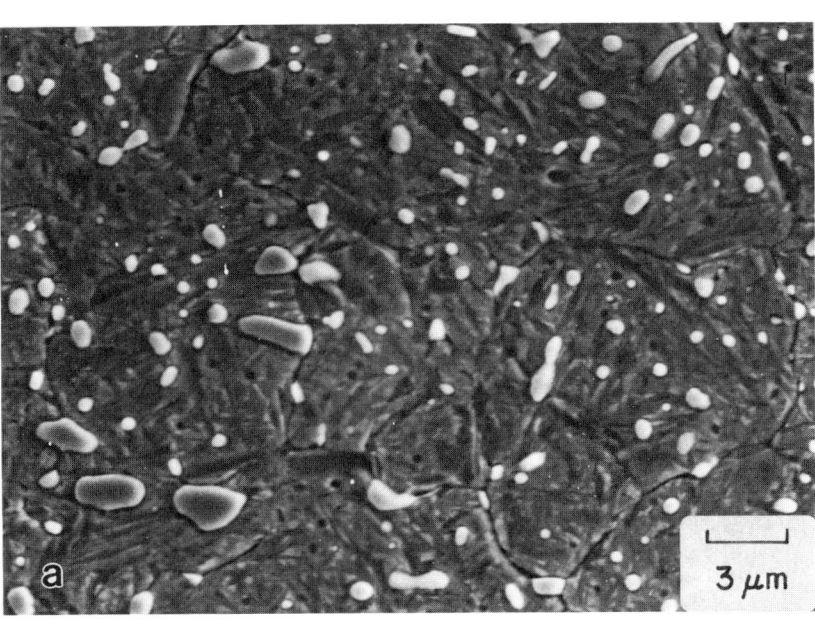

Fig. 13.12. Retained carbides and plate martensite formed in A2 tool steel. (a) Scanning electron micrograph. Courtesy of A. Wahid, Colorado School of Mines, Golden. (b) Transmission electron micrograph showing martensite, retained carbides, and retained austenite. Courtesy of J.R.T. Branco, Colorado School of Mines, Golden

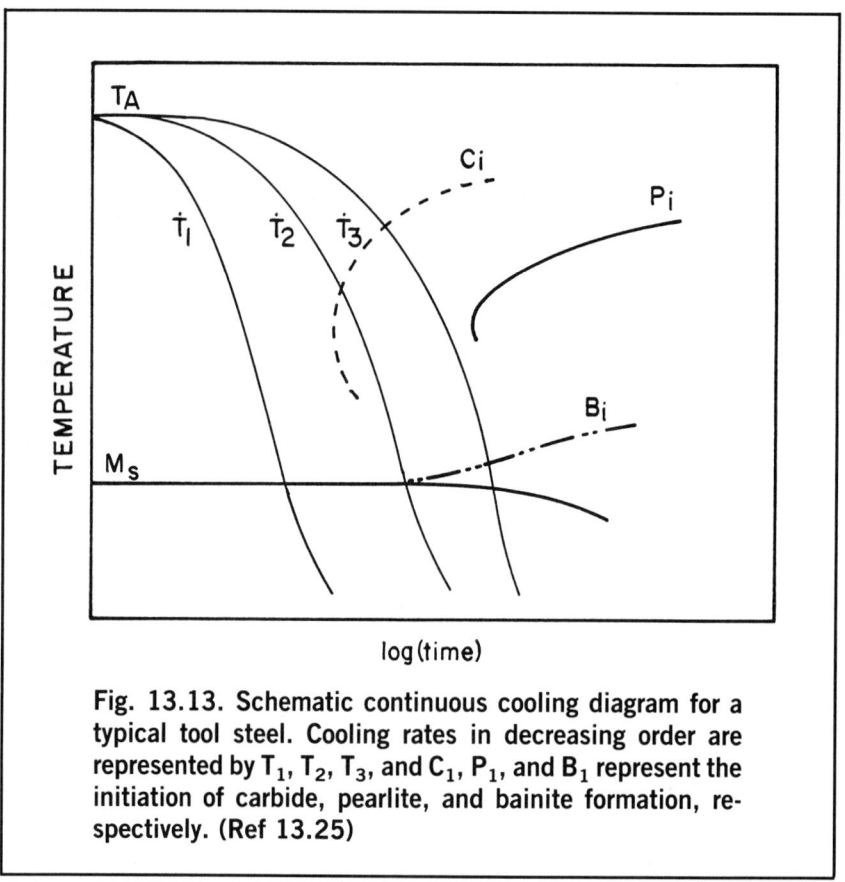

Fig. 13.13. Schematic continuous cooling diagram for a typical tool steel. Cooling rates in decreasing order are represented by T_1, T_2, T_3, and C_1, P_1, and B_1 represent the initiation of carbide, pearlite, and bainite formation, respectively. (Ref 13.25)

given austenite grain has the same orientation despite its quite dispersed appearance within the martensite. The austenite retained after quenching may be transformed to martensite by subzero cooling, as discussed relative to Fig. 13.10, or more commonly is transformed to carbides and ferrite during high-temperature tempering.

Grain Boundary Carbide Formation

Tool steels are susceptible to grain boundary carbide formation during relatively slow oil quenching or air cooling for hardening. Figure 13.13 shows schematically the effects of three cooling rates on the transformation of a typical tool steel. The high hardenability of tool steels effectively suppresses pearlite formation at all cooling rates. Bainite formation is also readily suppressed except in heavy sections, which cool slowly. However, the formation of small amounts of carbides on austenite

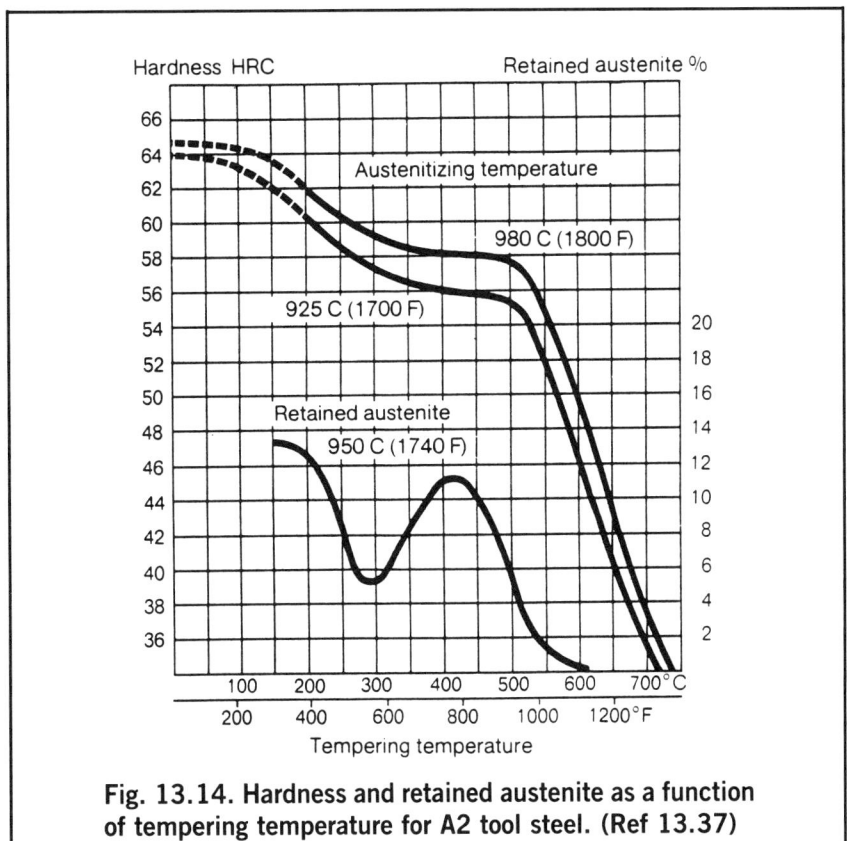

Fig. 13.14. Hardness and retained austenite as a function of tempering temperature for A2 tool steel. (Ref 13.37)

grain boundaries is difficult to suppress, as shown by the intersection of the two slower cooling rates with the carbide initiation curve, C_1. The small amounts of carbides do not significantly affect hardness but may lower tool steel fracture resistance, leading to quench cracking, intergranular fracture of overheated tool steels, or reduced performance of hot work die steels such as H-13 (Ref 13.30, 13.32).

Phosphorus segregates to austenite grain boundaries during austenitizing for hardening and contributes to fracture problems in high-carbon steel. The combination of segregated phosphorus and high carbon content leads to cementite grain boundary allotriomorph formation even during oil quenching (Ref 13.33, 13.34), and the phosphorus and carbides lower the fracture strength of the prior austenite grain boundaries (Ref 13.35, 13.36). As a result, if quenching or service stresses are high enough, failure by intergranular cracking occurs, especially in steels with coarse austenitic grain sizes. The susceptibility to grain-boundary cracking is reduced by maintaining recommended aus-

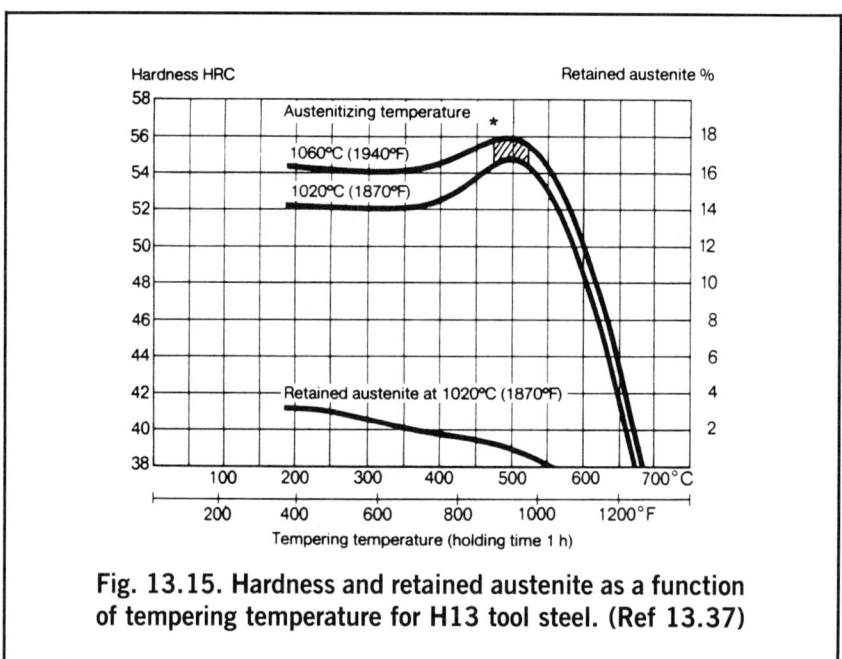

Fig. 13.15. Hardness and retained austenite as a function of tempering temperature for H13 tool steel. (Ref 13.37)

tenitizing temperatures, which in most tool steels of high carbon content are designed to produce fine austenitic grain sizes. The latter microstructures, when hardened and tempered at low temperatures, fracture by transgranular shear or ductile fracture associated with microvoid formation around the retained carbide particles (Ref 13.36).

Tempering of Tool Steels

Tempering is the final heat treatment step applied to tool steels (Fig. 13.6). As in the lower alloy steels discussed in Chapter 8, tempering has the important function of improving toughness in both low-alloy and tool steels. However, secondary hardening or the precipitation of alloy carbides at high tempering temperatures is much more important in tool steels than in low-alloy steels. Also, double or even triple tempering steps are applied to tool steels to ensure that toughness is improved after microstructural changes are induced by the first tempering steps.

Figures 13.14, 13.15, and 13.16 show examples of hardness as a function of tempering temperature for A-2, H-13, and several high-speed tool steels (Ref 13.38), respectively. The curves for the A-2 and H-13 are based on double tempering and that of the high-speed steels

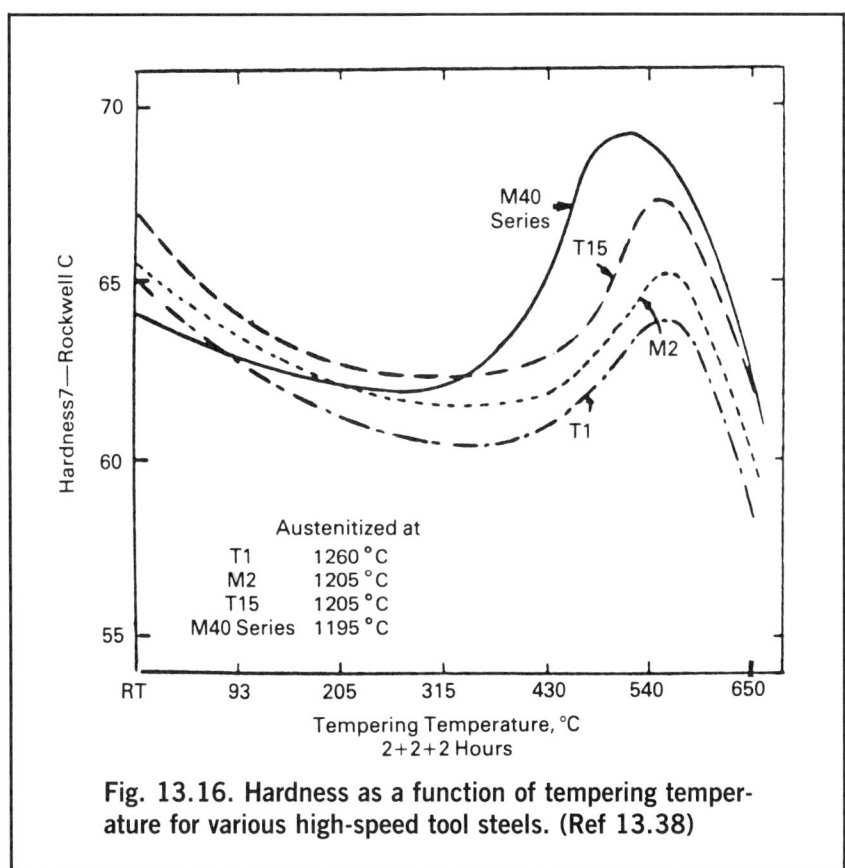

Fig. 13.16. Hardness as a function of tempering temperature for various high-speed tool steels. (Ref 13.38)

on triple tempering. Each tempering treatment was at least 2 h in duration. The peak hardness associated with secondary hardening around 500 °C (932 °F) increases with increasing alloy content and depends, as discussed earlier, on the balance of retained carbides, retained austenite, and the composition of the martensite in the as-quenched condition. In the A-2 and H-13 steels, higher martensite carbon content offsets the effects of increased retained austenite on hardness in the specimens austenitized at the higher temperatures.

The formation of alloy carbides during tempering requires the diffusion of carbide-forming elements. The atoms of the latter mostly diffuse substitutionally through the bcc iron lattice of the tempered martensite, a process characterized by low diffusion coefficients. The sluggish diffusion makes effective diffusion distances very short, leading to very fine, closely space alloy-carbide precipitates. The same sluggish diffusion retards carbide coarsening during high-temperature

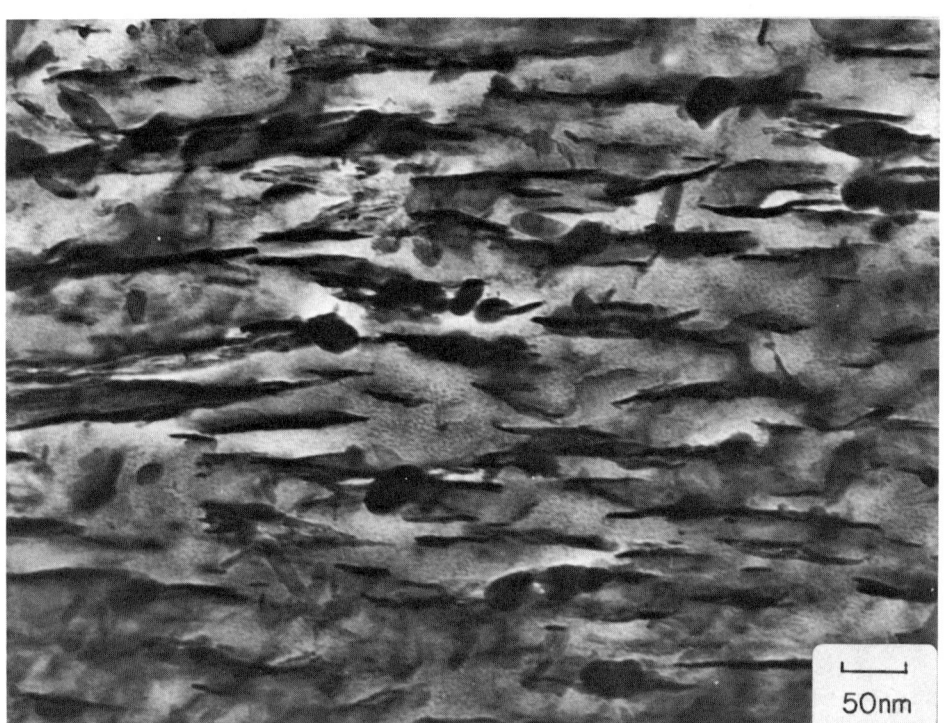

Fig. 13.17. Alloy carbides in a lath of martensite in H13 tool steel tempered 100 h at 530 °C (986 °F). Transmission electron micrograph. Courtesy of J.R.T. Branco, Colorado School of Mines, Golden

service, and makes tool steels resistant to softening during hot forging, die casting, and high-speed cutting operations.

At low tempering temperatures, transition iron carbides and cementite (M_3C) form as discussed in Chapter 8. At higher tempering temperatures, because of increased diffusivity of the alloying elements, alloy carbides precipitate. Honeycombe (Ref 13.39) has reviewed alloy carbide formation on tempering and shown that many of the alloy carbides form as fine discs or needles on preferred crystallographic habit planes within the plates or laths of tempered martensite. Figure 13.17 shows fine precipitates formed in a lath of tempered martensite in H-13 steel tempered at 550 °C (1022 °F).

The crystallography and composition of the alloy carbides which form during tempering are very sensitive to the specific alloying elements present. In tool steels containing chromium, the sequence of

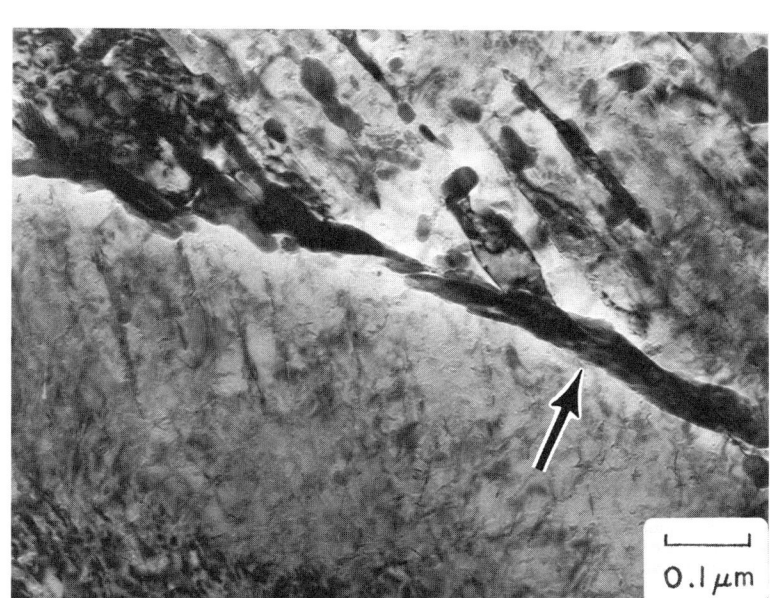

Fig. 13.18. Interlath carbides (arrow) formed in H13 tool steel tempered at 600 °C (1112 °F) for 2 h. Transmission electron micrograph. Courtesy of J.R.T. Branco, Colorado School of Mines, Golden. (Ref 13.25)

precipitation with increasing tempering may be M_3C, then M_7C_3, followed by $M_{23}C_6$ (Ref 13.40), while in tool steels rich in molybdenum, the sequence might be M_3C, followed by M_2C, followed by M_6C (Ref 13.41). Binary carbides, i.e., those formed between a single element and carbon, coarsen readily during tempering or in high-temperature service. Carbides composed of multiple alloying elements coarsen at lower rates. Multiple alloying might also change the sequence of carbide precipitation noted above. For example, $M_{23}C_6$ may form after M_2C and prior to M_6C in steels with high ratios of chromium and molybdenum to carbon (Ref 13.25).

Retained Austenite Transformation and Double Tempering in Tool Steels

Retained austenite transforms to ferrite and cementite during tempering. In low-alloy steels retained austenite transforms to cementite and ferrite between 200 and 300 °C (392 and 572 °F) (Chapter

8). The austenite in highly alloyed tool steels, however, is much more stable and does not fully transform until temperatures in excess of 500 °C (932 °F) are attained, as shown in Fig 13.14 and 13.15. Figure 13.18 shows interlath carbides which have formed in H-13 steel tempered at 600 °C (1112 °F). The carbides are rather coarse and planar, and similar carbide distributions may be responsible for the drop in impact toughness noted in H-13 steels tempered between 475 and 535 °C (887 and 995 °F). Double tempering would tend to spheroidize and render less harmful interlath carbides formed by the transformation of retained austenite. Double tempering is also believed to temper any martensite that may have formed by transformation of retained austenite to martensite during cooling after the first tempering treatment.

Summary

This chapter has reviewed the effects of alloying and processing on the properties and structure of tool steels. Tool steels have evolved to perform machining and the most difficult hot and cold forming operations on other materials, and considerable practical information about selection, processing, and performance of tool steels has been generated over the years (Ref 13.1-13.19). However, greater and greater production demands are being placed on tool steels and much deeper understanding of mechanisms of wear and fracture and the alloying-processing-structure relationships which optimize performance of tool steels is required. This next level of effort is in progress (Ref 13.8, 13.42), especially with respect to factors which control strength and toughness of tool steels (Ref 13.43-13.45).

References

13.1 G.A. Roberts and R.A. Cary, *Tool Steels*, 4th ed., American Society for Metals, Metals Park, OH, 1980
13.2 R. Wilson, *Metallurgy and Heat Treatment of Tool Steels*, McGraw-Hill, London, 1975
13.3 P. Payson, *The Metallurgy of Tool Steels*, Wiley, New York, 1962
13.4 M.G.H. Wells and L.W. Lherbier (Eds.), *Processing and Properties of High Speed Tool Steels*, TMS-AIME, Warrendale, PA, 1980
13.5 G. Hoyle, *High Speed Steels*, Butterworths, London, 1988
13.6 K-E. Thelning, *Steel and Its Heat Treatment*, 2nd ed., Butterworths, London, 1984
13.7 Tool Materials, in *Metals Handbook, Desk Edition*, American Society for Metals, Metals Park, OH, 1985

13.8 G. Krauss and H. Nordberg (Eds.), *Tool Materials for Molds and Dies*, Colorado School of Mines Press, Golden, 1987

13.9 *Metals Handbook*, Vol 3, 9th ed., *Properties and Selection: Stainless Steels, Tool Materials, and Special-Purpose Metals*, American Society for Metals, Metals Park, OH, 1980

13.10 P.D. Harvey (Ed.), *Heat Treatment of Tool Steels*, Metals Engineering Institute, American Society for Metals, Metals Park, OH, 1981

13.11 S. Floreen, The Physical Metallurgy of Maraging Steels, *Met Rev*, Vol 12, 1968, p 115-128

13.12 R.K. Wilson (Ed.), *Maraging Steels—Recent Developments and Applications*, TMS-AIME, Warrendale, PA, 1988

13.13 L.E. Toth, *Transition Metal Carbides and Nitrides*, Academic Press, New York, 1971

13.14 D.H. Jack and K.N. Jack, Carbides and Nitrides in Steels, *Mater Sci Eng*, Vol 11, 1973, p 1-27

13.15 H. Brandis, E. Haberling, and H.H. Weigard, Metallurgical Aspects of Carbides in High Speed Steels, in *Processing and Properties of High Speed Tool Steels*, M.G.H. Wells and L.W. Lherbier (Eds.), TMS-AIME, Warrendale, PA, 1980, p 1-18

13.16 B. Uhrenius and H. Harvig, A Thermodynamic Evaluation of Carbide Solubilities in the Fe-Mo-C, Fe-W-C, and Fe-Mo-W-C Systems at 1000 °C, *Met Sci*, Vol 9, 1975, p 67-81

13.17 T. Wada and E.K. Ohriner, Phase Equilibrium in the Fe-Mo-W-C Systems Containing $(Mo_{0.7}W_{0.3})C$ Carbide, *CALPHAD*, Vol 8, 1984, p 69-74

13.18 H. Wada, Thermodynamics of the Fe-Cr-C System at 985 K, *Met Trans A*, Vol 16A, 1985, p 1479-1490

13.19 S. Hertzman, A Study of Equilibria in the Fe-Cr-Ni-Mo-C-N System at 1273 K, *Met Trans A*, Vol 18A, 1987, p 1767-1778

13.20 J-O Andersson, A Thermodynamic Evaluation of the Fe-Cr-C System, *Met Trans A*, Vol 19A, p 627-636

13.21 P. Gustafson, An Experimental Study and a Thermodynamic Evaluation of the Cr-Fe-W System, *Met Trans A*, Vol 19A, 1988, p 2531-2546

13.22 P. Gustafson, A Thermodynamic Evaluation of the C-Cr-Fe-W System, *Met Trans A*, Vol 19A, 1988, p 2547-2554

13.23 L.R. Woodyatt and G. Krauss, Iron-Chromium-Carbon System at 870 °C, *Met Trans A*, Vol 7A, 1976, p 983-989

13.24 K. Bungardt, E. Kunze, and E. Horn, Investigation of the Structure of the Iron-Chromium-Carbon System, *Archiv Eisenhütt*, Vol 29, 1958, p 193

13.25 J.R.T. Branco and G. Krauss, Heat Treatment and Microstructure of Tool Steels for Molds and Dies, in *Tool Materials for Molds and Dies*, G. Krauss and H. Nordberg (Eds.), Colorado School of Mines Press, Golden, 1987, p 94-117

13.26 R.K. Barkalow, R.W. Kraft, and J.I. Goldstein, Solidification of M2 High Speed Steel, *Met Trans A*, Vol 3, 1972, p 919-926

13.27 W. Roberts, Dynamic Changes That Occur During Hot Working and Their Significance Regarding Microstructural Development and Hot Workability, in *Deformation, Processing, and Structure*, G. Krauss (Ed.), American Society for Metals, Metals Park, OH, 1984, p 109-184

13.28 E. Kula and M. Cohen, Grain Growth in High Speed Steel, *ASM Trans*, Vol 46, 1954, p 727-798

13.29 C.F. Jatczak, Hardenability in High Carbon Steels, *Met Trans*, Vol 4, 1973, p 2267-2277
13.30 H. Nilsson, O. Sandberg, and W. Roberts, Influence of Austenitization Temperature and the Cooling Rate after Austenitization on the Mechanical Properties of the Hot Work Tool Steel H-11 and H-13, in *Tools for Die Casting*, Uddeholm and Swedish Institute for Metals Research, 1983, p 51-70
13.31 M.L. Schmidt, Effect of Austenitizing Temperature on Laboratory Treated and Large Section Sizes of H-13 Tool Steel, in *Tool Materials for Molds and Dies*, G. Krauss and H. Nordberg (Eds.), Colorado School of Mines Press, Golden, 1987, p 118-164
13.32 D.L. Cocks, Longer Die Life From H-13 Die Casting Dies by the Practical Applications of Recent Research Results, in *Tool Materials for Molds and Dies*, G. Krauss and H. Nordberg (Eds.), Colorado School of Mines Press, Golden, 1987, p 340-350
13.33 H.M. Obermeyer and G. Krauss, Toughness and Intergranular Fracture of a Simulated Carburized Case in Ex-24 Type Steel, *J Heat Treat*, Vol 1 (No. 3), 1980, p 31-39
13.34 T. Ando and G. Krauss, The Effect of Phosphorus Content on Grain Boundary Cementite Formation in AISI 52100 Steel, *Met Trans A*, Vol 12A, 1981, p 1283-1290
13.35 D.L. Yaney, The Effects of Phosphorus and Tempering on the Fracture of AISI 52100 Steel, M.S. thesis, Colorado School of Mines, Golden, 1981
13.36 G. Krauss, The Relationship of Microstructure to Fracture Morphology and Toughness of Hardened Hypereutectoid Steels, in *Case-Hardened Steels: Microstructural and Residual Stress Effects*, D.E. Diesburg (Ed.), TMS-AIME, Warrendale, PA, 1983, p 33-58
13.37 Data Sheets: Cold Work Tool Steel (AISI A2) and Hot Work Tool Steel (AISI H 13), Uddeholm Steel Division, Hagfors, Sweden
13.38 S.G. Fletcher and C.R. Wendell, *ASM Metals Eng Quart*, 1 Feb 1966, in Ref 13.5, p 146
13.39 R.W.K. Honeycombe, *Steels: Microstructure and Properties*, Edward Arnold Ltd and American Society for Metals, Metals Park, OH, 1982
13.40 R.G. Baker and J. Nutting, The Tempering of 2.25%Cr-1%Mo Steel for Quenching and Normalizing, *J Iron Steel Inst*, Vol 192, 1959, p 257-268
13.41 K.H. Kuo and C.L. Jia, Crystallography of $M_{23}C_6$ and M_6C Precipitate in a Low Alloy Steel, *Acta Met*, Vol 33, 1985, p 991-996
13.42 H. Berns (Ed.), *Proceedings, Tool Steel Conference*, Bochum, FRG, Sept 1989
13.43 F.B. Pickering, The Properties of Tool Steels for Mold and Die Applications, in *Tool Materials for Molds and Dies*, G. Krauss and H. Nordberg (Eds.), Colorado School of Mines Press, Golden, 1987, p 3-32
13.44 H. Berns, Strength and Toughness of Hot Working Tool Steels, in *Tool Materials for Molds and Dies*, G. Krauss and H. Nordberg (Eds.), Colorado School of Mines Press, Golden, 1987, p 45-65
13.45 R.M. Hemphill and D.E. Wert, Impact and Fracture Toughness Testing of Common Grades of Tool Steels, in *Tool Materials for Molds and Dies*, G. Krauss and H. Nordberg (Eds.), Colorado School of Mines Press, Golden, 1987, p 66-91

CHAPTER 14

Cast Irons

Cast irons are a large family of alloys based on the Fe-C system. The carbon content of cast irons is significantly higher than that of steels, 2 wt.% or greater, and as a result cast irons have unique solidification and microstructural characteristics. In particular, graphite in flake and nodular form is a microstructural component found only in cast irons. The solid-state transformations on which cast iron heat treatments are based, however, are similar to those applied to steels. This chapter describes the various types and microstructures of cast irons as influenced by alloying, solidification, and heat treatment.

Phase Relationships

Cast irons contain more than 2 wt.% carbon and therefore solidify at temperatures well below the solidification range of steels. The Fe-C phase diagram, Fig. 1.1 (Chapter 1), shows the low-melting eutectic reaction, centered on a liquid composition of 4.3 wt.% carbon, which accounts for the excellent castability of cast irons. Primary-phase solidification and the eutectic reaction may involve the formation of either cementite or graphite. For example, the eutectic reaction may take place as:

$$L\ (4.30\%\ C) = A\ (2.11\%\ C) + Cementite\ (6.67\%\ C) \quad (Eq\ 14.1)$$

or

$$L\ (4.26\%\ C) = A\ (2.08\%\ C) + Graphite\ (100\%\ C) \quad (Eq\ 14.2)$$

The forms and morphology of the carbon-rich phase then determine the nature of a given cast iron. Cast irons may be hypoeutectic (carbon content below the eutectic carbon content), hypereutectic (carbon content above the eutectic carbon content), or eutectic. The eutectic

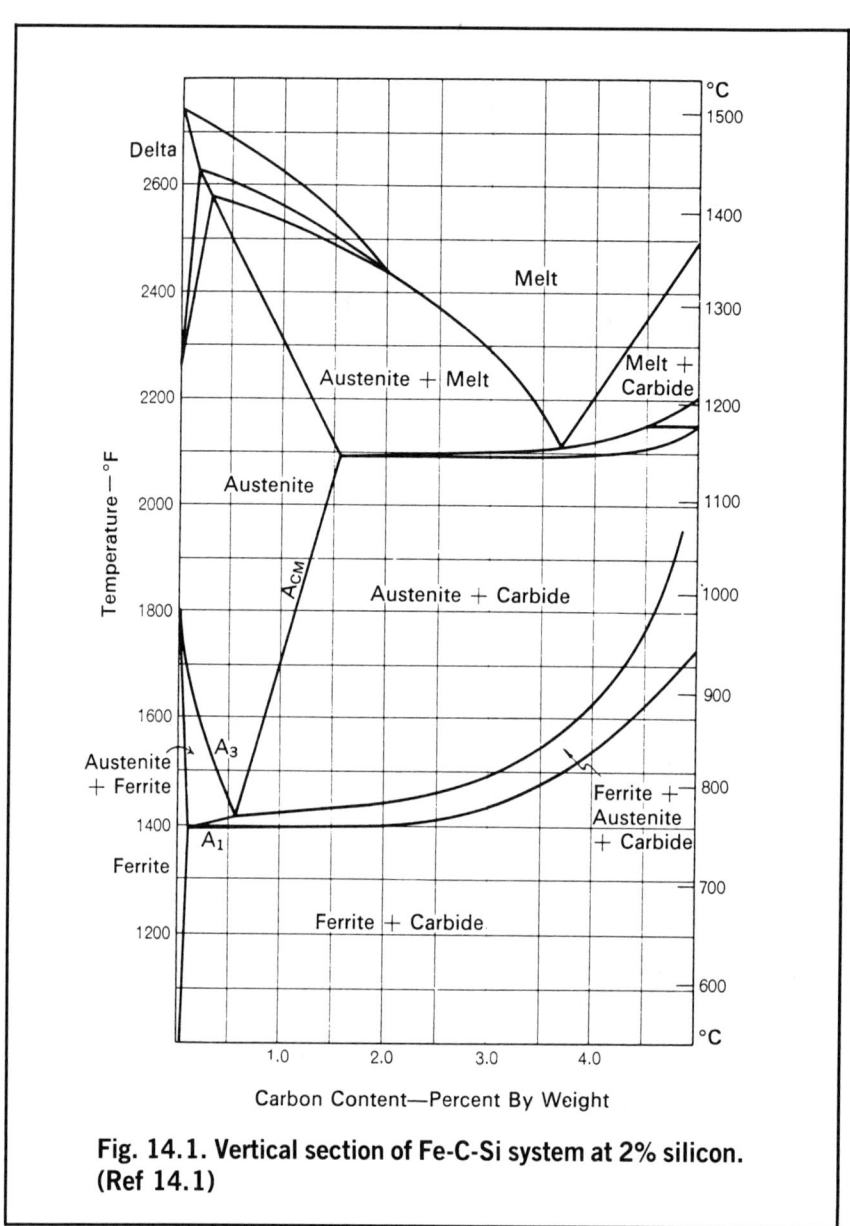

Fig. 14.1. Vertical section of Fe-C-Si system at 2% silicon. (Ref 14.1)

structure of austenite and cementite is given the special name ledeburite.

Cast irons contain elements which significantly shift the boundaries of the phase fields in the Fe-C diagram, and ternary isothermal or vertical sections must be used to approximate the phase relationships. Silicon is a major alloying element in cast irons, and Fig. 14.1

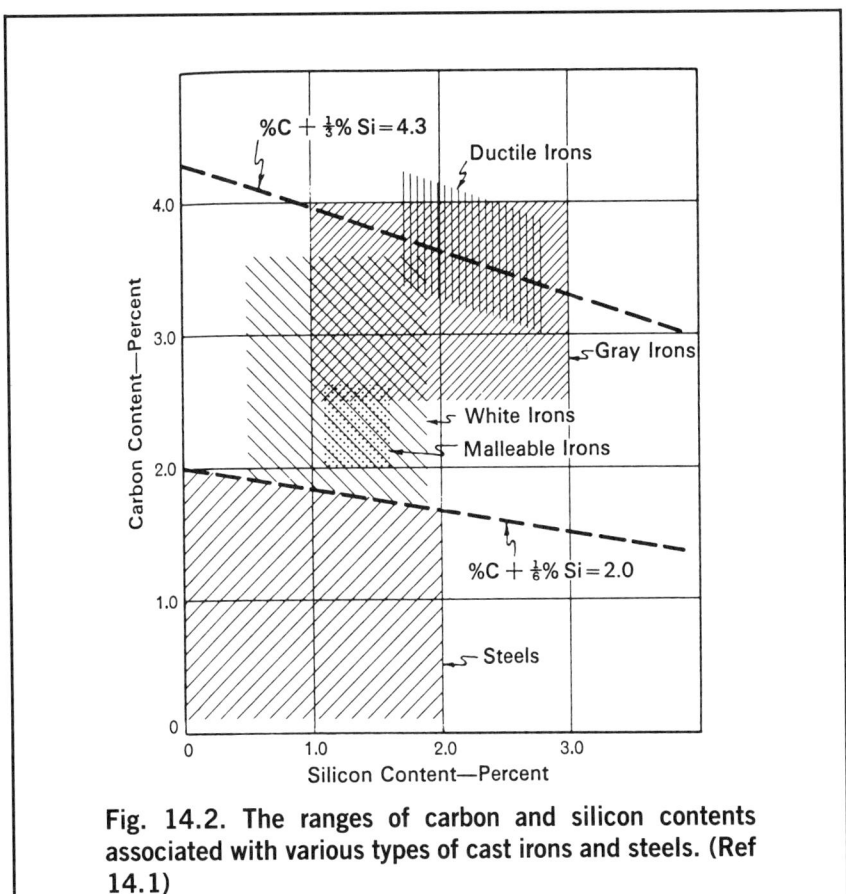

Fig. 14.2. The ranges of carbon and silicon contents associated with various types of cast irons and steels. (Ref 14.1)

shows a vertical section through the Fe-C-Si system at 2.00% silicon (Ref 14.1). The eutectoid and eutectic compositions are shifted to lower carbon contents, and the three-phase fields exist over a range of temperatures. Figure 14.1 shows carbon is present in the form of cementite. Silicon, however, is an alloying element which strongly promotes graphite formation, as do aluminum, boron (<0.15%), copper, nickel, and titanium (<0.25%) (Ref 14.2). Other elements such as bismuth, boron (>0.15%), chromium, manganese, molybdenum, tellurium, vanadium, and titanium (>0.25%), tend to decrease the graphitization potential of cast irons.

Figure 14.2 shows the ranges of carbon and silicon contents associated with steels and the various types of cast irons (Ref 14.1). Gray and ductile irons, which solidify directly to graphite and austenite, generally have higher silicon and carbon contents than white and

Table 14.1. Composition Ranges for Unalloyed Cast Irons (Ref 14.1)

Element	Gray iron, %	White iron, %	Malleable iron (cast white), %	Ductile iron, %
Carbon	2.5–4.0	1.8–3.6	2.00–2.60	3.0–4.0
Silicon	1.0–3.0	0.5–1.9	1.10–1.60	1.8–2.8
Manganese	0.25–1.0	0.25–0.80	0.20–1.00	0.10–1.00
Sulfur	0.02–0.25	0.06–0.20	0.04–0.18	0.03 max
Phosphorus	0.05–1.0	0.06–0.18	0.18 max	0.10 max

malleable cast irons, which solidify as austenite and cementite. As noted in reference to Fig. 14.1, silicon lowers the carbon content of the eutectic liquid. This is quantitatively recognized by a factor referred to as the carbon equivalent (C.E.), defined as (Ref 14.1):

$$\text{C.E.} = \%C + 1/3\%Si \qquad (Eq\ 14.3)$$

or if significant amounts of phosphorus are present:

$$\text{C.E.} = \%C + \frac{\%Si + \%P}{3} \qquad (Eq\ 14.4)$$

A C.E. of 4.3 indicates that an alloy is of the eutectic composition, as is the case for an Fe-4.3C alloy or any alloy where the sum of iron, silicon, and phosphorus contents add up to 4.3 according to Eq 14.3 and 14.4. Values of C.E. less and greater than 4.3 indicate hypoeutectic or hypereutectic alloys, respectively.

Cooling rates also affect solidification, with slower cooling favoring graphite formation and rapid cooling favoring cementite formation. Therefore, a cast iron with the same silicon and carbon content may form graphite in heavy, slow-cooled sections and cementite in thin, more rapidly cooled sections.

Table 14.1 shows typical composition ranges for silicon, manganese, sulfur, and phosphorus in the various types of cast irons. Low sulfur in solid solution, either as a result of restricting sulfur content in the melt or combination with manganese in the form of MnS, promotes matrix ferrite formation. High sulfur in solid solution, or manganese over the amount required to tie up sulfur, promote matrix pearlite formation (Ref 14.1). Phosphorus, in amounts above 0.2%, forms a low-melting eutectic structure, referred to as steadite, which consists of iron phosphide and ferrite or iron phosphide, cementite, and ferrite (Ref 14.2). Steadite forms in the last liquid to freeze, promoting fluidity, and is a hard, relatively brittle microstructural component of gray irons. Other elements are also added to the various cast irons for specific reasons, and will be discussed in the following sections on the specific types of cast irons. A large literature on cast irons exists, and Ref 14.1 through 14.5 have been selected as good sources for additional theory and technology regarding the production and use of cast irons.

Gray Cast Irons

Gray cast irons solidify by the formation of graphite with a flake-like morphology. When a gray cast iron is fractured, the graphite on the surface produces the gray appearance which gives gray cast irons their name.

During eutectic solidification, graphite flakes and austenite grow into the liquid (Ref 14.2). The graphite crystal structure is hexagonal, and the flakes grow by extension along the "a" axis. The graphite plates lead the austenite into the liquid. As a result, the leading edges of the flakes are not constrained and the flakes turn and bend, producing the nonparallel flake arrays typically seen on two-dimensional metallographic surfaces. Branching of the flakes also occurs in response to solidification conditions, and all of the graphite in a given eutectic cell tends to be interconnected.

A typical gray cast iron microstructure, unetched and etched, is shown in Fig. 14.3. The curved, irregular thin cross-sections of the graphite flakes are shown. The austenite which formed on solidification with the graphite has transformed to pearlite on cooling through the eutectoid temperature. Figure 14.4 shows the fracture surface of the same gray iron broken at room temperature. The three-dimensional plate morphology of the graphite flakes and the tendency for the secondary fractures to open along the flake-matrix interface are shown.

A number of variants of flake graphite distributions may develop in gray cast iron, depending on cooling rate and composition (Ref 14.4-14.5). Figure 14.5 shows five types of flake graphite as classified according to AFS-ASTM specifications. These sketches have been reduced for publication, but are available from ASTM as charts which represent graphite distributions at 100×.

The mechanical properties of gray cast irons are directly related to their microstructures. The graphite flakes have almost no strength, and this fact, together with the interconnection of the flakes, produces very low fracture resistance under tensile or bending loads. Therefore, ultimate tensile strengths and ductilities of gray cast irons with ferritic microstructures may be quite low, 22 ksi (152 MPa) and well

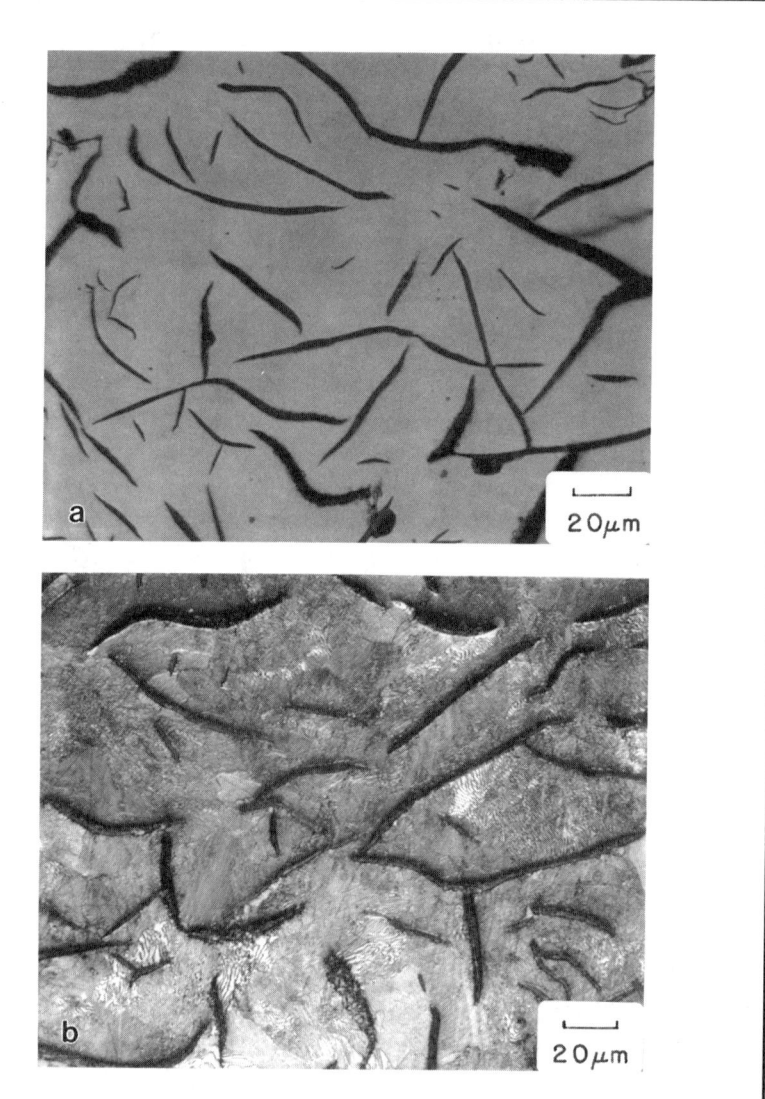

Fig. 14.3. Microstructure of gray cast iron. Graphite is in flake form. (a) Unetched. (b) Etched with nital. Light micrographs. Courtesy of J. McClain, Colorado School of Mines, Golden

below 1.0%, respectively (Ref 14.4, 14.5). Tensile strengths and hardness, however, can be significantly increased by causing the matrix to transform to pearlite instead of ferrite and by producing graphite flake distributions which reduce continuous fracture paths through the microstructure.

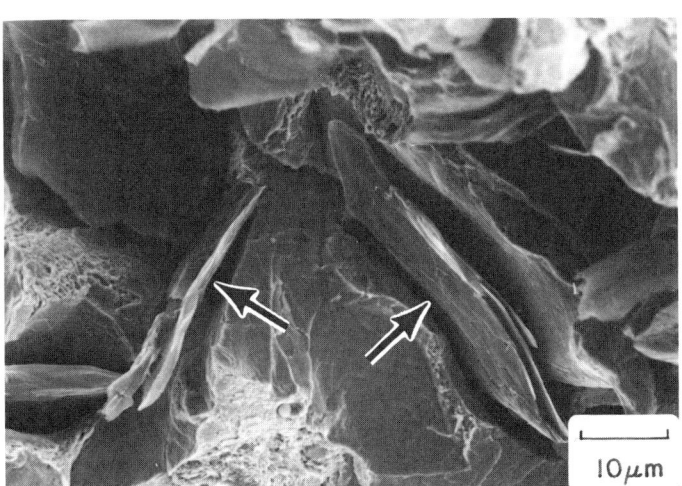

Fig. 14.4. Fracture surface of gray cast iron. Arrows point to graphite flakes. Scanning electron micrograph. Courtesy of S. Diets and R. McGrew, Colorado School of Mines, Golden

The compressive strengths of gray cast irons are high, and this characteristic, plus the excellent vibration damping capacity contributed by the graphite network, makes gray cast irons well suited for heavy equipment and machine tool foundations. Castings, of course, can make complicated shapes of a variety of sizes which would be difficult or impossible to make by fabrication with wrought materials. Most castings must receive some finish machining, and the machinability of gray cast irons, because of their low strength and fracture resistance, is generally quite good.

The heat treatments applied to gray cast irons are most commonly either stress relief heat treatments required because of nonuniform cooling of castings with irregular shapes or annealing heat treatments required to improve machinability. Subcritical heating is used for both stress relief and annealing. Temperatures between 550 and 650 °C (1022 and 1202 °F), are used to relieve stresses without significantly lowering strength and hardness, while higher temperatures, between 700 and 760 °C (1292 and 1400 °F), are used to lower hardness for improved machinability by spheroidizing matrix pearlite. Much higher annealing temperatures in the austenite-carbide-graphite phase field, 900 to 955 °C (1652 to 1751 °F), are used to transform very hard, massive carbides to graphite. Microstructures with extensive

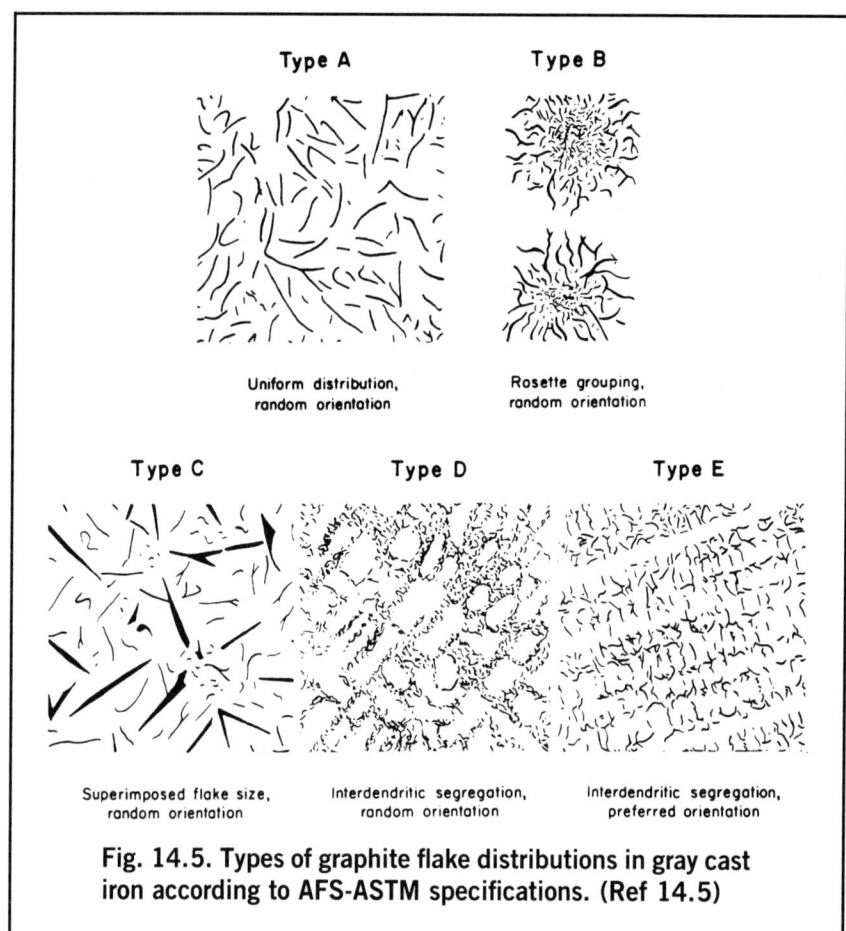

Fig. 14.5. Types of graphite flake distributions in gray cast iron according to AFS-ASTM specifications. (Ref 14.5)

massive carbides, i.e., white cast iron microstructures, may form in thin sections of gray cast iron. These microstructures are almost impossible to machine, and where possible should be avoided by casting and alloy design (Ref 14.1-14.5) rather than heat treatment.

Ductile Cast Iron

Ductile cast irons solidify by the formation of graphite in a spherical morphology. Other terms for ductile cast iron are nodular, spherulitic, and spheroidal cast iron. The growth of spheres of graphite is caused by innoculation of cerium or magnesium into molten cast iron. The graphite spheres nucleate within the liquid independent of austenite formation, and grow outwards until they contact the solidifying austenite. The mechanism of spherical growth is still controversial, and Elliot (Ref 14.2) has reviewed the various explanations for the effect of cerium, magnesium, and elements such as oxygen and sulfur on graphite growth.

Figure 14.6 shows the unetched and etched microstructure of a ductile cast iron. The spherical shape of the graphite is readily inferred from the circular cross sections of the graphite particles. The etched microstructure of the ductile cast iron shows that the austenite matrix has transformed to ferrite and pearlite. This mixed microstructure, with its alternate black and white circular appearance, is often referred to as a bull's-eye structure. Figure 14.7 shows the spherical three-dimensional morphology of graphite nodules on the fracture surface of a ductile cast iron.

The transformation of the austenite to ferrite, pearlite, or both structures, is a function of alloying and cooling rate. Once a gray or ductile cast iron has solidified to graphite and austenite, the decreasing solubility of carbon in austenite with decreasing temperature (Fig. 1.1 and 14.1) is accommodated by precipitation of additional graphite on the flakes or nodules. Then on further cooling through the eutectoid transformation temperature range, the austenite may form ferrite and graphite, with the graphite again growing on the existing flakes or spheres. This process is favored by silicon, aluminum, and small amounts of titanium, and since it depends on carbon diffusion from the austenite through the ferrite to the graphite, is favored by slow cooling. The austenite may also transform to pearlite. This process is promoted by more rapid cooling and alloying elements such as manganese, chromium, and molybdenum as well as many of the other elements found in cast irons. Table 14.2 lists the effects of a number of alloying elements on graphite and carbide formation during solidification and eutectoid transformation in cast irons (Ref 14.1). Thus, depending on alloying and processing conditions, the matrix microstructure of gray and ductile cast irons may be entirely ferritic, entirely pearlitic, or a combination of ferrite and pearlite.

The dispersion of the graphite into separate spherical particles dramatically improves ultimate tensile strengths, ductility, and toughness of ductile cast irons relative to gray cast irons. For example, a ductile cast iron with a ferrite matrix will have a tensile strength around 414 MPa (60 ksi) and an elongation around 18% (Ref 14.5). Higher strengths and lower ductilities are produced by increasing matrix strength. The fracture resistance of the ductile cast irons is much more dependent on the matrix microstructures rather than that of gray cast irons with their interconnected network of graphite flakes. As a result of the beneficial dispersion of graphite, castings from ductile cast iron compete favorably for many applications where forged steels may also be used.

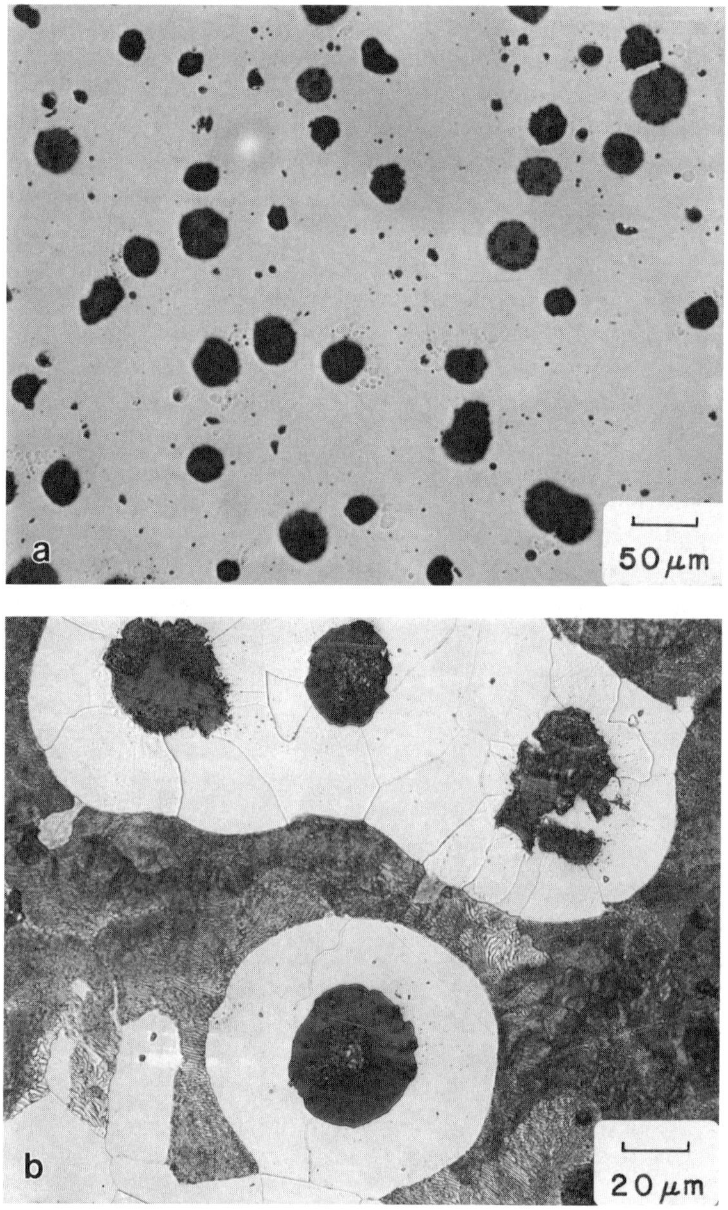

Fig. 14.6. Microstructure of ductile cast iron. Graphite is in nodular form. (a) Unetched. (b) Etched with nital. Light micrographs. Courtesy of J. McClain, Colorado School of Mines, Golden

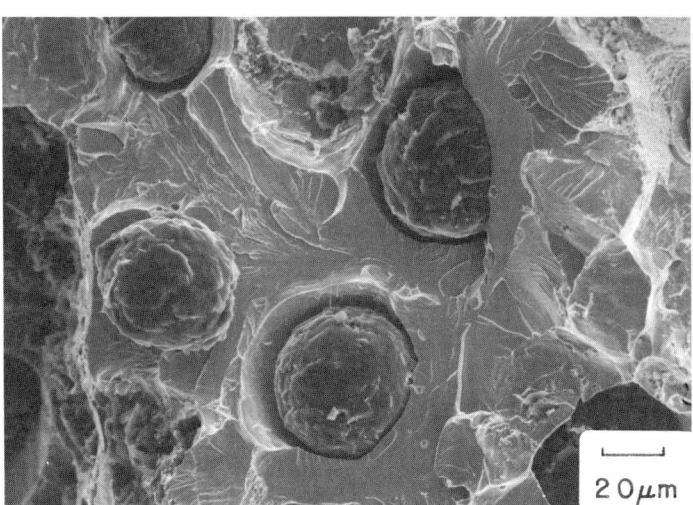

Fig. 14.7. Fracture surface of ductile cast iron showing spherical morphology of graphite nodules. Scanning electron micrograph. Courtesy of S. Diets and R. McGrew, Colorado School of Mines, Golden

Heat treatments of ductile cast irons include not only stress relief and annealing, but also those which are typically applied to steels (Ref 14.4, 14.5). The most ductile microstructures are produced by completely transforming the matrix to ferrite. Ferritizing heat treatments are accomplished by austenitizing at 900 °C (1652 °F), then holding at 700 °C (1292 °F) in order to completely transform the austenite to ferrite and graphite. Normalizing and tempering treatments produce higher strength, wear-resistant pearlite matrix microstructures, and quenching and tempering heat treatments produce the highest strength ductile cast iron microstructures (Ref 14.1-14.5). Austempering is being increasingly applied to ductile cast irons; however, special requirements are necessary. These requirements are described in a later section of this chapter.

Malleable Cast Iron

Malleable cast irons have graphite morphologies and distributions similar to ductile cast irons, and therefore are produced with ductility and toughness which merit the name "malleable." The process by

Table 14.2. Effects of Various Elements on Microstructure Formation in Cast Irons (Ref 14.1)

Element	Effect during solidification	Effect during eutectoid reaction
Aluminum	Strong graphitizer	Promotes ferrite and graphite formation
Antimony	Little effect in amounts used	Strong pearlite retainer
Bismuth	Carbide promoter, but not carbide former	Very mild pearlite stabilizer
Chromium	Strong carbide former. Forms complex carbides which are very stable.	Strong pearlite former
Copper	Mild graphitizer	Promotes pearlite formation
Manganese	Mild carbide former	Pearlite former
Molybdenum	Mild carbide former	Strong pearlite former
Nickel	Graphitizer	Mild pearlite promoter
Silicon	Strong graphitizer	Promotes ferrite and graphite formation
Tellurium	Very strong carbide promoter, but not stabilizer	Very mild pearlite stabilizer
Tin	Little effect with amount used	Strong pearlite retainer
Titanium under 0.25%	Graphitizer	Promotes graphite formation
Vanadium	Strong carbide former	Strong pearlite former

which malleable cast irons are produced, however, is quite different from that used to produce ductile cast irons. First white cast iron is produced, followed by heat treatment to transform the carbides to graphite. The requirement of white cast iron as the starting microstructure explains the range of carbon and silicon contents blocked for all malleable cast irons in Fig. 14.2. Reduced carbon and silicon contents relative to gray and ductile cast irons promote carbide formation on cooling, but minimum amounts of carbon are necessary to maintain good casting fluidity and a reasonable rate of graphitization during heat treatment.

The graphite produced during malleabilizing heat treatment is roughly spherical but much more irregular than the graphite nodules produced from the melt in ductile cast irons. The irregular graphite particles are traditionally referred to as temper carbon. Figure 14.8 shows the microstructure of a malleable cast iron in which the temper carbon is distributed in a matrix of ferrite. This structure typically has tensile strengths of 40 ksi (275 MPa) and elongations of 5%. The conversion of the as-cast white cast iron microstructure to malleable cast iron is accomplished in a two-stage annealing cycle. In the first

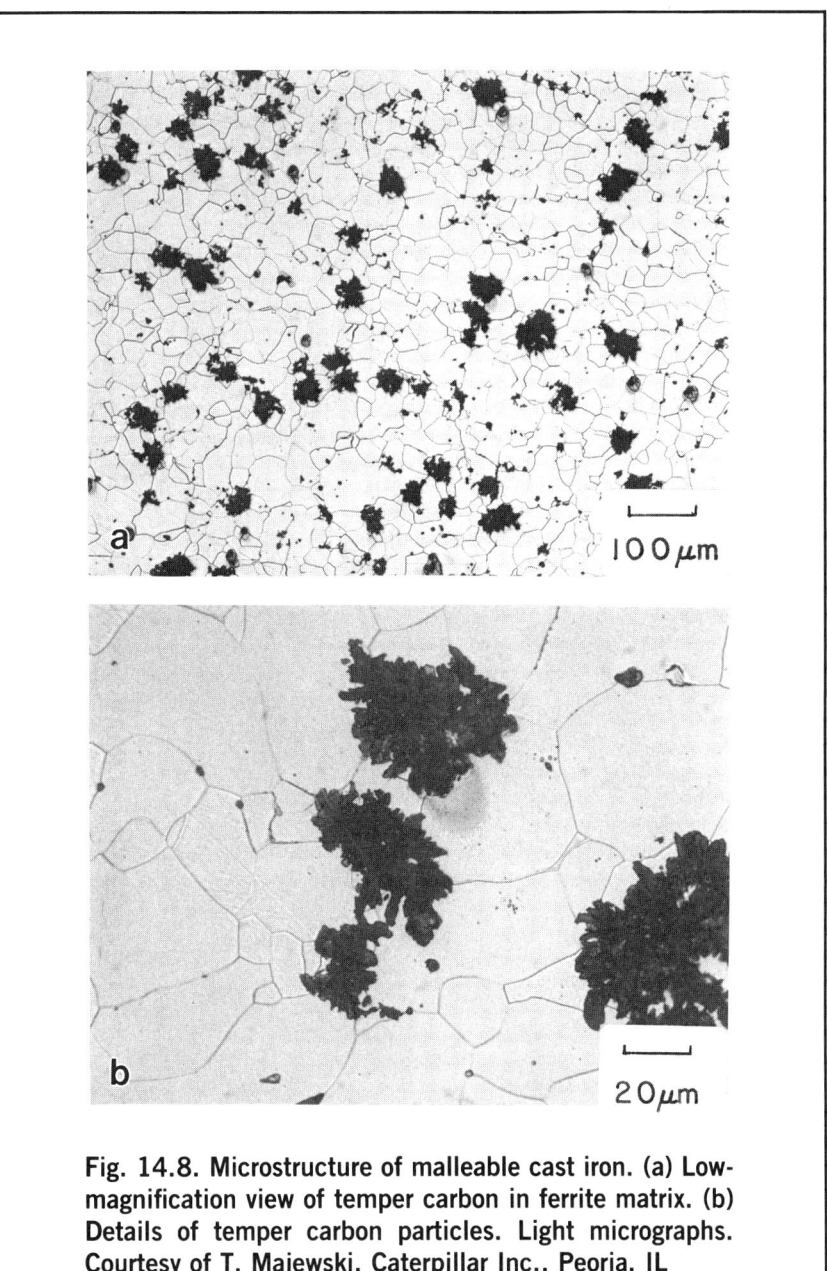

Fig. 14.8. Microstructure of malleable cast iron. (a) Low-magnification view of temper carbon in ferrite matrix. (b) Details of temper carbon particles. Light micrographs. Courtesy of T. Majewski, Caterpillar Inc., Peoria, IL

stage the cast iron is heated into the austenite-cementite phase field, commonly at temperatures around 950 °C (1740 °F). Graphite nucleates at the austenite-carbide interfaces, becomes surrounded by austenite, and grows by the diffusion of carbon from the dissolving carbide particles to the graphite through the intervening austenite. Eventually the metastable austenite-cementite structure is completely replaced by the stable austenite-graphite structure. The diffusion-dependent first stage of graphitization takes anywhere from 3 to 20 h, depending on composition and the spacing of the growing graphite nuclei (Ref 14.4). High carbon and silicon contents promote graphitization, but elements such as chromium stabilize carbides and reduce the rates of graphitization. The second stage of annealing, in the case of ferritic malleable cast irons, is designed to transform all of the austenite to ferrite and graphite. The graphite forms on the temper carbon particles both during cooling from the first-stage annealing temperature to the cast eutectoid temperature range and during cooling through the eutectoid reaction when all of the austenite must be converted to graphite and ferrite. This conversion to temper carbon and ferrite is accomplished by moderate rates of cooling to about 750 °C (1380 °F), followed by very slow cooling at rates of 3 to 11 °C (5 to 20 °F) per hour from that temperature (Ref 14.5). Conversion of the austenite to a pearlite matrix after the first-stage graphitization is accomplished by air cooling castings from about 870 °C (1600 °F).

Malleable cast iron can be heat treated to essentially the same microstructures as ductile cast irons, and since the same favorable distribution of dispersed graphite particles is developed, malleable cast irons can be used for applications which require higher tensile strengths, ductilities, and toughness than gray cast irons. Section sizes of microstructures on solidification, and the long malleabilizing heat treatment must be weighed relative to the selection of the ductile cast irons which produce nodular graphite directly on solidification without heat treatment.

White Cast Irons

White cast irons solidify directly to microstructures of carbides and austenite. The carbides are hard and brittle and produce flat, reflective fracture surfaces which give white cast irons their name. As discussed earlier, relatively low carbon and silicon contents (Fig. 14.2), rapid cooling, and small section sizes favor the formation of white cast iron microstructures.

Figure 14.9 shows a micrograph of hypoeutectic white cast iron. Solidification occurred by formation of primary dendrites of austenite and an interdendritic eutectic structure of cementite and austenite. The hardness of a white cast iron is very high, 50 HRC or higher depending on volume fraction of carbides and matrix microstructure.

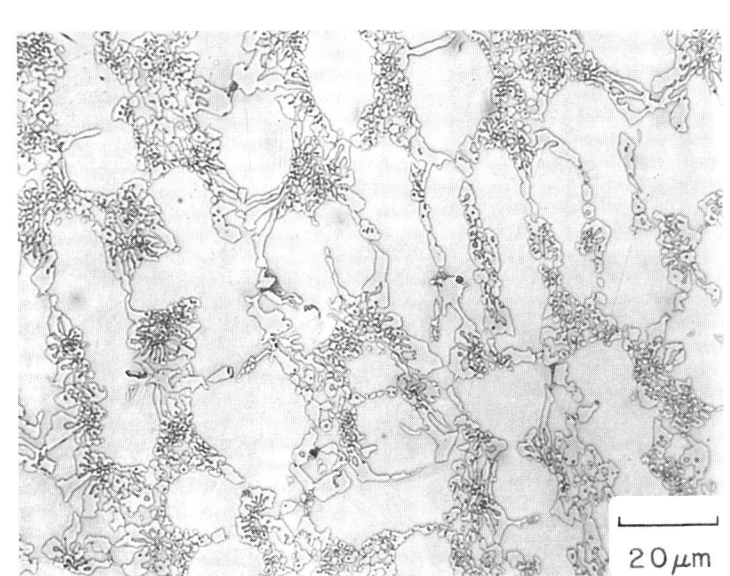

Fig. 14.9. Microstructure of white cast iron which solidified as primary dendrites of austenite and interdendritic eutectic structure of austenite and cementite. Light micrograph. Courtesy of J. McClain, Colorado School of Mines, Golden

Therefore, white cast irons are virtually nonmachinable and any surface finishing is often done by grinding. Although the high hardness of a white cast iron microstructure is generally undesirable in gray cast irons, it is sometimes incorporated in areas of gray iron castings which require high wear resistance. This is accomplished by the introduction of chills into casting molds. The chills are metal or graphite mold inserts which cause localized rapid heat transfer and solidification to microstructures with carbides instead of graphite.

Despite the low toughness of white cast irons, they are extensively used for applications which require high wear, corrosion, and oxidation resistance. Silicon, nickel, and chromium are the major alloying elements which provide corrosion protection, heat resistance, and oxidation resistance not only to white cast irons but also to a variety of special cast irons (Ref 14.2).

Three types of white cast irons are produced. Unalloyed or low-alloy white cast irons have microstructures of cementite and pearlite and are quite brittle. Somewhat improved toughness is obtained by alloying with moderate additions of nickel and chromium, for example, Ni-hard irons containing around 4% nickel and 2.5% chromium. These alloys have microstructures of carbides and martensite with increased strength, corrosion resistance, and oxidation resistance.

The most highly alloyed cast irons contain from 15 to 30% chromium and are used for ore crushing and other demanding applications in corrosive media. The high-chromium cast irons solidify as austenite and chromium-rich M_7C_3 carbides. Volume fractions of carbides, austenite stability, and hardenability are determined by the ratio of chromium to carbon and alloying elements such as molybdenum. An atlas of continuous cooling diagrams and microstructures of high Cr-Mo (11 to 26% chromium, 0 to 3.4% molybdenum) white cast irons has been published (Ref 14.6). This atlas shows that the matrix structures may range from pearlite to mixtures of austenite and martensite depending on composition and cooling rate. Figure 14.10 shows some examples of high Cr-Mo white cast iron microstructures.

Several studies (Ref 14.7-14.9) have been directed to optimizing microstructure and fracture in high-chromium cast irons for severe abrasive and corrosive service. Typically the white cast iron microstructures of massive alloy carbides and matrices of austenite or martensite show improved fracture resistance relative to white cast iron with pearlitic matrices. Stable austenite is produced by high chromium to carbon ratios and molybdenum additions. Martensite is produced by heating high Cr-Mo white cast irons with stable austenite matrices to temperatures between 900 and 1000 °C (1652 to 1832 °F), where alloy carbides precipitate, lower the austenite stability, and cause martensite to form on cooling. The dynamic fracture toughness of both austenitic and martensitic high Cr-Mo white cast irons decreases with increasing volume fraction of alloy carbides and tends to be higher in austenitic alloys (17 to 25 ksi$\sqrt{\text{in.}}$, or 19 to 27 MPa$\sqrt{\text{m}}$) than in martensitic alloys (13 to 19 ksi$\sqrt{\text{in.}}$, or to 21 MPa$\sqrt{\text{m}}$). Nevertheless, apparently because of higher hardness, the martensitic white cast irons show better abrasion resistance in application where high dynamic fracture toughness is not required.

Austempered Ductile Irons

Austempered ductile irons are solidified as ductile cast irons and heat treated by austempering (i.e., austenitizing and isothermal trans-

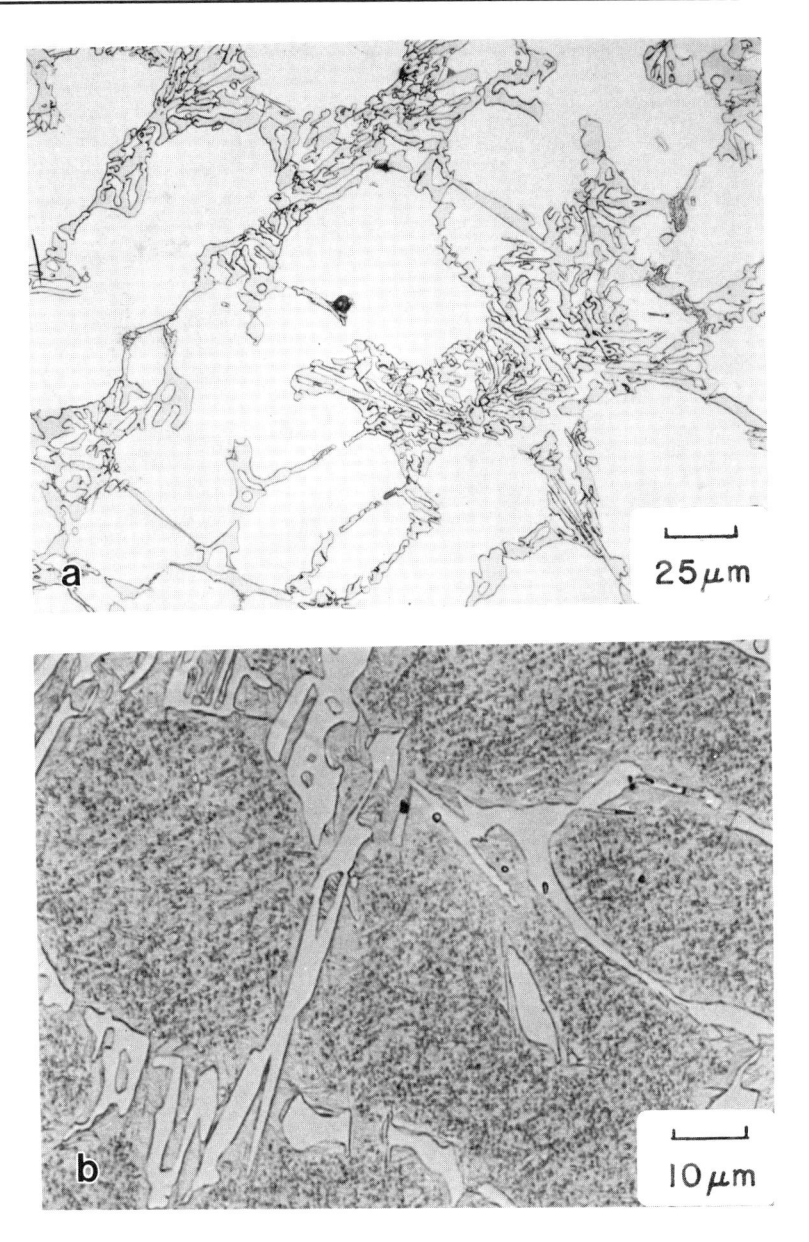

Fig. 14.10. Microstructures of high Cr-Mo white cast irons. (a) White cast iron containing 2.7C-19.7Cr-0.73Mo with microstructure of austenite and carbides. (b) White cast iron containing 2.3C-12.0Cr-1.4Mo with matrix microstructure containing secondary carbides and martensite after heat treatment at 980 °C (1796 °F). Light micrographs. (Ref 14.9)

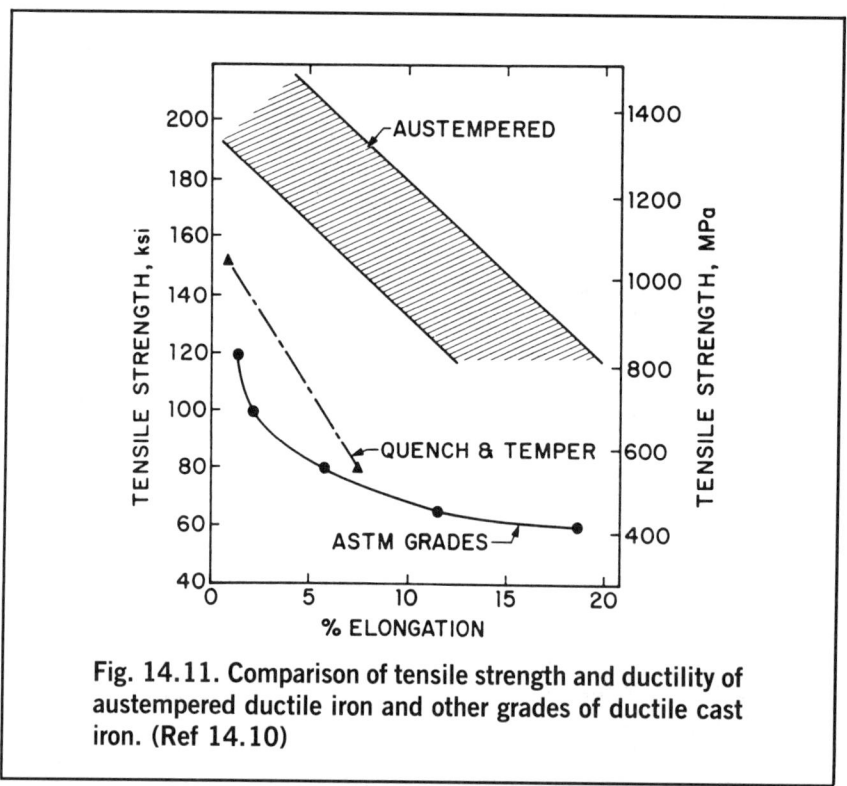

Fig. 14.11. Comparison of tensile strength and ductility of austempered ductile iron and other grades of ductile cast iron. (Ref 14.10)

formation) to bainitic microstructures. These cast irons have higher strengths with good ductility and toughness compared to standard quenched and tempered grades of ductile irons (Fig. 14.11) (Ref 14.10). Alloying elements such as manganese, nickel, and molybdenum must be added to the austempered ductile irons in order to prevent pearlite formation on cooling to isothermal transformation temperatures. Transformation to pearlite would significantly lower toughness of the austempered ductile irons.

The austenite in the ductile irons transforms in two steps during isothermal treatment. The first step consists of the formation of ferrite plates and the necessary rejection of carbon into the austenite adjacent to the plates. This high-carbon austenite is quite stable and becomes a component of the austempered ductile iron bainitic microstructures which have high strength and toughness.

The bainite, a mixture of ferrite and austenite, is different from the classical ferrite-carbide bainitic structures produced in low-alloy medium-carbon steels. The suppression of cementite is apparently caused by the high silicon content of austempered ductile irons. Silicon has very low solubility in the crystal structure of cementite and

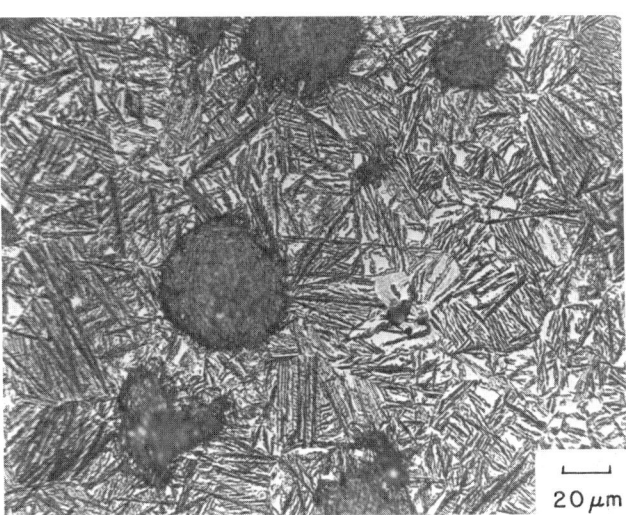

Fig. 14.12. Microstructure of austempered ductile iron. Light micrograph. Nital etch. Courtesy of M. Shea, General Motors Corp., Warren, MI

therefore must be rejected from areas of the cementite growth (Ref 14.11). Thus cementite formation is initially retarded because of the required silicon diffusion. High silicon content also retards cementite formation during the tempering of steels (Ref 14.12), and stable austenite is a characteristic of bainitic microstructures formed in high-silicon steels (Ref 14.13).

The second step of transformation during austempering consists of the decomposition of the interplate austenite into ferrite and carbide (Ref 14.10). Apparently this transformation develops after longer times make possible silicon diffusion from the growing cementite. The formation of ferrite-carbide microstructures lowers ductility and toughness of austempered ductile irons, and therefore isothermal transformation should be terminated prior to this stage.

Austenitizing temperatures in excess of 900 °C (1652 °F) may also lead to embrittlement of austempered ductile iron (Ref 14.14). Phosphorus, perhaps due to the dissolution of a Mg-P precipitate, has been found to cause severe embrittlement and intergranular fracture at austenite grain boundaries of specimens austenitized at 1070 °C (1958 °F).

Figure 14.12 shows the microstructure of an austempered ductile iron containing 3.72% carbon, 2.32% silicon, 0.42% manganese, and

0.05% magnesium, and heat treated by austenitizing at 900 °C (1652 °F) for 1 h and austempering in salt at 370 °C (698 °F) for 1 h. The matrix microstructure between the graphite nodules consists of bainitic ferrite and austenite. The ferrite is in the form of plates, and the interplate austenite, in the amount of 37%, appears white. There are two temperature ranges which are used for austempering (Ref 14.10, 14.15). The higher temperature range, around 400 °C (750 °F), produces relatively coarse microstructures of ferrite and austenite with hardness between 30 and 35 HRC. The lower temperature range, around 300 °C (570 °F), produces a finer structure of ferrite and austenite with hardness between 45 and 50 HRC.

The intensive development of austempered ductile irons has led to the availability of a new material based on foundry and heat treatment technology. Although questions remain about composition and microstructural control (Ref 14.15), austempered ductile irons provide a high-strength material with good toughness for consideration for gears and other structural components (Ref 14.10).

Cast Iron Surface Modification

Cast irons can be coated and surface modified by many techniques for appearance, increased surface hardness, wear resistance, corrosion protection, and high-temperature oxidation resistance. The *Gray and Ductile Iron Castings Handbook* (Ref 14.1) has an excellent chapter reviewing the various coatings for cast irons, and Table 14.3 from that reference lists a few of the techniques used to improve the surface characteristics of cast irons. Induction and flame hardening are also often used to harden surfaces of ductile and malleable cast irons (Ref 14.14).

In addition to established surface modification techniques, newer approaches to surface treatment are being applied to cast irons. An especially active area is the use of lasers to produce surface solid-state transformation or hardening (Ref 14.16) and surface melting (Ref 14.17) of cast irons. For example, laser hardening has been applied to selective hardening of cylinder liners, valve seats, and crankshaft fillets. In these applications, laser technology permits highly directed laser beams from remote sources to effectively harden critical areas.

In both laser surface hardening and melting, the high carbon content of the cast irons and the very high heating and cooling rates associated with laser heating produce localized surface structures of substantially increased hardness relative to bulk gray, ductile, or malleable cast irons. Laser melting effectively acts as a chill for

Table 14.3. Diffusion Coatings Applied to Cast Irons (Ref 14.1)

Type	Coating structure	Properties	Uses
Calorized	Metallic aluminum introduced into surface layer forming aluminum-iron alloy	High-temperature oxidation resistance	Chemical processes, steam superheaters, and heat transfer
Chromized	Chromium carbide case formed on surface	High hardness and wear resistance	Combustion and mechanical equipment
Cyanided, Carbonitrided	Carbon-nitrogen compound formed by diffusion into surface	Wear resistance for surfaces combined with core toughness	Gears, cams, pawls, and shafts
Nickel-phosphorus	Ammonium phosphate and nickel oxide products reduced and diffused	Corrosion resistance comparable to austenitic irons. Poor wear resistance	Chemical process pipe and fittings
Nitrided	Nitrogen introduced into surface by contact of ammonia or other nitrogenous material	Wear and corrosion resistance at elevated temperatures	Same as for carbonitrided
Sheradized	Zinc introduced into surface	Corrosion resistance	Atmospheric corrosion resistance

graphitic cast iron, and converts surface zones into very fine solidified zones of metastable austenite and cementite (Ref 14.17).

Summary

Cast irons are a large group of ferrous engineering alloys with unique structures and properties, and offer the very great advantage of fabrication by casting or near net shape processing. Casting of a shape requires no processing steps beyond finish machining and heat treatment. Therefore, casting and cast irons often offer the most economical approach to manufacturing, and in the case of alloys which are very difficult or impossible to form or machine, the only method of manufacture. Castings, however, are subject to segregation and porosity, which in wrought materials are homogenized by hot working. Considerable foundry technology must therefore be applied to produce sound castings of specified composition, structure, and properties (Ref 14.1-14.5). As in the case of steels, the versatility of cast irons can be considerably extended by both established and newer techniques of surface modification.

References

14.1 C.F. Walton (Ed.), *Gray and Ductile Iron Castings Handbook*, Gray and Ductile Iron Founders' Society, Cleveland, 1971

14.2 R. Elliott, *Cast Iron Technology*, Butterworths, London, 1988

14.3 I. Minkoff, *The Physical Metallurgy of Cast Iron*, John Wiley, New York, 1983

14.4 Cast Irons, in *Metals Handbook*, Vol 1, *Properties and Selection: Irons and Steels*, American Society for Metals, Metals Park, OH, 1978, p 1-106

14.5 Cast Irons, in *Metals Handbook, Desk Edition*, American Society for Metals, Metals Park, OH, 1985, p 5-1 to 5-20

14.6 F. Maratray and R. Usseglio-Nanot, *Atlas: Transformation Characteristics of Chromium and Chromium-Molybdenum White Irons*, Climax Molybdenum S.A., Paris

14.7 K-H. Zum Gahr and D.V. Doane, Optimizing Fracture Toughness and Abrasion Resistance in White Cast Irons, *Met Trans A*, Vol 11A, 1980, p 613-620

14.8 J. Dodd, R.B. Gundlach, and J.L. Parks, Some Factors Affecting the Metallurgy and Production of High Chromium Molybdenum Irons, in *Proceedings of Third International Congress on Heat Treatment of Materials*, Book Number 310, The Metals Society, London, 1984, p 5.48-5.60

14.9 R.H. Frost, T. Majewski, and G. Krauss, Impact Fracture Behavior of High Chromium Molybdenum White Cast Irons, *AFS Trans*, Vol 56, 1986, p 297-322

14.10 R.B. Gundlach and J.F. Janowak, A Review of Austempered Ductile Iron Metallurgy, in *First International Conference on Austempered Ductile Iron*, American Society for Metals, Metals Park, OH, 1984, p 1-12

14.11 S.J. Barnard, G.D.W. Smith, A.J. Garrett-Reed, and J. Vander Sande, Atom Probe Studies: 1) The Role of Silicon in the Tempering of Steel and 2) Low Temperature Chromium Diffusivity in Bainite, in *Solid-Solid Phase Transformations*, H.I. Aaronson et al. (Eds.), TMS-AIME, Warrendale, PA, 1982, p 881-885

14.12 W.S. Owen, The Effect of Silicon on the Kinetics of Tempering, *Trans ASM*, Vol 46, 1954, p 812-829

14.13 H.K.D.H. Bhadeshia and D.V. Edwards, The Bainite Transformation in a Silicon Steel, *Met Trans A*, Vol 10A, 1979, p 895-907

14.14 R.C. Klug, M.B. Hintz, and K.B. Rundman, Embrittlement of Austempered Nodular Irons: Grain Boundary Phosphorus Enrichment Resulting from Precipitate Decomposition, *Met Trans A*, Vol 16A, 1985, p 797-803

14.15 C.R. Loper, Jr., Conference Summary, in *First International Conference on Austempered Ductile Iron*, American Society for Metals, Metals Park, OH, 1984, p 153-156

14.16 P.A. Molian, Engineering Applications and Analysis of Hardening Data for Laser Heat Treated Ferrous Alloys, *Surface Eng*, Vol 2, 1986, p 19-28

14.17 H.W. Bergmann, Current Status of Laser Surface Melting of Cast Iron, *Surface Eng*, Vol 1, 1985, p 137-155

APPENDIX 1

Glossary of Selected Terms

A_{cm}, A_1, A_3, A_4. Same as Ae_{cm}, Ae_1, Ae_3 and Ae_4.

Ac_{cm}, Ac_1, Ac_3, Ac_4. Defined under *transformation temperature*.

acicular ferrite. A highly substructured nonequiaxed ferrite that forms upon continuous cooling by a mixed diffusion and shear mode of transformation that begins at a temperature slightly higher than the temperature transformation range for upper bainite. It is distinguished from bainite in that it has a limited amount of carbon available; thus, there is only a small amount of carbide present.

Ae_{cm}, Ae_1, Ae_3, Ae_4. Defined under *transformation temperature*.

age hardening. Hardening by aging, usually after rapid cooling or cold working. See also *aging*.

age softening. Spontaneous decrease of strength and hardness that takes place at room temperature in certain strain hardened alloys, especially those of aluminum.

aging. A change in the properties of certain metals and alloys that occurs at ambient or moderately elevated temperatures after hot working or a heat treatment (quench aging in ferrous alloys, natural or artificial aging in ferrous and nonferrous alloys) or after a cold working operation (strain aging). The change in properties is often, but not always, due to a phase change (precipitation), but never involves a change in chemical composition of the metal or alloy. See also *age hardening, artificial aging, interrupted aging, natural aging, overaging, precipitation hardening, precipitation heat treatment, progressive aging, quench aging, step aging.*

air-hardening steel. A steel containing sufficient carbon and other alloying elements to harden fully during cooling in air or other gaseous mediums from a temperature above its transformation range. The term should be restricted to steels that are capable of being hardened by cooling in air in fairly large sections, about 2 in. (50 mm) or more in diameter. Same as self-hardening steel.

alpha ferrite. See *ferrite*.

alpha iron. The body-centered cubic form of pure iron, stable below 910 °C (1670 °F).

Alumel. A nickel-base alloy containing about 2.5% Mn, 2% Al and 1% Si used chiefly as a component of pyrometric thermocouples.

annealing. A generic term denoting a treatment, consisting of heating to and holding at a suitable temperature followed by cooling at a suitable rate, used primarily to soften metallic materials, but also to simultaneously produce desired changes in other properties or in microstructure. The purpose of such changes may be, but is not confined to: improvement of machinability, facilitation of cold work, improvement of mechanical or electrical properties, and/or increase in stability of dimensions. When the term is used without qualification, full annealing is implied. When applied only for the relief of stress, the process is properly called stress relieving or stress-relief annealing.

In ferrous alloys, annealing usually is done above the upper critical temperature, but the time-temperature cycles vary widely in both maximum temperature attained and in cooling rate employed, depending on composition, material condition, and results desired. When applicable, the following commercial process names should be used: black annealing, blue annealing, box annealing, bright annealing, cycle annealing, flame annealing, full annealing, graphitizing, in-process annealing, isothermal annealing, malleabilizing, orientation annealing, process annealing, quench annealing, spheroidizing, subcritical annealing.

In nonferrous alloys, annealing cycles are designed to: (1) remove part or all of the effects of cold working (recrystallization may or may not be involved); (2) cause substantially complete coalescence of precipitates from solid solution in relatively coarse form; or (3) both, depending on composition and material condition. Specific process names in commercial use are final annealing, full annealing, intermediate annealing, partial annealing, recrystallization annealing, stress-relief annealing, anneal to temper.

annealing carbon. Fine, apparently amorphous carbon particles formed in white cast iron and certain steels during prolonged annealing. Also called temper carbon.

annealing twin. A twin form in a crystal during recrystallization.

anneal to temper. A final partial anneal that softens a cold worked nonferrous alloy to a specified level of hardness or tensile strength.

Ar_{cm}, Ar_1, Ar_3, Ar_4, Ar', Ar''. Defined under *transformation temperature*.

artificial aging. Aging above room temperature. See also *aging*. Compare with *natural aging*.

athermal transformation. A reaction that proceeds without benefit of thermal fluctuations; that is, thermal activation is not required. Such reactions are diffusionless and can take place with great speed when the driving force is sufficiently high. For example, many martensitic transformations occur athermally on cooling, even at relatively low temperatures, because of the progressively increasing driving force. In contrast, a reaction that occurs at constant temperature is an *isothermal transformation*; thermal activation is necessary in this case and the reaction proceeds as a function of time.

ausforming. Hot deformation of metastable austenite within controlled ranges of temperature and time that avoids formation of nonmartensitic transformation products.

austempering. A heat treatment for ferrous alloys in which a part is quenched from the austenitizing temperature at a rate fast enough to avoid formation of ferrite or pearlite and then held at a temperature just above M_s until transformation to bainite is complete.

austenite. A solid solution of one or more elements in face-centered cubic iron. Unless otherwise designated (such as nickel austenite), the solute is generally assumed to be carbon.

austenitic grain size. The size attained by the grains of steel when heated to the austenitic region; may be revealed by appropriate etching of cross sections after cooling to room temperature.

austenitizing. Forming austenite by heating a ferrous alloy into the transformation range (partial austenitizing) or above the transformation range (complete austenitizing). When used without qualification, the term implies complete austenitizing.

bainite. A metastable aggregate of ferrite and cementite resulting from the transformation of austenite at temperatures below the pearlite range but above M_s. Its appearance is feathery if formed in the upper part of the bainite transformation range; acicular, resembling tempered martensite, if formed in the lower part.

baking. (1) Heating to a low temperature to remove gases. (2) Curing or hardening surface coatings such as paints by exposure to heat. (3) Heating to drive off moisture, as in the baking of sand cores after molding.

bark. The decarburized layer just beneath the scale that results from heating steel in an oxidizing atmosphere.

black annealing. Box annealing or pot annealing ferrous alloy sheet, strip, wire. See also *box annealing*.

black oxide. A black finish on a metal produced by immersing it in hot oxidizing salts or salt solutions.

blank carburizing. Simulating the carburizing operation without introducing carbon. This is usually accomplished by using an inert material in place of the carburizing agent, or by applying a suitable protective coating to the ferrous alloy.

blank nitriding. Simulating the nitriding operation without introducing nitrogen. This is usually accomplished by using an inert material in place of the nitriding agent or by applying a suitable protective coating to the ferrous alloy.

block brazing. An obsolete brazing process in which the joint was heated using hot blocks.

blue annealing. Heating hot rolled ferrous sheet in an open furnace to a temperature within the transformation range and then cooling in air, in order to soften the metal. The formation of a bluish oxide on the surface is incidental.

blue brittleness. Brittleness exhibited by some steels after being heated to some temperature within the range of about 200 to 370 °C (400 to 700 °F), particularly if the steel is worked at the elevated temperature. Killed steels are virtually free of this kind of brittleness.

bluing. Subjecting the scale-free surface of a ferrous alloy to the action of air, steam, or other agents at a suitable temperature, thus forming a thin blue film of oxide and improving the appearance and resistance to corrosion. *Note*: This term is ordinarily applied to sheet, strip, or finished parts. It is used also to denote the heating of springs after fabrication to improve their properties.

box annealing. Annealing a metal or alloy in a sealed container under conditions that minimize oxidation. In box annealing a ferrous alloy, the charge is usually heated slowly to a temperature below the transformation range, but sometimes above or within it, and is then cooled slowly; this process is also called close annealing or pot annealing. See also *black annealing*.

brazing. A group of welding processes that join solid materials together by heating them to a suitable temperature and by using a filler metal having a liquidus above 450 °C (840 °F) and below the solidus of the base materials. The filler metal is distributed between the closely fitted surfaces of the joint by capillary attraction.

brazing alloy. See preferred term *brazing filler metal*.

brazing filler metal. A nonferrous filler metal used in *brazing* and braze welding.

brazing sheet. Brazing filler metal in sheet form or flat-rolled metal clad with brazing filler metal on one or both sides.

breaks. Creases or ridges usually in "untempered" or in aged material where the yield point has been exceeded. Depending on the origin of the break, it may be termed a cross break, a coil break, an edge break, or a sticker break.

bright annealing. Annealing in a protective medium to prevent discoloration of the bright surface.

burning. (1) Permanently damaging a metal or alloy by heating to cause either incipient melting or intergranular oxidation. See also *overheating*. (2) In grinding, getting the work hot enough to cause discoloration or to change the microstructure by tempering or hardening.

calorizing. Imparting resistance to oxidation to an iron or steel surface by heating in aluminum powder at 800 to 1000 °C (1472 to 1832 °F).

carbonitriding. A case hardening process in which a suitable ferrous material is heated above the lower transformation temperature in a gaseous atmosphere of such composition as to cause simultaneous absorption of carbon and nitrogen by the surface and, by diffusion, create a concentration gradient. The process is completed by cooling at a rate that produces the desired properties in the workpiece.

carbonization. Conversion of an organic substance into elemental carbon. (Should not be confused with carburization).

carbon potential. A measure of the ability of an environment containing active carbon to alter or maintain, under prescribed conditions, the carbon level of the steel. *Note*: In any particular environment, the carbon level attained will depend on such factors as temperature, time, and steel composition.

carbon restoration. Replacing the carbon lost in the surface layer from previous processing by carburizing this layer to substantially the original carbon level. Sometimes called recarburizing.

carburizing. Absorption and diffusion of carbon into solid ferrous alloys by heating, to a temperature usually above Ac_3, in contact with a suitable carbonaceous material. A form of *case hardening* that produces a carbon gradient extending inward from the surface, enabling the surface layer to be hardened either by quenching directly from the carburizing temperature or by cooling to room temperature, then reaustenitizing and quenching.

carburizing flame. A gas flame that will introduce carbon into some heated metals, as during a gas welding operation. A carburizing flame is a *reducing flame*, but a reducing flame is not necessarily a carburizing flame.

case. That portion of a ferrous alloy, extending inward from the surface, whose composition has been altered so that it can be *case hardened*. Typically considered to be the portion of the alloy (1) whose composition has been measurably altered from the original composition, (2) that appears dark on an etched cross section, or (3) that has a hardness, after hardening, equal to or greater than a specified value. Contrast with *core*.

case hardening. A generic term covering several processes applicable to steel that change the chemical composition of the surface layer by absorption of carbon, nitrogen, or a mixture of the two and, by diffusion, create a concentration gradient. The processes commonly used are carburizing and quench hardening; cyaniding; nitriding; and carbonitriding. The use of the applicable specific process name is preferred.

cementation. The introduction of one or more elements into the outer portion of a metal object by means of diffusion at high temperature.

cementite. A compound of iron and carbon, known chemically as iron carbide and having the approximate chemical formula Fe_3C. It is characterized by an orthorhombic crystal structure. When it occurs as a phase in steel, the chemical composition will be altered by the presence of manganese and other carbide-forming elements.

checks. Numerous, very fine cracks in a coating or at the surface of a metal part. Checks may appear during processing or during service and are most often associated with thermal treatment or thermal cycling. Also called check marks, checking, *heat checks*.

Chromel. (1) A 90Ni-10Cr alloy used in thermocouples. (2) A series of nickel-chromium alloys, some with iron, used for heat-resistant applications.

close annealing. Same as *box annealing*.

coarsening. An increase in the grain size, usually, but not necessarily, by *grain growth*.

coherency. The continuity of lattice of precipitate and parent phase (solvent) maintained by mutual strain and not separated by a phase boundary.

coherent precipitate. A crystalline precipitate that forms from solid solution with an orientation that maintains continuity between the crystal lattice of the precipitate and the lattice of the matrix, usually accompanied by some strain in both lattices. Because the lattices fit at the interface between precipitate and matrix, there is no discernible phase boundary.

cold treatment. Exposing to suitable subzero temperatures for the purpose of obtaining desired conditions or properties such as dimensional or microstructural stability. When the treatment involves the transformation of retained austenite, it is usually followed by tempering.

columnar structure. A coarse structure of parallel elongated grains formed by unidirectional growth, most often observed in castings, but sometimes in structures resulting from diffusional growth accompanied by a solid-state transformation.

combined carbon. The part of the total carbon in steel or cast iron that is present as other than *free carbon*.

conditioning heat treatment. A preliminary heat treatment used to prepare a material for desired reaction to a subsequent heat treatment. For the term to be meaningful, the exact heat treatment must be specified.

congruent transformation. An isothermal or isobaric phase change in which both of the phases concerned have the same composition throughout the process.

constantan. A group of copper-nickel alloys containing 45 to 60% copper with minor amounts of iron and manganese and characterized by relatively constant electrical resistivity irrespective of temperature; used in resistors and thermocouples.

constitution diagram. A graphical representation of the temperature and composition limits of phase fields in an alloy system as they actually exist under the specific conditions of heating or cooling (synonymous with phase diagram). A constitution diagram may be an equilibrium diagram, an approximation to an equilibrium diagram, or a representation of metastable conditions or phases. Compare with *equilibrium diagram*.

continuous precipitation. Precipitation from a supersaturated solid solution in which the precipitate particles grow by long-range diffusion without recrystallization of the matrix. Continuous precipitates grow from nuclei distributed more or less uniformly throughout the matrix. They usually are randomly oriented, but may form a Widmanstätten structure. Also called general precipitation. Compare with *discontinuous precipitation, localized precipitation*.

controlled cooling. Cooling from an elevated temperature in a predetermined manner, to avoid hardening, cracking, or internal damage, or to produce desired microstructure or mechanical properties.

cooling curve. A curve showing the relation between time and temperature during the cooling of a material.

cooling stresses. Residual stresses resulting from nonuniform distribution of temperature during cooling.

core. In a ferrous alloy prepared for *case hardening*, that portion of the alloy that is not part of the *case*. Typically considered to be the portion that (1) appears light on an etched cross section, (2) has an essentially unaltered chemical composition, or (3) has a hardness, after hardening, less than a specified value.

critical cooling rate. The rate of continuous cooling required to prevent undesirable transformation. For steel, it is the minimum rate at which austenite must be continuously cooled to suppress transformations above the M_s temperature.

critical point. (1) The temperature or pressure at which a change in crystal structure, phase or physical properties occurs. Same as *transformation temperature*. (2) In an equilibrium diagram, that specific value of composition, temperature, and pressure, or combinations thereof, at which the phases of a heterogeneous system are in equilibrium.

critical strain. The strain just sufficient to cause *recrystallization*; because the strain is small, usually only a few percent, recrystallization takes place from only a few nuclei, which produces a recrystallized structure consisting of very large grains.

critical temperature. (1) Synonymous with *critical point* if the pressure is constant. (2) The temperature above which the vapor phase cannot be condensed to liquid by an increase in pressure.

critical temperature ranges. Synonymous with *transformation ranges*, which is the preferred term.

Curie temperature. The temperature of magnetic transformation below which a metal or alloy is ferromagnetic and above which it is paramagnetic.

cyaniding. A case hardening process in which a ferrous material is heated above the lower transformation range in a molten salt containing cyanide to cause simultaneous absorption of carbon and nitrogen at the surface and, by diffusion, create a concentration gradient. Quench hardening completes the process.

cycle annealing. An annealing process employing a predetermined and closely controlled time-temperature cycle to produce specific properties or microstructures.

dead soft. A *temper* of nonferrous alloys and some ferrous alloys corresponding to the condition of minimum hardness and tensile strength produced by *full annealing*.

decalescence. A phenomenon, associated with the transformation of alpha iron to gamma iron on the heating (superheating) of iron or steel, revealed by the darkening of the metal surface owing to the sudden decrease in temperature caused by the fast absorption of the latent heat of transformation. Contrast with *recalescence*.

decarburization. Loss of carbon from the surface layer of a carbon-containing alloy due to reaction with one or more chemical substances in a medium that contacts the surface.

degrees of freedom. The number of independent variables (such as temperature, pressure, or concentration within the phases present) that may be altered at will without causing a phase change in an alloy system at equilibrium; or the number of such variables that must be fixed arbitrarily to define the system completely.

delta ferrite. See *ferrite*.

differential heating. Heating that intentionally produces a temperature gradient within an object such that, after cooling, a desired stress distribution or variation in properties is present within the object.

diffusion. (1) Spreading of a constituent in a gas, liquid, or solid, tending to make the composition of all parts uniform. (2) The spontaneous movement of atoms or molecules to new sites within a material.

diffusion coefficient. A factor of proportionality representing the amount of substance diffusing across a unit area through a unit concentration gradient in unit time.

direct quenching. (1) Quenching carburized parts directly from the carburizing operation. (2) Also used for quenching pearlitic malleable parts directly from the malleabilizing operation.

discontinuous precipitation. Precipitation from a supersaturated solid solution in which the precipitate particles grow by short-range diffusion, accompanied by recrystallization of the matrix in the region of precipitation. Discontinuous precipitates grow into the matrix from nuclei near grain boundaries, forming cells of alternate lamellae of precipitate and depleted (and recrystallized) matrix. Often referred to as cellular or nodular precipitation. Compare with *continuous precipitation, localized precipitation*.

double aging. Employment of two different aging treatments to control the type of precipitate formed from a supersaturated matrix in order to obtain the desired properties. The first aging treatment, sometimes referred to as intermediate or stabilizing, is usually carried out at higher temperature than the second.

double tempering. A treatment in which a quench-hardened ferrous metal is subjected to two complete tempering cycles, usually at substantially the same temperature, for the purpose of ensuring completion of the tempering reaction and promoting stability of the resulting microstructure.

drawing. A misnomer for *tempering*.

dry cyaniding. (obsolete) Same as *carbonitriding*.

embrittlement. Reduction in the normal ductility of a metal due to a physical or chemical change. Examples include *blue brittleness, hydrogen embrittlement, and temper brittleness*.

enantiotropy. The relation of crystal forms of the same substance in which one form is stable above a certain temperature and the other form stable below that temperature. Ferrite and austenite are enantiotropic in ferrous alloys, for example.

end-quench hardenability test. A laboratory procedure for determining the hardenability of a steel or other ferrous alloy; widely referred to as the *Jominy test*. Hardenability is determined by heating a standard specimen above the upper critical temperature, placing the hot specimen in a fixture so that a stream of cold water impinges on one end, and, after cooling to room temperature is completed, measuring the hardness near the surface of the specimen at regularly spaced intervals along its length. The data are normally plotted as hardness versus distance from the quenched end.

equilibrium diagram. A graphical representation of the temperature, pressure, and composition limits of phase fields in an alloy system as they exist under conditions of complete equilibrium. In metal systems, pressure is usually considered constant.

eutectic. (1) An isothermal reversible reaction in which a liquid solution is converted into two or more intimately mixed solids on cooling, the number of solids formed being the same as the number of components in the system. (2) An alloy having the composition indicated by the eutectic point on an equilibrium diagram. (3) An alloy structure of intermixed solid constituents formed by a eutectic reaction.

eutectic carbide. Carbide formed during freezing as one of the mutually insoluble phases participating in the eutectic reaction of ferrous alloys.

eutectic melting. Melting of localized microscopic areas whose composition corresponds to that of the eutectic in the system.

eutectoid. (1) An isothermal reversible reaction in which a solid solution is converted into two or more intimately mixed solids on cooling, the number of solids formed being the same as the number of components in the system. (2) An alloy having the composition indicated by the eutectoid point on an equilibrium diagram. (3) An alloy structure of intermixed solid constituents formed by a eutectoid reaction.

extra hard. A *temper* of nonferrous alloys and some ferrous alloys characterized by tensile strength and hardness about one-third of the way from *full hard* to *extra spring* temper.

extra spring. A *temper* of nonferrous alloys and some ferrous alloys corresponding approximately to a cold worked state above *full hard* beyond which further cold work will not measurably increase the strength and hardness.

ferrite. A solid solution of one or more elements in body-centered cubic iron. Unless otherwise designated (for instance, as chromium ferrite), the solute is generally assumed to be carbon. On some equilibrium diagrams, there are two ferrite regions separated by an austenite area. The lower area is alpha ferrite; the upper, delta ferrite. If there is no designation, alpha ferrite is assumed.

ferritizing anneal. A treatment given as-cast gray or ductile (nodular) iron to produce an essentially ferritic matrix. For the term to be meaningful, the final microstructure desired or the time-temperature cycle used must be specified.

file hardness. Hardness as determined by the use of a file of standardized hardness on the assumption that a material that cannot be cut with the file is as hard as, or harder than, the file. Files covering a range of hardnesses may be employed.

final annealing. An imprecise term used to denote the last anneal given to a nonferrous alloy prior to shipment.

finishing temperature. The temperature at which hot working is completed.

flame annealing. Annealing in which the heat is applied directly by a flame.

flame hardening. A process for hardening the surfaces of hardenable ferrous alloys in which an intense flame is used to heat the surface layers above the upper transformation temperature, whereupon the workpiece is immediately quenched.

flame straightening. Correcting distortion in metal structures by localized heating with a gas flame.

fog quenching. Quenching in a fine vapor or mist.

free carbon. The part of the total carbon in steel or cast iron that is present in elemental form as graphite or temper carbon. Contrast with *combined carbon*.

free ferrite. Ferrite that is formed directly from the decomposition of hypoeutectoid austenite during cooling, without the simultaneous formation of cementite. Also proeutectoid ferrite.

freezing range. That temperature range between liquidus and solidus temperatures in which molten and solid constituents coexist.

full annealing. An imprecise term that denotes an annealing cycle to produce minimum strength and hardness. For the term to be meaningful, the composition and starting condition of the material and the time-temperature cycle used must be stated.

full hard. A *temper* of nonferrous alloys and some ferrous alloys corresponding approximately to a cold worked state beyond which the material can no longer be formed by bending. In specifications, a full hard temper is commonly defined in terms of minimum hardness or minimum tensile strength (or, alternatively, a range of hardness or strength) corresponding to a specific percentage of cold reduction following a full anneal. For aluminum, a full hard temper is equivalent to a reduction of 75% from *dead soft*; for austenitic stainless steels, a reduction of about 50 to 55%.

furnace brazing. A mass-production *brazing* process in which the filler metal is preplaced on the joint, then the entire assembly is heated to brazing temperature in a furnace. Usually, a protective furnace atmosphere is required, and wetting of the joint surfaces is accomplished without using a brazing flux.

fusion. A change of state from solid to liquid; melting.

gamma iron. The face-centered cubic form of pure iron, stable from 910 to 1400 °C (1670 to 2550 °F).

gas cyaniding. A misnomer for *carbonitriding*.

grain growth. An increase in the average size of the grains in polycrystalline metal, usually as a result of heating at elevated temperature.

grain size. For metals, a measure of the areas or volumes of grains in a polycrystalline material, usually expressed as an average when the individual sizes are fairly uniform. In metals containing two or more phases, the grain size refers to that of the matrix unless otherwise specified. Grain sizes are reported in terms of number of grains per unit area or volume, average diameter, or as a grain-size number derived from area measurements.

graphitization. Formation of graphite in iron or steel. Where graphite is formed during solidification, the phenomenon is called primary graphitization; where formed later by heat treatment, secondary graphitization.

graphitizing. Annealing a ferrous alloy in such a way that some or all of the carbon is precipitated as graphite.

growth. In cast iron, a permanent increase in dimensions resulting from repeated or prolonged heating at temperatures above 480 °C (900 °F) due either to graphitizing of carbides or to oxidation.

half hard. A *temper* of nonferrous alloys and some ferrous alloys characterized by tensile strength about midway between that of *dead soft* and *full hard* tempers.

hardenability. The relative ability of a ferrous alloy to form martensite when quenched from a temperature above the upper critical temperature. Hardenability is commonly measured as the distance below a quenched surface where the metal exhibits a specific hardness (50 HRC, for example) or a specific percentage of martensite in the microstructure.

hardening. Increasing hardness by suitable treatment, usually involving heating and cooling. When applicable, the following more specific terms should be used: *age hardening, case hardening, flame hardening, induction hardening, precipitation hardening* and *quench hardening*.

hard temper. Same as *full hard* temper.

heat-resisting alloy. An alloy developed for very high temperature service where relatively high stresses (tensile, thermal, vibratory, or shock) are encountered and where oxidation resistance is frequently required.

heat tinting. Coloration of a metal surface through oxidation by heating to reveal details of the microstructure.

heat treatable alloy. An alloy that can be hardened by heat treatment.

heat treating film. A thin coating or film, usually an oxide, formed on the surface of metals during heat treatment.

heat treatment. Heating and cooling a solid metal or alloy in such a way as to obtain desired conditions or properties. Heating for the sole purpose of hot working is excluded from the meaning of this definition.

homogeneous carburizing. Use of a carburizing process to convert a low-carbon ferrous alloy to one of uniform and higher carbon content throughout the section.

homogenizing. Holding at high temperature to eliminate or decrease chemical segregation by diffusion.

hot quenching. An imprecise term used to cover a variety of quenching procedures in which a quenching medium is maintained at a prescribed temperature above 70 °C (160 °F).

hydrogen brazing. A term sometimes used to denote brazing in a hydrogen-containing atmosphere, usually in a furnace; use of the appropriate process name is preferred.

hydrogen embrittlement. A condition of low ductility in metals resulting from the absorption of hydrogen.

induction brazing. *Brazing* in which the required heat is generated by subjecting the workpiece to electromagnetic induction.

induction hardening. A surface-hardening process in which only the surface layer of a suitable ferrous workpiece is heated by electromagnetic induction to above the upper critical temperature and immediately quenched.

induction heating. Heating by combined electrical resistance and hysteresis losses induced by subjecting a metal to the varying magnetic field surrounding a coil carrying alternating current.

intermediate annealing. Annealing wrought metals at one or more stages during manufacture and before final treatment.

interrupted aging. Aging at two or more temperatures, by steps, and cooling to room temperature after each step. See also *aging*. Compare with *progressive aging* and *step aging*.

interrupted quenching. A quenching procedure in which the workpiece is removed from the first quench at a temperature substantially higher than that of the quenchant and is then subjected to a second quenching system having a different cooling rate than the first.

isothermal annealing. Austenitizing a ferrous alloy and then cooling to and holding at a temperature at which austenite transforms to a relatively soft ferrite carbide aggregate.

isothermal transformation. A change in phase that takes place at a constant temperature. The time required for transformation to be completed, and in some instances the time delay before transformation begins, depends on the amount of supercooling below (or superheating above) the equilibrium temperature for the same transformation.

Jominy test. See *end-quench hardenability test*.

ledeburite. The eutectic of the iron-carbon system, the constituents being austenite and cementite. The austenite decomposes into ferrite and cementite on cooling below the Ar_1.

liquid phase sintering. *Sintering* a powder metallurgy compact under conditions that maintain a liquid metallic phase within the compact during all or part of the sintering schedule. The liquid phase may be derived from a component of the green compact or may be infiltrated into the compact from an outside source.

localized precipitation. Precipitation from a supersaturated solid solution similar to *continuous precipitation*, except that the precipitate particles form at preferred locations, such as along slip planes, grain boundaries, or incoherent twin boundaries.

malleabilizing. Annealing white cast iron in such a way that some or all of the combined carbon is transformed to graphite or, in some instances, part of the carbon is removed completely.

maraging. A precipitation-hardening treatment applied to a special group of iron-base alloys to precipitate one or more intermetallic compounds in a matrix of essentially carbon-free martensite. *Note*: The first developed series of maraging steels contained, in addition to iron, more than 10% nickel and one or more supplemental hardening elements. In this series, aging is done at 480 °C (900 °F).

marquenching. See *martempering*.

martempering. (1) A hardening procedure in which an austenitized ferrous workpiece is quenched into an appropriate medium whose temperature is maintained substantially at the M_s of the workpiece, held in the medium until its temperature is uniform throughout—but not long enough to permit bainite to form—and then cooled in air. The treatment is frequently followed by tempering. (2) When the process is applied to carburized material, the controlling M_s temperature is that of the case. This variation of the process is frequently called marquenching.

martensite. A generic term for microstructures formed by diffusionless phase transformation in which the parent and product phases have a specific crystallographic relationship. Martensite is characterized by an acicular pattern in the microstructure in both ferrous and nonferrous alloys. In alloys where the solute atoms occupy interstitial positions in the martensitic lattice (such as carbon in iron), the structure is hard and highly strained; but where the solute atoms occupy substitutional positions (such as nickel in iron), the martensite is soft and ductile. The amount of high-temperature phase that transforms to martensite on cooling depends to a large extent on the lowest temperature attained, there being a rather distinct beginning temperature (M_s) and a temperature at which the transformation is essentially complete (M_f).

martensite range. The temperature interval between M_s and M_f.

martensitic transformation. A reaction that takes place in some metals on cooling, with the formation of an acicular structure called *martensite*.

McQuaid-Ehn test. A test to reveal grain size after heating into the austenitic temperature range. Eight standard McQuaid-Ehn grain sizes rate the structure, No. 8 being finest, No. 1 coarsest.

metallurgy. The science and technology of metals and alloys. Process metallurgy is concerned with the extraction of metals from their ores and with the refining of metals; physical metallurgy, with the physical and mechanical properties of metals as affected by composition, processing, and environmental conditions; and mechanical metallurgy, with the response of metals to applied forces.

M_f temperature. For any alloy system, the temperature at which martensite formation on cooling is essentially finished. See *transformation temperature* for the definition applicable to ferrous alloys.

microhardness. The hardness of a material as determined by forcing an indenter such as a Vickers or Knoop indenter into the surface of a material under very light load; usually, the indentations are so small that they must be measured with a microscope. Capable of determining hardnesses of different microconstituents within a structure, or of measuring steep hardness gradients such as those encountered in case hardening.

mill scale. The heavy oxide layer formed during hot fabrication or heat treatment of metals.

monotropism. The ability of a solid to exist in two or more forms (crystal structures), but in which one form is the stable modification at all temperatures and pressures. Ferrite and martensite are a monotropic pair below Ac_1 in steels, for example. May also be spelled monotrophism.

M_s temperature. For any alloy system, the temperature at which martensite starts to form on cooling. See *transformation temperature* for the definition applicable to ferrous alloys.

natural aging. Spontaneous aging of a supersaturated solid solution at room temperature. See also *aging*. Compare with *artificial aging*.

neutral flame. A gas flame in which there is no excess of either fuel or oxygen in the inner flame. Oxygen from ambient air is used to complete the combustion of CO_2 and H_2 produced in the inner flame.

nitriding. Introducing nitrogen into the surface layer of a solid ferrous alloy by holding at a suitable temperature (below Ac_1 for ferritic steels) in contact with a nitrogenous material, usually ammonia or molten cyanide of appropriate composition. Quenching is not required to produce a hard case.

nitrocarburizing. Any of several processes in which both nitrogen and carbon are absorbed into the surface layers of a ferrous material at temperatures below the lower critical temperature and, by diffusion, create a concentration gradient. Nitrocarburizing is done mainly to provide an antiscuffing surface layer and to improve fatigue resistance. Compare with *carbonitriding*.

normalizing. Heating a ferrous alloy to a suitable temperature above the transformation range and then cooling in air to a temperature substantially below the transformation range.

optical pyrometer. An instrument for measuring the temperature of heated material by comparing the intensity of light emitted with a known intensity of an incandescent lamp filament.

overaging. Aging under conditions of time and temperature greater than those required to obtain maximum change in a certain property, so that the property is altered in the direction of the initial value. See also *aging*.

overheating. Heating a metal or alloy to such a high temperature that its properties are impaired. When the original properties cannot be restored by further heat treating, by mechanical working, or by a combination of working and heat treating, the overheating is known as *burning*.

oxidizing flame. A gas flame produced with excess oxygen in the inner flame.

packing material. Any material in which powder metallurgy compacts are embedded during the presintering or sintering operations.

partial annealing. An imprecise term used to denote a treatment given cold worked material to reduce the strength to a controlled level or to effect stress relief. To be meaningful, the type of material, the degree of cold work, and the time-temperature schedule must be stated.

patenting. In wiremaking, a heat treatment applied to medium-carbon or high-carbon steel before the drawing of wire or between drafts. This process consists of heating to a temperature above the transformation range and then cooling to a temperature below Ae_1 in air or in a bath of molten lead or salt.

pearlite. A metastable lamellar aggregate of ferrite and cementite resulting from the transformation of austenite at temperatures above the bainite range.

postheating. Heating weldments immediately after welding, for tempering, for stress relieving, or for providing a controlled rate of cooling to prevent formation of a hard or brittle structure.

pot annealing. Same as *box annealing*.

precipitation hardening. Hardening caused by the precipitation of a constituent from a supersaturated solid solution. See also *age hardening* and *aging*.

precipitation heat treatment. *Artificial aging* in which a constituent precipitates from a supersaturated solid solution.

preheating. Heating before some further thermal or mechanical treatment. For tool steel, heating to an intermediate temperature immediately before final austenitizing. For some nonferrous alloys, heating to a high temperature for a long time, to homogenize the structure before working. In welding and related processes, heating to an intermediate temperature for a short time immediately before welding, brazing, soldering, cutting, or thermal spraying.

presintering. The heating of a powder metallurgy compact to a temperature lower than the normal temperature for final sintering, usually to increase the ease of handling or forming the compact or to remove a lubricant or binder before sintering.

process annealing. An imprecise term denoting various treatments used to improve workability. For the term to be meaningful, the condition of the material and the time-temperature cycle used must be stated.

progressive aging. Aging by increasing the temperature in steps or continuously during the aging cycle. See also *aging*. Compare with *interrupted aging* and *step aging*.

pseudocarburizing. See *blank carburizing*.

pseudonitriding. See *blank nitriding*.

pusher furnace. A type of continuous furnace in which parts to be heated are periodically charged into the furnace in containers, which are pushed along the hearth against a line of previously charged containers thus advancing the containers toward the discharge end of the furnace, where they are removed.

quarter hard. A *temper* of nonferrous alloys and some ferrous alloys characterized by tensile strength about midway between that of *dead soft* and *half hard* tempers.

quench-age embrittlement. Embrittlement of low-carbon steel evidenced by a loss of ductility on aging at room temperature following rapid cooling from a temperature below the lower critical temperature.

quench aging. Aging induced by rapid cooling after *solution heat treatment*.

quench annealing. Annealing an austenitic ferrous alloy by *solution heat treatment* followed by rapid quenching.

quench cracking. Fracture of a metal during quenching from elevated temperature. Most frequently observed in hardened carbon steel, alloy steel, or tool steel parts of high hardness and low toughness. Cracks often emanate from fillets, holes, corners, or other stress raisers and result from high stresses due to the volume changes accompanying transformation to martensite.

quench hardening. (1) Hardening suitable alpha-beta alloys (most often certain copper or titanium alloys) by solution treating and quenching to develop a martensitic-like structure. (2) In ferrous alloys, hardening by austenitizing and then cooling at a rate such that a substantial amount of austenite transforms to martensite.

quenching. Rapid cooling. When applicable, the following more specific terms should be used: *direct quenching, fog quenching, hot quenching, interrupted quenching, selective quenching, spray quenching,* and *time quenching*.

recalescence. A phenomenon, associated with the transformation of gamma iron to alpha iron on the cooling (supercooling) of iron or steel, revealed by the brightening (reglowing) of the metal surface owing to the sudden increase in temperature caused by the fast liberation of the latent heat of transformation. Contrast with *decalescence*.

recarburize. (1) To increase the carbon content of molten cast iron or steel by adding carbonaceous material, high-carbon pig iron, or a high-carbon alloy. (2) To carburize a metal part to return surface carbon lost in processing; also known as carbon restoration.

recovery. Reduction or removal of work-hardening effects, without motion of large-angle grain boundaries.

recrystallization. (1) The formation of a new, strain-free grain structure from that existing in cold worked metal, usually accomplished by heating. (2) The change from one crystal structure to another, as occurs on heating or cooling through a critical temperature.

recrystallization annealing. Annealing cold worked metal to produce a new grain structure without phase change.

recrystallization temperature. The approximate minimum temperature at which complete recrystallization of a cold worked metal occurs within a specified time.

recuperator. Equipment for transferring heat from gaseous products of combustion to incoming air or fuel. The incoming material passes through pipes surrounded by a chamber through which the outgoing gases pass.

reducing flame. A gas flame produced with excess fuel in the inner flame.

refractory. (1) A material of very high melting point with properties that make it suitable for such uses as furnace linings and kiln construction. (2) The quality of resisting heat.

refractory alloy. (1) A heat-resistant alloy. (2) An alloy having an extremely high melting point. See also *refractory metal*. (3) An alloy difficult to work at elevated temperatures.

refractory metal. A metal having an extremely high melting point; for example, tungsten, molybdenum, tantalum, niobium (columbium), chromium, vanadium, and rhenium. In the broad sense, it refers to metals having melting points above the range of iron, cobalt, and nickel.

regenerator. Same as *recuperator* except the gaseous products of combustion heat brick checkerwork in a chamber connected to the exhaust side of the furnace while the incoming air and fuel are being heated by the brick checkerwork in a second chamber, connected to the entrance side. At intervals, the gas flow is reversed so that incoming air and fuel contact hot checkerwork while that in the second chamber is being reheated by exhaust gases.

resist. (1) A material applied to a part of the surface of an article to prevent reaction of metal from that area during chemical or electrochemical processes. (2) A material applied to prevent the flow of brazing filler metal into unwanted area.

resistance brazing. Brazing by resistance heating, the joint being part of the electrical circuit.

reverberatory furnace. A furnace with a shallow hearth, usually unregenerative, having a roof that deflects the flame and radiates heat toward the hearth or the surface of the charge.

Rockwell hardness test. An indentation hardness test based on the depth of penetration of a specified penetrator into the specimen under certain arbitrarily fixed conditions.

rotary furnace. A circular furnace constructed so that the hearth and workpieces rotate around the axis of the furnace during heating.

selective heating. Intentionally heating only certain portions of a workpiece.

selective quenching. Quenching only certain portions of an object.

self-hardening steel. See preferred term, *air-hardening steel*.

shrink forming. Forming metal wherein the inner fibers of a cross section undergo a reduction in a localized area by the application of heat, cold upset, or mechanically induced pressures.

siliconizing. Diffusing silicon into solid metal, usually steel, at an elevated temperature.

sinter. To heat a mass of fine particles for a prolonged time below the melting point, usually to cause agglomeration.

sintering. The bonding of adjacent surfaces in a mass of particles by molecular or atomic attraction on heating at high temperatures below the melting temperature of any constituent in the material. Sintering strengthens a powder mass and normally produces densification and, in powdered metals, recrystallization. See also *liquid phase sintering*.

slack quenching. The incomplete hardening of steel due to quenching from the austenitizing temperature at a rate slower than the critical cooling rate for the particular steel, resulting in the formation of one or more transformation products in addition to martensite.

slot furnace. A common batch furnace where stock is charged and removed through a slot or opening.

snap temper. A precautionary interim stress-relieving treatment applied to high-hardenability steels immediately after quenching to prevent cracking because of delay in tempering them at the prescribed higher temperature.

soaking. Prolonged holding at a selected temperature to effect homogenization of structure or composition.

soft temper. Same as *dead soft* temper.

solution heat treatment. Heating an alloy to a suitable temperature, holding at that temperature long enough to cause one or more constituents to enter into solid solution, and then cooling rapidly enough to hold these constituents in solution.

sorbite. (obsolete) A fine mixture of ferrite and cementite produced either by regulating the rate of cooling of steel or by tempering steel after hardening. The first type is very fine pearlite difficult to resolve under the microscope; the second type is tempered martensite.

spheroidite. An aggregate of iron or alloy carbides of essentially spherical shape dispersed throughout a matrix of ferrite.

spheroidizing. Heating and cooling to produce a spheroidal or globular form of carbide in steel. Spheroidizing methods frequently used are: (1) Prolonged holding at a temperature just below Ae_1. (2) Heating and cooling alternately between temperatures that are just above and just below Ae_1. (3) Heating to a temperature above Ae_1 or Ae_3 and then cooling very slowly in the furnace or holding at a temperature just below Ae_1. (4) Cooling at a suitable rate from the minimum temperature at which all carbide is dissolved, to prevent the reformation of a carbide network, and then reheating in accordance with method 1 or 2 above. (Applicable to hypereutectoid steel containing a carbide network.)

spinodal structure. A fine homogeneous mixture of two phases that form by the growth of composition waves in a solid solution during suitable heat treatment. The phases of a spinodal structure differ in composition from each other and from the parent phase but have the same crystal structure as the parent phase.

spray quenching. Quenching in a spray of liquid.

spring temper. A *temper* of nonferrous alloys and some ferrous alloys characterized by tensile strength and hardness about two-thirds of the way from *full hard* to *extra spring* temper.

stabilizing treatment. (1) Before finishing to final dimensions, repeatedly heating a ferrous or nonferrous part to or slightly above its normal operating temperature and then cooling to room temperature to ensure dimensional stability in service. (2) Transforming retained austenite in quenched hardenable steels, usually by *cold treatment*. (3) Heating a solution-treated stabilized grade of austenitic stainless steel to 870 to 900 °C (1600 to 1650 °F) to precipitate all carbon as TiC, NbC, or TaC so that *sensitization* is avoided on subsequent exposure to elevated temperature.

Stead's brittleness. A condition of brittleness that causes transcrystalline fracture in the coarse grain structure that results from prolonged annealing of thin sheets of low-carbon steel previously rolled at a temperature below about 705 °C (1300 °F). The fracture usually occurs at about 45° to the direction of rolling.

step aging. Aging at two or more temperatures, by steps, without cooling to room temperature after each step. See also *aging*. Compare with *interrupted aging* and *progressive aging*.

stoking. (obsolete) Presintering, or sintering, in such a way that powder metallurgy compacts are advanced through the furnace at a fixed rate by manual or mechanical means; also called continuous sintering.

stop-off. See *resist*.

stopping off. (1) Applying a *resist*. (2) Depositing a metal (copper, for example) in localized areas to prevent carburization, decarburization, or nitriding in those areas.

strain-age embrittlement. A loss in ductility accompanied by an increase in hardness and strength that occurs when low-carbon steel (especially rimmed or capped steel) is aged following plastic deformation. The degree of embrittlement is a function of aging time and temperature, occurring in a matter of minutes at about 200 °C (400 °F) but requiring a few hours to a year at room temperature.

stress relieving. Heating to a suitable temperature, holding long enough to reduce residual stresses, and then cooling slowly enough to minimize the development of new residual stresses.

subcritical annealing. A process anneal performed on ferrous alloys at a temperature below Ac_1.

superalloy. See *heat-resisting alloy*.

supercooling. Cooling below the temperature at which an equilibrium phase transformation can take place, without actually obtaining the transformation.

superheating. Heating above the temperature at which an equilibrium phase transformation should occur without actually obtaining the transformation.

surface hardening. A generic term covering several processes applicable to a suitable ferrous alloy that produces, by quench hardening only, a surface layer that is harder or more wear resistant than the core. There is no significant alteration of the chemical composition of the surface layer. The processes commonly used are induction hardening, flame hardening, and shell hardening. Use of the applicable specific process name is preferred.

temper. (1) In heat treatment, reheating hardened steel or hardened cast iron to some temperature below the eutectoid temperature for the purpose of decreasing hardness and increasing toughness. The process also is sometimes applied to normalized steel. (2) In tool steels, temper is sometimes used, but inadvisedly, to denote the carbon content. (3) In nonferrous alloys and in some ferrous alloys (steels that cannot be hardened by heat treatment), the hardness and strength produced by mechanical or thermal treatment, or both, and characterized by a certain structure, mechanical properties, or reduction in area during cold working.

temper brittleness. Brittleness that results when certain steels are held within, or are cooled slowly through, a certain range of temperature below the transformation range. The brittleness is manifested as an upward shift in ductile-to-brittle transition temperature, but only rarely produces a low value of reduction of area in a smooth-bar tension test of the embrittled material.

temper carbon. Same as *annealing carbon*.

temper color. A thin, tightly adhering oxide skin (only a few molecules thick) that forms when steel is tempered at a low temperature, or for a short time, in air or a mildly oxidizing atmosphere. The color, which ranges from straw to blue depending on the thickness of the oxide skin, varies with both tempering time and temperature.

thermocouple. A device for measuring temperatures, consisting of lengths of two dissimilar metals or alloys that are electrically joined at one end and connected to a voltage-measuring instrument at the other end. When one junction is hotter than the other, a thermal electromotive force is produced that is roughly proportional to the difference in temperature between the hot and cold junctions.

thermomechanical working. A general term covering a variety of processes combining controlled thermal and deformation treatments to obtain synergistic effects such as improvement in strength without loss of toughness. Same as thermal-mechanical treatment.

three-quarters hard. A *temper* of nonferrous alloys and some ferrous alloys characterized by tensile strength and hardness about midway between those of *half hard* and *full hard* tempers.

time quenching. Interrupted quenching in which the time in the quenching medium is controlled.

total carbon. The sum of the free and combined carbon (including carbon in solution) in a ferrous alloy.

transformation-induced plasticity. A phenomenon, occurring chiefly in certain highly alloyed steels that have been heat treated to produce metastable austenite or metastable austenite plus martensite, whereby, on subsequent deformation, part of the austenite undergoes strain-induced transformation to martensite. Steels capable of transforming in this manner, commonly referred to as TRIP steels, are highly plastic after heat treatment, but exhibit a very high rate of strain hardening and thus have high tensile and yield strengths after plastic deformation at temperatures between about 20 and 500 °C (70 and 930 °F). Cooling to −195 °C (−320 °F) may or may not be required to complete the transformation to martensite. Tempering usually is done following transformation.

transformation ranges. Those ranges of temperature within which a phase forms during heating and transforms during cooling. The two ranges are distinct, sometimes overlapping, but never coinciding. The limiting temperatures of the ranges depend on the composition of the alloy and on the rate of change of temperature, particularly during cooling. See also *transformation temperature*.

transformation temperature. The temperature at which a change in phase occurs. The term is sometimes used to denote the limiting temperature of a transformation range. The following symbols are used for iron and steels:

Ac_{cm}. In hypereutectoid steel, the temperature at which the solution of cementite in austenite is completed during heating.

Ac_1. The temperature at which austenite begins to form during heating.

Ac_3. The temperature at which transformation of ferrite to austenite is completed during heating.

Ac_4. The temperature at which austenite transforms to delta ferrite during heating.

Ae_{cm}, Ae_1, Ae_3, Ae_4. The temperatures of phase changes at equilibrium.

Ar_{cm}. In hypereutectoid steel, the temperature at which precipitation of cementite starts during cooling.

Ar_1. The temperature at which transformation of austenite to ferrite or to ferrite plus cementite is completed during cooling.

Ar_3. The temperature at which austenite begins to transform to ferrite during cooling.

Ar_4. The temperature at which delta ferrite transforms to austenite during cooling.

Ar'. The temperature at which transformation of austenite to pearlite starts during cooling.

M_f. The temperature at which transformation of austenite to martensite finishes during cooling.

M_s (or Ar''). The temperature at which transformation of austenite to martensite starts during cooling.

Note: All these changes except the formation of martensite occur at lower temperatures during cooling than during heating, and depend on the rate of change of temperature.

TRIP steel. A commercial steel product exhibiting *transformation-induced plasticity*.

troostite. (obsolete). A previously unresolvable rapidly etching fine aggregate of carbide and ferrite produced either by tempering martensite at low temperature or by quenching a steel at a rate slower than the critical cooling rate. Preferred terminology for the first product is tempered martensite; for the latter, fine pearlite.

undercooling. Same as *supercooling*.

APPENDIX 2

Temperature Conversions

The general arrangement of this conversion table was devised by Sauveur and Boylston. The middle columns of numbers (in **boldface** type) contain the temperature readings (°F or °C) to be converted. When converting from degrees Fahrenheit to degrees Celsius, read the Celsius equivalent in the column headed "C". When converting from Celsius to Fahrenheit, read the Fahrenheit equivalent in the column headed "F".

F		C	F		C	F		C	F		C
.....	−458	−272.22	.....	−378	−227.78	.....	−298	−183.33	−360.4	−218	−138.89
.....	−456	−271.11	.....	−376	−226.67	.....	−296	−182.22	−356.8	−216	−137.78
.....	−454	−270.00	.....	−374	−225.56	.....	−294	−181.11	−353.2	−214	−136.67
.....	−452	−268.89	.....	−372	−224.44	.....	−292	−180.00	−349.6	−212	−135.56
.....	−450	−267.78	.....	−370	−223.33	.....	−290	−178.89	−346.0	−210	−134.44
.....	−448	−266.67	.....	−368	−222.22	.....	−288	−177.78	−342.4	−208	−133.33
.....	−446	−265.56	.....	−366	−221.11	.....	−286	−176.67	−338.8	−206	−132.22
.....	−444	−264.44	.....	−364	−220.00	.....	−284	−175.56	−335.2	−204	−131.11
.....	−442	−263.33	.....	−362	−218.89	.....	−282	−174.44	−331.6	−202	−130.00
.....	−440	−262.22	.....	−360	−217.78	.....	−280	−173.33	−328.0	−200	−128.89
.....	−438	−261.11	.....	−358	−216.67	.....	−278	−172.22	−324.4	−198	−127.78
.....	−436	−260.00	.....	−356	−215.56	.....	−276	−171.11	−320.8	−196	−126.67
.....	−434	−258.89	.....	−354	−214.44	.....	−274	−170.00	−317.2	−194	−125.56
.....	−432	−257.78	.....	−352	−213.33	−457.6	−272	−168.89	−313.6	−192	−124.44
.....	−430	−256.67	.....	−350	−212.22	−454.0	−270	−167.78	−310.0	−190	−123.33
.....	−428	−255.56	.....	−348	−211.11	−450.4	−268	−166.67	−306.4	−188	−122.22
.....	−426	−254.44	.....	−346	−210.00	−446.8	−266	−165.56	−302.8	−186	−121.11
.....	−424	−253.33	.....	−344	−208.89	−443.2	−264	−164.44	−299.2	−184	−120.00
.....	−422	−252.22	.....	−342	−207.78	−439.6	−262	−163.33	−295.6	−182	−118.89
.....	−420	−251.11	.....	−340	−206.67	−436.0	−260	−162.22	−292.0	−180	−117.78
.....	−418	−250.00	.....	−338	−205.56	−432.4	−258	−161.11	−288.4	−178	−116.67
.....	−416	−248.89	.....	−336	−204.44	−428.8	−256	−160.00	−284.8	−176	−115.56
.....	−414	−247.78	.....	−334	−203.33	−425.2	−254	−158.89	−281.2	−174	−114.44
.....	−412	−246.67	.....	−332	−202.22	−421.6	−252	−157.78	−277.6	−172	−113.33
.....	−410	−245.56	.....	−330	−201.11	−418.0	−250	−156.67	−274.0	−170	−112.22
.....	−408	−244.44	.....	−328	−200.00	−414.4	−248	−155.56	−270.4	−168	−111.11
.....	−406	−243.33	.....	−326	−198.89	−410.8	−246	−154.44	−266.8	−166	−110.00
.....	−404	−242.22	.....	−324	−197.78	−407.2	−244	−153.33	−263.2	−164	−108.89
.....	−402	−241.11	.....	−322	−196.67	−403.6	−242	−152.22	−259.6	−162	−107.78
.....	−400	−240.00	.....	−320	−195.56	−400.0	−240	−151.11	−256.0	−160	−106.67
.....	−398	−238.89	.....	−318	−194.44	−396.4	−238	−150.00	−252.4	−158	−105.56
.....	−396	−237.78	.....	−316	−193.33	−392.8	−236	−148.89	−248.8	−156	−104.44
.....	−394	−236.67	.....	−314	−192.22	−389.2	−234	−147.78	−245.2	−154	−103.33
.....	−392	−235.56	.....	−312	−191.11	−385.6	−232	−146.67	−241.6	−152	−102.22
.....	−390	−234.44	.....	−310	−190.00	−382.0	−230	−145.56	−238.0	−150	−101.11
.....	−388	−233.33	.....	−308	−188.89	−378.4	−228	−144.44	−234.4	−148	−100.00
.....	−386	−232.22	.....	−306	−187.78	−374.8	−226	−143.33	−230.8	−146	−98.89
.....	−384	−231.11	.....	−304	−186.67	−371.2	−224	−142.22	−227.2	−144	−97.78
.....	−382	−230.00	.....	−302	−185.56	−367.6	−222	−141.11	−223.6	−142	−96.67
.....	−380	−228.89	.....	−300	−184.44	−364.0	−220	−140.00	−220.0	−140	−95.56

(continued on the next page)

Temperature Conversions (continued)

F		C	F		C	F		C	F		C
−216.4	−138	−94.44	+35.6	+2	−16.67	287.6	142	61.11	539.6	282	138.89
−212.8	−136	−93.33	+39.2	+4	−15.56	291.2	144	62.22	543.2	284	140.00
−209.2	−134	−92.22	+42.8	+6	−14.44	294.8	146	63.33	546.8	286	141.11
−205.6	−132	−91.11	+46.4	+8	−13.33	298.4	148	64.44	550.4	288	142.22
−202.0	−130	−90.00	+50.0	+10	−12.22	302.0	150	65.56	554.0	290	143.33
−198.4	−128	−88.89	+53.6	+12	−11.11	305.6	152	66.67	557.6	292	144.44
−194.8	−126	−87.78	+57.2	+14	−10.00	309.2	154	67.78	561.2	294	145.56
−191.2	−124	−86.67	+60.8	+16	−8.89	312.8	156	68.89	564.8	296	146.67
−187.6	−122	−85.56	+64.4	+18	−7.78	316.4	158	70.00	568.4	298	147.78
−184.0	−120	−84.44	+68.0	+20	−6.67	320.0	160	71.11	572.0	300	148.89
−180.4	−118	−83.33	+71.6	+22	−5.56	323.6	162	72.22	575.6	302	150.00
−176.8	−116	−82.22	+75.2	+24	−4.44	327.2	164	73.33	579.2	304	151.11
−173.2	−114	−81.11	+78.8	+26	−3.33	330.8	166	74.44	582.8	306	152.22
−169.6	−112	−80.00	+82.4	+28	−2.22	334.4	168	75.56	586.4	308	153.33
−166.0	−110	−78.89	+86.0	+30	−1.11	338.0	170	76.67	590.0	310	154.44
−162.4	−108	−77.78	+89.6	+32	±0.00	341.6	172	77.78	593.6	312	155.56
−158.8	−106	−76.67	+93.2	+34	+1.11	345.2	174	78.89	597.2	314	156.67
−155.2	−104	−75.56	+96.8	+36	+2.22	348.8	176	80.00	600.8	316	157.78
−151.6	−102	−74.44	+100.4	+38	+3.33	352.4	178	81.11	604.4	318	158.89
−148.0	−100	−73.33	+104.0	+40	+4.44	356.0	180	82.22	608.0	320	160.00
−144.4	−98	−72.22	107.6	42	5.56	359.6	182	83.33	611.6	322	161.11
−140.8	−96	−71.11	111.2	44	6.67	363.2	184	84.44	615.2	324	162.22
−137.2	−94	−70.00	114.8	46	7.78	366.8	186	85.56	618.8	326	163.33
−133.6	−92	−68.89	118.4	48	8.89	370.4	188	86.67	622.4	328	164.44
−130.0	−90	−67.78	122.0	50	10.00	374.0	190	87.78	626.0	330	165.56
−126.4	−88	−66.67	125.6	52	11.11	377.6	192	88.89	629.6	332	166.67
−122.8	−86	−65.56	129.2	54	12.22	381.2	194	90.00	633.2	334	167.78
−119.2	−84	−64.44	132.8	56	13.33	384.8	196	91.11	636.8	336	168.89
−115.6	−82	−63.33	136.4	58	14.44	388.4	198	92.22	640.4	338	170.00
−112.0	−80	−62.22	140.0	60	15.56	392.0	200	93.33	644.0	340	171.11
−108.4	−78	−61.11	143.6	62	16.67	395.6	202	94.44	647.6	342	172.22
−104.8	−76	−60.00	147.2	64	17.78	399.2	204	95.56	651.2	344	173.33
−101.2	−74	−58.89	150.8	66	18.89	402.8	206	96.67	654.8	346	174.44
−97.6	−72	−57.78	154.4	68	20.00	406.4	208	97.78	658.4	348	175.56
−94.0	−70	−56.67	158.0	70	21.11	410.0	210	98.89	662.0	350	176.67
−90.4	−68	−55.56	161.6	72	22.22	413.6	212	100.00	665.6	352	177.78
−86.8	−66	−54.44	165.2	74	23.33	417.2	214	101.11	669.2	354	178.89
−83.2	−64	−53.33	168.8	76	24.44	420.8	216	102.22	672.8	356	180.00
−79.6	−62	−52.22	172.4	78	25.56	424.4	218	103.33	676.4	358	181.11
−76.0	−60	−51.11	176.0	80	26.67	428.0	220	104.44	680.0	360	182.22
−72.4	−58	−50.00	179.6	82	27.78	431.6	222	105.56	683.6	362	183.33
−68.8	−56	−48.89	183.2	84	28.89	435.2	224	106.67	687.2	364	184.44
−65.2	−54	−47.78	186.8	86	30.00	438.8	226	107.78	690.8	366	185.56
−61.6	−52	−46.67	190.4	88	31.11	442.4	228	108.89	694.4	368	186.67
−58.0	−50	−45.56	194.0	90	32.22	446.0	230	110.00	698.0	370	187.78
−54.4	−48	−44.44	197.6	92	33.33	449.6	232	111.11	701.6	372	188.89
−50.8	−46	−43.33	201.2	94	34.44	453.2	234	112.22	705.2	374	190.00
−47.2	−44	−42.22	204.8	96	35.56	456.8	236	113.33	708.8	376	191.11
−43.6	−42	−41.11	208.4	98	36.67	460.4	238	114.44	712.4	378	192.22
−40.0	−40	−40.00	212.0	100	37.78	464.0	240	115.56	716.0	380	193.33
−36.4	−38	−38.89	215.6	102	38.89	467.6	242	116.67	719.6	382	194.44
−32.8	−36	−37.78	219.2	104	40.00	471.2	244	117.78	723.2	384	195.56
−29.2	−34	−36.67	222.8	106	41.11	474.8	246	118.89	726.8	386	196.67
−25.6	−32	−35.56	226.4	108	42.22	478.4	248	120.00	730.4	388	197.78
−22.0	−30	−34.44	230.0	110	43.33	482.0	250	121.11	734.0	390	198.89
−18.4	−28	−33.33	233.6	112	44.44	485.6	252	122.22	737.6	392	200.00
−14.8	−26	−32.22	237.2	114	45.56	489.2	254	123.33	741.2	394	201.11
−11.2	−24	−31.11	240.8	116	46.67	492.8	256	124.44	744.8	396	202.22
−7.6	−22	−30.00	244.4	118	47.78	496.4	258	125.56	748.4	398	203.33
−4.0	−20	−28.89	248.0	120	48.89	500.0	260	126.67	752.0	400	204.44
−0.4	−18	−27.78	251.6	122	50.00	503.6	262	127.78	755.6	402	205.56
+3.2	−16	−26.67	255.2	124	51.11	507.2	264	128.89	759.2	404	206.67
+6.8	−14	−25.56	258.8	126	52.22	510.8	266	130.00	762.8	406	207.78
+10.4	−12	−24.44	262.4	128	53.33	514.4	268	131.11	766.4	408	208.89
+14.0	−10	−23.33	266.0	130	54.44	518.0	270	132.22	770.0	410	210.00
+17.6	−8	−22.22	269.6	132	55.56	521.6	272	133.33	773.6	412	211.11
+21.2	−6	−21.11	273.2	134	56.67	525.2	274	134.44	777.2	414	212.22
+24.8	−4	−20.00	276.8	136	57.78	528.8	276	135.56	780.8	416	213.33
+28.4	−2	−18.89	280.4	138	58.89	532.4	278	136.67	784.4	418	214.44
+32.0	±0	−17.78	284.0	140	60.00	536.0	280	137.78	788.0	420	215.56

TEMPERATURE CONVERSIONS / 471

F		C	F		C	F		C	F		C
791.6	**422**	216.67	1130.0	**610**	321.11	2390.0	**1310**	710.00	3650.0	**2010**	1098.9
795.2	**424**	217.78	1148.0	**620**	326.67	2408.0	**1320**	715.56	3668.0	**2020**	1104.4
798.8	**426**	218.89	1166.0	**630**	332.22	2426.0	**1330**	721.11	3686.0	**2030**	1110.0
802.4	**428**	220.00	1184.0	**640**	337.78	2444.0	**1340**	726.67	3704.0	**2040**	1115.6
806.0	**430**	221.11	1202.0	**650**	343.33	2462.0	**1350**	732.22	3722.0	**2050**	1121.1
809.6	**432**	222.22	1220.0	**660**	348.89	2480.0	**1360**	737.78	3740.0	**2060**	1126.7
813.2	**434**	223.33	1238.0	**670**	354.44	2499.0	**1370**	743.33	3758.0	**2070**	1132.2
816.8	**436**	224.44	1256.0	**680**	360.00	2516.0	**1380**	748.89	3776.0	**2080**	1137.8
820.4	**438**	225.56	1274.0	**690**	365.56	2534.0	**1390**	754.44	3794.0	**2090**	1143.3
824.0	**440**	226.67	1292.0	**700**	371.11	2552.0	**1400**	760.00	3812.0	**2100**	1148.9
827.6	**442**	227.78	1310.0	**710**	376.67	2570.0	**1410**	765.56	3830.0	**2110**	1154.4
831.2	**444**	228.89	1328.0	**720**	382.22	2588.0	**1420**	771.11	3848.0	**2120**	1160.0
834.8	**446**	230.00	1346.0	**730**	387.78	2606.0	**1430**	776.67	3866.0	**2130**	1165.6
838.4	**448**	231.11	1364.0	**740**	393.33	2624.0	**1440**	782.22	3884.0	**2140**	1171.1
842.0	**450**	232.22	1382.0	**750**	398.89	2642.0	**1450**	787.78	3902.0	**2150**	1176.7
845.6	**452**	233.33	1400.0	**760**	404.44	2660.0	**1460**	793.33	3920.0	**2160**	1182.2
849.2	**454**	234.44	1418.0	**770**	410.00	2678.0	**1470**	798.89	3938.0	**2170**	1187.8
852.8	**456**	235.56	1436.0	**780**	415.56	2696.0	**1480**	804.44	3956.0	**2180**	1193.3
856.4	**458**	236.67	1454.0	**790**	421.11	2714.0	**1490**	810.00	3974.0	**2190**	1198.9
860.0	**460**	237.78	1472.0	**800**	426.67	2732.0	**1500**	815.56	3992.0	**2200**	1204.4
863.6	**462**	238.89	1490.2	**810**	432.22	2750.0	**1510**	821.11	4010.0	**2210**	1210.0
867.2	**464**	240.00	1508.0	**820**	437.78	2768.0	**1520**	826.67	4028.0	**2220**	1215.6
870.8	**466**	241.11	1526.0	**830**	443.33	2786.0	**1530**	832.22	4046.0	**2230**	1221.1
874.4	**468**	242.22	1544.0	**840**	448.89	2804.0	**1540**	837.78	4064.0	**2240**	1226.7
878.0	**470**	243.33	1562.0	**850**	454.44	2822.0	**1550**	843.33	4082.0	**2250**	1232.2
881.6	**472**	244.44	1580.0	**860**	460.00	2840.0	**1560**	848.89	4100.0	**2260**	1237.8
885.2	**474**	245.56	1598.0	**870**	465.56	2858.0	**1570**	854.44	4118.0	**2270**	1243.3
888.8	**476**	246.67	1616.0	**880**	471.11	2876.0	**1580**	860.00	4136.0	**2280**	1248.9
892.4	**478**	247.78	1634.0	**890**	476.67	2894.0	**1590**	865.56	4154.0	**2290**	1254.4
896.0	**480**	248.89	1652.0	**900**	482.22	2912.0	**1600**	871.11	4172.0	**2300**	1260.0
899.6	**482**	250.00	1670.0	**910**	487.78	2930.0	**1610**	876.67	4190.0	**2310**	1265.6
903.2	**484**	251.11	1688.0	**920**	493.33	2948.0	**1620**	882.22	4208.0	**2320**	1271.1
906.8	**486**	252.22	1706.0	**930**	498.89	2966.0	**1630**	887.78	4226.0	**2330**	1276.7
910.4	**488**	253.33	1724.0	**940**	504.44	2984.0	**1640**	893.33	4244.0	**2340**	1282.2
914.0	**490**	254.44	1742.0	**950**	510.00	3002.0	**1650**	898.89	4262.0	**2350**	1287.8
917.6	**492**	255.56	1760.0	**960**	515.56	3020.0	**1660**	904.44	4280.0	**2360**	1293.3
921.2	**494**	256.67	1778.0	**970**	521.11	3038.0	**1670**	910.00	4298.0	**2370**	1298.9
924.8	**496**	257.78	1796.0	**980**	526.67	3056.0	**1680**	915.56	4316.0	**2380**	1304.4
928.4	**498**	258.89	1814.0	**990**	532.22	3074.0	**1690**	921.11	4334.0	**2390**	1310.0
932.0	**500**	260.00	1832.0	**1000**	537.78	3092.0	**1700**	926.67	4352.0	**2400**	1315.6
935.6	**502**	261.11	1850.0	**1010**	543.33	3110.0	**1710**	932.22	4370.0	**2410**	1321.1
939.2	**504**	262.22	1868.0	**1020**	548.89	3128.0	**1720**	937.78	4388.0	**2420**	1326.7
942.8	**506**	263.33	1886.0	**1030**	554.44	3146.0	**1730**	943.33	4406.0	**2430**	1332.2
946.4	**508**	264.44	1904.0	**1040**	560.00	3164.0	**1740**	948.89	4424.0	**2440**	1337.8
950.0	**510**	265.56	1922.0	**1050**	565.56	3182.0	**1750**	954.44	4442.0	**2450**	1343.3
953.6	**512**	266.67	1940.0	**1060**	571.11	3200.0	**1760**	960.00	4460.0	**2460**	1348.9
957.2	**514**	267.78	1958.0	**1070**	576.67	3218.0	**1770**	965.56	4478.0	**2470**	1354.4
960.8	**516**	268.89	1976.0	**1080**	582.22	3236.0	**1780**	971.11	4496.0	**2480**	1360.0
964.4	**518**	270.00	1994.0	**1090**	587.78	3254.0	**1790**	976.67	4514.0	**2490**	1365.6
968.0	**520**	271.11	2012.0	**1100**	593.33	3272.0	**1800**	982.22	4532.0	**2500**	1371.1
971.6	**522**	272.22	2030.0	**1110**	598.89	3290.0	**1810**	987.78	4550.0	**2510**	1376.7
975.2	**524**	273.33	2048.0	**1120**	604.44	3308.0	**1820**	993.33	4568.0	**2520**	1382.2
978.8	**526**	274.44	2066.0	**1130**	610.00	3326.0	**1830**	998.89	4586.0	**2530**	1387.8
982.4	**528**	275.56	2084.0	**1140**	615.56	3344.0	**1840**	1004.4	4604.0	**2540**	1393.3
986.0	**530**	276.67	2102.0	**1150**	621.11	3362.0	**1850**	1010.0	4622.0	**2550**	1398.9
989.6	**532**	277.78	2120.0	**1160**	626.67	3380.0	**1860**	1015.6	4640.0	**2560**	1404.4
993.2	**534**	278.89	2138.0	**1170**	632.22	3398.0	**1870**	1021.1	4658.0	**2570**	1410.0
996.8	**536**	280.00	2156.0	**1180**	637.78	3416.0	**1880**	1026.7	4676.0	**2580**	1415.6
1000.4	**538**	281.11	2174.0	**1190**	643.33	3434.0	**1890**	1032.2	4694.0	**2590**	1421.1
1004.0	**540**	282.22	2192.0	**1200**	648.89	3452.0	**1900**	1037.8	4712.0	**2600**	1426.7
1007.6	**542**	283.33	2210.0	**1210**	654.44	3470.0	**1910**	1043.3	4730.0	**2610**	1432.2
1011.2	**544**	284.44	2228.0	**1220**	660.00	3488.0	**1920**	1048.9	4748.0	**2620**	1437.8
1014.8	**546**	285.56	2246.0	**1230**	665.56	3506.0	**1930**	1054.4	4766.0	**2630**	1443.3
1018.4	**548**	286.67	2264.0	**1240**	671.11	3524.0	**1940**	1060.0	4784.0	**2640**	1448.9
1022.0	**550**	287.78	2282.0	**1250**	676.67	3542.0	**1950**	1065.6	4802.0	**2650**	1454.4
1040.0	**560**	293.33	2300.0	**1260**	682.22	3560.0	**1960**	1071.1	4820.0	**2660**	1460.0
1058.0	**570**	298.89	2318.0	**1270**	687.78	3578.0	**1970**	1076.7	4838.0	**2670**	1465.6
1076.0	**580**	304.44	2336.0	**1280**	693.33	3596.0	**1980**	1082.2	4856.0	**2680**	1471.1
1094.0	**590**	310.00	2354.0	**1290**	698.89	3614.0	**1990**	1087.8	4874.0	**2690**	1476.7
1112.0	**600**	315.56	2372.0	**1300**	704.44	3632.0	**2000**	1093.3	4892.0	**2700**	1482.2

(continued on the next page)

Temperature Conversions (continued)

F	C		F	C		F	C		F	C	
4910.0	2710	1487.8	5270.0	2910	1598.9	6152.0	3400	1871.1	8402.0	4650	2565.5
4928.0	2720	1493.3	5288.0	2920	1604.4	6242.0	3450	1898.8	8492.0	4700	2593.3
4946.0	2730	1498.9	5306.0	2930	1610.0	6332.0	3500	1926.6	8582.0	4750	2621.1
4964.0	2740	1504.4	5324.0	2940	1615.6	6422.0	3550	1954.4	8672.0	4800	2648.8
4982.0	2750	1510.0	5342.0	2950	1621.1	6512.0	3600	1982.2	8762.0	4850	2676.6
5000.0	2760	1515.6	5360.0	2960	1626.7	6602.0	3650	2010.0	8852.0	4900	2704.4
5018.0	2770	1521.1	5378.0	2970	1632.2	6692.0	3700	2037.7	8942.0	4950	2732.2
5036.0	2780	1526.7	5396.0	2980	1637.8	6782.0	3750	2065.5	9032.0	5000	2760.0
5054.0	2790	1532.2	5414.0	2990	1643.3	6872.0	3800	2093.3	9122.0	5050	2787.7
5072.0	2800	1537.8	5432.0	3000	1648.9	6962.0	3850	2121.1	9212.0	5100	2815.5
5090.0	2810	1543.3	5450.0	3010	1654.4	7052.0	3900	2148.8	9302.0	5150	2843.3
5108.0	2820	1548.9	5468.0	3020	1660.0	7142.0	3950	2176.6	9392.0	5200	2871.1
5126.0	2830	1554.4	5486.0	3030	1665.6	7232.0	4000	2204.4	9482.0	5250	2898.8
5144.0	2840	1560.0	5504.0	3040	1671.1	7322.0	4050	2232.2	9572.0	5300	2926.6
5162.0	2850	1565.6	5522.0	3050	1676.7	7412.0	4100	2260.0	9662.0	5350	2954.4
5180.0	2860	1571.1	5540.0	3060	1682.2	7502.0	4150	2287.7	9752.0	5400	2982.2
5198.0	2870	1576.7	5558.0	3070	1687.8	7592.0	4200	2315.5	9842.0	5450	3010.0
5216.0	2880	1582.2	5576.0	3080	1693.3	7682.0	4250	2343.3	9932.0	5500	3037.7
5234.0	2890	1587.8	5594.0	3090	1698.9	7772.0	4300	2371.1	10022.0	5550	3065.5
5252.0	2900	1593.3	5612.0	3100	1704.4	7862.0	4350	2398.8	10112.0	5600	3093.3
			5702.0	3150	1732.2	7952.0	4400	2426.6			
			5792.0	3200	1760.0	8042.0	4450	2454.4			
			5882.0	3250	1787.7	8132.0	4500	2482.2			
			5972.0	3300	1815.5	8222.0	4550	2510.0			
			6062.0	3350	1843.3	8312.0	4600	2537.7			

Index

Emphasis in the index, as in the book, is placed on the crystallographic, microstructural, and phase transformation phenomena underlying heat treatment processes, rather than on specific methods or specific materials. The letter (D), (M), or (T) following an entry signifies that information on the subject is presented in a diagram, micrograph, or table.

A_1, A_3, and A_{cm} temperatures (D), 14–15
A2 tool steel
  composition (T), 403
  effect of tempering temperature on hardness and retained austenite (D), 423, 424–425
  effect on increasing austenitizing temperature (D), 417, 418
  M_7C_3 carbides contained in equilibrium, 409–410
  plate martensite formation and retained carbides (M), 419, 421
A-286 stainless steel. *See* AISI 600 stainless steel
Ac_{cm}, Ac_1, Ac_3, and Ac_4 temperatures (D), 14–15, 468
Acicular ferrite, 453
Acicular structure of austenite, in duplex stainless steels, 393–394
Acicular structure of bainite, 80
Acicular structure of martensite (M), 47, 48, 67, 71
Adatoms, adsorption during physical vapor deposition, 331, 333, 334
Ae_{cm}, Ae_1, Ae_3, and Ae_4 temperatures (D), 14–15, 468
Age hardening, 453
Age softening, 453
Aging (D), 125–131, 453
  artificial, 454
  double, 458
  effect on austenite, 126
  effect on hardness (D), 148, 149
  effect on martensite strength, 151
  effect on precipitation-strengthening stainless steels (M), 387, 390
  interrupted, 461
  natural, 462
  overaging, 463
  progressive, 463
  quench (D), 126–130, 131, 464
  step, 466
  strain, 125–130
Air cooling
  after normalizing, 113, 114
  effect on cast irons, 444
  effect on maraging steels, 405, 419
  microstructures produced as related to bar diameter (D,M), 99–105
  residual stress source, 130
  and spheroidization (D), 115, 117
Air-hardening steel, 453
AISI 200 austenitic stainless steels, compositions (T), 372
AISI 201 stainless steel, composition (T), 372
AISI 202 stainless steel, composition (T), 372
AISI 216 stainless steel, composition and minimum mechanical properties (T), 373
AISI 300 austenitic stainless steels, compositions (T), 359–361
AISI 301 stainless steel
  composition (T), 359
  strength enhanced by strain-induced martensite formation, 368
  stress-strain curves (D), 367, 368
  varying austenite stability relative to martensite formation during cold work, 361
AISI 302 stainless steel
  composition (T), 359
  strength enhanced by strain-induced martensite formation, 368

AISI 302 stainless steel *(continued)*
  varying austenite stability relative to martensite formation during cold work, 361
AISI 304 stainless steel
  cast alloy CF-8, 371
  chromium carbide precipitation on various boundaries (M), 364, 365
  chromium depletion as function of distance from grain boundaries (D), 363
  composition (T), 359
  effect of chromium carbide precipitation at grain boundaries (M), 361–362
  effect of temperature on strain-induced martensite formation (D), 367, 368
  $M_{23}C_6$ carbide precipitation kinetics (D), 364, 365
  stress-strain curves (D), 367, 368
  stress-strain curves obtained in constant temperature baths (D), 368, 369, 370
  varying austenite stability relative to martensite formation during cold work, 361
AISI 304L stainless steel
  alloying approach to minimize chromium carbide precipitation, 365
  carbon reduction, 361
  cast alloy CF-3, 371
  chromium carbide formation eliminated, 361
  composition (T), 359
  delta ferrite (M), 369, 370
  intergranular corrosion eliminated, 361
  sensitive to strain-induced martensite formation, 368
AISI 309 stainless steel
  composition (T), 359
  heat-resisting HH, 372
  high-temperature strength by alloying with chromium and nickel, 361
  scaling resistance by alloying with chromium and nickel, 361
AISI 310 stainless steel
  composition (T), 359
  heat-resisting HK, 372
  high-temperature strength by alloying with chromium and nickel, 361
  scaling resistance by alloying with chromium and nickel, 361
AISI 316 stainless steel
  cast alloy CF-8M, 371
  composition (T), 359
  pitting resistance increased by alloying with molybdenum, 361
AISI 316L stainless steel
  alloying approach to minimize chromium carbide precipitation, 365
  carbon reduction, 361
  cast alloy CF-3M, 371
  chromium carbide formation eliminated, 361
  composition (T), 359
  effect of annealing on microstructure (M), 360, 361
  intergranular corrosion eliminated, 361
AISI 321 stainless steel
  carbon reduction, 361
  chromium carbide formation eliminated, 361
  composition (T), 359
  intergranular corrosion eliminated, 361
  intragranular sigma formation, 371
  stabilized grade, 365
AISI 347 stainless steel
  carbon reduction, 361
  chromium carbide formation eliminated, 361
  composition (T), 359
  intergranular corrosion eliminated, 361
  stabilized grade, 365
AISI 400 ferritic stainless steels, compositions (T), 374, 375
AISI 400 martensitic stainless steels, compositions (T), 381
AISI 403 stainless steel
  comparable to type 410 stainless steel, 382
  composition (T), 381
  hardening effect on lath martensite microstructure (M), 384
  special-quality grade for turbine blade applications, 382
  spheroidized carbides dispersed in ferrite after annealing (M), 382, 383
AISI 410 stainless steel
  comparable to type 403 stainless steel, 382
  composition (T), 381
  mechanical property changes after tempering (D), 384, 385
AISI 416 stainless steel
  composition (T), 381
  spherical carbides dispersed in ferrite after annealing (M), 382, 383
AISI 420 stainless steel

INDEX / 475

composition (T), 381
for dies to mold plastics, 402
AISI 430 stainless steel, composition (T), 374
AISI 430 F stainless steel, composition (T), 374
AISI 430 F Se stainless steel, composition (T), 374
AISI 431 stainless steel, composition (T), 381
AISI 440A stainless steel
  composition (T), 381
  hardening effect on microstructure, 384
  higher hardness, 382
AISI 440B stainless steel
  composition (T), 381
  hardening effect on microstructure, 384
  higher hardness, 382
AISI 440C stainless steel
  bearing application of ion implantation, 326
  composition (T), 381, 382
  hardening effect on microstructure, 384
  higher hardness, 382
AISI 446 stainless steel
  composition (T), 374
AISI 600 stainless steel
  composition (T), 386, 387
  heavily alloyed to stabilize austenite, 390
  trade name A-286 (T), 386
AISI 630 stainless steel
  composition (T), 386, 387
  trade name 17.4 PH (T), 386
AISI 631 stainless steel
composition (T), 386, 387
effect of solution treatment and heat treatments (M), 389, 390
surface tilting caused by martensite formation (M), 389, 390
trade name 17.7 PH (T), 386
AISI 633 stainless steel
  composition (T), 386, 387
  trade name AM 350 (T), 386
AISI 635 stainless steel
  composition (T), 386, 387
  trade name Stainless W (T), 386
Al 2205 stainless steel
  composition (T), 391
  effect of temperature on stress-strain curves (D), 394, 395
  ferrite and austenite grains in microstructure (M), 393, 394
Allotriomorph formation, cementite grain boundary, 423
Alloying elements
  effect on carbon solubility during carburizing, 285–286, 290
  effect on cast irons (T), 432–433, 438–439, 442, 445, 448
  effect on continuous cooling transformations (D,M), 95–104
  effect on crystal structure on Fe-C alloys, 10–11
  effect on eutectoid composition and temperature (D), 11–15
  effect on grain size control (D), 199–201
  effect on hardenability (D), 154, 167, 169, 170
  effect on laser surface heat treatment, 345
  effect on martensitic transformation (D), 52–54
  effect on pearlite formation (D), 29, 30, 31
  effect on precipitation-hardening stainless steels, 386, 388
  effect on stainless steels (D), 351–353, 354, 365–366
  effect on tempering (D), 212–218, 224, 236–238, 247
  effect on tool steels (M), 402, 405–410, 411, 413, 415, 417, 419
  for grain size control, 136–138
  in high-strength low-alloy (HSLA) or microalloyed steels, 34
  low-temperature solid solubility of intermetallic phases, 377
  partitioning for sigma formation, 371
  responsible for stabilization of carbides, 248
Alloy steels
  nitriding of, 308
  for ferritic nitrocarburizing, 312
Alpha ferrite, 453
Alpha iron, 3, 4, 5, 6, 453 (*see also* Ferrite)
Alpha phase, ferritic stainless steels, 379–380
Alpha prime phase, ferritic stainless steels, 380–381
Alumel, 453
Aluminum
  effect on bainite formation, 80
  effect on grain size (D), 193, 198–199
  use in nitriding steels, 306
Aluminum bronze, effect on embrittlement, 241
Aluminum nitride, effect on austenitic grain size (D), 193, 198–199, 200
AM 350 stainless steel. *See* AISI 633 stainless steel

Ammonia gas, nitriding and carbonitriding reactions (D), 308–309, 310, 311
Annealing (*see also* Black annealing; Blue annealing; Box annealing; Bright annealing; Continuous annealing; Cycle annealing; Final annealing; Flame annealing; Full annealing; Homogenizing heat treatment; Intermediate annealing; Isothermal annealing; Partial annealing; Process annealing; Quench annealing; Recrystallization annealing; Subcritical annealing)
  austenite stainless steels, 373
  definition, 453
  effect on cast irons, 437, 441, 444
  effect on duplex stainless steels, 393, 394
  effect on intermetallic phases of ferritic stainless steels (M), 378
  effect on tool steels (M), 405–406, 413–415
  ferritic stainless steels, 375
  intercritical (D), 274–277, 278
Annealing carbon (temper carbon) (M), 442, 443, 444, 454
Annealing twin, 454
  in duplex stainless steels, 394
Anneal to temper, 454
Antimony, effect on temper embrittlement, 236, 239
Ar_{cm}, Ar_1, Ar_3, Ar_4, Ar', and Ar'' temperatures (D), 14–15, 468
Argon-oxygen-decarburization (AOD), 377
  effect on duplex stainless steels, 392
Arsenic, effect on temper embrittlement, 236, 239
Athermal transformation kinetics (D,M), 43–44, 47, 49, 50, 51, 55, 56, 454
Atomic percent, 3
Auger electron spectroscopy (AES), for composition analysis related to temper embrittlement, 237
Ausforming, 271, 454
Austempering (D,T), 263, 265, 267–269, 454
  effect on cast irons, 441, 446–450
  effect on tool steels, 416
Austenite
  carbon addition effect (D), 2, 6–7, 9
  crystal structure (D), 5–6, 7–8, 10
  definition, 454
  formation from cast irons, 433, 435, 443–444, 446, 447
  homogeneous, 91
  homogenizing (M), 109, 110, 112
  in duplex stainless steels (M), 390, 392, 393, 394, 395
  phase field in the Fe-C system (D), 4, 11, 13
Austenite formation
  effect of cold working of steels (M), 122, 127
  from ferrite and spheroidized cementite (D), 182, 184–185, 186
  from martensite (D,M), 185–186, 187–188, 189, 190–194
  from pearlite (D), 182–183, 184, 185, 186
  nucleation sites (D), 184–186
Austenite-stabilizing elements (D), 10–11, 12, 100, 311, 353–354, 358
Austenite transformation (*see also* Continuous cooling transformation; Eutectoid transformation; Isothermal transformation diagrams; Martensitic transformation)
  during tempering, 220–221
  effect of alloying elements on transformation temperature (D), 12–14, 30, 31
  effect of double tempering, 428
  in slowly cooled steels, 17
  isothermal, 92
  to bainite, 91
  to cementite, 91
  to ferrite, 91, 92
  to martensite (D,M,T), 43–84
  to pearlite (D), 21–30, 91, 92–93
  to proeutectoid ferrite and cementite (M), 34–41
  volume expansion and contraction effects, 263–264
Austenitic nitrocarburizing (D), 310, 311
Austenitic stainless steels. See Stainless steels, austenitic
Austenitizing, 454
  of cast irons, 441, 449–450
  effect on microstructures from continuous cooling transformation, 105
  of tool steels, 405–406, 416–418, 419
Austenitizing temperature
  determining carbide dissolution and better corrosion resistance and strength, 382–384
  effect on fracture morphologies and particle distribution, 249
  effect on isothermal transformation, 91
  effect on M_s temperature of precipitation-hardening stainless steels (D), 388–389
  effect on rate of austenite formation (D), 182, 184, 186, 194–195, 197, 198, 201

effect on tool steels, 417, 423
  for annealing and hardening, 109
  for induction heating, 284, 285
  for normalizing, 113, 136
  for spheroidizing (D), 116, 118
Autocatalysis, 55–57
Autotempering of martensite, 75–76, 95, 151
Bainite, 454
  crystallographic orientation relationships, 81
  effect on hardenability, 165
  formation and structure, 22, 78
  formation as product of continuous cooling transformation, 95
  formation by austempering, 268
  formation on continuous cooling (D,M), 84, 95, 98–99, 101
  hardness, 92
  in cast irons, 448, 449
  isothermal formation, 91
  isothermal transformation (M), 78, 79, 81
  lower (M), 78, 80–81, 82
  lower with midrib (M), 80–81, 83
  spheroidization, 115
  upper (M), 78, 79, 80, 81
  Widmanstätten formation, 58
Bainite transformation, isothermal, 89, 93
Bain strain (D), 58, 59, 60–61
Baking, 454
Bar diameter
  correlation with hardness distribution (D), 152–156, 164–167
  critical size and ideal size in hardenability testing (D,T), 164–169
  effect on continuous cooling transformations (D), 102, 103, 104–105
  effect on severity of quench (D), 160–161
  effect on strength properties, 208
Bark, 454
Bending fatigue tests (D), 303, 305
Black annealing, 454 (see also Annealing)
Black oxide, 455
Blank carburizing, 455
Blank nitriding, 455
Block brazing, 455
Blue annealing, 455 (see also Annealing)
Bluing, 455
Body-centered cubic structure (D), 4–5, 7, 9–10, 352
Body-centered tetragonal structure (D), 44–45, 58–59, 151, 370

Box annealing (D), 122–124, 128, 129, 130, 455 (see also Annealing)
Brass, effect on embrittlement, 241
Brazing, 455
Brazing alloy, 455
Brazing filler metal, 455
Brazing sheet, 455
Breaks, 455
Bright annealing, 455 (see also Annealing)
Brittle fracture, 375
  effect of section size on ferritic stainless steels, 376
  mechanism of, 231, 241
  stress-controlled mode, 247
Brittleness (see also Stead's brittleness)
  blue, 455
  reduction by tempering, 205
Bull's-eye structure, 439
Burning, 455
Cadmium, effect on embrittlement, 241, 243
Calorizing, 455
Carbide-forming elements, 10
  effect on laser surface melting (M), 343, 344, 345
  effect on retardation of softening during tempering (D), 213–215, 216
Carbides (see also Cementite)
  alloys found in tool steels, characteristics (T), 406, 407
  chi (M), 222–223, 225
  effect on control of grain size (D), 199, 200–201
  formation during tempering, 251
  formation in austenitic stainless steels, 371
  formation in cast irons (M), 442, 444, 446, 447
  formation in tool steels (M,T), 405–410, 415, 416, 417, 422, 425–428
  formation on tempering (M), 95, 205, 207, 213–216, 219–222, 224, 227, 234, 236
  hardness comparisons (D), 406–407, 408
  identification in complex tool steels, 407
  in bainite (M), 78, 80, 82
  network formation at grain boundaries (M), 109, 110, 113
  production by physical vapor deposition, 331
  spheroidization (M), 109, 110, 112, 115
  transition (epsilon and eta) (M), 219–220, 221, 222, 248, 253–254, 256
  transition metal (T), 406, 407, 426

Carbon
   activity coefficient, 286, 287
   as austenite-stabilizing element, 358
   control in carburizing (D), 285–291, 295
   diffusion coefficient, 23, 117
   diffusion during austenite transformation, 22–23, 28, 29–30, 38
   diffusion during bainite formation, 81–83
   diffusion in carburizing, 286, 289–290
   effect on absorption of CVN impact energy (D), 250
   effect on annealing, 108, 109, 114
   effect on bainite formation (M), 78, 81, 82–84
   effect on cast irons (D,T), 433–434, 439, 442, 444, 450
   effect on continuous cooling transformations and hardenability (D), 98, 99, 102, 103, 104
   effect on Fe-C system (D), 6–10
   effect on ferrite-pearlite mechanical properties (D), 132, 133–134, 138
   effect on gamma loop in iron-chromium alloys (D), 381
   effect on hardness of martensite (D), 145–152, 167, 169
   effect on hardness of martensite (D,M), 207–208, 214, 217, 219–220
   effect on hardness, strength, and impact toughness of tempered steels (D), 250, 252
   effect on impact toughness (D), 206, 207
   effect on intergranular fracture after carburizing (M), 299–301
   effect in isothermal transformation (D), 89–90, 91
   effect on martensite formation and morphology (D,M), 44, 45, 50–55, 61–62, 65, 67, 72
   effect on microstructures of martensitic stainless steels, 354
   effect on M_s and M_f temperatures, 90, 95
   effect on normalizing, 114
   effect on softening during tempering, 212, 219
   effect on structural features of tempered steels (D,M), 253, 254, 256
   effect on tensile strength, strain hardening, and elongation in tempered steels (D), 252, 254, 255, 256
   effect on tool steels (D), 402, 405, 407, 409, 410, 422
   free, 459
   habit plane a function of content, 58
   in heat-resistant stainless steels, 372
   in precipitation-hardening stainless steels, 387
   in proeutectoid phases, 34–36, 41
   restricted in ferritic stainless steels, 375
   solubility in austenite, 285
   steel matrix, hardness comparison (D), 408
   stress-strain curve after intercritical heat treatment (D), 275
   total, 467
Carbon content of eutectoid structures, effect on ferrite-pearlite microstructure (D), 134, 136
Carbon equivalent (C.E.) factor, 434
Carbonitriding (D), 310–311, 455
Carbonization, 455
Carbon monoxide and carbon dioxide in carburizing reactions (D), 287–289
Carbon potential, 289, 455–456
Carbon restoration, 456
Carburizing, 456
   effect on grain size (D), 195–198
   fatigue fracture observations (D), 302–305
   homogeneous, 460–461
   processing principles (D), 285–291
   properties and structure (D,M,T), 291–302
   reaction equations, 286–289
   temperatures for, 287
Carburizing flame, 456
Carburized steels
   compositions (T), 292–293
   microstructures (T), 291–292
Case, 456
Case depth
   in carburizing (T), 289, 290, 291, 292, 293
   in ferritic nitrocarburizing, 312
   in flame hardening (D), 281, 282
   in induction heating, 283
   in ion implantation, 325, 326
   in laser transformation hardening, 342
   in plasma carburizing, 323
Case hardening, 456
Cast irons (D,M,T), 431–451
   austempered ductile irons (D,M), 446–450
   classification of alloys by carbon content, 3
   composition ranges for unalloyed (T), 434
   crystal structure, 435

diffusion coatings (T), 451
ductile irons (M,T), 433, 434, 438–441, 442, 444, 450
  effect of alloying elements (T), 432–433, 438–439, 442, 445, 448
  eutectic structure (M), 431–432, 433, 434, 435, 445
  eutectoid structure, 433, 439, 444
  gray irons (D,M,T), 433, 434, 435–438, 439, 442, 444, 445, 450
  hypereutectic structure, 431, 434
  hypoeutectic structure (M), 431, 434, 444, 445
  malleable cast irons (M,T), 433, 434, 441, 450
  nodular. See Cast irons, ductile irons
  phase relationships, 431–435
  spheroidal. See Cast irons, ductile irons
  spherulitic. See Cast irons, ductile irons
  surface modification (T), 450–451
  white cast irons (M,T), 433, 434, 438, 442, 444–446, 447
CC diagrams. See Continuous cooling transformation diagrams
"C" curve kinetics, 377–379
Cementation, 346, 456
Cementite
  austenite formation from (D), 182, 184–185
  crystal structures (D), 7, 10, 11
  definition, 456
  formation after quench aging, 129
  formation during ferritic nitrocarburizing, 314
  formation on tempering of martensite (D,M), 209–212, 213, 218, 219, 221–223, 224, 225, 231–234
  hardness comparison of tool steels (D), 406–407, 408, 425
  heat treatments, 107
  in bainite, 78, 80
  in carburized steel (D), 297, 301
  in cast irons, 431–432, 433, 434, 442–443, 444, 445, 448–449
  in pearlitic structure (D), 19–21, 22–23, 24, 29–30
  isothermal formation, 91
  precipitated in martensite during quenching, 75–76
  proeutectoid (D,M), 18, 34, 36, 37, 90, 95, 193

  proeutectoid network formation (D,M), 109, 110, 111
  spheroidization (D), 109, 111, 116, 119
  transformation to austenite (D), 12–14
Charpy impact test. See Impact properties
Checks, 456
Chemical vapor deposition (CVD), 337
Chills, 445, 450–451
Chi phase (D,T), 371, 376, 377, 378
Chromel, 456
Chromium
  as a carbide former, 10, 30
  effect on continuous cooling transformations and hardenability (D), 97, 98, 101, 103
  effect on corrosion resistance of stainless steels (D), 351–353, 365
  effect on duplex stainless steels, 392, 395
  effect on hardness during tempering, 217
  effect on mechanical properties, ferrite-pearlite microstructures, 139
  effect on M_s temperatures, 54
  effect on M_s temperature during martensite formation, 366
  effect on pearlite formation (D), 30
  effect on phase equilibria of stainless steels (D), 353–356, 359
  effect on phase formation in Fe-C system (D), 10–11, 13
  effect on properties of cast irons, 446
  effect on rate of softening during tempering, 212, 213–214
  effect on strengthening of stainless steels, 368
  effect on surface oxidation, 295
  effect on susceptibility to severe intergranular corrosive attack (D), 362–363
  effect on temper embrittlement, 236, 238
  effect on tool steels, 409
  as ferrite stabilizer, 10
  higher contents in martensitic stainless steels, 382
  in cast alloys, 371
  in production of alloy carbide layers, 338–339
  major alloying element in ferritic stainless steels, 375
  major component of sigma phase, 370–371
  use in nitriding steels, 306
Chromium carbide precipitation, stainless steels (D,M), 361–362, 364, 365, 373, 374
Chromium carbonitride coatings (M), 338

Cladding, 346
Cleavage fracture. *See* Fracture, cleavage
Close annealing. *See* Box annealing
Coarsening, 456
Cobalt
　effect on deformation during tempering, 247
　effect on M_s temperatures, 54
　effect on tool steels, 405
Coherency, 456
Coherent precipitate, 456
Cold treatment, 456
Cold working, 373
　dislocation substructure (M), 122, 123
　effect of quench aging (D), 130
　effect of strain aging, 129
　process and recrystallization annealing (D,M), 118–125
　stress relieving (D), 118, 130–133
Columnar structure, 456
Combined carbon, 456
Compressive stresses, production during physical vapor deposition, 333–334
Computers, use in hardenability calculations, 174–177
Conditioning heat treatment, 456–457
Congruent transformation, 457
Constantan, 457
Constitution diagram, 457
Continuous annealing (D), 122, 124, 128, 129 (*see also* Annealing)
Continuous cooling transformations (D), 92–104
　effect of alloying elements (D,M), 95–104
　effect of bar diameter or section size (D), 102, 103, 104–105
Continuous precipitation, 457
Controlled cooling, 457
Convective heating, effect on plasma nitriding, 322
Cooling curve, 457
Cooling rates
　calculation of hardness values of carburizing steels, 293
　critical, 457
　effect of quenching medium (D,T), 156, 157, 159, 161–163, 176
　effect of section size (bar diameter) (D), 102, 103, 104–105, 114, 264
　effect of thermal diffusivity, 156–157, 159

　effect on cast iron solidification, 434, 435, 450
　effect on continuous cooling transformations (D,M), 92–104
　effect on hardenability evaluation (D), 171–172, 175, 176
　effect on intercritical annealing (D,M), 276–277, 278
　effect on pearlite formation (D), 24–25
　factors affecting (D), 156–159, 163
　for full annealing and normalizing (D), 109, 112, 124
　in tool steels (D), 422
Cooling stresses, 457
Copper
　effect on continuous cooling transformations and hardenability, 100
　effect on embrittlement, 241
Core, 457
Corrosion resistance
　improved by nitrogen ion implantation, 325
　in duplex stainless steels, 395
Critical cooling rate, 457
Critical point, 457
Critical size in hardenability testing. *See* Bar diameter
Critical strain, 457
Critical temperature ranges, 457
Critical temperatures, Fe-C alloys (D), 2, 11–15
Crystallographic orientation relationships
　in bainite formation, 81
　in ferrite formation (M), 38–41
　in martensite formation (D), 57–61, 74
Crystal structures
　of austenite (D), 5–6, 7–8
　of cast irons, 435
　of cementite (D), 7, 10, 11
　of Fe-C alloys (D), 7–10, 11, 12
　of ferrite (D), 4–5, 7, 8
　of martensite (D), 10, 44
CT diagrams. *See* Continuous cooling transformation diagrams
Curie temperature, 457
CVD. *See* Chemical vapor deposition
Cyaniding, 457
Cycle annealing, 457 (*see also* Annealing)
D2 tool steel
　composition (T), 403
　effect of annealing (M), 415

INDEX / 481

M_7C_3 carbides contained in equilibrium, 409–410
DBTT. *See* Ductile-to-brittle transition temperature
Dead soft, 457–458
Decalescence, 458
Decarburization, 458
Degrees of freedom, 458
Delong constitution diagram, weld metal (D), 357, 359
Delta ferrite. *See* Ferrite
Delta iron (M), 4, 6, 369, 370, 371 (*see also* Ferrite)
Dew point, 289
Differential heating, 458
Diffusion
  and austenite grain boundaries in proeutectoid phases, 37, 38, 39
  bulk or volume vs. grain boundary, 29–30
  and critical temperatures, 14
  definition, 458
  effect on hardenability, 100
  in bainite formation, 78
  massive or short range (SRD), 41
  suppressed in martensitic transformation, 44, 58
Diffusion coatings, application to cast irons (T), 451
Diffusion coefficient
  definition, 458
  tempering of tool steels, 425
Diffusion coefficient for carbon in austenite (D), 23, 289–290, 291
Diffusion of carbon
  during spheroidizing treatments, 117–118
  effect on cast irons, 439, 443–444
  in carburizing reactions, 287
  in salt bath coating process, 339
Dilatometry
  use for determination of continuous cooling transformations, 95, 98, 104
  use for determination of isothermal transformations, 92
Diode ion plating. *See* Ion plating
Direct quenching, 458
Discontinuous precipitation, 458
Dislocation density, 75
Dislocations
  effect of aging, 126, 127, 129
  effect of tempering (M), 209–210, 211–212, 213, 224, 227–228

  effect on martensite yield strength, 149–150, 151
  in bainite, 78
  in cold worked steels (D,M), 121–124, 127, 129
  in martensitic formation of austenite, 188
  in martensitic structure (D,M), 59, 60, 61, 63–66, 74–76
Double tempering. *See* Tempering
DPH hardness of microstructures (D,M), 98, 99, 100, 101
DPI stainless steel, composition (T), 391
Drawing, 458
Dry cyaniding, 458
Dual-phase microstructure from intercritical annealing (D), 126–127, 275, 276
Dubé classification system of proeutectoid phases, 36–37
Ductile-to-brittle transition temperature (DBTT) (D), 134, 136, 139
  effect of section thickness in ferritic stainless steels (D), 375–376
  equations for, 137–138
  factors of ferritic stainless steels, 376–377
  increase by change in fracture appearance, 381
  shifts in (D), 229–230, 236, 237
Ductility, 107
  of cast irons (D), 435, 439, 444, 448, 449
  of duplex stainless steels, 392, 395
  effect of aging phenomena, 126
  effect of carbon content, 133–134, 138
  effect of embrittlement, 241–242
  effect of normalizing, 114
  effect of subcritical annealing, 118–119, 124
  effect of tempering, 208
  ferritic stainless steels, 375, 376–377
  of precipitation-hardening stainless steels, 390
  restoration in austenitic stainless steels, 373
E-Brite 26–1 stainless steel, effect of annealing on microstructure (M), 374
Elastic limits, after tempering (D,T), 252, 253, 254, 255
Electrodeposition, 346
Electron beam hardening, 339–340, 342–343
Electron diffraction, identification of aluminum nitride particles, 240

Embrittlement (*see also* Temper embrittlement; Tempered martensite embrittlement)
  by aluminum nitride (M), 229, 239–241
  by hydrogen (D), 229, 241–244, 461
  by liquid metals (D), 229, 241, 242
  definition, 458
  quench-age, 464
  types and characteristics (D), 229–231
Enantiotropy, 458
End-quench test. *See* Jominy test
Entropy and enthalpy in minimum free energy principle, 23–25
Epsilon carbide, 80
Epsilon carbonitride (D), 310, 311, 312, 314
Epsilon iron, 3
Equilibrium, hierarchy of (T), 342–343
Equilibrium diagram, definition, 458–459
Eutectic, 459
Eutectic carbide, 459
Eutectic melting, 459
Eutectoid, 459
Eutectoid carbon content, effect on ferrite-pearlite microstructure, 134
Eutectoid steels
  austenitizing (D), 182, 184
  cast irons, 431
  continuous cooling transformation (D), 92, 94
  isothermal transformation (D), 21, 23, 25–26, 28, 29–31
  isothermal transformation diagrams (D), 89–91, 93
  use for high hardness and wear resistance, 138
Eutectoid transformation of austenite (D), 17–19, 21, 23, 25–26, 28
  effect of alloying elements (D), 10–14, 17–19, 30, 31
Evaporative-source physical vapor deposition, 330
EX 24 (Exchange grade 24) steel
  carburizing and diffusion effect on microstructure (M), 293, 294
  grain boundary carbides formation after carburizing (D), 295, 297
Exchange grade (EX) steels, 293, 294, 295, 297
Extra hard (temper), 459
Extra spring (temper), 459
Face-centered cubic crystal structure (D), 5–6, 7, 9, 353

Fatigue properties
  effect of austenitic grain size and retained austenite in carburized steel (D), 307
  effect of carburizing (D), 295, 297–300, 301
  effect of hydrogen embrittlement (D), 243–244
  effect of nitriding, 310
Fe-C equilibrium diagram. *See* Iron-carbon equilibrium diagram
Ferralium 255 stainless steel, composition (T), 391
Ferrite
  carbon addition effect (D), 2, 6–7, 8, 9
  cold working and recrystallization (D,M), 119, 121, 126, 127
  converted from malleable cast iron, 444
  crystal structure (D), 4–5, 7, 8, 10
  crystal structure and solid solution strengthening, 138
  definition, 459
  delta (M), 6, 369, 370, 371, 388, 390, 411
  effect of austenitic grain size on formation, 182–183
  effect of quench aging, 126–127
  effect of strain aging, 125–126
  effect of temperature in duplex stainless steels (M), 390–391, 392, 393, 394, 395
  fine structure, 136
  formation as occurrence of continuous cooling transformation, 95, 98–99
  formation on tempering of martensite (D), 221–222, 228
  free, 459
  hardness, 92, 100, 101
  heat treatments to produce (D), 107, 112, 117, 118–119
  in bainite, 78
  in pearlitic structure (D), 19–21, 22–23, 24, 29–30
  isothermal formation, 91
  nucleation and growth (M), 36–41
  proeutectoid (D,M), 18, 34–36, 38, 90–91, 95, 100–101, 109, 113, 193
  solubility for phosphorus and sulfur, 358
  solubility of carbon in (D), 7, 8, 10
  transformation to austenite, 12–13
  Widmanstätten formation, 58
Ferrite-austenite microstructure, in stainless steels, 356–358
Ferrite-martensite microstructure

production by intercritical annealing (M), 276, 277
Ferrite number (FN) (D), 357, 358, 359
Ferrite-pearlite microstructure
  cold work and recrystallization (M), 122, 125
  effect of carburizing, 291–292
  effect of continuous cooling transformation, 104–105
  full annealing for (D), 109, 112
  hardness as function of carbon content (D), 145, 146
  mechanical properties (D), 132, 133–139
  microalloyed steels for bars and forgings (M), 139–142
  normalizing for, 113
  yield strength, 200
Ferrite-stabilizing elements, 10, 353–354, 358
Ferritic nitrocarburizing. See Nitrocarburizing, ferritic
Ferritizing anneal, 459
File hardness, 459
Final annealing, 459 (see also Annealing)
Finishing temperature, 459
Flame annealing, 459 (see also Annealing)
Flame hardening (D), 281–282, 341, 459
  applied to cast irons, 450
Flame straightening, 459
FN. See Ferrite number
Forging, homogenizing for, 110–111
Forgings, overheating of, 244–245
Formality, use of proeutectoid ferrite, 35–36
Fracture
  brittle. See Brittle fracture
  cleavage (D,M), 134, 135, 139, 233, 235, 375
  intergranular (D,M), 180–182, 232–233, 237, 239, 240–242
  intergranular, in cast irons, 449
  intergranular (M), effect of carburizing, 299, 301
  intergranular, effect of improved matrix fracture resistance, 249
  intergranular, in tool steels, 417, 422, 423
  "rock-candy," 240
  transgranular, 231, 234, 299
Fracture toughness, effect of austenitizing temperature, 249
Fracture toughness testing, 181–182
Frank-van der Merwe growth, 331
Free carbon, 459
Free energy principle for stability of phases and microstructures, 23–24

Free ferrite, 459
Freezing range, 459
Frenkel defects, formation by ion implantation, 325–326
Fretting corrosion, effect of ion mixing, 328
Friction coefficient, reduced by titanium addition, 326
Full annealing (D,M), 107–112, 113, 115, 116, 119, 120–121, 125 (see also Annealing)
  compared to normalizing, 113–114
  definition, 459
  temperature ranges (D), 90
Full hard (temper), 460
Furnace brazing, 460
Fusion, 460
Gamma iron, 3, 4, 5, 460 (see also Austenite)
Gamma loops (D), 352, 353, 375, 381, 382
Gas carburizing (D,M,T), 286–303, 323, 324
Gas cyaniding, 460
Gas nitriding, 307
  vs. plasma nitriding, 321
Glide-plane decohesion, 244
GM 980X, stress-strain curve after intercritical heat treatment (D), 275–276
Grain boundaries
  aluminum nitride precipitation, 239
  austenitic (D,M), 31, 32, 36–40, 180, 185, 192–194
  austenitic, containing segregated phosphorus, 237
  austenitic, effect of tempering, 225, 231, 232–233
  austenitic, intergranular fracture, 239
  austenitic, of tool steels, 422
  effect of annealing on type 316L stainless steel (M), 360, 361
  effect of chromium carbide precipitation on type 304 stainless steel (D), 361–362
Grain boundary allotriomorphs, 36, 37, 39
Grain boundary embrittlement, 181–182
Grain growth, definition, 460
Grain size
  ASTM numbers (M,T), 188–193
  austenitic (D,M,T), 66, 91, 113, 138–139, 167, 179, 188–201
  austenitic, definition, 454
  austenitic, effect on fatigue limits (D), 307
  austenitic, effect on hardenability (D), 169, 179, 180, 181
  austenitic, hardness, 148, 152
  austenitic, in tool steels, 417, 423

Grain size *(continued)*
  of carburizing steels (T), 292
  control (D,M,T), 136–138, 193–201
  definition, 460
  effect on fatigue fracture of carburized steels, 304
  effect on mechanical properties (D), 137–138
  ferritic (D,M), 113–114, 119, 121, 127, 136, 138, 142, 180, 200–201
  refinement by normalizing, 113
Graphite
  in cast irons (M), 431
  in ductile cast irons, 438, 439
  in gray cast irons, 433, 434, 435, 436, 437
  in iron-carbon equilibrium diagram (D), 3
  in malleable cast irons, 442, 443–444
  in white cast irons, 445
Graphitization, 460
Graphitizing, 460
Growth, definition, 460
H13 tool steel
  alloy carbides in lath martensite (M), 426
  composition (T), 404
  interlath carbide formation (M), 427–428
  lath martensite formation (M), 419, 420
  M_7C_3 carbides contained in equilibrium, 409–410
  tempering temperature effect on hardness and retained austenite (D), 424, 425, 426
Habit planes
  for bainite formation, 81
  for martensite formation (D,M), 47, 55–58, 60, 64–66, 68–70, 74, 75
Half hard (temper), 460
Hall-Petch plots (D), 148, 150
Hardenability
  application in matching steel composition to section size and cooling rates, 265
  of carbonitriding steels, 311
  of carburizing steels (T), 292–293
  computer calculations, 174–177
  definitions, 145, 152, 460
  effect of alloying elements (D,M), 93, 95–102
  effect of bar diameter (D,T), 164–169
  effect of carbon content (D), 167, 169, 170, 174
  effect of intercritical heat treatment, 276
  as function of bar diameter, 104–105
  Grossmann-Bain approach, 163–167
  in tool steels, 402, 405–406, 415, 416–417, 419, 422
  Jominy specimen and test (D), 92–93, 94, 96, 169–173
  martensitic stainless steels, 382
  necessary for steels suitable for austempering, 268
  quantitative (D), 163–167
  ranges for various steels (T), 169, 171
Hardenability bands (D), 170–171, 175
Hardening
  definition, 460
  secondary. *See* Secondary hardening
Hardfacing, 346
Hardness
  of austempered vs. quenched and tempered steel (D), 269
  of carburizing steels, 293
  of cast irons, 436, 450
  comparisons of alloy carbides, cementite, and a carbon steel matrix (D), 408
  distribution along quenched bar (D), 152–156, 164–166
  effect of nitrogen, 326
  effect of tempering (D), 205, 206, 207–208, 209, 214, 216
  effect of tempering temperature on tool steel (D), 423–425
  as function of bar diameter, 104–105
  as function of carbon content of martensite (D), 145–149, 164, 166
  as function of distance from a carburized surface, 293
  high-speed steel comparisons (D), 408
  measurements for determination of isothermal transformations, 91–92, 93
  produced by ferritic nitrocarburizing (D), 312
  use of proeutectoid pearlite, 35–36
Hard temper, 460
Heat-resisting alloy, 460
Heat tinting, 460
Heat treatable alloy, 460
Heat treating film, 460
Heat treatment
  definition, 460
  effect on austenitic stainless steels, 373–374
  effect on cast irons, 437, 441, 442, 444
  effect on tool steels (D,M), 410–413
  intercritical (D,M), 274–279
  scientific basis, 1, 2–3

special types, 263–265
thermomechanical (D,T), 269–274
Hereditary treatments (D), 272
High-deposition-rate electron beam evaporation, 346
High-strength low-alloy (HSLA) steels (D), 34, 136, 137, 180
   bars and forgings (M), 139–142
   grain size control (D), 199–201
   intercritical annealing (D), 274–276, 279
   thermomechanical heat treatments (D,M), 275–276, 279
"Hollow cathode" effect, 322
Homogeneous carburizing, 460–461
Homogenizing heat treatment (D), 107–112, 461
   compared to normalizing, 113
   effect on segregation produced during solidification, 6
Hot dipping, 345
Hot stage cinephotomicrography, 50
Hot strip mill processing, 121
Hot working
   effect on duplex stainless steels, 393, 394
   effect on grain size, 113, 142, 198–199, 201
   effect on segregation produced during solidification, 6
   effect on tool steels, 405–406
   homogenizing for (D), 108, 110–111
HSLA steels. See High-strength low-alloy steels
H-steels, hardenability data (T), 169, 171
Hydrogen brazing, 461
Hydrogen embrittlement. See Embrittlement, by hydrogen
Hydrogen flaking, 238
Hypereutectic steels, cast irons, 431, 434
Hypereutectoid steels, 34
   annealing, 108, 109
   bainite formation, 80–81
   isothermal transformation diagrams, 90
   normalizing, 112–113
Hypoeutectoid steels
annealing, 108, 109, 113
cast irons (M), 431, 434, 444, 445
and continuous cooling transformations, 95
isothermal transformation, 91
isothermal transformation diagram (D), 90
normalizing, 112–113
Ideal size in hardenability testing. See Bar diameter

Impact properties (see also Toughness)
   comparison of austempered and quenched-and-tempered carbon steel (D), 269
   effect of austenitic grain size (D), 180, 181–182
   effect of austenitizing temperature, 249
   effect of carbon (D), 250, 252
   effect of phosphorus (D), 250, 251
   effect of tempering (D), 205–207, 208, 231, 232, 233, 235
   energy absorption as a function of tempering temperature (D), 250
   of ferrite-pearlite microstructures, 134
   in martensitic stainless steels, 385
   tempered steels, 246–247
Impact transition temperature. See Ductile-to-brittle transition temperature
Inclusions
   effect on the tempering of steels, 246–247
   shape control, 247
   steelmaking innovations for reduction of, 247
Induction brazing, 461
Induction hardening, 461
Induction heating (D), 282–285, 341, 461
   applied to cast irons, 450
Initial layer-by-layer growth, 331
Intercritical heat treatments (D,M), 126–128, 130, 274–279
Intermediate annealing, 461 (see also Annealing)
Intermetallic phases, characteristics in ferritic stainless steels (T), 376
Interphase precipitation of carbides (M), 31–34
   effect of aging (D), 126, 129, 131
   in microalloyed bars and forgings (M), 141, 142
Interstitial solid solutions of carbon in iron, 7, 10
Interstitial voids in fcc and bcc iron (D), 7–10
Ion-beam mixing. See Ion mixing
Ion implantation (D), 325–328
Ion mixing (D), 325–328
Ion nitriding. See Plasma nitriding
Ion plating, 330–331
Iron. See Cast irons; Ferrite
Iron carbide. See Carbides; Cementite
Iron-carbon equilibrium diagram (D), 1–4, 7, 17–19

Iron-carbon equilibrium diagram *(continued)*
    temperature ranges for various heat treatments (D), 108, 116, 118
Iron-carbon-silicon system, vertical section (D), 432–433
Iron-chromium-carbon system, isothermal and vertical sections (D), 409, 410
Iron-chromium equilibrium diagram, 370
Iron-chromium equilibrium phase diagram (D), 352
Iron-chromium-nickel ternary system (D), 354–356
Iron-graphite equilibrium diagram (D), 2–3
Iron-iron carbide equilibrium diagram (D), 128
Iron-nickel equilibrium phase diagram (D), 353, 354
Iron-nitrogen phase diagram (D), 308, 320
Island growth, 331
Isothermal annealing, 461 (*see also* Annealing)
Isothermal formation of martensite (D), 56–57, 78
Isothermal formation of pearlite, 78
Isothermal transformation, 461
    carbon content effect (D), 89–90, 91
    of cast irons, 449
    and spheroidizing, 115
Isothermal transformation (IT) diagrams (D), 21–22, 25, 89–90, 91
Isothermal transformation kinetics, 25, 28, 56–57
IT diagrams. *See* Isothermal transformation diagrams
JBK-75 stainless steel, precipitates formed (M), 390, 391
Jominy specimen (D), 92–93, 94, 96, 170, 172, 173, 175
    used in calculating hardness gradients from carbon gradients, 293
Jominy test for hardenability (end-quench hardenability test) (D), 169–173, 174, 175, 177, 458
Lamellar structure and spacing of pearlite (D), 19, 21, 23, 24–25, 26, 28, 29, 34
    effect of normalizing, 113–114
    effect of spheroidizing, 109, 114, 115
    effect on yield strength, 138–139
Laser and electron beam surface modification (D,M,T), 339–346

Laser glazing, 345
Laser hardening, effect on cast irons, 450
Lasers, 339
    CO_2 (carbon dioxide), 340
    excimer, 340
    Nd:YAG (neodymium dissolved in yttrium aluminum garnet), 340
Laser surface alloying, 345
Laser surface heat treatment (M,T), 339–341, 346
    effect of alloying elements, 345
Laser transformation hardening, 339, 340–341, 346
Lath martensite
    effect of tempering (M), 225–228
    interrelationship of packet size, strength, and austenitic grain size (D), 148–151
Lattice correspondence, 59–60
Lattice invariant deformation (D), 59, 60, 61, 64
Laves phase (T), 371, 376, 377
Lead, effect on embrittlement (D), 241, 242
Lead-tin solders, effect on embrittlement, 241
Ledeburite, 431–432, 461
Lever rule, 19–21, 34, 36
Light metallography, not able to reveal lath width distribution (D), 70, 74
Light microscopy
    for resolving plate martensites, 62–63
    to determine the amount of retained austenite, 55
Liquid phase sintering, 461
Liquidus surfaces, stainless steels (D), 354–355
Lithium, effect on embrittlement, 241
Localized precipitation, 461
Low-alloy high-strength steels. *See* High-strength low-alloy steels
Lüders bands, 127–128, 130
Lüders strain (D), 127–128, 129
M2 tool steel
    composition (T), 404
    effect of solidification (M), 411, 413
    feathery, herringbone, and MC eutectics (M), 412, 414
    primary dendrite with axis normal to the plane of polish (M), 413
M42 tool steel
    composition (T), 404
    effect of laser surface melting (M), 343, 344, 345

laser-melted dendritic structure (M), 343, 345
McQuaid-Ehn test for austenite grain size, 193, 462
Magnetic fields
 application during sputtering process (D), 330
 for induction heating (D), 283–285
Magnetron sputtering (D), 330
Malleabilizing, 461
Manganese
 as austenite stabilizer, 10, 358
 composition ranges for unalloyed cast irons (T), 434
 effect on bainite formation, 84
 effect on continuous cooling transformations, and hardenability (D), 98, 99, 100, 102, 104
 effect on hardness during tempering, 217, 218
 effect on martensite strength, 152
 effect on mechanical properties, 138, 139
 effect on M_s temperature, 54
 effect on pearlite formation (D), 29–30, 31
 effect on surface oxidation, 295
 effect on temper embrittlement, 236, 238, 239, 245
 substituted for nickel in Type 200 stainless steels, 372
Maraging, 461
Marquenching. See Martempering
Martempering (D), 263, 265, 266–267, 416, 462
Martensite
 alpha prime, 366–368
 athermal formation (D,M), 43–44, 47, 49–51, 54–56, 57, 76
 austenite formation from (D,M), 185–186, 187–188, 189, 190–194
 body-centered cubic alpha prime, 366–368
 crystal structure (D), 10, 44, 149–150, 151
 definition, 462
 effect of interphase precipitation, 31
 epsilon, 366, 367–368
 50% level for hardenability testing (D), 164–166, 167
 fine structure (M), 47, 58, 61, 63–67, 70, 73–76
 hardness, 92, 93, 100
 hardness affected by packet size (D), 148–150, 151, 152
 hardness as function of carbon content (D), 145–148
 hardness comparison (D), 406–407, 408
 hexagonal close-packed epsilon, 366, 367–368
 lath and plate composition ranges (D), 52
 lath and plate morphologies (D,M), 61–77, 254–255
 microcracking (D,M), 46, 66, 67, 70, 75, 295, 297–300
 modification by thermomechanical treatments (D), 266, 270, 271, 276, 277
 packet size related to austenitic grain size (D), 180, 181
 reasons for strength (D), 149–152, 180
 spheroidization (D,M), 114, 115, 117
 susceptibility to quench cracking, 264
 tempering. See Tempering of steel
 transformations on tempering (D,M), 214, 216, 217, 219–220, 226, 227
 yield strength, 151
Martensite formation (D), 26, 36
 after quenching, 126
 effect of chromium content on M_s temperature in stainless steels, 365
 effect of laser surface heat treatment, 340–341
 in austenitic stainless steels, 366–370
 in cast irons (M), 446, 447
 in 17–7PH stainless steel (M), 389, 390
 in tool steels (M), 417, 419, 420, 421
 involving residual stresses, 132
 strain-induced as function of strain at various temperatures (D), 367, 368, 370
 strain-induced for strengthening by cold working, 373
Martensite range, 462
Martensitic transformation, 462
 athermal (D,M), 43–44, 47, 49, 50, 51, 54–55, 57, 76
 burst and stabilization phenomena, 55–56
 crystallographic theory of (D), 57–61, 74
 isothermal (D), 56–57, 91
 kinetics (D,M), 47–57
Matrix alpha phase, 380
M_D temperature, 368
Mechanical deformation, in austenite, 6
Metallographic examination to establish the microstructure produced during a cooling sequence, 96

Metallographic method for determination of isothermal transformations, 91, 92
Metallurgy, definition, 462
Metal nitride, production by physical vapor deposition, 331
M_f temperature
  definition, 462, 468
  effect of carbon content, 52, 56
Microalloyed steels. *See* High-strength low-alloy steels
Microcracks, development during carburizing (D,M), 295, 297–298, 299–300
Microhardness, definition, 462
Mill scale, 462
Mixed-mode growth, 331
Molybdenum
  as carbide former, 10, 30, 31
  effect on bainite formation, 84
  effect on continuous cooling transformation and hardenability (D,M), 95, 97, 98, 99, 101, 102, 103
  effect on corrosion resistance of stainless steels (D), 352, 353
  effect on duplex stainless steels, 392, 395
  effect on hardness during tempering, 217
  effect on mechanical properties, 139
  effect on M_s temperatures, 54
  effect on pearlite formation (D), 30
  effect on softening during tempering (D), 212, 213–214, 215
  effect on thermal conductivity, 265
  effect on tool steels, 409
  as ferrite stabilizer, 358
  in crystal structure with iron and chromium, 370
  reducing susceptibility to temper embrittlement, 236, 238
  use in nitriding steels, 306
Monotropism, 462
7-Mo PLUS stainless steel
  composition (T), 391
  microstructure (M), 392, 393
M_s temperature (D,T), 50–52, 53, 54–55, 56, 75–76
  definition, 462, 468
  effect of austenite grain size, 179
  effect of austenitizing temperature on precipitation-hardening stainless steels (D), 388–389
  effect of carbon content, 90
  effect of carbon in carburized steels, 301

effect of chromium in austenitic stainless steels, 365
effect of cooling rate, 93, 95, 102–103
effect on isothermal transformations, 91
effect on tool steels, 417
equation for austenitic stainless steels, 366
and yield strength of martensite, 150, 151
Neutral flame, 462
Nickel
  as austenite stabilizer, 10
  effect on austenite formation, 187
  effect on continuous cooling transformations and hardenability (D,M), 98, 99, 100, 101, 102, 103
  effect on duplex stainless steels, 392, 395
  effect on hardness during tempering, 217, 218
  effect on intermetallic phases of ferritic stainless steels, 377
  effect on lath and plate morphology of martensite (D), 65, 69
  effect on martensite strength, 151
  effect on M_s temperature, 54, 63
  effect on M_s temperature during martensite formation, 366
  effect on pearlite formation (D), 29–30, 31
  effect on phase equilibria of stainless steels (D), 353–356, 359
  effect on precipitation-hardening stainless steels, 387
  effect on strengthening of stainless steels, 368
  effect on temper embrittlement, 237–238
  effect on tool steels, 405
  in cast alloys, 371
  substitutions made in Type 200 austenitic stainless steels, 372
Niobium
  as carbide former, 10, 31, 34
  effect on austenitic grain size, 199–200
  effect on chromium carbide precipitation in stainless steels, 365
  effect on grain size control, 136
  effect on mechanical properties of ferrite-pearlite microstructures (M), 139, 141, 142
  as ferrite stabilizer, 10, 358
  in production of alloy carbide layers (M), 338–339
  in steel for intercritical annealing (M), 276, 277

interphase precipitation of carbides (M), 141, 142
  stabilizing element for ferritic stainless steels, 377
Niobium carbide coatings (M), 338–339
Nitralloy 135 Modified, white layer and diffusion zone (M), 309
Nitralloy N, increased hardness due to diffusion zone, 312
Nitrides
  formation in austenitic stainless steels, 371
  formation in nitriding and carbonitriding, 307–311
Nitriding (D,M), 305–310, 462–463
Nitrocarburizing
  definition, 463
  ferritic (D), 311–314
Nitrogen
  as austenite stabilizer, 358
  effect on gamma loop in iron-chromium alloys (D), 381
  effect on hardness, 326
  substituted for nickel in Type 200 stainless steels, 372
Nitronic stainless steels (T), 372–373
Normalizing (D), 107, 108, 111, 112–114
  definition, 463
  effect on cast irons, 441
  temperature ranges (D), 90
Oil quenching (see also Quenching)
  effect on hardness distribution in hardenability evaluation (D), 153–157, 164, 166
  effect on tool steels, 419
  equivalent cooling rates (D), 176
  for water-hardening tool steels, 402
  microstructures produced as related to bar diameter (D,M), 99–105
  severity of quench (D,T), 159, 160–161
Optical pyrometer, 463
Orientation relationships. See Crystallographic orientation relationships
Overaging. See Aging
Overheating, 244–245, 463
Oxidation
  during gas carburizing (M), 293–295, 296, 297
  in ion implantation, 325
  in plasma carburizing, 323
Oxidation resistance, effect of ion mixing, 328

Oxide ceramic coatings, production by physical vapor deposition, 331
Oxidizing flame, 463
Packing material, 463
Partial annealing, 463 (see also Annealing)
Patenting, 463
Pearlite
  austenite formation from (D), 182–183, 184, 185, 186
  definition, 463
  effect of continuous cooling transformation, 100, 101
  formation on continuous cooling, 95
  hardness, 92, 93, 100, 101
  heat treatments to produce (D), 107, 109, 111, 112, 114–116, 118, 119, 122
  isothermal formation, 89, 91
  isothermal transformation (D), 21–22, 25, 26–28
  proeutectoid (M), 34, 35, 109, 113
  structure (D), 19–21
15.5PH stainless steel, ion mixing for resistance to fretting corrosion, 328
17.4PH stainless steel. See AISI 630 stainless steel
17.7PH stainless steel. See AISI 631 stainless steel
PH-13.8 Mo stainless steel
  composition (T), 386, 387
  effect of solution treating and aging (M), 387
Phases in steel (D), 1–15
Phase transformations, 1, 4, 7
  critical temperatures (D), 13–15
  effect of alloying elements (D), 10–13
Phosphorus
  composition ranges for unalloyed cast irons (T), 434, 449
  effect on absorption of CVN impact energy (D), 250–251
  effect on hardness during tempering (D,M), 217, 231, 232, 233, 235–236
  effect on impact toughness, 180
  effect on impact toughness of tempered steels (D), 250, 251
  effect on intergranular fracture after carburizing (M), 299–301
  effect on temper embrittlement, 236
  effect on tool steels, 423
  segregation in austenitic grain boundaries, 237, 238, 239
  solubility in ferrite, 358

Physical vapor deposition (PVD)
 coating microstructures (D,M), 331–336
 processing, 328–331, 346
Plasma-assisted physical vapor deposition, 330
Plasma carburizing (M), 322–325
Plasma nitriding (D), 320–322, 346
 vs. plasma carburizing, 323
Plastic deformation
 accompanying martensite formation, 46, 75
 effect of strain aging, 127
 in thermomechanical treatments, 264
Plasticity, transformation-induced, 467
Polygonization (D), 228, 272
Postheating, 463
Pot annealing. *See* Box annealing
Precipitation hardening, definition, 463
Precipitation heat treatment, 463
Preheating, 463
Presintering, 463
Process annealing (D), 118–124, 133, 463 (*see also* Annealing)
 martensitic stainless steels, 382
Proeutectoid phases (D), 18, 34–41
 avoided in formation of hard martensitic microstructures, 146
 and continuous cooling transformations, 100
 formation and morphology (M), 37–41
 formation at grain boundaries, 179, 180
 formation on continuous cooling, 95
 isothermal transformation (D), 90
Pseudocarburizing. *See* Blank carburizing
Pseudonitriding. *See* Blank nitriding
Pusher furnace, 464
PVD. *See* Physical vapor deposition
Quarter hard, 464
Quench annealing, 464 (*see also* Annealing)
Quench cracking, 173, 464
 causes and relief of, 264–265
 in tool steels, 419, 422
Quench hardening, 464
Quenching, 464 (*see also* Oil quenching; Water quenching)
 fog, 459
 hot, 461
 interrupted, 461
 selective, 465
 severity of (D,T), 159–163, 167, 173
 slack, 465
 spray, 466
 time, 467

Quenching media
 cooling rates (D,T), 153, 158, 159, 161–163
 effect on critical size in hardenability testing (D), 167, 168
 for austempering, 269
 for martempering, 266–267
 severity of quench (D,T), 159–163, 167, 173, 276
Quench tempering, martensite formation, 75–76
Rail steel, properties as affected by microstructure, 138, 139
Rare earth metals (REM), effect on deformation during tempering, 247
Recalescence, 464
 during continuous cooling, 103–104
Recarburize, 464
Recovery, 464
 of cold worked steels, 119, 121, 122, 133
 of tool steels, 412, 416
Recrystallization
 austenite during annealing, 188
 austenitic stainless steels, 373
 definition, 464
 during annealing of austenitic stainless steels, 361
 effect of recovery mechanisms, 229
 effect of thermomechanical treatments (D), 272
 effect on duplex stainless steels, 394–395
 in cold worked steels (D,M), 118–122, 124, 126, 127, 133
 secondary, 195, 201
 of tool steels, 412, 416
Recrystallization annealing (D,M), 118–127, 133, 464 (*see also* Annealing)
 ferritic stainless steels, 375
Recrystallization temperature, 464
Recuperator, 464
Reducing flame, 464
Refractory, 464
Refractory alloy, 464
Refractory metal, 464–465
Regenerator, 465
Residual stresses
 cause, effect, and relief of, 107, 114, 130–133, 205, 263–265
 produced by carburizing (D), 301–302
 produced by plasma carburizing, 323
 produced in tool steels, 416

production during physical vapor deposition, 331, 333, 334–335
Resist, 465
Resistance brazing, 465
Retained austenite
  after bainite transformation (M), 81
  after martensitic transformation (D), 52–53, 55, 62–63, 68
  austenite formation (M), 192
  decomposition in tempering, 231–232
  effect of carbon content in tempered steels, 254
  effect of fatigue fracture of carburized steels (D), 304, 306
  effect on fatigue limits in carburized steel (D), 307
  effect on hardness, 146–148, 181
  in martensitic microstructure (D,M), 220–221, 223, 224, 231–232
  in tool steels (D), 417, 419, 423, 424–425, 427–428
  martensite formation in carburizing steels, 293, 295, 297–298, 305
Reverberatory furnace, 465
Rockwell hardness test, 465
Rotary furnace, 465
SAE 950X steel, stress-strain curve after intercritical heat treatment (D), 275–276
SAE 980X steel, stress-strain curve after intercritical heat treatment (D), 275–276
SAE 1015 steel, austenitic grain size after carburizing (D), 197
SAE 1040 steel, effect of laser transformation hardening, 341
SAE 1042 steel, effect of heating rate and microstructure on austenitizing temperature (D), 284–285
SAE 1045 steel
  chromium carbonitride coating deposited by salt bath process (M), 338
  hardening response to quenching (D), 153, 154, 156
  hardness distribution along quenched bar (D,T), 153–156
SAE 1050 steel
  effect of flame speed on depth of hardening (D), 281–282
  effect of laser transformation hardening, 341
SAE 1070 steel, effect of laser transformation hardening, 341
SAE 1080 steel
  isothermal transformation (D), 21, 22, 89
  pearlite formation growth kinetics (D), 21, 22
SAE 3140 steel
  center hardness as a function of bar diameter (D), 164, 166
  embrittlement detectable by increase in impact transition temperature (D), 237
  severity of oil and water quenching (D), 159, 160
SAE 4130 steel
  low-phosphorus fracture surface after tempering (M), 235–236
  mechanical properties after tempering (T), 252, 253
  microvoids and extracted particles on fracture surface (M), 248
  stress-strain curves after quenching and tempering (D), 252, 253
  transgranular fracture mode, 234
  transition carbides and retained austenite presence (M), 254, 256
SAE 4140 steel
  continuous cooling transformation (D), 96, 97, 98–99
  greater fracture modes for higher carbon steels, 236
  isothermal transformation (D), 97, 98
  mechanical properties after tempering (T), 252, 253
  retained austenite and cementite as a function of tempering temperature (D), 224
  retained austenite presence after tempering, 254
  retained austenite present in as-quenched specimens, 221
  stress-strain curves after quenching and tempering (D), 252, 253
SAE 4145 steel, tensile properties as a function of tensile test temperature (D), 241, 242
SAE 4150 steel
  greater fracture modes for higher carbon steels, 236
  mechanical properties after tempering (T), 252, 253
  retained austenite presence after tempering, 254
  stress-strain curves after quenching and tempering (D), 252, 253

SAE 4150 steel *(continued)*
  transition carbides precipitated in a martensite lath (M), 256
SAE 4340 steel
  anisotropy in properties and fracture, 247
  effect of hydrogen embrittlement (D), 243–244
  effect of laser transformation hardening, 341
  effect of phosphorus on impact toughness (D,M), 231, 232–234
  effect of tempering on mechanical properties (T), 211
  effect of tempering temperature (D), 208, 209, 210
  formation of austenite in tempered martensite (M), 185–186, 187–188
  fracture toughness and impact properties varying with austenitizing temperature, 249
  intergranular fracture (D), 232
  retained austenite and cementite as a function of tempering temperature (D), 224
  retained austenite present in as-quenched specimens, 221
  shear along the slip line field at notch root (D), 251
  toughness limited by void sheets, 249
SAE 4360 steel
  lower bainite with fine carbides within plates (M), 80, 82
  lower bainite transformed at 300 °C (M), 80, 82
  interlath cementite transformed to upper bainite (M), 78, 80
  upper bainite formation (M), 78, 79
SAE 4615 steel, austenitic grain size after carburizing (D), 197
SAE 6140 steel
  hardening response to quenching (D), 153, 157
  hardness distribution along quenched bar (D,T), 153–155, 157
SAE 8620 steel
  carburizing effect on microstructure (M), 293, 294
  effect of gas carburizing on retained austenite content (D), 295, 296
  microcracking (D,M), 295, 297–298, 299–300
  overload case fracture surfaces after carburizing (M), 299–301
SAE 8650 steel, Jominy test for hardenability method (D), 175
SAE 8650H steel, hardenability band (D), 175
SAE 8719 steel
  effect of austenite grain size and retained austenite on fatigue limits (D), 307
  fatigue limits as function of austenitic grain size (D), 304
  fatigue limits as function of retained austenite in carburized steel (D), 306
  plasma-carburized vs. gas carburized (M), 324
  stress vs. cycles for bending fatigue after carburizing (D), 303, 305
SAE 52100 steel
  carbide network formation in (M), 109, 110
  continuous cooling transformation characteristic, 95
  CVN impact energy absorption as a function of tempering temperature, 250–251
  density of void-initiating carbide particles increased by carbon content, 252
  effect of laser transformation hardening, 341
  proeutectoid cementite network after normalization (D), 109, 111
  tool steel applications of ion implantation, 326
Salt bath coating process (M), 338–339
Schaeffler constitution diagram, stainless steel weld metal (D), 357, 359
Secondary hardening (D), 213, 215, 216, 217–218, 219, 223, 224
  of tool steels, 417, 424
Section size *(see also* Bar diameter)
  effect on cooling rate, 114, 264
  effect on tool steel hardenability, 419
Selected area diffraction techniques, 64
Selective heating, 465
Self-hardening steel. *See* Air-hardening steel
Sensitization, heat treatments for prevention, 373
Shear initiation zone, 248–249
Shear mechanism
  along the slip line field at root of CVN specimen (D), 251
  associated with Widmanstätten ferrite formation, 39
  of austenite formation, 187, 188

bainite formation, 78, 81
  ferrite formation by continuous cooling, 101
  martensite formation (D), 45–46, 50–52, 55, 56, 57, 58, 61, 179
Shrink forming, 465
Siewart, McCowan, Olson constitution diagram, stainless steels (D), 358, 359
Sigma phase (D,T), 370–371, 376, 377, 378, 391–392, 395
Silicon
  effect on austenite grain growth, 193–194
  effect on bainite formation (M), 78, 80, 81
  effect on cast irons (D,T), 432–434, 442, 444, 448–449
  effect on continuous cooling transformations and hardenability (D), 98, 99, 101, 104
  effect on mechanical properties, 138, 139
  effect on M_s temperatures, 54
  effect on softening during tempering, 217, 218
  effect on surface oxidation, 295
  effect on temper embrittlement, 236, 239
  as ferrite stabilizer, 10, 358
  in cast alloys, 371
Siliconizing, 465
Sinter, 465
Sintering, 465
Slip (dislocation movement) in martensitic structure (D), 59, 60, 61
Slot furnace, 465
Snap temper, 465
Soaking, 465
Softening of martensite during tempering (D), 205–216
  effect of alloying elements (D), 212–218
Soft temper. See Dead soft temper
Solidification, effect on tool steels, 405–406
Solid solution strengthening, 373
Solidus surfaces, stainless steels (D), 354–355
Solution heat treatment, 465
  precipitation-hardening stainless steels, 390
Sorbite, 465
Spheroidite, 465
Spheroidized structure
  austenite formation from (D), 183, 184–185, 186
  hardness as function of carbon content (D), 114, 116, 145, 146
Spheroidizing (D,M), 107, 109, 111, 114–118, 119, 465–466
  for ductility, 114

rate of coarsening of structure, 117
  temperatures for (D), 90, 116, 118
Spinodal decomposition, 380
Spinodal structure, 466
Spring temper, 466
Sputtering (D), 329–330, 331, 333
  bias, 334
Stabilization, 377
  in precipitation-hardening stainless steels, 387
Stabilizing heat treatments, 365–366, 466
Stainless steel 21-6-9, 371
Stainless steels (D), 351–396 (see also specific stainless steels under AISI; Al 2205 stainless steel; DPI stainless steel; Ferralium 255 stainless steel; JBK-75 stainless steel; 7-Mo PLUS stainless steel; PH 13-8 Mo stainless steel; U45 stainless steel)
  alloy design (D), 351–353
  austenitic, 351, 353, 356, 359–361
  austenitic, heat treatment of, 373–374
  austenitic, intergranular carbides in, 361–366
  austenitic, martensite formation, 366–370
  austenitic, other phases, 370–371
  austenitic, physical metallurgical principles, 361
  austenitic, stabilized, 365–366
  austenitic, strain-induced martensite formation, 368
  austenitic, welded, 365
  austenitic, wrought, 365
  cast alloys, 371
  desirable features, 351
  duplex (D,M,T), 356, 359, 390–395
  effect of alloying elements (D), 351–353, 354, 365–366
  ferritic, 352–353, 375–377
  ferritic, compositions of some common types (T), 374–375
  ferritic, ductility, 376–377
  ferritic, effect of annealing on E-Brite 26-1 (M), 374
  ferritic, effect of section thickness on DBTT (D), 375–376
  ferritic, 475 °C (885 °F) embrittlement (D,T), 379–381
  ferritic, intermetallic phases (D,M,T), 376, 377–379

Stainless steels *(continued)*
  ferritic, intermetallic phase time-temperature-transformation diagrams (D), 377–379
  ferritic, toughness, 376–377
  heat-resisting grades, 371–372
  high-strength manganese austenitic, compositions and minimum mechanical properties (T), 373
  martensitic (T), 351, 352–353, 354, 382–385
  martensitic, NbC coating deposited by salt bath process (M), 338
  phase equilibria (D), 351, 353–358
  physical vapor deposition metal nitride coating deposits (D,M), 335, 336
  precipitation-hardening (D,M,T), 386–390
  precipitation-hardening, alloying elements, 386, 388
  precipitation-hardening, compositions (T), 386
  precipitation-hardening, strengthening, 386
  precipitation-hardening, strengthening precipitates formed (M), 390, 391
  precipitation-hardening, three classes, 386
  solidification ranges, 359
Stainless W. *See* AISI 635 stainless steel
Steadite, 434
Stead's brittleness, 466
Stoking, 466
Stop-off. *See* Resist
Stopping off, 466
Strain-age embrittlement, 466
Strain hardening, 368, 370
Stranski-Krastanov growth, 331
Stress relieving (D), 118, 130–133, 264, 265, 466
  effect on austenite stainless steels, 373, 374
  effect on cast irons, 437, 441
  effect on tool steels, 416
Stress-strain curves
  effect of continuous yielding and discontinuous yielding (D), 129
  effect of quenching and tempering (D), 252, 253
  effect of temperature on Al 2205 stainless steel (D), 394, 395
  effect of tempering (D), 208–209, 212
  effect of temper rolling, 130
  for type 301 austenitic stainless steel (D), 367, 368
  for type 304 austenitic stainless steel (D), 367, 368
  for type 304 stainless steel at various temperatures (D), 368, 369, 370
Stretcher strains. *See* Lüders bands
Subcritical annealing, 466
Subcritical heating, 382
  effect on cast irons, 437
Subzero cooling
  effect on martensite formation, 419
  of precipitation-hardening stainless steels, 390
Sulfur
  composition ranges for unalloyed cast irons (T), 434
  deliberate addition for machinability of martensitic stainless steels, 382
  major source of inclusions, 239, 245
  solubility in ferrite, 358
Superalloy. *See* Heat-resisting alloy
Supercooling, 466
Superhardness (D), 188, 295
Superheating, 466
Surface engineering, definition, 320
Surface hardening (D,M,T), 281–315, 466–467
Surface melting and alloying, 340
Surface modifications (D,M,T), 319–346
Surface relief. *See* Surface tilting
Surface tilting in martensite formation (D,M), 46, 47, 48, 50, 51, 63, 64, 188, 189, 389, 390
Tantalum, effect on chromium carbide precipitation in stainless steels, 365
TD process. *See* Toyota diffusion coating process
Temper, definition, 467
Temperature conversions (T), 469–472
Temper brittleness, 467
Temper carbon. *See* Annealing carbon
Temper color, 467
Tempered martensite embrittlement (D), 206–207, 208, 229, 231–233
  effect of impurities and phosphorus (D), 231–232, 238
  intergranular fracture related to quench cracking, 264
Temper embrittlement (D), 206–207, 229–230, 236–239
  of cast irons, 449
  compositional factors, 236
  of duplex stainless steels, 395

fracture modes, 231
  heat treatment factors (D), 236, 237
  of martensitic stainless steels, 385
Tempering of steel (D,M,T), 205–255
  of cast irons, 441
  comparison of mechanical properties to those produced by austempering (T), 268
  components of microstructure, 245–246
  deleterious effect of inclusions on fracture, 246–247
  double, 458
  effect of alloying elements (D), 212–218, 224, 236–239, 247
  effect on continuous cooling transformation of martensite, 93, 95
  effect on embrittlement, 206–207
  effect on hardness (D), 145, 205, 206, 207–208, 209, 212–218
  effect on hardness of tool steels (D,M), 417, 418, 427–428
  effect on impact toughness (D), 205–207, 208
  effect on mechanical properties (D,T), 205–212
  effect on mechanical property changes in type 410 stainless steel (D), 384, 385
  effect on precipitation-hardening stainless steels, 387–388
  effect on stress-strain curve and work hardening (D), 209–210, 212
  effect on tool steels, 424–427
  effect on transformation of retained austenite (D,M), 218–224
  embrittlement phenomena (D,M), 229–244
  fracture modes (D,M,T), 245–255
  microstructural effects (D,M), 218–229
  three stages of, 218–219
  time-temperature relationships (D), 215–216
  tensile test results (T), 252, 253
Temper rolling, 130
Tenelon, composition and minimum mechanical properties (T), 373
Tensile strength
  of annealed stainless steels, 361
  of cast irons (D), 435, 439, 442, 444, 448
  effect of microstructure (D,M), 121–122, 123, 124
  effect of tempering (D,T), 206, 208–210, 252, 253
  effect of tensile test temperature (D), 242
  of ferrite-pearlite microstructure, 133–134, 138
  of ferritic stainless steels, 375
  microalloyed bar and forging steels, 139
  of tool steels, 405
  ultimate, of stainless steels, 368
Thermal evaporation, 329, 330
  during physical vapor deposition, 333
Thermal spraying, 346
Thermocouple, definition, 467
Thermodynamics of phase transformations, 23–26
Thermomechanical glow discharge surface treatments (D), 319–325
Thermomechanical treatments (D,T), 269–274
  classification (D,T), 270–273
  Soviet classification (D,T), 271, 272, 273
Thermomechanical working, 467
Three-dimensional island growth, 331
Three-quarters hard, 467
Tilting surface relief. See Surface tilting
Time-temperature-transformation (TTT) diagrams, 89
  ferritic stainless steels (D), 377, 379
  for isothermal martensite formation (D), 56
  for pearlite formation (D), 27, 28
  for thermomechanical treatments (D), 270
Tin, effect of temper embrittlement, 236, 239
Titanium
  as carbide former, 10, 31
  effect on austenite grain size, 199–200
  effect on chromium carbide precipitation in stainless steels, 365
  effect on coefficient of friction, 326
  effect on deformation during tempering, 247
  effect on mechanical properties of ferrite-pearlite microstructures, 139
  effect on surface oxidation, 295
  in production of alloy carbide layers, 338–339
  stabilizing element for ferritic stainless steels, 377
Titanium aluminum nitride coatings, physical vapor deposition application (D,M), 334, 335–336
Titanium nitride, physical vapor deposition coating, 335
Titanium nitride coatings, chemical vapor deposition application, 337
Tool steels (D,M,T), 401–428

Tool steels *(continued)*
  air-hardening medium-alloy cold work (AISI type A) (T), 403
  alloy design (M), 405–410, 411, 413, 415, 417, 419
  classification system (T), 401–405
  effect of tempering (D,M), 417, 418, 424–428
  grain boundary carbide formation, 422
  high-carbon high-chromium cold work (AISI type D) (T), 402, 403
  high-speed, 411
  hot work (AISI type H) (T), 402, 404
  low-alloy special-purpose (AISI type L) (T), 402, 403
  maraging (AISI Grades 90, 110, 125) (T), 405
  mold steels (AISI type P) (T), 402, 403
  molybdenum high-speed (AISI type M) (T), 404, 405
  molybdenum hot work (AISI type H) (T), 402–405
  oil-hardening cold work (AISI type O) (T), 402, 403
  primary processing (D,M), 410–413
  proprietary hot work (AISI 6G, 6F2, 6F3) (T), 404, 405
  shock-resisting (AISI type S) (T), 402, 403
  tungsten high-speed (AISI type T) (T), 404, 405
  ultrahard high-speed (AISI M41, M42, M43, M44, M46, M47) (T), 404–405
  water-hardening (AISI type W) (T), 402, 403
Total carbon, 467
Toughness
  of austempered vs. quenched and tempered steel (D), 267–269
  of cast irons, 439, 444, 445–446, 448, 449–450
  of duplex stainless steels, 392, 395
  effect of embrittlement (D), 229–230
  effect of phosphorus (D,M), 231–235
  effect of tempering (D), 205–206
  of ferrite-pearlite microstructures, 134, 138
  of ferritic stainless steels, 375, 376–377
  improved by low sulfur content, 239
  limited by void sheets, 249
  and proeutectoid cementite network, 36
  of tool steels, 402, 405, 424
Toyota diffusion coating process (TD process), 338–339
Transformation-induced plasticity, 467

Transformation-induced plasticity (TRIP) steels, 271, 468
Transformation ranges, 467
Transformation temperature, 467
Tribological properties, improved by nitrogen ion implantation, 325
Triode ion plating (D,M), 331, 334, 335
TRIP steels. *See* Transformation-induced plasticity steels
Troostite, 468
TTT diagrams. *See* Time-temperature-transformation diagrams
Tungsten
  as carbide former, 31
  effect on thermal conductivity, 265
Twinning
  annealing of austenitic stainless steels, 361
  in austenite, 6
  in martensitic structure (D), 59, 60, 61, 63, 150, 218, 223
  in martensitic structure, deformation, 64, 65, 66, 67, 76
Two-dimensional layer-by-layer growth, 331
U45 stainless steel, composition (T), 391
Undercooling. *See* Supercooling
Unified Numbering System (UNS) alloy numbers, 361
Vacancy-interstitial pairs, formation by ion implantation, 325–326
Vacuum degassing, for reduction of hydrogen flaking, 238
Vacuum melting, 377
Vanadium
  as carbide former, 31, 34
  effect on austenitic grain size, 199–200
  effect on corrosion resistance of stainless steels (D), 352, 353
  effect on grain size control, 136
  effect on hardness during tempering, 217
  effect on mechanical properties on ferrite-pearlite microstructures (M), 139–140, 141, 142
  effect on softening during tempering, 212
  effect on temper embrittlement, 238
  effect on tool steels, 409
  in production of alloy carbide layers, 338–339
  interphase precipitation of carbides (M), 140, 142
  use in nitriding steels, 306
Voids. *See* Interstitial voids

Void sheets, 249
Voltage-current relationships, nitrogen-hydrogen mixtures (D), 321
Volmer-Weber growth, 331
Water quenching (*see also* Quenching)
  effect on hardness distribution in hardenability evaluation (D), 153–156, 159, 164, 166
  equivalent cooling rates (D), 176
  following annealing of austenitic stainless steels, 361
  for water-hardening tool steels, 402
  microstructures produced as related to bar diameter (D,M), 99–105
  severity of quench (D,T), 159, 160–161
  Widmanstätten side plate formation (M), 39, 40
Weight percent, 3
Widmanstätten
  bainite, 58, 101–102
  ferrite, 58, 101–102
  side plates (M), 36–37, 38, 39, 40
  structure of austenite, in duplex stainless steels, 393–394
Workability, and proeutectoid cementite network, 36
Work hardening effect of tempering (D), 209–210, 212
Work hardening rate, increased by pearlite, 134
X-ray analysis, for determining amount of retained austenite, 55
X-ray diffraction techniques
  for determining crystallographic orientation relationships, 57
  to measure retained austenite content, 52
Yield strength
  effect of annealing on stainless steels, 361
  effect of grain size (D), 137
  effect of martensite packet size (D), 148, 150
  effect of microstructure (D), 127–130
  effect of tempering (D), 208, 209, 210
  effect of thermomechanical treatments (D), 273
  equations for determination, 137–138
  of ferrite-pearlite microstructures, 133–134
  of ferritic stainless steels, 375
  of high-strength low-alloy steels, 200
  of microalloyed bars and forgings, 140
  of Nitronic stainless steels, 372
  vs. elongation of thermomechanically treated steels (D), 273
Zinc, effect on embrittlement, 241
Zirconium, effect on deformation during tempering, 247
Zirconium nitride coating, physical vapor deposition application (M), 336–337